Excel

Get the Results You Want!

SUCCESS ONE® HSC TOPIC-BY-TOPIC

MATHEMATICS STANDARD 2

Past HSC questions arranged into topics with worked answers **2001–2021**

BRAND NEW SERIES

PASCAL PRESS

ISBN 978 1 74125 732 8

Pascal Press
PO Box 250
Glebe NSW 2037
(02) 8585 4044
www.pascalpress.com.au

Publisher: Vivienne Joannou
Project editor: Rosemary Peers
Edited by Rosemary Peers
Answers checked by Peter Little
Cover and page design by Sonia Woo
Page layout and typesetting by lj Design (Julianne Billington)
Printed by Vivar Printing/Green Giant Press

CONTENTS

ABOUT THIS BOOK

Revise the smart way and use this guide during the year to complete HSC questions whenever you have completed a topic in class and need to study for a topic test, an Assessment Task or lastly for the HSC Exam.

- This book contains over 1200 questions from 21 years of HSC past papers arranged into syllabus topics.
- Below each question you will find the HSC paper it is taken from and its original question number.

Each topic:

- has questions arranged in order with the most recent HSC papers first
- has questions marked with one of three levels of difficulty: Easy, Medium and Hard
- has questions relevant only to the current course (the Mathematics Standard 2 course replaced the Mathematics General 2 course and was first examined in 2019).

Some questions have additional information:

- CQ This symbol is next to questions that are common to both the HSC Mathematics Standard 2 and the Mathematics Advanced papers.
- **Bonus question** Some questions have been written to give extra practice on a topic. These are usually new syllabus topics where there are only past HSC questions from 2019 onwards.
- Formulas in square brackets have been added to some questions. In the original HSC paper the formulas would have been provided in the HSC Formulae Sheet but for the current syllabus they are provided in the HSC Exam. For example, page 2, Question 17:

[$BAC_{female} = \dfrac{10N - 7.5H}{5.5M}$

where N is number of standard drinks consumed
H is number of hours of drinking
M is person's mass in kilograms]

Worked answers:

- The worked answers contained in this publication are examples of answers which the authors believe would score full marks.
- They are not necessarily ideal or model answers, nor are they the only answers which would score full marks.
- They are not endorsed by NESA.
- Worked answers for the 2001–2009 HSC Examination papers were written by Barbara D'Angelo, MA, Dip. Ed.
- Worked answers for the 2010–2021 HSC Examination papers were written by Lyn Baker.
- Bonus questions and answers were written by Allyn Jones.
- The suggested mark allocation for working in solutions is provided as a guide only and is not endorsed by NESA.

Acknowledgements:

- Source of photo on page 26: www.nyeinparis.files.wordpress.com/2010/01/dscf1250.jpg.
- Image of energy rating label on page 39 adapted from Department of the Environment and Energy.
- Photo of windfarm on page 120 reproduced with the kind permission of Miguel Saavedra.
- Diagrams on pages 137 and 157 reproduced from Google Maps.

SELF-ASSESSMENT SUMMARY

YEAR 11	EASY	MEDIUM	HARD	MASTERY
ALGEBRA				
Formulae and equations	/22	/25	/11	1 2 3 4 5
Linear relationships	/18	/29	/6	1 2 3 4 5
MEASUREMENT				
Practicalities of measuring		/15	/3	1 2 3 4 5
Perimeter, area and volume	/24	/49	/15	1 2 3 4 5
Units of energy and mass	/3	/7	/1	1 2 3 4 5
Working with time	/11	/13	/5	1 2 3 4 5
FINANCIAL MATHEMATICS				
Interest and depreciation	/10	/25	/8	1 2 3 4 5
Earning and managing money	/30	/20		1 2 3 4 5
Budgeting and household expenses	/9	/12		1 2 3 4 5
STATISTICAL ANALYSIS				
Classifying and representing data	/17	/2	/1	1 2 3 4 5
Summary statistics	/42	/54	/19	1 2 3 4 5
Relative frequency and probability	/51	/48	/17	1 2 3 4 5

YEAR 12	EASY	MEDIUM	HARD	MASTERY
ALGEBRA				
Simultaneous linear equations	/6	/12	/7	1 2 3 4 5
Non-linear relationships	/19	/46	/26	1 2 3 4 5
MEASUREMENT				
Non-right-angled trigonometry	/35	/39	/31	1 2 3 4 5
Rates and ratio	/20	/32	/9	1 2 3 4 5
FINANCIAL MATHEMATICS				
Investments	/11	/19	/2	1 2 3 4 5
Depreciation and loans	/22	/24	/9	1 2 3 4 5
Annuities	/4	/14	/10	1 2 3 4 5
STATISTICAL ANALYSIS				
Bivariate data analysis	/22	/26	/10	1 2 3 4 5
The normal distribution	/12	/31	/9	1 2 3 4 5
NETWORK CONCEPTS				
Network concepts	/24	/3	/1	1 2 3 4 5
Shortest paths	/9	/21	/7	1 2 3 4 5
Critical path analysis	/6	/52	/28	1 2 3 4 5

1 Suppose $a = \frac{b}{7}$, where $b = 22$.

What is the value of a, correct to three significant figures?

A 3.14 **B** 3.15
C 3.142 **D** 3.143 *(1 mark)*

(Q6, **2021 HSC**) Medium

2 A student is thinking of a number. Let the number be x.

When the student subtracts 8 from this number and multiplies the result by 3, the answer is 2 more than x.

Which equation can be used to find x?

A $3(x - 8) = 2x$ **B** $3x - 8 = 2x$
C $3(x - 8) = x + 2$ **D** $3x - 8 = x + 2$

(1 mark)

(Q9, **2021 HSC**) Medium

3 Solve $x + \frac{x-1}{2} = 9$. *(2 marks)*

CQ (Q29, **2021 HSC**) Hard

4 When Jake stops drinking alcohol at 10:30 pm, he has a blood alcohol content (BAC) of 0.08375.

The number of hours required for a person to reach zero BAC after they stop consuming alcohol is given by the formula

$$\text{Time} = \frac{BAC}{0.015}.$$

At what time on the next day should Jake expect his BAC to be 0.05?

A 12:45 am **B** 1:50 am
C 2:15 am **D** 4:05 am *(1 mark)*

(Q13, **2020 HSC**) Hard

5 Which of the following correctly expresses y as the subject of the formula $3x - 4y - 1 = 0$?

A $y = \frac{3}{4}x - 1$ **B** $y = \frac{3}{4}x + 1$

C $y = \frac{3x - 1}{4}$ **D** $y = \frac{3x + 1}{4}$ *(1 mark)*

(Q11, **2019 HSC**) Medium

6 The formula below is used to calculate an estimate for blood alcohol content (BAC) for females.

$$BAC_{\text{Female}} = \frac{10N - 7.5H}{5.5M}$$

The number of hours required for a person to reach zero BAC after they stop consuming alcohol is given by the following formula.

$$\text{Time} = \frac{BAC}{0.015}$$

The number of standard drinks in a glass of wine and a glass of spirits is shown.

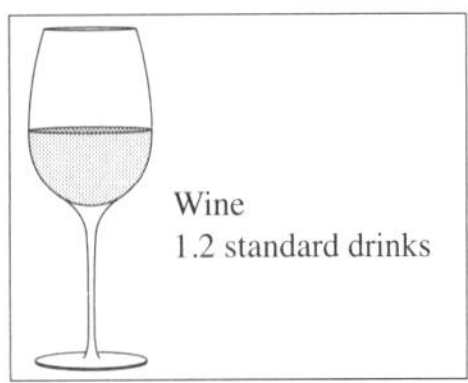

Hannah weighs 60 kg. She consumed 3 glasses of wine and 4 glasses of spirits between 6:15 pm and 12:30 am the following day. She then stopped drinking alcohol.

Using the given formulae, calculate the time in the morning when Hannah's BAC should reach zero. *(4 marks)*

(Q28, **2019 HSC**) Medium

7 Which of the following expresses v as the subject of $k = \frac{1}{2}mv^2$?

A $v = \pm\sqrt{\frac{2k}{m}}$ **B** $v = \pm\frac{\sqrt{k}}{2m}$

C $v = \pm\sqrt{\frac{k}{2m}}$ **D** $v = \pm\frac{2\sqrt{k}}{m}$ *(1 mark)*

(Q25, **2018 HSC**) Medium

8 Clark's formula, given below, is used to determine the dosage of medicine for children.

$$\text{Dosage} = \frac{\text{weight in kg} \times \text{adult dosage}}{70}$$

For a particular medicine, the adult dosage is 325 mg and the correct dosage for a specific child is 90 mg.

How much does the child weigh, to the nearest kg? *(2 marks)*

(Q26b, **2018 HSC**) Easy

9 Solve the equation $\frac{2x}{5} + 1 = \frac{3x + 1}{2}$, leaving your answer as a fraction. *(3 marks)*

(Q28b, **2018 HSC**) Hard

10 Sophie is driving at 70 km/h. She notices a branch on the road ahead and decides to apply the brakes. Her reaction time is 1.5 seconds. Her braking distance (D metres) is given by $D = 0.01v^2$, where v is speed in km/h.

What is Sophie's stopping distance, to the nearest metre?

[stopping distance = reaction-time distance + braking distance] *(3 marks)*

(Q28e, **2018 HSC**) **Hard**

11 It is given that $I = \frac{3}{2}MR^2$.

What is the value of I when $M = 26.55$ and $R = 3.07$, correct to two decimal places?

A 375.35 **B** 3246.08
C 9965.45 **D** 14 948.18 *(1 mark)*

(Q7, **2017 HSC**) **Easy**

12 What is the value of x in the equation $\frac{5-x}{3} = 6$?

A −13 **B** −3
C 3 **D** 13 *(1 mark)*

(Q9, **2017 HSC**) **Easy**

13 Young's formula, shown below, is used to calculate the dosage of medication for children aged 1–12 years based on the adult dosage.

$$D = \frac{yA}{y+12}$$

where D = dosage for children aged 1–12 years
y = age of child (in years)
A = adult dosage

A child's dosage is calculated to be 20 mg, based on an adult dosage of 40 mg.

How old is the child in years?

A 6 **B** 8
C 10 **D** 12 *(1 mark)*

(Q19, **2017 HSC**) **Medium**

14 Rhys is drinking low alcohol beer at a party over a five-hour period. He reads on the label of the low alcohol beer bottle that it is equivalent to 0.8 of a standard drink.

Rhys weighs 90 kg.

What is the maximum number of complete bottles of the low alcohol beer he can drink to remain under a Blood Alcohol Content (BAC) of 0.05?

[$BAC_{male} = \frac{10N - 7.5H}{6.8M}$

where N is number of standard drinks consumed
H is number of hours of drinking
M is person's mass in kilograms] *(4 marks)*

(Q27e, **2017 HSC**) **Medium**

15 Make y the subject of the equation $x = \sqrt{yp - 1}$. *(2 marks)*

(Q28d, **2017 HSC**) **Medium**

16 Which of the following equations has $x = 5$ as the solution?

A $x - 5 = 10$ **B** $5 - x = 10$
C $\frac{x}{2} = 10$ **D** $2x = 10$ *(1 mark)*

(Q2, **2016 HSC**) **Easy**

17 Caroline drinks two small bottles of wine over a three-hour period. Each of these bottles contains 2.3 standard drinks. Caroline weighs 53 kg.

What is her approximate blood alcohol content (BAC) at the end of this period?

A 0.081 **B** 0.065
C 0.0017 **D** 0.0014

[$BAC_{female} = \frac{10N - 7.5H}{5.5M}$

where N is number of standard drinks consumed
H is number of hours of drinking
M is person's mass in kilograms] *(1 mark)*

(Q10, **2016 HSC**) **Medium**

18 Which of the following correctly expresses Q as the subject of $e = iR + \frac{Q}{C}$?

A $Q = Ce + CiR$ **B** $Q = Ce - CiR$
C $Q = \frac{e + iR}{C}$ **D** $Q = \frac{e - iR}{C}$ *(1 mark)*

(Q24, **2016 HSC**) **Medium**

19 The number of 'standard drinks' in various glasses of wine is shown.

Number of standard drinks

White Wine		*Red Wine*	
small glass	*large glass*	*small glass*	*large glass*
0.9	1.4	1.0	1.5

A woman weighing 62 kg drinks three small glasses of white wine and two large glasses of red wine between 8 pm and 1 am.

What would be her blood alcohol content (BAC) estimate at 1 am, correct to three decimal places?

A 0.030 **B** 0.037
C 0.046 **D** 0.057

[$BAC_{female} = \frac{10N - 7.5H}{5.5M}$

where N is number of standard drinks consumed
H is number of hours of drinking
M is person's mass in kilograms] *(1 mark)*

(Q23, **2015 HSC**) **Medium**

20 Consider the equation $\frac{2x}{3} - 4 = \frac{5x}{2} + 1$.
Which of the following would be a correct step in solving this equation?

A $\frac{2x}{3} - 3 = \frac{5x}{2}$ **B** $\frac{2x}{3} = \frac{5x}{2} + 5$

C $2x - 4 = \frac{15x}{2} + 3$ **D** $\frac{4x}{6} - 8 = 5x + 2$ *(1 mark)*

(Q24, **2015 HSC**) **Medium**

21 Clark's formula is used to determine the dosage of medicine for children.

$$\text{Dosage} = \frac{\text{weight in kg} \times \text{adult dosage}}{70}$$

The adult daily dosage of a medicine contains 3150 mg of a particular drug.

A child who weighs 35 kg is to be given tablets each containing 525 mg of this drug.

How many tablets should this child be given daily? *(2 marks)*

(Q26b, **2015 HSC**) **Easy**

22 The formula $C = \frac{5}{9}(F - 32)$ is used to convert temperatures between degrees Fahrenheit (F) and degrees Celsius (C).

Convert 3°C to the equivalent temperature in Fahrenheit. *(2 marks)*

(Q28d, **2015 HSC**) **Easy**

23 Claire is driving on a motorway at a speed of 110 kilometres per hour and has to brake suddenly. She has a reaction time of 2 seconds and a braking distance of 59.2 metres.

Calculate her stopping distance.

[stopping distance = reaction-time distance + braking distance] *(2 marks)*

(Q30d, **2015 HSC**) **Hard**

24 Young's formula below is used to calculate the required dosages of medicine for children aged 1–12 years.

$$\text{Dosage} = \frac{\text{age of child (in years)} \times \text{adult dosage}}{\text{age of child (in years)} + 12}$$

How much of the medicine should be given to an 18-month-old child in a 24-hour period if each adult dosage is 45 mL? The medicine is to be taken every 6 hours by both adults and children.

A 5 mL **B** 20 mL
C 27 mL **D** 30 mL *(1 mark)*

(Q4, **2014 HSC**) **Medium**

25 Solve the equation $\frac{5x+1}{3} - 4 = 5 - 7x$. *(3 marks)*

(Q26c, **2014 HSC**) **Medium**

26 What is the maximum number of standard drinks that a male weighing 84 kg can consume over 4 hours in order to maintain a blood alcohol content (BAC) of less than 0.05?

[$BAC_{male} = \frac{10N - 7.5H}{6.8M}$

where N is number of standard drinks consumed
H is number of hours of drinking
M is person's mass in kilograms] *(3 marks)*

(Q29b, **2014 HSC**) **Medium**

27 Which equation correctly shows r as the subject of $S = 800(1 - r)$?

A $r = \frac{800 - S}{800}$ **B** $r = \frac{S - 800}{800}$

C $r = 800 - S$ **D** $r = S - 800$ *(1 mark)*

(Q21, **2013 HSC**) **Hard**

28 Sarah tried to solve this equation and made a mistake in Line 2.

$\frac{W+4}{3} - \frac{2W-1}{5} = 1$ ...... Line 1
$5W + 20 - 6W - 3 = 15$ ...... Line 2
$17 - W = 15$ ...... Line 3
$W = 2$ ...... Line 4

Copy the equation in Line 1 into your writing booklet and continue your solution to solve this equation for W. Show all lines of working. *(2 marks)*

(Q29a, **2013 HSC**) **Hard**

29 Which of the following correctly expresses c as the subject of $E = mc^2 + p$?

A $c = \pm\sqrt{\frac{E - p}{m}}$ **B** $c = \pm\frac{\sqrt{E - p}}{m}$

C $c = \pm\sqrt{\frac{E}{m}} - p$ **D** $c = \pm\sqrt{\frac{E}{m} - p}$ *(1 mark)*

(Q21, **2012 HSC**) **Medium**

30 Which of the following correctly expresses a as the subject of $s = ut + \frac{1}{2}at^2$?

A $a = \frac{2(s - ut)}{t^2}$ **B** $a = \frac{2s - ut}{t^2}$

C $a = \frac{\frac{1}{2}(s - ut)}{t^2}$ **D** $a = \frac{\frac{1}{2}s - ut}{t^2}$ *(1 mark)*

(Q18, **2011 HSC**) **Hard**

31 If $M = -9$, what is the value of $\dfrac{3M^2 + 5M}{6}$?

A -250.5 **B** -48
C 33 **D** 235.5 *(1 mark)*

(Q7, **2010 HSC**) **Medium**

32 Which of the following correctly expresses x as the subject of $a = \dfrac{nx}{5}$?

A $x = \dfrac{an}{5}$ **B** $x = \dfrac{5a}{n}$

C $x = \dfrac{a-5}{n}$ **D** $x = 5a - n$ *(1 mark)*

(Q18, **2010 HSC**) **Medium**

33 Fred tried to solve this equation and made a mistake in Line 2.

$$\begin{aligned} 4(y+2) - 3(y+1) &= -3 && \text{Line 1} \\ 4y + 8 - 3y + 3 &= -3 && \text{Line 2} \\ y + 11 &= -3 && \text{Line 3} \\ y &= -14 && \text{Line 4} \end{aligned}$$

Copy the equation in Line 1 into your writing booklet.

i Rewrite Line 2 correcting his mistake. *(1 mark)* **Easy**

ii Continue your solution showing the correct working for Lines 3 and 4 to solve this equation for y. *(1 mark)* **Easy**

(Q24a, **2010 HSC**)

34 Which of the following correctly expresses n as the subject of $v = \dfrac{3mn^2}{r}$.

A $n = \pm\dfrac{\sqrt{rv}}{3m}$ **B** $n = \pm r\sqrt{\dfrac{v}{3m}}$

C $n = \pm\dfrac{r\sqrt{v}}{3m}$ **D** $n = \pm\sqrt{\dfrac{rv}{3m}}$ *(1 mark)*

(Q15, **2009 HSC**) **Medium**

35 What is the value of $\sqrt{\dfrac{x + 2y}{8y}}$ if $x = 5.6$ and $y = 3.1$, correct to 2 decimal places?

A 0.69 **B** 2.62
C 2.83 **D** 4.77 *(1 mark)*

(Q9, **2008 HSC**) **Easy**

36 Solve $\dfrac{5x + 1}{2} = 4x - 7$. *(3 marks)*

(Q23d, **2008 HSC**) **Medium**

37 Which of the following correctly expresses T as the subject of $B = 2\pi\left(R + \dfrac{T}{2}\right)$?

A $T = \dfrac{B}{\pi} - 2R$ **B** $T = \dfrac{B}{\pi} - R$

C $T = 2R - \dfrac{B}{\pi}$ **D** $T = \dfrac{B}{4\pi} - \dfrac{R}{2}$ *(1 mark)*

(Q19, **2007 HSC**) **Hard**

38 The distance in kilometres (D) of an observer from the centre of a thunderstorm can be estimated by counting the number of seconds (t) between seeing the lightning and first hearing the thunder.

Use the formula $D = \dfrac{t}{3}$ to estimate the number of seconds between seeing the lightning and hearing the thunder if the storm is 1.2 km away. *(1 mark)*

(Q24b, **2007 HSC**) **Easy**

39 If $V = \dfrac{4}{3}\pi r^3$, what is the value of V when $r = 2$, correct to two decimal places?

A 8.38 **B** 12.57
C 25.13 **D** 33.51 *(1 mark)*

(Q2, **2006 HSC**) **Easy**

40 What is the formula for q as the subject of $4p = 5t + 2q^2$?

A $q = \pm\dfrac{\sqrt{5t - 4p}}{2}$ **B** $q = \pm\dfrac{\sqrt{4p - 5t}}{2}$

C $q = \pm\sqrt{\dfrac{5t - 4p}{2}}$ **D** $q = \pm\sqrt{\dfrac{4p - 5t}{2}}$ *(1 mark)*

(Q18, **2006 HSC**) **Hard**

41 What is the value of $\dfrac{a - b}{4}$, if $a = 240$ and $b = 56$?

A 4 **B** 46
C 226 **D** 736 *(1 mark)*

(Q2, **2005 HSC**) **Easy**

42 Using the formula $d = 5t^3 - 2$, Marcia tried to find the value of t when $d = 137$.

Here is her solution. She has made one mistake.

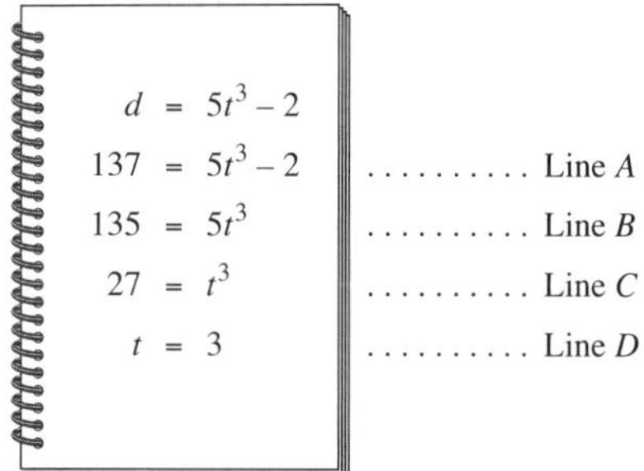

Which line does NOT follow correctly from the previous line?

A Line A **B** Line B
C Line C **D** Line D *(1 mark)*

(Q14, **2005 HSC**) Easy

43 The formula $D = \frac{2A}{15}$ is used to calculate the dosage of Hackalot cough medicine to be given to a child.

- D is the dosage of Hackalot cough medicine in millilitres (mL).
- A is the age of the child in months.

i If George is nine months old, what dosage of Hackalot cough medicine should he be given? *(1 mark)* Easy

ii The correct dosage of Hackalot cough medicine for Sam is 4 mL. What is the difference in the ages of Sam and George, in months? *(3 marks)* Hard

(Q24b, **2005 HSC**)

44 Make L the subject of the equation $T = 2\pi L^2$. *(2 marks)* Medium

(Q24c, **2005 HSC**)

45 If $K = Ft^3$, $F = 5$ and $t = 0.715$, what is the value of K correct to three significant figures?

A 1.82 **B** 1.827
C 1.828 **D** 1.83 *(1 mark)*

(Q3, **2004 HSC**) Easy

46 If $d = 6t^2$, what is a possible value of t when $d = 2400$?

A 0.05 **B** 20
C 120 **D** 400 *(1 mark)*

(Q11, **2004 HSC**) Easy

47 Kirbee is shopping for computer software. *Novirus* costs $115 more than *Funmaths*. Let x dollars be the cost of *Funmaths*.

i Write an expression involving x for the cost of Novirus. *(1 mark)* Easy

ii *Novirus* and *Funmaths* together cost $415. Write an equation involving x and solve it to find the cost of *Funmaths*. *(2 marks)* Medium

(Q23b, **2004 HSC**)

48 A health rating, R, is calculated by dividing a person's weight, w, in kilograms by the square of the person's height, h, in metres.

i Fred is 150 cm and weighs 72 kg. Calculate Fred's health rating. *(1 mark)* Easy

ii Over several years, Fred expects to grow 10 cm taller. By this time he wants his health rating to be 25. How much weight should he gain or lose to achieve his aim? Justify your answer with mathematical calculations. *(2 marks)* Medium

(Q28a, **2004 HSC**)

49 If $d = \sqrt{\frac{h}{5}}$, what is the value of d, correct to one decimal place, when $h = 28$?

A 1.1 **B** 2.4
C 2.8 **D** 5.6 *(1 mark)*

(Q4, **2003 HSC**) Easy

50 Kylie and Danny work in a music store. The weekly wage $\$W$ of an employee at the store is given by

$$W = 0.75\,n + 50,$$

where n is the number of CDs the employee sells.

If Kylie sells two more CDs than Danny in one week, how much more will she earn?

A $0.75 **B** $1.50
C $50.75 **D** $51.50 *(1 mark)*

(Q15, **2003 HSC**) Easy

51 **i** While Sandra is on holiday she visits countries where the Fahrenheit temperature scale is used. She knows that the correct way to convert from Celsius to Fahrenheit is:

'Multiply the Celsius temperature by 1.8, then add 32.'

Find the value of A in the following table.

Celsius	*Fahrenheit*
5	41
15	59
A	77

(1 mark) Medium

ii Peter uses the following method to approximate the conversion from Celsius to Fahrenheit:

'Add 12 to the Celsius temperature, then double your result.'

Express Peter's rule as an algebraic equation. Use C for the Celsius temperature and F for the approximate Fahrenheit temperature. *(2 marks)* Medium

(Q28c, **2003 HSC**)

52 If $w = 2y^3 - 1$, what is the value of y when $w = 13$?

A $\frac{\sqrt[3]{14}}{2}$ **B** $\sqrt[3]{6}$

C $\sqrt[3]{7}$ **D** $\sqrt[3]{14}$ *(1 mark)*

(Q16, **2002 HSC**) Easy

53 If $w = \frac{15y}{y + 12}$, and $y = 7$, find the value of w (correct to two decimal places).

A 5.53 **B** 8.26

C 15.75 **D** 27.00 *(1 mark)*

(Q2, **2001 HSC**) Easy

Year 11 Formulae and equations—Worked answers

1 $a = \frac{b}{7}$

When $b = 22$,

$a = \frac{22}{7}$

$= 3.142\,857\,14\ldots$

$= 3.14$ (3 significant figures)

Answer A

2 The number is x.

Subtracting 8 gives $x - 8$.

Then multiplying by 3 gives $3(x - 8)$.

2 more than x is $x + 2$.

So the equation is $3(x - 8) = x + 2$.

Answer C

3 $x + \frac{x-1}{2} = 9$

Multiplying by 2:

$2x + x - 1 = 18$ ✓

$3x - 1 = 18$

$3x = 19$

$x = 6\frac{1}{3}$ ✓ *(2 marks)*

4 $\text{Time} = \frac{BAC}{0.015}$

If $BAC = 0.083\,75$

$\text{Time} = \frac{0.083\,75}{0.015}$

$= 5$ h 35 min

So, according to the formula, Jake's BAC will be zero 5 h 35 min after 10:30 pm or at 4:05 am.

If $BAC = 0.05$,

$\text{Time} = \frac{0.05}{0.015}$

$= 3$ h 20 min

According to the formula, BAC will reach 0 from 0.05 after 3 h 20 min.

So Jake might assume that his BAC is 0.05 3 h 20 min before 4:05 am, so at 12:45 am.

Answer A

5 $3x - 4y - 1 = 0$

$4y = 3x - 1$

$y = \frac{3x-1}{4}$

Answer C

6 $M = 60$

$N = 3 \times 1.2 + 4$

$= 7.6$ ✓

From 6:15 pm until 12:30 am is $6\frac{1}{4}$ hours.

So $H = 6.25$

$BAC = \frac{10N - 7.5H}{5.5M}$

$= \frac{10 \times 7.6 - 7.5 \times 6.25}{5.5 \times 60}$

$= 0.088\,257\,5757\ldots$ ✓

$\text{Time} = \frac{BAC}{0.015}$

$= 5.883\,838\,38\ldots$ h

$= 5$ h 53 min (nearest minute) ✓

Now 5 h 53 min after 12:30 am is 6:23 am.

So, using these formulae, Hannah's BAC should reach zero at 6:23 am. ✓ *(4 marks)*

7 $k = \frac{1}{2}mv^2$

$2k = mv^2$

$v^2 = \frac{2k}{m}$

$v = \pm\sqrt{\frac{2k}{m}}$

Answer A

8 Dosage = 90 mg, adult dosage = 325 mg

Dosage = $\frac{\text{weight in kg} \times \text{adult dosage}}{70}$

$90 = \frac{\text{weight in kg} \times 325}{70}$ ✓

$6300 = \text{weight in kg} \times 325$

Weight = 19.384 6153… kg

= 19 kg (nearest kg) ✓

(2 marks)

9 $\frac{2x}{5} + 1 = \frac{3x+1}{2}$

$10 \times \frac{2x}{5} + 10 \times 1 = \frac{10(3x+1)}{2}$ ✓

$4x + 10 = 15x + 5$ ✓

$5 = 11x$

$x = \frac{5}{11}$ ✓ *(3 marks)*

10 70 km/h = 70 000 m/h

= 19.444 4444… m/s

In 1.5 s Sophie travels 1.5 × 19.444 4444… m

So reaction time distance

= 29.166 6666… m

= 29 m (nearest metre) ✓

Braking distance:

$D = 0.01v^2$

$= 0.01 \times 70^2$

$= 49$ m ✓

Stopping distance

= (29 + 49) m = 78 m

Sophie's stopping distance is 78 m, to the nearest metre. ✓ *(3 marks)*

11 $I = \frac{3}{2}MR^2$

When $M = 26.55$ and $R = 3.07$,

$I = \frac{3}{2} \times 26.55 \times 3.07^2$

$= 375.346\,642\ldots$

$= 375.35$ (2 d.p.)

Answer A

12 $\frac{5-x}{3} = 6$

$5 - x = 18$

$-x = 13$

$x = -13$

Answer A

13 $D = \frac{yA}{y+12}$

When $A = 40$, $D = 20$

$20 = \frac{40y}{y+12}$

$20y + 240 = 40y$

$240 = 20y$

$12 = y$

Answer D

14 $BAC_{\text{male}} = \frac{10N - 7.5H}{6.8M}$

If $BAC = 0.05$, $H = 5$ and $M = 90$,

$0.05 = \frac{10N - 7.5 \times 5}{6.8 \times 90}$ ✓

$= \frac{10N - 37.5}{612}$

$30.6 = 10N - 37.5$

$68.1 = 10N$

$6.81 = N$ ✓

So the number of standard drinks must be less than 6.81.

Number of bottles = 6.81 ÷ 0.8

= 8.5125 ✓

So the maximum number of complete bottles Rhys can drink in 5 hours to stay under 0.05 is 8. ✓ *(4 marks)*

15 $x = \sqrt{yp - 1}$

$x^2 = yp - 1$ ✓

$x^2 + 1 = yp$

$\frac{x^2+1}{p} = y$

$y = \frac{x^2+1}{p}$ ✓ *(2 marks)*

16 $2x = 10$

$x = 5$

Answer D

17 $N = 2 \times 2.3 = 4.6$

$H = 3$ $M = 53$

$BAC_{\text{female}} = \dfrac{10N - 7.5H}{5.5M}$

$= \dfrac{10 \times 4.6 - 7.5 \times 3}{5.5 \times 53}$

$= 0.080\,617\,4957\ldots$

$= 0.081$ (2 sig. figs.)

Answer A

18 $e = iR + \dfrac{Q}{C}$

$\dfrac{Q}{C} = e - iR$ ✓

$Q = Ce - CiR$

Answer B

19 Number of standard drinks

$= 3 \times 0.9 + 2 \times 1.5$

$= 5.7$

$BAC_{\text{female}} = \dfrac{10N - 7.5H}{5.5M}$

$= \dfrac{10 \times 5.7 - 7.5 \times 5}{5.5 \times 62}$

$= 0.057\,184\,75\ldots$

$= 0.057$ (3 d.p.)

Answer D

20 $\dfrac{2x}{3} - 4 = \dfrac{5x}{2} + 1$

Add 4 to both sides:

$\dfrac{2x}{3} = \dfrac{5x}{2} + 5$

Answer B

21 Dosage

$= \dfrac{\text{weight in kg} \times \text{adult dosage}}{70}$

$= \dfrac{35 \times 3150}{70}$

$= 1575$ mg ✓

Number of tablets $= 1575 \div 525$

$= 3$

The child should be given 3 tablets daily. ✓ *(2 marks)*

22 $C = \dfrac{5}{9}(F - 32)$

$3 = \dfrac{5}{9}(F - 32)$ ✓

$5.4 = F - 32$

$37.4 = F$

So 3 °C is equivalent to 37.4 °F. ✓ *(2 marks)*

23 Speed = 110 km/h

= 110 000 m/h

= 30.555 55… m/s ✓

Reaction time distance

$= 2 \times 30.5555\ldots$

$= 61.1111\ldots$

$= 61.1$ m (1 d.p.)

Stopping distance

$= (61.1 + 59.2)$ m

$= 120.3$ m ✓ *(2 marks)*

24 Each dose $= \dfrac{1.5 \times 45}{1.5 + 12}$

$= 5$ mL

Number of doses $= 24 \div 6$

$= 4$

Total medicine $= 4 \times 5$ mL

$= 20$ mL

Answer B

25 $\dfrac{5x + 1}{3} - 4 = 5 - 7x$

$5x + 1 - 12 = 15 - 21x$ ✓

$26x = 26$ ✓

$x = 1$ ✓ *(3 marks)*

26 $BAC_{\text{male}} = \dfrac{10N - 7.5H}{6.8M}$

$0.05 = \dfrac{10N - 7.5 \times 4}{6.8 \times 84}$ ✓

$= \dfrac{10N - 30}{571.2}$

$10N - 30 = 28.56$ ✓

$10N = 58.56$

$N = 5.856$

So the maximum number of standard drinks to stay under 0.05 would be 5. ✓ *(3 marks)*

27 $S = 800(1 - r)$

$\dfrac{S}{800} = 1 - r$

$r = 1 - \dfrac{S}{800}$

$= \dfrac{800}{800} - \dfrac{S}{800}$

$= \dfrac{800 - S}{800}$

Answer A

28 $\dfrac{W + 4}{3} - \dfrac{2W - 1}{5} = 1$

$5W + 20 - 6W + 3 = 15$ ✓

$23 - W = 15$

$W = 8$ ✓

(2 marks)

29 $E = mc^2 + p$

$mc^2 + p = E$

$mc^2 = E - p$

$c^2 = \dfrac{E - p}{m}$

$c = \pm\sqrt{\dfrac{E - p}{m}}$

Answer A

30 $s = ut + \dfrac{1}{2}at^2$

$s - ut = \dfrac{1}{2}at^2$

$2(s - ut) = at^2$

$\dfrac{2(s - ut)}{t^2} = a$

So $a = \dfrac{2(s - ut)}{t^2}$

Answer A

31 If $M = -9$,

$\dfrac{3M^2 + 5M}{6}$

$= \dfrac{3 \times (-9)^2 + 5 \times (-9)}{6}$

$= \dfrac{243 - 45}{6}$

$= 33$

Answer C

32 $a = \dfrac{nx}{5}$

[Multiply both sides by 5.]

$5a = nx$

[Divide both sides by n.]

$\dfrac{5a}{n} = x$

So $x = \dfrac{5a}{n}$

Answer B

33 $4(y + 2) - 3(y + 1) = -3$

i $4y + 8 - 3y - 3 = -3$ *(1 mark)*

ii $y + 5 = 3$

$y = -8$ *(1 mark)*

34 $v = \dfrac{3mn^2}{r}$

$rv = 3mn^2$

$\dfrac{rv}{3m} = n^2$

$\therefore n = \pm\sqrt{\dfrac{rv}{3m}}$

Answer D

35 $\sqrt{\dfrac{x + 2y}{8y}} = \sqrt{\dfrac{5.6 + 2 \times 3.1}{8 \times 3.1}}$

$= 0.6897\ldots$ (by calc.)

$\doteqdot 0.69$.

Answer A

36 $\dfrac{5x + 1}{2} = 4x - 7$

$5x + 1 = 2(4x - 7)$ ✓

$5x + 1 = 8x - 14$

$5x - 8x = -14 - 1$ ✓

$-3x = -15$

$x = \dfrac{-15}{-3}$

$= 5$ ✓ *(3 marks)*

37 $B = 2\pi\left(R + \frac{T}{2}\right)$

$\therefore \frac{B}{2\pi} = R + \frac{T}{2}$

$\therefore \frac{B}{2\pi} - R = \frac{T}{2}$

$\therefore T = 2\left(\frac{B}{2\pi} - R\right)$

$= \frac{B}{\pi} - 2R$

Answer A

38 $D = \frac{t}{3}$

$\therefore 1.2 = \frac{t}{3}$

$\therefore t = 3.6$ seconds

(1 mark)

39 $V = \frac{4}{3}\pi r^3$

$\therefore V = \frac{4}{3} \times \pi \times 2^3$

$= 33.5103\ldots$

$= 33.51$ to 2 decimal places

Answer D

40 $4p = 5t + 2q^2$

$2q^2 = 4p - 5t$

$q^2 = \frac{4p - 5t}{2}$

$q = \pm\sqrt{\frac{4p - 5t}{2}}$

Answer D

41 $\frac{a - b}{4} = \frac{240 - 56}{4}$

$= 46$

Answer B

42 $d = 5t^3 - 2$

$\therefore 137 = 5t^3 - 2$ … A ✓

$139 = 5t^3$ … B

$\therefore$ Line B is incorrect.

Answer B

43 **i** $D = \frac{2A}{15}$

If $A = 9$,

$D = \frac{2 \times 9}{15}$

$= 1.2$ mL

(1 mark)

ii If $D = 4$,

$4 = \frac{2A}{15}$ ✓

$4 \times 15 = 2A$

$60 = 2A$

$\therefore A = 30$

$\therefore$ Sam is 30 months ✓

$\therefore$ Difference in the ages

$= 30 - 9$

$= 21$ months ✓

(3 marks)

44 $T = 2\pi L^2$

$2\pi L^2 = T$

$L^2 = \frac{T}{2\pi}$ ✓

$\therefore L = \pm\sqrt{\frac{T}{2\pi}}$ ✓

(2 marks)

45 $K = Ft^3$

$= 5 \times 0.715^3$

$= 1.8276\ldots$

$= 1.83$ to 3 sig. figures

Answer D

46 $d = 6t^2$

$\therefore 2400 = 6t^2$

$t^2 = \frac{2400}{6}$

$= 400$

$\therefore t = \pm\sqrt{400}$

$= \pm 20$

Answer B

47 **i** Cost of Novirus

$= \$115$ more than x

$= \$115 + x$ *(1 mark)*

ii $x + 115 + x = 415$ ✓

$2x = 415 - 115$

$= 300$

$\therefore x = 150$

$\therefore$ Cost of Funmaths is \$150. ✓

(2 marks)

48 **i** $R = \frac{w}{h^2}$

If $h = 150$ cm $= 1.5$ m

and $w = 72$ kg

$\therefore R = \frac{72}{1.5^2}$

$= 32$

(1 mark)

ii If $R = 25$

and $h = 1.6$ m

$\therefore 25 = \frac{w}{1.6^2}$ ✓

$w = 25 \times 1.6^2$

$= 64$ kg

$\therefore$ Fred should lose 8 kg. ✓

(2 marks)

49 $d = \sqrt{\frac{h}{5}}$

If $h = 28$,

$d = \sqrt{\frac{28}{5}}$

$= \sqrt{5.6}$

$= 2.366\ldots$

$= 2.4$ to 1 decimal place

Answer B

50 If Danny sells 1 CD,

$W = \$0.75 \times 1 + \$50$

$= \$50.75$

If Kylie sells 3 CDs (2 more)

$W = \$0.75 \times 3 + \$50$

$= \$52.25$

$\therefore$ Kylie earns \$1.50 more.

Answer B

51 **i** $A = (77 - 32) \div 1.8$

$= 25$

(1 mark)

ii $F = 2(C + 12)$ ✓✓

(2 marks)

52 $w = 2y^3 - 1$

$\therefore 13 = 2y^3 - 1$

$2y^3 = 14$

$y^3 = 7$

$y = \sqrt[3]{7}$

Answer C

53 $w = \frac{15y}{y + 12}$

$\therefore w = \frac{15 \times 7}{7 + 12}$

$= 5.526\ldots$

$\doteqdot 5.53$ to 2 decimal places

Answer A

1 Suppose $y = -1 - 2x$.

When the value of x increases by 5, the value of y decreases by

A 1 **B** 2
C 5 **D** 10 *(1 mark)*

(Q6, **2020 HSC**) 

2 A plumber charges a call-out fee of \$90 as well as \$2 per minute while working.

Suppose the plumber works for t hours.

Which equation expresses the amount the plumber charges (\$$C$) as a function of time (t hours)?

A $C = 2 + 90t$ **B** $C = 90 + 2t$
C $C = 120 + 90t$ **D** $C = 90 + 120t$

(1 mark)

(Q10, **2020 HSC**) Medium

3 The relationship between British pounds (p) and Australian dollars (d) on a particular day is shown in the graph.

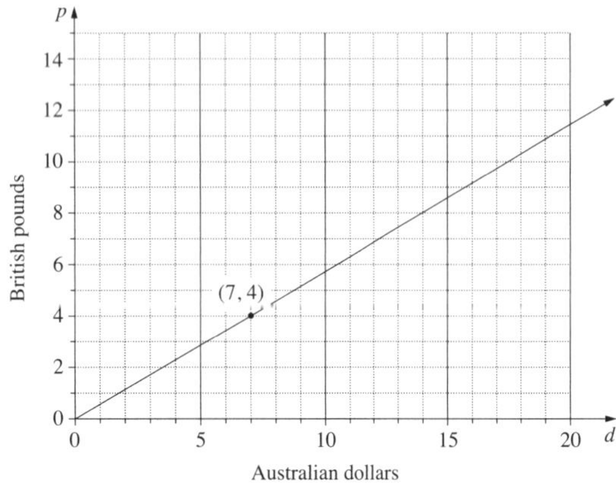

a Write the direct variation equation relating British pounds to Australian dollars in the form $p = md$. Leave m as a fraction. *(1 mark)* Medium

b The relationship between Japanese yen (y) and Australian dollars (d) on the same day is given by the equation $y = 76d$.

Convert 93 100 Japanese yen to British pounds. *(2 marks)* Medium

(Q34, **2019 HSC**)

4 The graph displays the cost (\$$c$) charged by two companies for the hire of a minibus for x hours.

Both companies charge \$360 for the hire of a minibus for 3 hours.

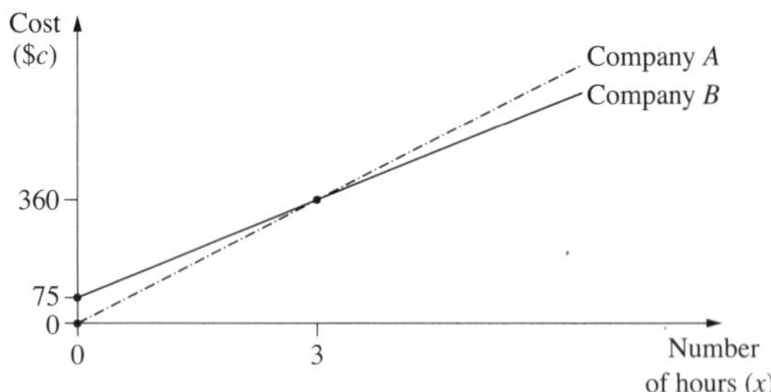

i What is the hourly rate charged by Company A? *(1 mark)* Easy

ii Company B charges an initial booking fee of \$75.

Write a formula, in the form of $c = mx + b$, for the cost of hiring a minibus from Company B for x hours. *(2 marks)* Easy

iii A minibus is hired for 5 hours from Company B.

Calculate how much cheaper this is than hiring from Company A. *(2 marks)* Easy

(Q27d, **2018 HSC**)

5 The graph shows the relationship between infant mortality rate (deaths per 1000 live births) and life expectancy at birth (in years) for different countries.

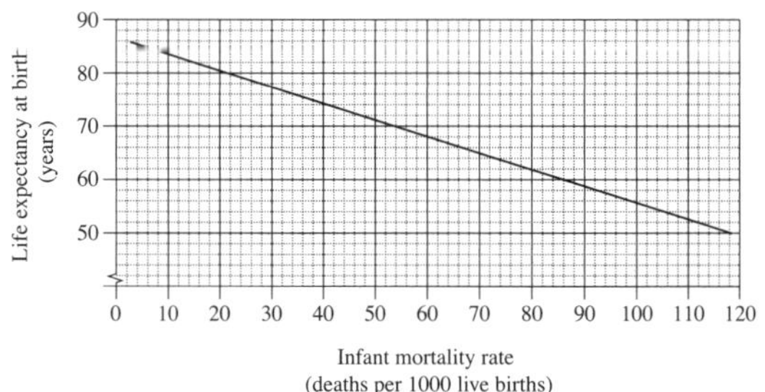

What is the life expectancy at birth in a country which has an infant mortality rate of 60?

A 68 years **B** 69 years
C 86 years **D** 88 years *(1 mark)*

(Q3, **2017 HSC**) Easy

6 A pentagon is created using matches.

By adding more matches, a row of two pentagons is formed.

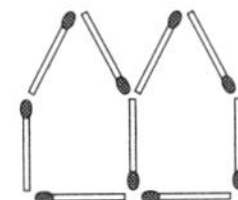

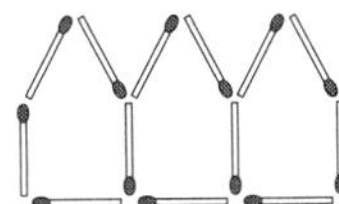

Continuing to add matches, a row of three pentagons can be formed.

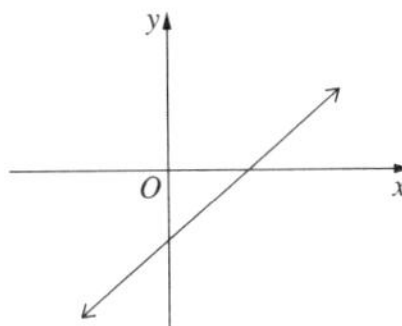

Continuing this pattern, what is the maximum number of complete pentagons that can be formed if 100 matches in total are available?

A 25 **B** 24
C 21 **D** 20 *(1 mark)*

(Q20, **2017 HSC**) Medium

7 The graph shows a line which has an equation in the form $y = mx + b$.

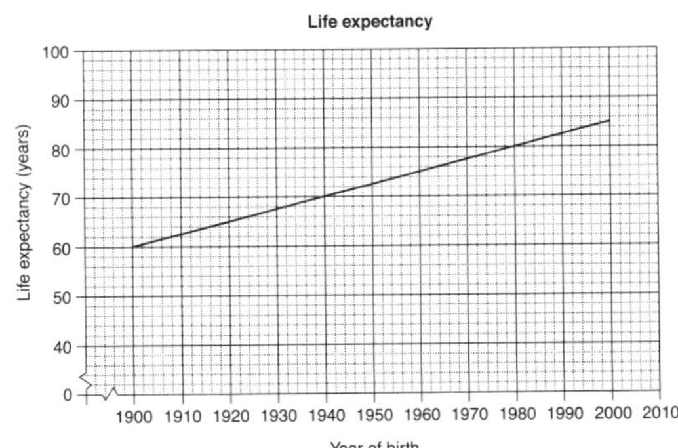

Which of the following statements is true?

A m is positive and b is negative
B m is negative and b is positive
C m and b are both positive
D m and b are both negative *(1 mark)*

(Q14, **2016 HSC**) Easy

8 The graph shows the life expectancy of people born between 1900 and 2000.

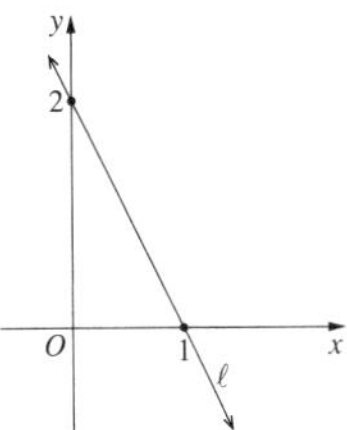

i According to the graph, what is the life expectancy of a person born in 1932? *(1 mark)* Easy

ii With reference to the value of the gradient, explain the meaning of the gradient in this context. *(2 marks)* Easy

(29e, **2016 HSC**)

9 What is the equation of the line ℓ?

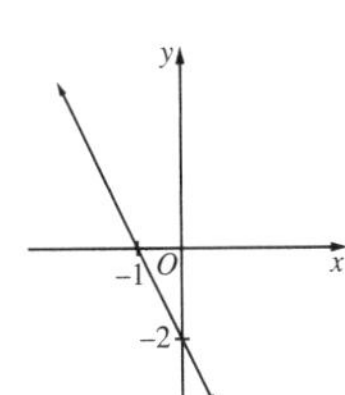

A $y = -2x + 2$ **B** $y = 2x + 2$
C $y = -\frac{x}{2} + 2$ **D** $y = \frac{x}{2} + 2$ *(1 mark)*

(Q13, **2015 HSC**) Medium

10 Which of the following is the graph of $y = 2x - 2$?

A

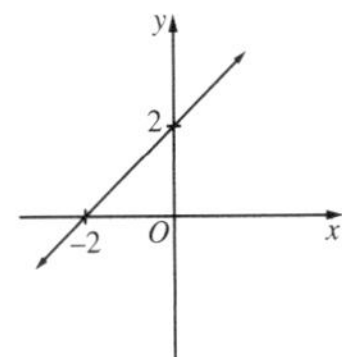

B

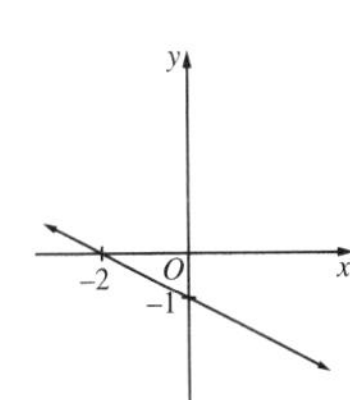

C

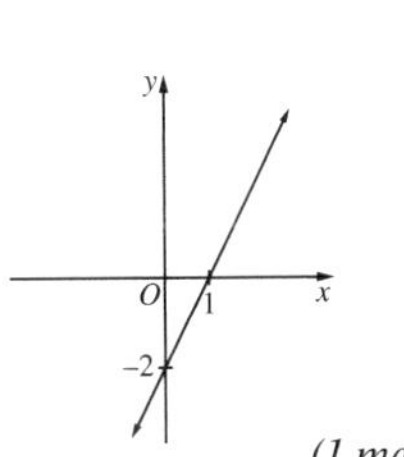

D *(1 mark)*

(Q7, **2014 HSC**) Medium

11 The weight of an object on the moon varies directly with its weight on Earth. An astronaut who weighs 84 kg on Earth weighs only 14 kg on the moon.

A lunar landing craft weighs 2449 kg when on the moon. Calculate the weight of this landing craft when on Earth. *(2 marks)*

(Q26f, **2014 HSC**) Easy

12 The line ℓ has intercepts p and q, where p and q are positive integers.

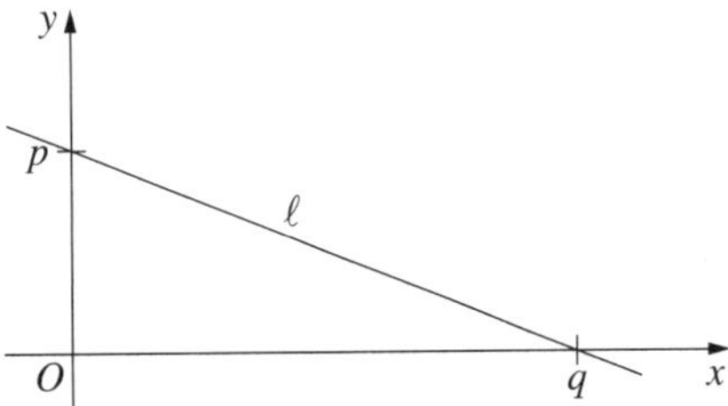

What is the gradient of line ℓ?

A $-\frac{p}{q}$ **B** $-\frac{q}{p}$
C $\frac{p}{q}$ **D** $\frac{q}{p}$ *(1 mark)*

(Q5, **2012 HSC**) Medium

13 Dots were used to create a pattern. The first three shapes in the pattern are shown.

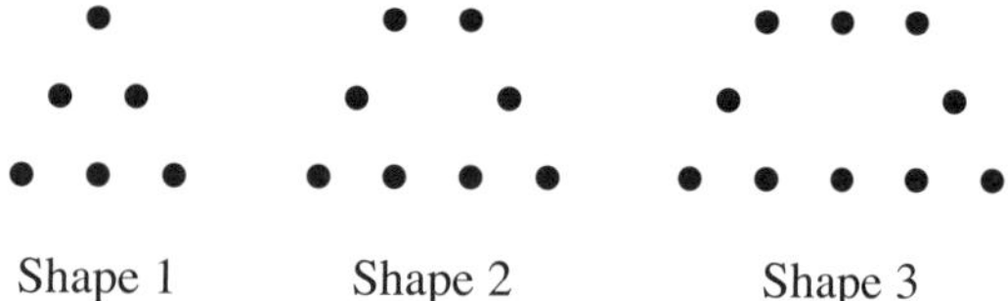

The number of dots used in each shape is recorded in the table.

Shape (S)	1	2	3
Number of dots (N)	6	8	10

How many dots would be required for Shape 156?

A 316 **B** 520
C 624 **D** 936 *(1 mark)*

(Q8, **2012 HSC**) Medium

14 Conversion graphs can be used to convert from one currency to another.

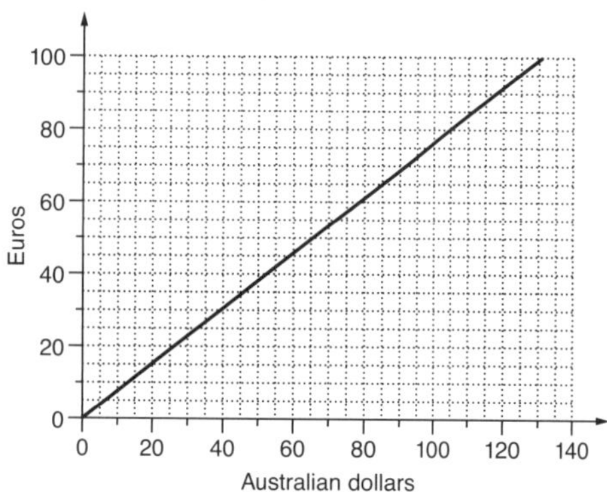

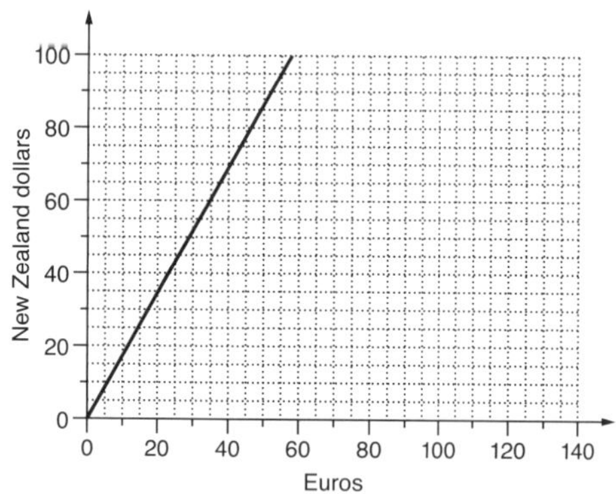

Sarah converted 60 Australian dollars into Euros. She then converted all of these Euros into New Zealand dollars.

How much money, in New Zealand dollars, should Sarah have?

A \$26 **B** \$45
C \$78 **D** \$135 *(1 mark)*

(Q13, **2012 HSC**) Easy

15 The graph shows tax payable against taxable income, in thousands of dollars.

i Use the graph to find the tax payable on a taxable income of \$21 000. *(1 mark)* Easy

ii Use suitable points from the graph to show that the gradient of the section of the graph marked **A** is $\frac{1}{3}$. *(1 mark)* Medium

iii How much of each dollar earned between \$21 000 and \$39 000 is payable in tax? *(1 mark)* Medium

iv Write an equation that could be used to calculate the tax payable, T, in terms of the taxable income, I, for taxable incomes between \$21 000 and \$39 000. *(2 marks)* Hard

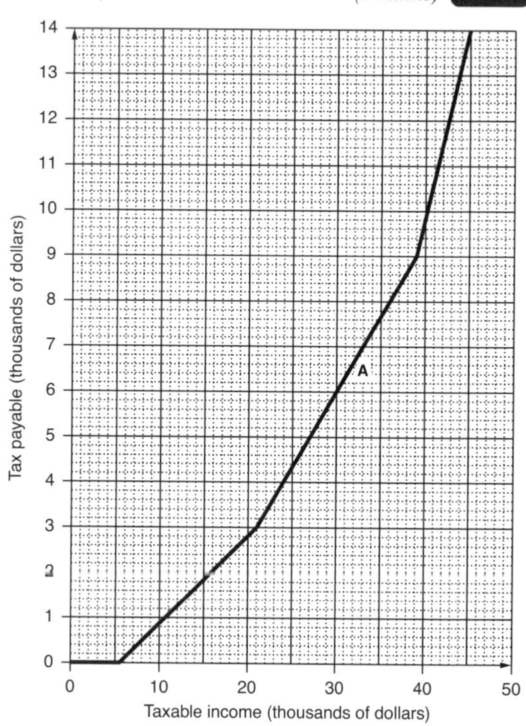

(Q27c, **2010 HSC**)

16 A factory makes boots and sandals. In any week

- the total number of pairs of boots and sandals that are made is 200
- the maximum number of pairs of boots made is 120
- the maximum number of pairs of sandals made is 150.

The factory manager has drawn a graph to show the numbers of pairs of boots (x) and sandals (y) that can be made.

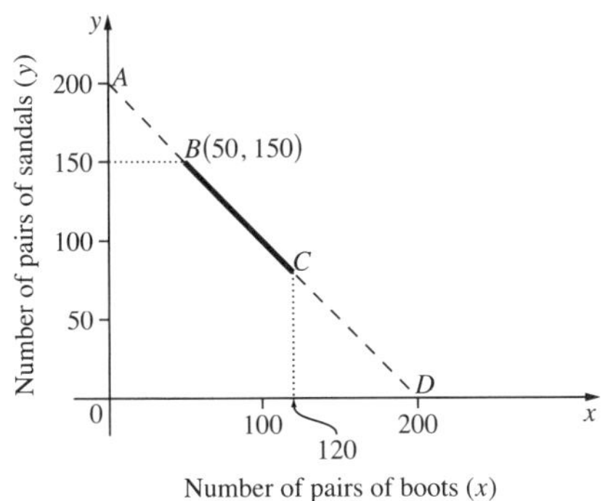

i Find the equation of the line AD. *(1 mark)* **Medium**

ii Explain why this line is only relevant between B and C for this factory. *(1 mark)* **Medium**

iii The profit per week, $\$P$, can be found by using the equation

$$P = 24x + 15y.$$

Compare the profits at B and C. *(2 marks)* **Hard**

(Q24d, **2009 HSC**)

17 Sandy travels to Europe via the USA. She uses this graph to calculate her currency conversions.

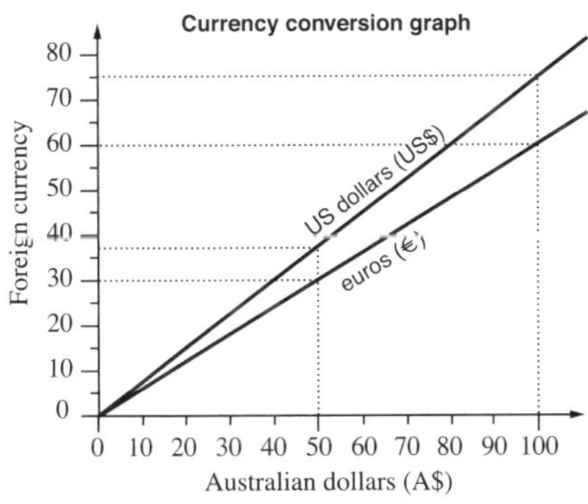

i After leaving the USA she has US\$150 to add to the A\$600 that she plans to spend in Europe. She converts all of her money to euros.
How many euros does she have to spend in Europe? *(3 marks)* **Medium**

ii If the value of the euro falls in comparison to the Australian dollar, what will be the effect on the gradient of the line used to convert Australian dollars to euros? *(1 mark)* **Medium**

(Q24c, **2007 HSC**)

18 A clubhouse uses four long-life light globes for five hours every night of the year. The purchase price of each light globe is \$6.00 and they each cost $\$d$ per hour to run.

i Write an equation for the total cost ($\$c$) of purchasing and running these four light globes for one year in terms of d. *(2 marks)* **Medium**

ii Find the value of d (correct to three decimal places) if the total cost of running these four light globes for one year is \$250. *(1 mark)* **Medium**

iii If the use of the light globes increases to ten hours per night every night of the year, does the total cost double? Justify your answer with appropriate calculations. *(1 mark)* **Hard**

(Q27b, **2007 HSC**)

19 Which equation represents the relationship between x and y in this table?

x	0	2	4	6	8
y	1	2	3	4	5

A $y = 2x + 1$ **B** $y = 2x - 2$
C $y = \frac{x}{2} - 2$ **D** $y = \frac{x}{2} + 1$ *(1 mark)*

(Q7, **2006 HSC**) **Easy**

20 The graph shows the amounts charged by Company A and Company B to deliver parcels of various weights.

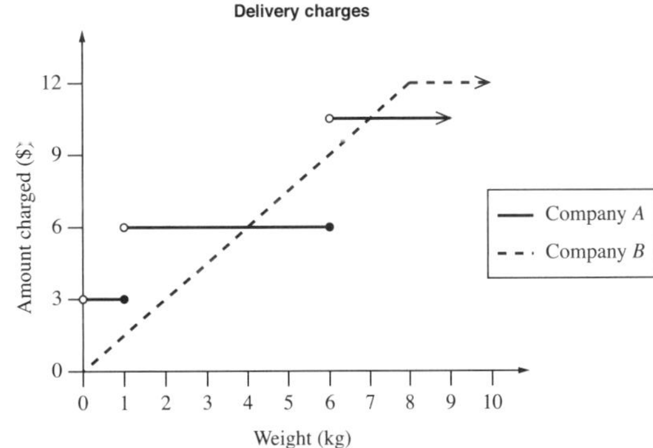

i How much does Company A charge to deliver a 3 kg parcel? *(1 mark)* **Easy**

ii Give an example of the weight of a parcel for which both Company A and Company B charge the same amount. *(1 mark)* **Easy**

iii For what weight(s) is it cheaper to use Company A? *(2 marks)* **Medium**

iv What is the rate per kilogram charged by Company B for parcels up to 8 kg? *(1 mark)* **Medium**

(Q23d, **2006 HSC**)

21 A new tunnel is built. When there is no toll to use the tunnel, 6000 vehicles use it each day. For each dollar increase in the toll, 500 fewer vehicles use the tunnel.

i Find the lowest toll for which no vehicles will use the tunnel. *(1 mark)* **Medium**

ii For a toll of \$5.00, how many vehicles use the tunnel each day and what is the total daily income from tolls? *(2 marks)* **Medium**

iii If d (dollars) represents the value of the toll, find an equation for the number of vehicles (v) using the tunnel each day in terms of d. *(2 marks)* **Hard**

iv Anne says 'A higher toll always means a higher total daily income'.

Show that Anne is incorrect and find the maximum daily income from tolls. (Use a table of values, or a graph, or suitable calculations.) *(3 marks)* **Hard**

(Q28b, **2006 HSC**)

22 The total cost, \$$C$, of a school excursion is given by $C = 2n + 5$, where n is the number of students.

If three extra students go on the excursion, by how much does the total cost increase?

A \$6 **B** \$11
C \$15 **D** \$16 *(1 mark)*

(Q17, **2005 HSC**) **Medium**

23 Susan drew a graph of the height of a plant.

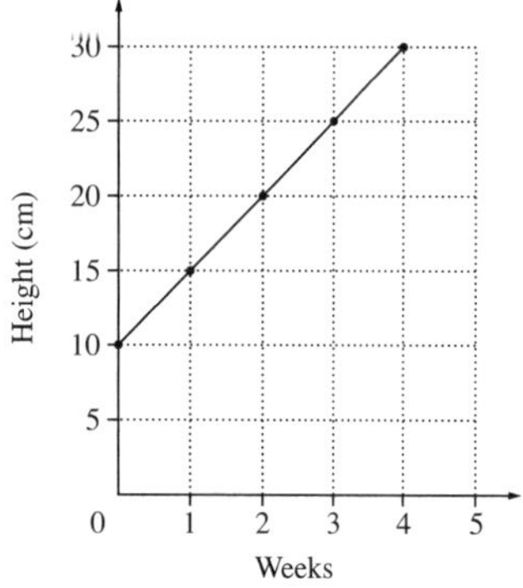

What is the gradient of the line?

A 1 **B** 5
C 7.5 **D** 10 *(1 mark)*

(Q2, **2004 HSC**) **Medium**

24 At a World Cup rugby match, the stadium was filled to capacity for the entire game. At the end of the game, people left the stadium at a constant rate.

The graph shows the number of people (N) in the stadium t minutes after the end of the game.

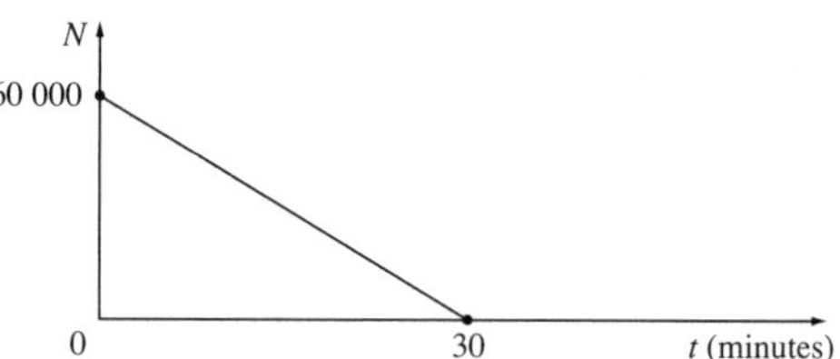

The equation of the line is of the form $N = a - bt$, where a and b are constants.

i Write down the value of a, and give an explanation of its meaning. *(2 marks)* **Medium**

ii **1** Calculate the value of b. *(1 mark)* **Easy**

2 What does the value of b represent in this situation? *(1 mark)* **Medium**

iii Rearrange the formula $N = a - bt$ to make t the subject. *(2 marks)* **Medium**

iv How long did it take 10 000 people to leave the stadium? *(1 mark)* **Medium**

v Copy or trace the graph of N against t shown above into your answer booklet.

Suppose that 15 minutes after the end of the game, several of the exits had been closed, reducing the rate at which people left.

On the same axes, carefully draw another graph of N against t that could represent this new situation. Your new graph should show N from $t = 15$ until all the people had left the stadium. *(2 marks)* **Hard**

(Q26a, **2003 HSC**)

25 Which one of the following could be the graph of $y = 3x + 1$?

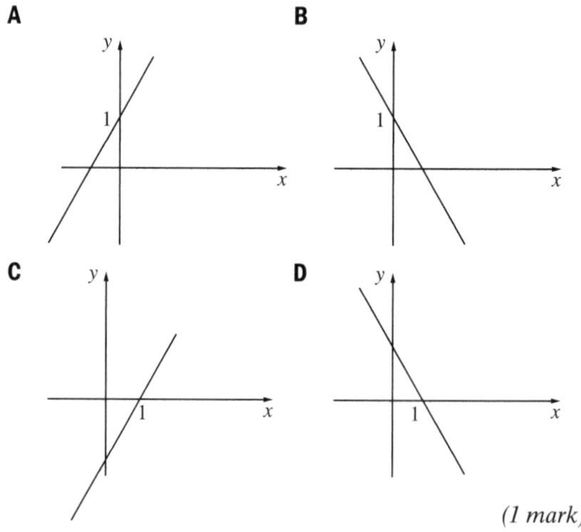

A

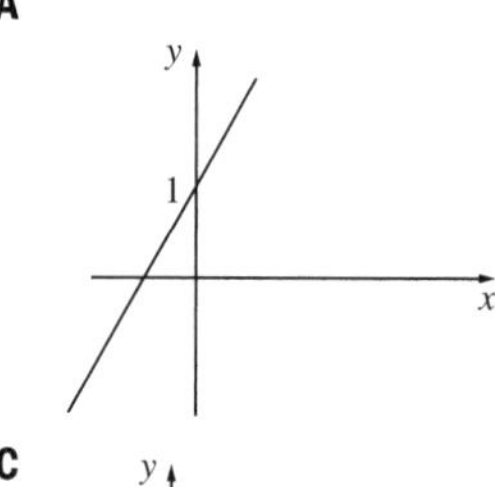

B

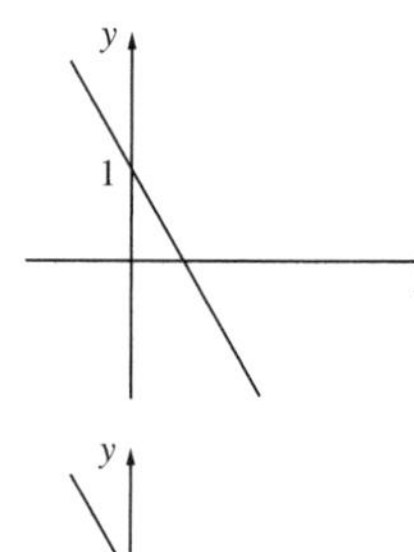

C

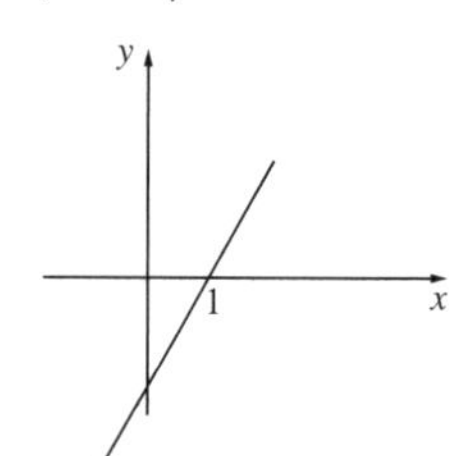

D

(1 mark)

(Q6, **2002 HSC**) **Easy**

26 What is the equation of the line ℓ?

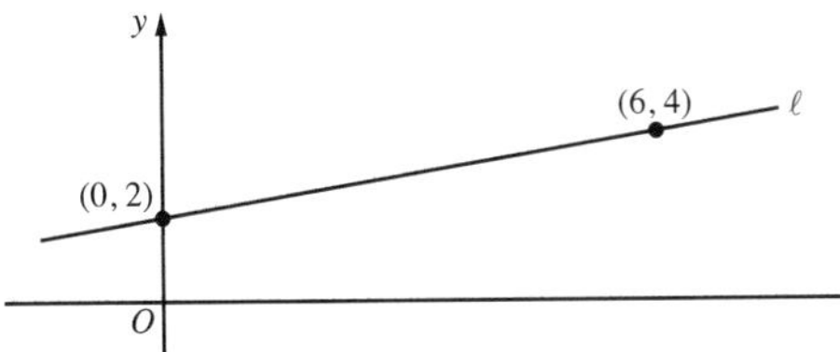

A $y = 6x + 2$ **B** $y = x + 2$

C $y = 3x + 2$ **D** $y = \frac{1}{3}x + 2$ *(1 mark)*

(Q13, **2001 HSC**) **Medium**

---

27 Otto is the manager of a weekend market in which there are 220 stalls for rent. From past experience, Otto knows that if he charges d dollars to rent a stall, then the number of stalls, s, that will be rented is given by:

$$s = 220 - 4d.$$

i How many stalls will be rented if Otto charges \$7.50 per stall? *(1 mark)* **Easy**

ii Copy and complete the following table for the function $s = 220 - 4d$. *(1 mark)* **Easy**

d	10	30	50
s			

iii Draw a graph of the function $s = 220 - 4d$.

Use your ruler to draw the axes. Label each axis, and mark a scale on each axis. *(2 marks)* **Easy**

iv Does it make sense to use the formula $s = 220 - 4d$ to calculate the number of stalls rented if Otto charges \$60 per stall? Explain your answer. *(1 mark)* **Medium**

(Q26a, **2001 HSC**)

Year 11 Linear relationships—Worked answers

1 $y = -1 - 2x$
If x becomes $x + 5$,
$y = -1 - 2(x + 5)$
$= -1 - 2x - 10$
So y decreases by 10.
Answer D

2 \$2 per minute = \$120 per hour
So $C = 90 + 120t$
Answer D

3 **a** $m = \frac{4}{7}$
So $p = \frac{4}{7}d$ *(1 mark)*

b $y = 76d$
When $y = 93\,100$
$76d = 93\,100$
$d = 1225$ ✓
When $d = 1225$
$p = \frac{4}{7} \times 1225$
$= 700$
So 93 100 yen is 700 British pounds. ✓ *(2 marks)*

4 **i** Rate for company A
$= \$360 \div 3$
$= \$120$ per hour *(1 mark)*

ii $m = \frac{\text{vertical change in position}}{\text{horizontal change in position}}$
$= \frac{360 - 75}{3}$
$= 95$ ✓
$b = 75$
$c = mx + b$
$c = 95x + 75$ ✓ *(2 marks)*

iii Cost from company B:
$c = 95x + 75$
When $x = 5$,
$c = 95 \times 5 + 75$
$= 550$ ✓
It costs \$550 to hire the minibus from company B.
Cost from company A
$= 5 \times \$120 = \$600$
So it is \$50 cheaper to hire the minibus from company B than from company A. ✓
(2 marks)

5 Life expectancy at birth = 68 years

Answer A

6 1 pentagon uses 5 matches.
2 pentagons use 9 matches.
3 pentagons use 13 matches.
Each extra pentagon uses an extra 4 matches.
So the rule for the number of matches is 4 times the number of pentagons plus 1.
Now $100 \div 4 = 25$.
So around 25 pentagons could be formed with 100 matches.
But $4 \times 25 + 1 = 101$ so there are not quite enough matches to make 25 pentagons.
$4 \times 24 + 1 = 97$
So 24 complete pentagons could be formed.
Answer B

7 The gradient, m, is positive.
The y-intercept, b, is negative.

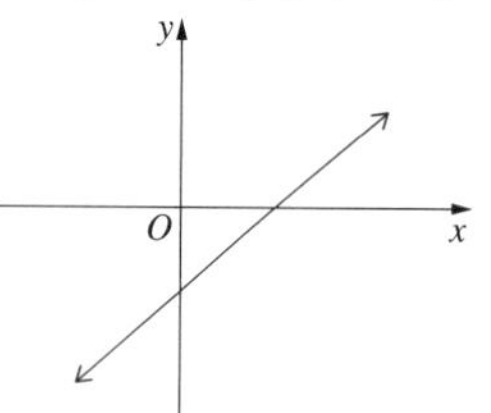

Answer A

8 **i** 68 years

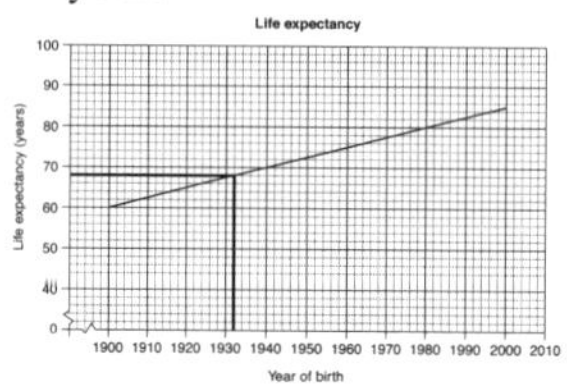

(1 mark)

ii Gradient $= \frac{20}{80} = \frac{1}{4}$ ✓
The gradient shows the number of extra years of life expectancy for every later year of birth. So for every 4 years later the year of birth is one extra year of life expectancy. ✓
(2 marks)

9 gradient $= \frac{-2}{1} = -2$
y-intercept = 2
Equation is $y = -2x + 2$
Answer A

10 $y = 2x - 2$ has y-intercept -2.
[So the graph could only be A or D.] The gradient is 2.
So the graph can only be D.
[A has a negative gradient.]

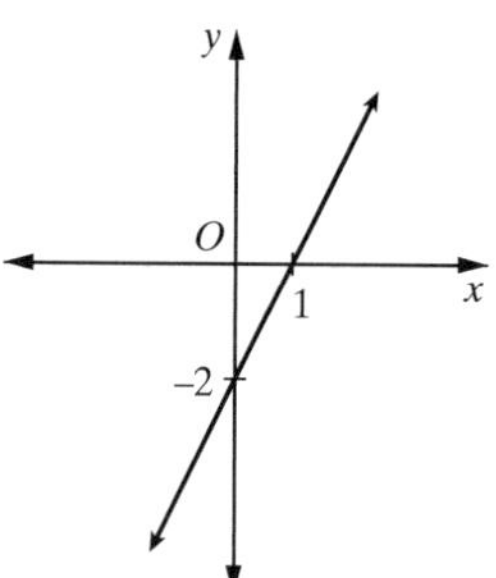

Answer D

11 Let m be the weight on the moon and e the weight on earth.
Now $m \propto e$
$m = ke$ ✓
When $e = 84, m = 14$
$14 = k \times 84$
$k = \frac{14}{84}$
$= \frac{1}{6}$
So $m = \frac{1}{6}e$
When $m = 2449$,
$2449 = \frac{1}{6}e$
$e = 2449 \times 6$
$= 14\,694$
On Earth, the weight of the landing craft would be 14 694 kg. ✓
(2 marks)

12

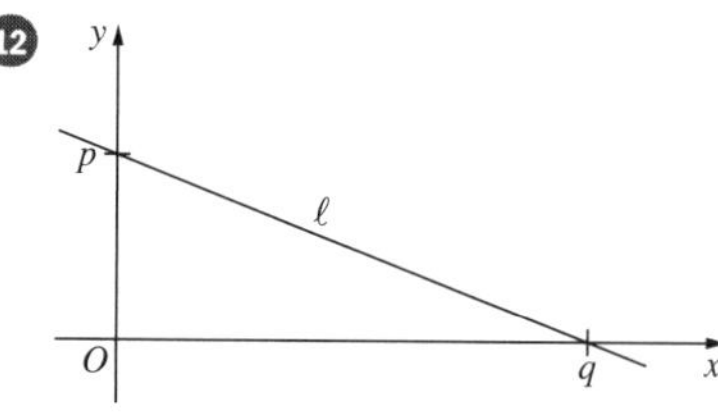

gradient
$= \frac{\text{vertical change in position}}{\text{horizontal change in position}}$
$= \frac{-p}{q}$
$= -\frac{p}{q}$
Answer A

13

Shape (S)	1	2	3
Number of dots (N)	6	8	10

The rule is $N = 2S + 4$
When $S = 156$,
$N = 2 \times 156 + 4$
$= 316$
Answer A

14 From the first graph,
60 Australian dollars is about 46 Euros.

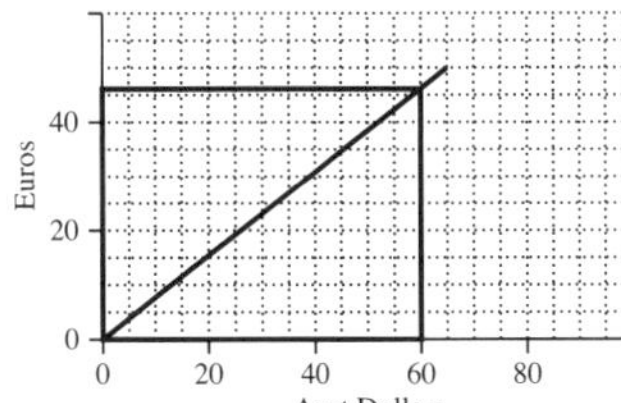

From the second graph 46 Euros is about 78 NZ dollars.

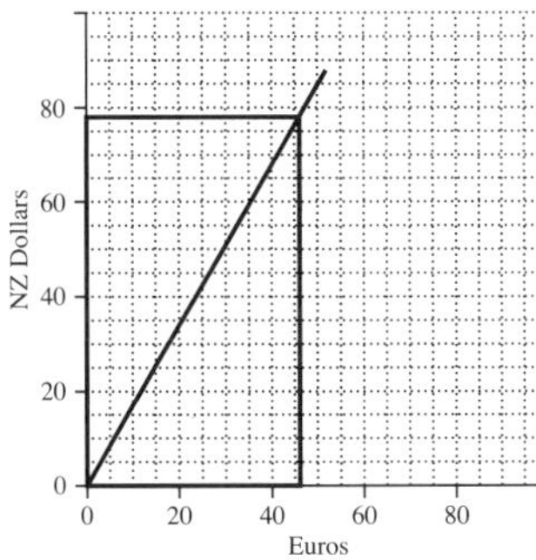

Answer C

15 **i** On \$21 000 the tax payable is \$3000.

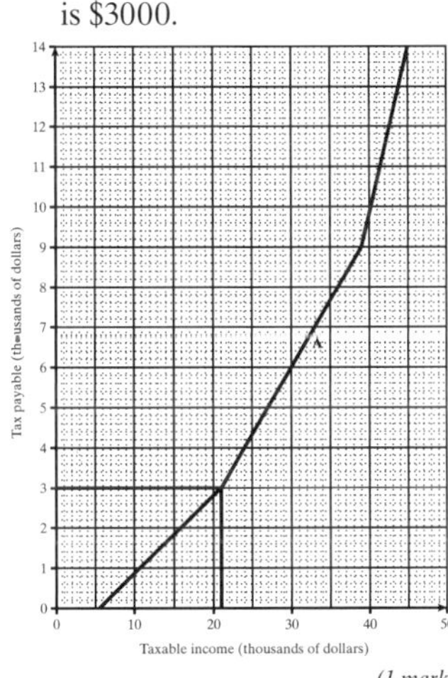

(1 mark)

ii On a taxable income of \$39 000, \$9000 tax is payable.

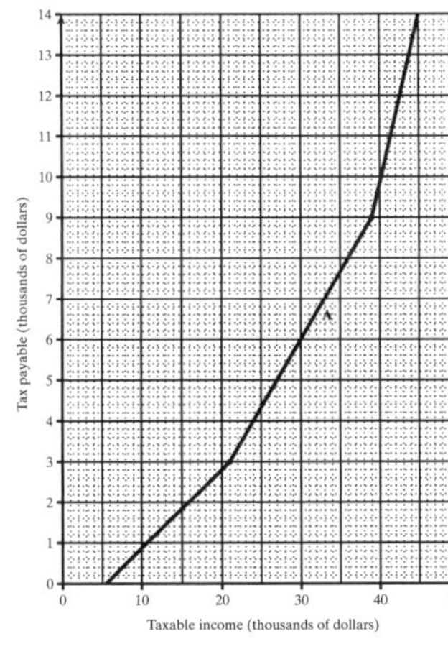

Gradient

$= \dfrac{\text{vertical change in position}}{\text{horizontal change in position}}$

$= \dfrac{9000 - 3000}{39000 - 21000}$

$= \dfrac{6000}{18000}$

$= \dfrac{1}{3}$

(1 mark)

iii $\frac{1}{3}$ of each dollar is payable in tax. So $33\frac{1}{3}$ cents in each dollar earned between \$21 000 and \$39 000 is payable in tax.

(1 mark)

iv The equation is of the form $y = mx + b$.

Now $m = \dfrac{1}{3}$

So $T = \dfrac{1}{3}I + b$

Now when $I = 21\,000$, $T = 3000$

$3000 = \dfrac{1}{3} \times 21\,000 + b$ ✓

$3000 = 7000 + b$

$b = 3000 - 7000$

$b = -4000$

$T = \dfrac{1}{3}I - 4000$ ✓

(2 marks)

16 **i** The number of pairs of boots (x) and the number of pairs of sandals (y), total 200 in a week.

$\therefore x + y = 200$

(1 mark)

ii $y \leqslant 150$ as the number of pairs of sandals must be 150 or less.

$x \leqslant 120$ as the number of pairs of boots must be 120 or less.

(1 mark)

iii $P = 24x + 15y$

At B, $x = 50$ and $y = 150$

$\therefore P = 24 \times 50 + 15 \times 150$
$= \$3450$ ✓

At C, $x = 120$ and $y = 80$

$\therefore P = 24 \times 120 + 15 \times 80$
$= \$4080$

∴ The profit is greater by \$630 when the number of pairs of boots made is the maximum number of pairs of boots that can be made. ✓

(2 marks)

17 **i** From the graph,

US\$75 = €60 = A\$100 ✓
∴ US\$150 = €120 = A\$200
and €360 = A\$600 ✓

∴ Sandy will have
€120 + €360 = €480. ✓

(3 marks)

ii If the value of the euro falls, then A\$100 will buy more euros than the current graph shows. Thus the line will become steeper and the gradient will be greater.

(1 mark)

18 **i** Cost of purchasing
1 globe = \$6
Cost of running
1 globe = \$$d$/h ✓
Running cost per year of 1 globe
$= 5 \times \$d \times 365$
$= \$1825d$

∴ Cost of purchasing and running 4 globes:
$\$c = 4(\$1825d + \$6)$. ✓

(2 marks)

ii $250 = 4(1825d + 6)$

$1825d + 6 = \dfrac{250}{4}$

$1825d = 62.5 - 6$
$1825d = 56.5$
$d = 0.030\,95\ldots$
$= 0.031$ to 3 decimal places
i.e. The cost of running each globe is \$0.031/h.

(1 mark)

iii The running cost will double but not the total cost.
Running cost per year
$= 10 \times \$d \times 365$
$= \$3650d$,
but purchase cost of each globe is still \$6.

(1 mark)

19 From the table,
when $x = 0$, $y = 1$
∴ A or D is possible.

Test another point, e.g. (4, 3):

(A) $y = 2x + 1$

When $x = 4$, $y = 2 \times 4 + 1$
$= 9$
$\neq 3$ ∴ not A.

(D) $y = \dfrac{x}{2} + 1$

When $x = 4$, $y = \dfrac{4}{2} + 1$
$= 3$ ∴ D

Answer D

20 i $6 *(1 mark)*

ii 4 kg or 7 kg *(1 mark)*

iii Weights > 4 kg and ≤ 6 kg, and weights > 7 kg. ✓✓ *(2 marks)*

iv From graph, for 8 kg, cost is $12

$\therefore$ Cost for 1 kg = $\frac{\$12}{8}$ = $1.50

$\therefore$ Rate is $1.50/kg. *(1 mark)*

21 i

Toll (d)	No. of vehicles (v)
$0	6000
$1	5500
$2	5000
$3	4500
⋮	⋮

Now 6000 ÷ 500 = 12

$\therefore$ When the toll is $12.00, no vehicles will use the tunnel. *(1 mark)*

ii Continuing the table above, when the toll is $5.00, 3500 vehicles use the tunnel. ✓

Total daily income = $5 × 3500
= $17 500 ✓ *(2 marks)*

iii Number of vehicles
= 6000 – 500 × cost of toll

$\therefore v = 6000 - 500d$ ✓✓ *(2 marks)*

iv Total income (I)
$= v \times d$
$= (6000 - 500d) \times d$
$= 6000d - 500d^2$ ✓

Looking at a table of values:

d	0	1	2	3	4
I	0	5500	10 000	13 500	16 000

d	5	6	7	8	10	12
I	17 500	18 000	17 500	16 000	10 000	0

Anne is incorrect as shown by the table of values. ✓ The maximum daily income is $18 000 when the toll is $6.00. As the toll increases above $6.00, the number of vehicles decreases as does the total income. ✓ *(3 marks)*

22 $C = 2n + 5$

Let $n = 3$,
$\therefore C = 2 \times 3 + 5$
= $11

If n increases by 3, then $n = 6$.
$\therefore C = 2 \times 6 + 5$
= $17

$\therefore$ Cost increases by $17 – $11 = $6.

Answer A

23 Gradient = $\frac{\text{difference in height}}{\text{difference in time}}$

$= \frac{15 - 10}{1 - 0}$

$= \frac{5}{1}$

= 5 cm/week

Answer B

24 i $a = 60\,000$ ✓

This is the number of people who were in the stadium before any started to leave at the end of the game. 60 000 is the capacity of the stadium. ✓ *(2 marks)*

ii 1 $b = \frac{\text{rise}}{\text{run}}$

$= \frac{60\,000 - 0}{30 - 0}$

= 2000 (by calc.) *(1 mark)*

2 b represents the rate at which the people left the stadium, i.e. 2000 people per minute. *(1 mark)*

iii $N = a - bt$
$\therefore N + bt = a$ ✓
$\therefore bt = a - N$
$\therefore t = \frac{a - N}{b}$ ✓ *(2 marks)*

iv If $a = 60\,000$ and $N = 50\,000$

$t = \frac{60\,000 - 50\,000}{2000}$

= 5 minutes *(1 mark)*

v

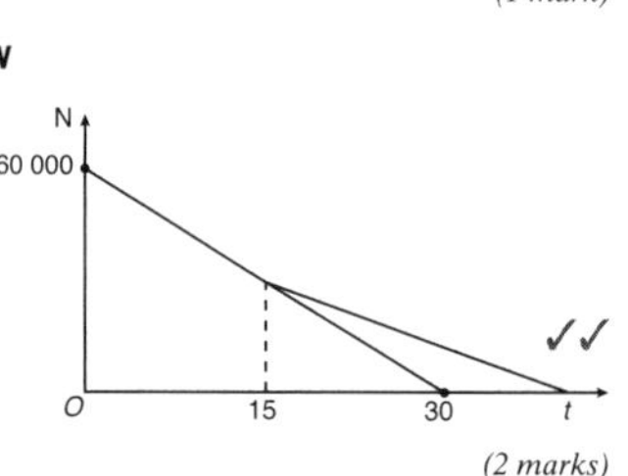

(2 marks)

25 $y = 3x + 1$ is in the form $y = mx + b$.

$\therefore$ The graph passes through 1 on the y axis, and has a positive slope of 3.

Answer A

26

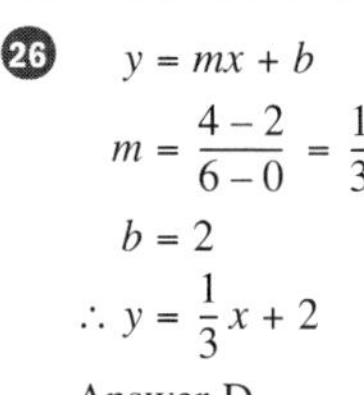

$y = mx + b$

$m = \frac{4 - 2}{6 - 0} = \frac{1}{3}$

$b = 2$

$\therefore y = \frac{1}{3}x + 2$

Answer D

27 i $s = 220 - 4d$
$= 220 - 4 \times 7.50$
$= 190$ *(1 mark)*

ii

d	10	30	50
s	180	100	20

When $d = 10$, $s = 220 - 4 \times 10$
= 180 etc. *(1 mark)*

iii

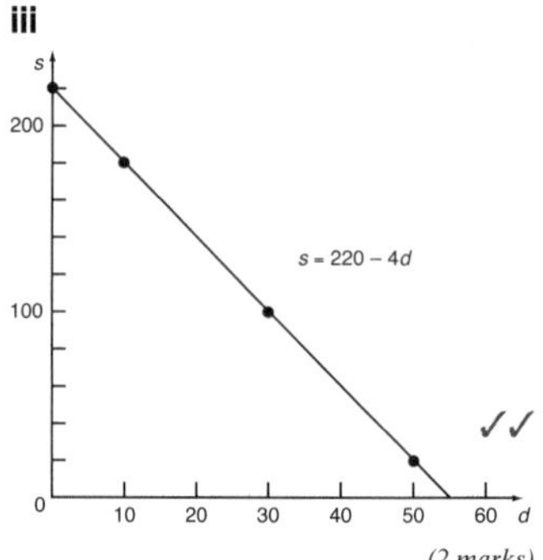

✓✓ *(2 marks)*

iv No, because $d = 60$ is not a possible value of d as it will give a negative number of stalls. *(1 mark)*

1 What is 0.002 073 expressed in standard form with two significant figures?

A 2.07×10^{-2} **B** 2.1×10^{-2}
C 2.07×10^{-3} **D** 2.1×10^{-3} *(1 mark)*

(Q2, **2020 HSC**) Medium

2 A plant stem is measured to be 16.0 cm, correct to one decimal place.

What is the percentage error in this measurement?

A 0.3125% **B** 0.625%
C 3.125% **D** 6.25% *(1 mark)*

(Q5, **2020 HSC**) Medium

3 A person's weight is measured as 79.3 kg.

What is the absolute error of this measurement?

A 10 grams **B** 50 grams
C 100 grams **D** 500 grams *(1 mark)*

(Q8, **2019 HSC**) Medium

4 The length of a window is measured as 2.4 m.

Which calculation will give the percentage error for this measurement?

A $\pm\left(\frac{0.05}{2.4}\right) \times 100$ **B** $\pm\left(\frac{0.05}{100}\right) \times 2.4$
C $\pm\left(\frac{0.5}{2.4}\right) \times 100$ **D** $\pm\left(\frac{0.5}{100}\right) \times 2.4$ *(1 mark)*

(Q18, **2018 HSC**) Medium

5 The length of a netball court is measured to be 30.50 metres, correct to the nearest centimetre.

What is the lower limit for the length of the netball court?

A 30.45 m **B** 30.49 m
C 30.495 m **D** 30.499 m *(1 mark)*

(Q21, **2017 HSC**) Medium

6 What is 208.345 correct to two significant figures?

A 208 **B** 210
C 208.34 **D** 208.35 *(1 mark)*

(Q1, **2016 HSC**) Medium

7 What is 1 560 200 km written in scientific notation correct to two significant figures?

A 1.56×10^4 km **B** 1.6×10^5 km
C 1.56×10^6 km **D** 1.6×10^6 km *(1 mark)*

(Q1, **2015 HSC**) Medium

8 The length of a fish was measured to be 49 cm, correct to the nearest cm.

What is the percentage error in this measurement, correct to one significant figure?

A ± 0.01% **B** ± 0.5%
C ± 1% **D** ± 2% *(1 mark)*

(Q12, **2015 HSC**) Medium

9 The top of the Sydney Harbour Bridge is measured to be 138.4 m above sea level.

What is the percentage error in this measurement?

A 0.036% **B** 0.050%
C 0.072% **D** 0.289% *(1 mark)*

(Q10, **2014 HSC**) Medium

10 A rectangular wooden chopping board is advertised as being 17 cm by 25 cm, with each side measured to the nearest centimetre.

i Calculate the percentage error in the measurement of the longer side. *(1 mark)* Hard

ii Between what lower and upper limits does the actual area of the top of the chopping board lie? *(2 marks)* Hard

(Q27d, **2013 HSC**)

11 The mass of a sample of microbes is 50 mg. There are approximately 2.5×10^6 microbes in the sample.

In scientific notation, what is the approximate mass in grams of one microbe? *(2 marks)*

(Q25b, **2009 HSC**) Hard

12 The capacity of a bottle is measured as 1.25 litres correct to the nearest 10 millilitres.

What is the percentage error for this measurement? *(1 mark)*

(Q23b, **2008 HSC**) Medium

13 What is 0.000 000 326 mm expressed in scientific notation?

A 0.326×10^{-6} mm **B** 3.26×10^{-7} mm
C 0.326×10^{6} mm **D** 3.26×10^{7} mm

(1 mark)

(Q1, **2007 HSC**) Medium

14 George measures the breadth and length of a rectangle to the nearest centimetre. His answers are 10 cm and 15 cm. Between what lower and upper values must the actual area of the rectangle lie?

A 10×15 cm^2 (lower) and 11×16 cm^2 (upper)
B 10×15 cm^2 (lower) and 10.5×15.5 cm^2 (upper)
C 9.5×14.5 cm^2 (lower) and 10×15 cm^2 (upper)
D 9.5×14.5 cm^2 (lower) and 10.5×15.5 cm^2 (upper) *(1 mark)*

(Q18, **2003 HSC**) **Medium**

---

15 Arrange the numbers 5.6×10^{-2}, 4.8×10^{-1}, 7.2×10^{-2} from smallest to largest.

A $5.6 \times 10^{-2}, 7.2 \times 10^{-2}, 4.8 \times 10^{-1}$
B $4.8 \times 10^{-1}, 5.6 \times 10^{-2}, 7.2 \times 10^{-2}$
C $7.2 \times 10^{-2}, 5.6 \times 10^{-2}, 4.8 \times 10^{-1}$
D $4.8 \times 10^{-1}, 7.2 \times 10^{-2}, 5.6 \times 10^{-2}$ *(1 mark)*

(Q14, **2002 HSC**) **Medium**

---

16 The number represented by a 1 followed by one hundred zeros is called a googol.

Which of the following is equal to a googol?

A 10^2 **B** 10^{10}
C 10^{99} **D** 10^{100} *(1 mark)*

(Q6, **2001 HSC**) **Medium**

---

17 Joyce measures the length of a piece of wood as 250 mm, correct to the nearest mm.

What is the percentage error in her measurement?

A $\pm 0.002\%$ **B** $\pm 0.004\%$
C $\pm 0.2\%$ **D** $\pm 0.4\%$ *(1 mark)*

(Q14, **2001 HSC**) **Medium**

Year 11 Practicalities of measuring—Worked answers

1 $0.002073 = 0.0021$ (2 sig. figs)
$= 2.1 \times 10^{-3}$
Answer D

2 Absolute error $= \frac{1}{2} \times 0.1$
$= 0.05$
Percentage error $= \frac{0.05}{16.0} \times 100\%$
$= 0.3125\%$
Answer A

3 Absolute error $= \frac{1}{2}$ of 0.1 kg
$= 0.05$ kg
$= 50$ g
Answer B

4 Error $= \pm 0.05$ m
Percentage error
$= \pm\left(\frac{0.05}{2.4}\right) \times 100\%$
So the required calculation
is $\pm\left(\frac{0.05}{2.4}\right) \times 100$
Answer A

5 Length = 30.50 m to the nearest centimetre
So length $= 30.50$ m $\pm$ 0.5 cm
Lower limit $= 30.50$ m $-$ 0.5 cm
$= (30.50 - 0.005)$ m
$= 30.495$ m
Answer C

6 $208.345 = 210$ (2 sig. figs.)
Answer B

7 1 560 200 km
$= 1\,600\,000$ km (2 sig. figs)
$= 1.6 \times 10^6$ km
Answer D

8 Error $= \pm 0.5$ cm
Percentage error
$= \pm\frac{0.5}{49} \times 100\%$
$= \pm 1.020408\ldots\ \%$
$= \pm 1\%$ (1 sig. fig.)
Answer C

9 Error $= \pm 0.05$ m
Percentage error
$= \pm\frac{0.05}{138.4} \times 100\%$
$= \pm 0.036127\ldots\%$
$= \pm 0.036\%$ (2 sig. figs)
Answer A

10 **i** Error $= \pm 0.5$ cm
Percentage error
$= \pm\frac{0.5}{25} \times 100\%$
$= \pm 2\%$
(1 mark)

ii Smallest area
$= (24.5 \times 16.5)\ \text{cm}^2$
$= 404.25\ \text{cm}^2$ ✓
Largest area
$= (25.5 \times 17.5)\ \text{cm}^2$
$= 446.25\ \text{cm}^2$
The actual area is greater than or equal to $404.25\ \text{cm}^2$ and less than or equal to $446.25\ \text{cm}^2$. ✓ *(2 marks)*

11 Mass of 2.5×10^6 microbes = 50 mg
∴ Mass of 1 microbe
$= \left(\frac{50}{2.5 \times 10^6}\right)$ mg ✓
$= 2 \times 10^{-5}$ mg
$= (2 \times 10^{-5} \div 1000)$ g
$= 2 \times 10^{-8}$ g ✓
(2 marks)

12 Percentage error
$= \frac{\text{error}}{\text{measurement}} \times 100$
$= \frac{5\text{ mL}}{1250\text{ mL}} \times 100$
$= 0.4\%$
(1 mark)

13 0.000 000 326 mm
$= 3.26 \times 10^{-7}$ mm
Answer B

14 Smallest unit = 1 cm
Error = 0.5 cm
∴ Length lies between 14.5 and 15.5 cm. Width lies between 9.5 and 10.5 cm.
∴ Area lies between
$(9.5 \times 14.5)\ \text{cm}^2$
and $(10.5 \times 15.5)\ \text{cm}^2$
Answer D

15 $5.6 \times 10^{-2} = 0.056$
$4.8 \times 10^{-1} = 0.48$
$7.2 \times 10^{-2} = 0.072$

Arranged from smallest to largest:
5.6×10^{-2}, 7.2×10^{-2}, 4.8×10^{-1}

Answer A

16 Imagine 1 followed by 3 zeros: $1000 = 10^3$
∴ 1 followed by 100 zeros is 10^{100}.
Answer D

17 Error = ± 0.5 mm
Percentage error $= \pm\frac{0.5}{250} \times 100$
$= \pm 0.2\%$

Answer C

1 Which of the following shapes has the largest perimeter?

A

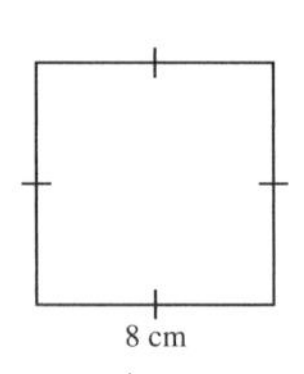

B

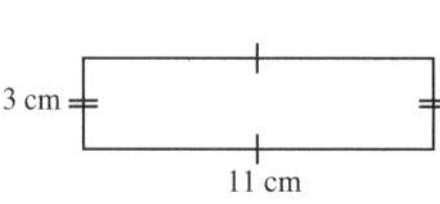

C

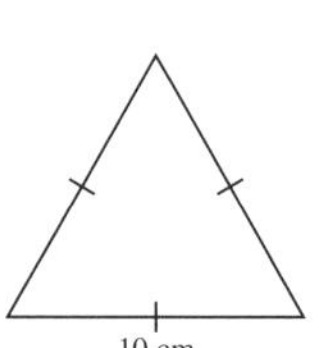

D

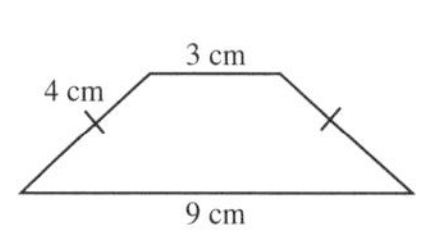

NOT TO SCALE

(1 mark)

(Q1, **2021 HSC**) Easy

2 A block of land is represented by the shaded region on the number plane. All measurements are in kilometres.

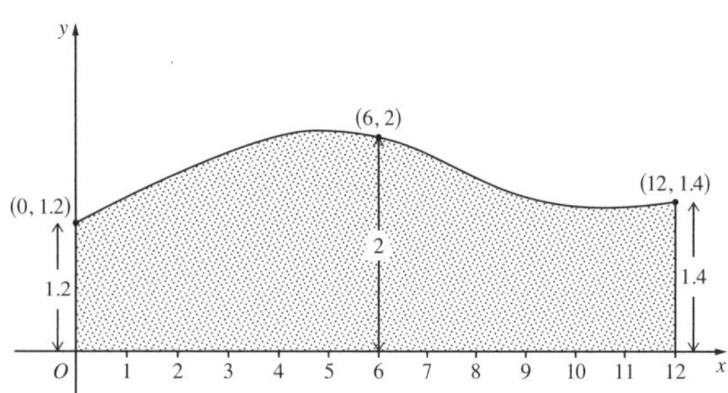

Which of the following is the approximation for the area of this block of land in square kilometres, using two applications of the trapezoidal rule?

A 9.9 **B** 19.8

C 39.6 **D** 72 *(1 mark)*

(Q12, **2021 HSC**) Easy

3 The volume, V, of a sphere is given by the formula

$$V = \frac{4}{3}\pi r^3,$$

where r is the radius of the sphere.

A tank consists of the bottom half of a sphere of radius 2 metres, as shown.

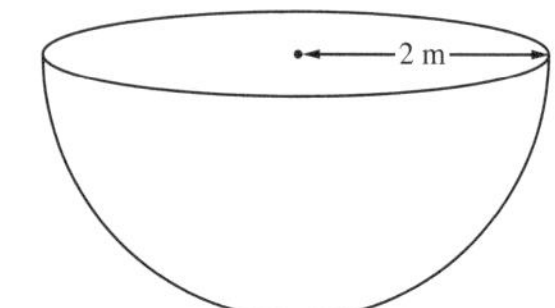

Find the volume of the tank in cubic metres, correct to one decimal place. *(2 marks)*

(Q16, **2021 HSC**) Easy

4 A composite solid consists of a triangular prism which fits exactly on top of a cube, as shown.

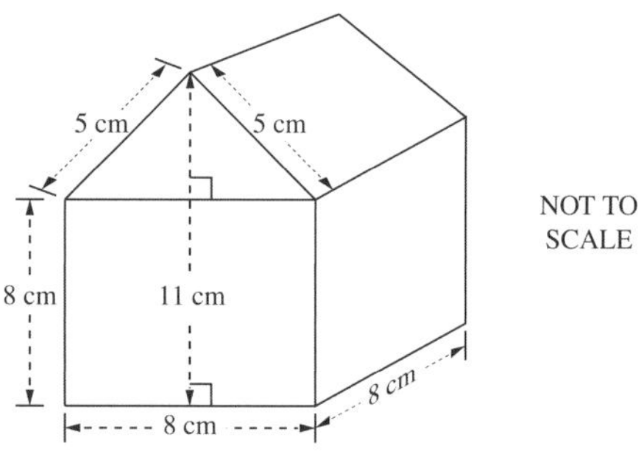

Find the surface area of the composite solid. *(3 marks)*

(Q25, **2020 HSC**) Medium

5 Which of the following shapes has a perimeter of 12 cm?

A

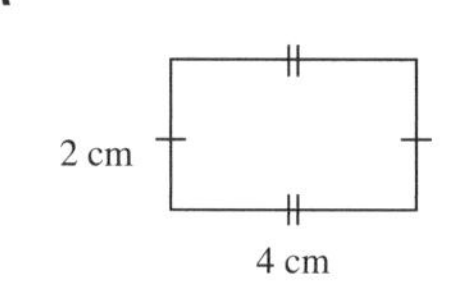

B

C

D

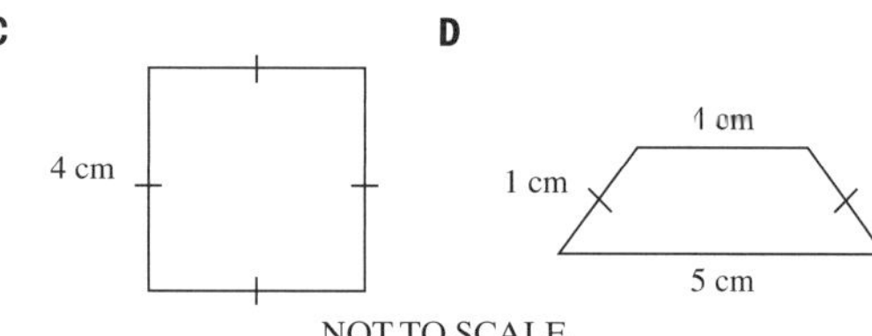

NOT TO SCALE

(1 mark)

(Q1, **2019 HSC**) Easy

6 A bowl is in the shape of a hemisphere with a diameter of 16 cm.

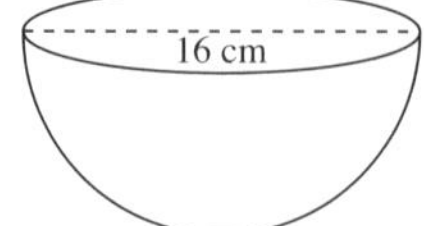

What is the volume of the bowl, correct to the nearest cubic centimetre? *(2 marks)*

(Q16, **2019 HSC**) Medium

7 In order to find the area of a dam, Madi took some measurements in metres and drew the following diagram (not to scale).

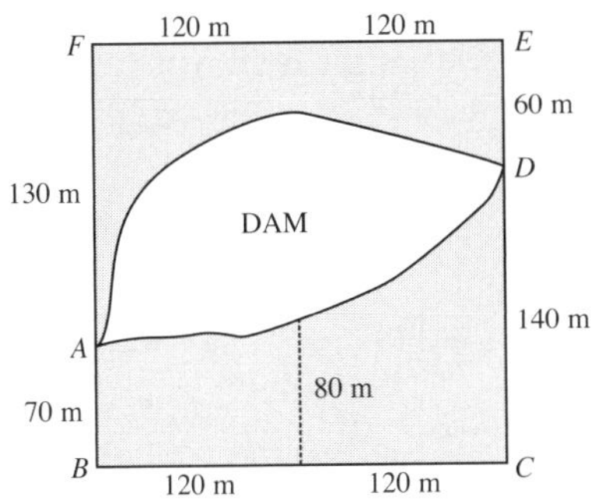

a Use applications of the trapezoidal rule to find an approximation for the shaded area $ABCD$. *(2 marks)* **Medium**

b It is known that the area of the shaded section $AFED$ is 13 800 m^2. If the average depth of water in the dam is 3.6 metres, what is the amount of water in the dam, to the nearest megalitre? *(2 marks)* **Hard**

Bonus question (see page iv)

8 A piece of timber is used as a bookend. It has a uniform cross-section which has been shaded in the diagram.

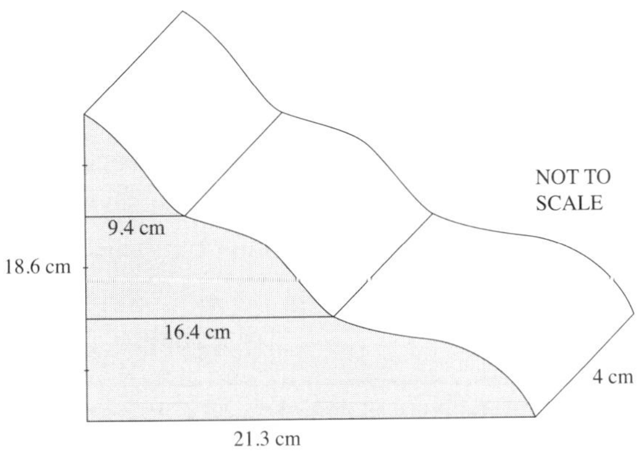

a Using applications of the trapezoidal rule find an approximate area of the cross-section. *(2 marks)* **Medium**

b The area of the curved section is 128.7 cm^2. What is the approximate surface area of the entire shape, to the nearest square centimetre? *(2 marks)* **Hard**

Bonus question

9 A river has a cross-section as shown in the diagram, where all measurements are in metres. Calculate the approximate area of this cross-section using three applications of the trapezoidal rule. *(2 marks)*

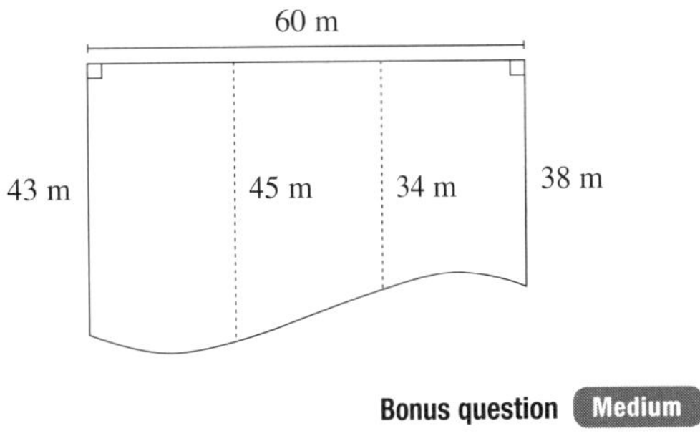

Bonus question **Medium**

10 The diagram shows a paddock with an existing road as one of the boundaries.

Steve used two applications of the trapezoidal rule to find the area of the paddock.

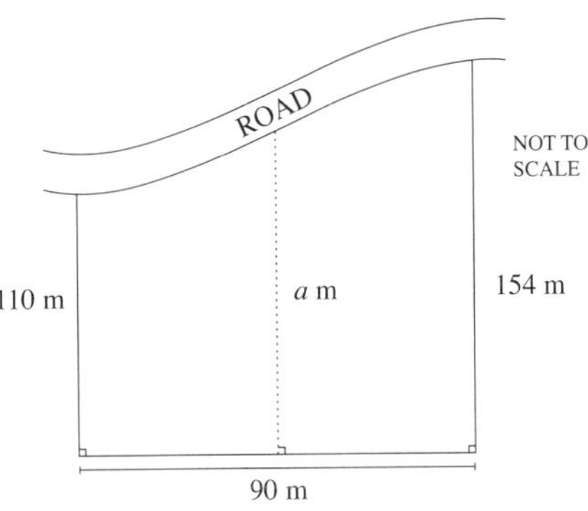

If the area is 1.215 ha what is the value of a? *(2 marks)*

Bonus question **Hard**

11 The solid is comprised of a cylinder and a hemisphere.

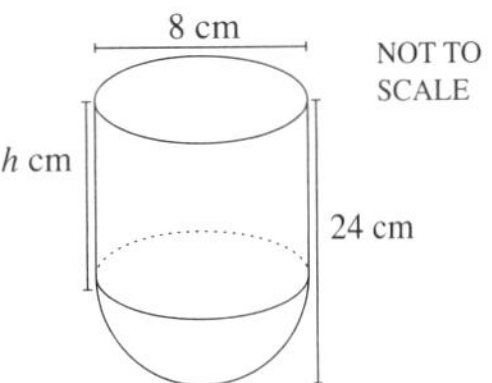

Find the volume and the surface area of the solid, correct to three significant figures. *(4 marks)*

Bonus question **Hard**

12 There is a lake inside the rectangular grass picnic area $ABCD$, as shown in the diagram.

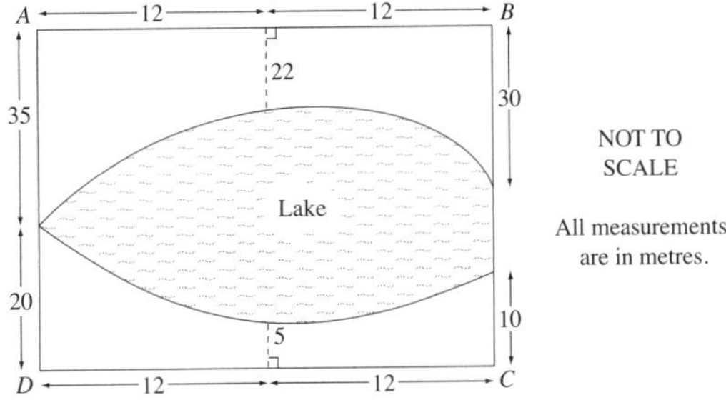

a Use applications of the trapezoidal rule to find the approximate area of the lake's surface. *(2 marks)* **Medium**

b The lake is 60 cm deep. Bozo the clown thinks he can empty the lake using a 4-L bucket. How many times would he have to fill his bucket from the lake in order to empty the lake? (Note that $1 \text{ m}^3 = 1000$ L). *(2 marks)* **Hard**

Bonus question

13 A large oil filter is in the shape of a cylinder of height 24 cm sitting on top of a hemisphere of radius 14 cm.

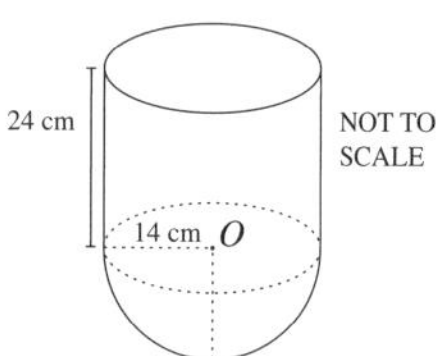

Calculate the volume of the filter, correct to the nearest cubic centimetre. *(2 marks)*

Bonus question **Hard**

14 The diagram shows the base of a pool with a consistent depth of 1.8 m.

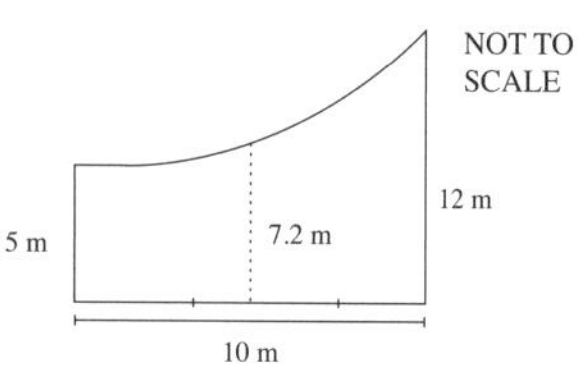

a Use two applications of the trapezoidal rule to estimate the volume of water in the pool. *(2 marks)* **Medium**

b On a hot and windy day the level of water drops by 16 mm. If a hose can fill a 10-litre bucket in 8 seconds, how long will it take the hose to fill the pool back to a depth of 1.8 m? Give your answer to the nearest minute. *(3 marks)* **Hard**

Bonus question

15 A rectangular pyramid has base side lengths $3x$ and $4x$. The perpendicular height of the pyramid is $2x$. All measurements are in metres.

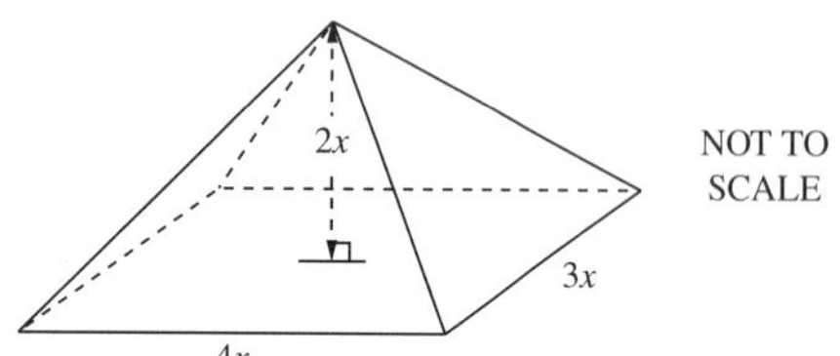

What is the volume of the pyramid in cubic metres?

A $8x^3$ **B** $9x^3$
C $12x^3$ **D** $24x^3$ *(1 mark)*

(Q13, **2018 HSC**) **Easy**

16 A shape consisting of a quadrant and a right-angled triangle is shown.

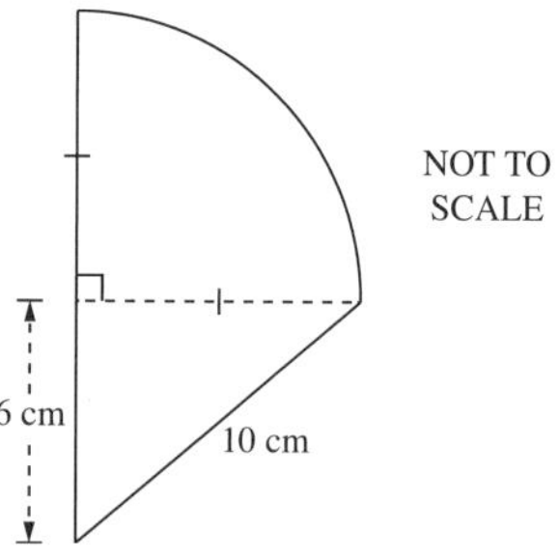

What is the perimeter of this shape, correct to one decimal place?

A 28.6 cm **B** 36.6 cm
C 66.3 cm **D** 74.3 cm *(1 mark)*

(Q22, **2018 HSC**) **Easy**

17 A shade shelter is to be constructed in the shape of half a cylinder with open ends. The diameter is 3.8 m and the length is 10 m.

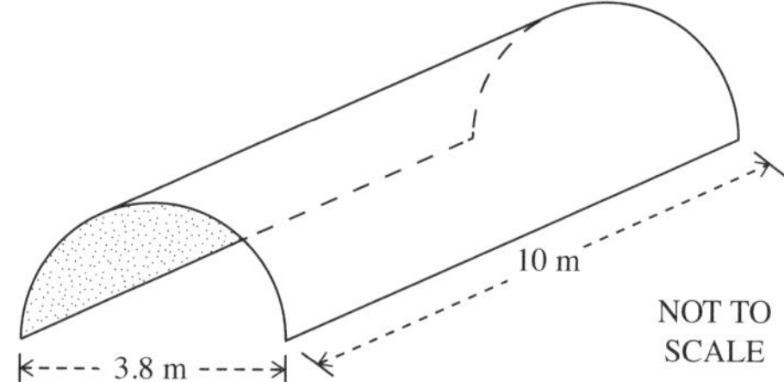

The curved roof is to be made of plastic sheeting.

What area of plastic sheeting is required, to the nearest m^2? *(2 marks)*

(Q27c, **2018 HSC**) **Medium**

18 A cylindrical water tank has a radius of 9 metres and a capacity of 1.26 megalitres.

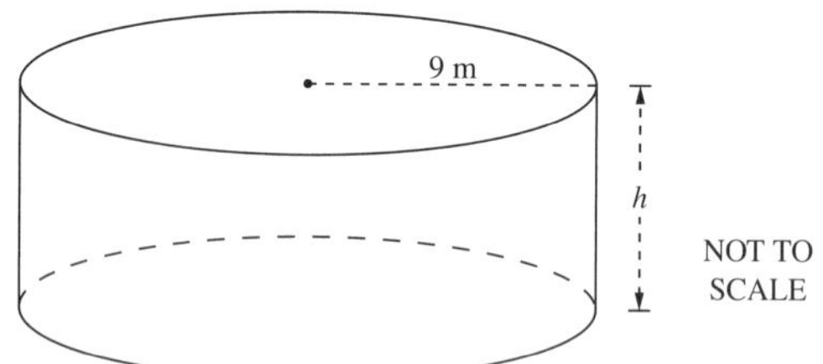

What is the height of the water tank? Give your answer in metres, correct to two decimal places. *(3 marks)*

(Q30a, **2018 HSC**) **Medium**

19 A skip bin is in the shape of a trapezoidal prism, with dimensions as shown.

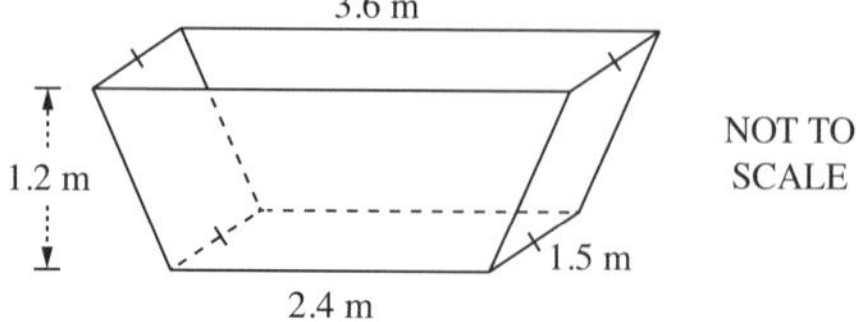

What is the volume of the skip bin?

A 5.4 m^3 **B** 7.776 m^3
C 10.8 m^3 **D** 15.552 m^3 *(1 mark)*

(Q18, **2017 HSC**) **Medium**

20 A concrete water pipe is manufactured in the shape of an annular cylinder. The dimensions are shown in the diagrams.

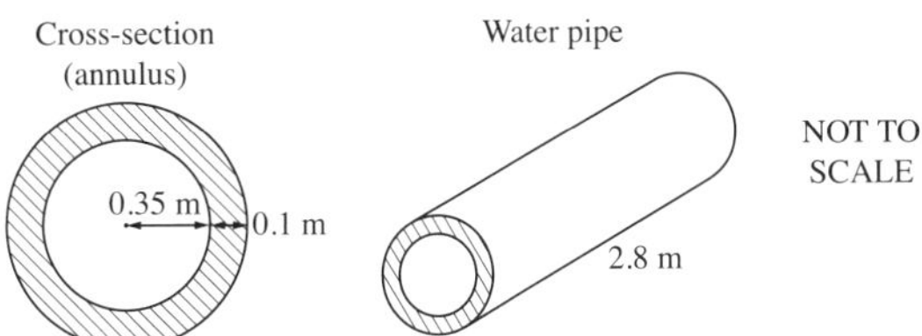

What is the approximate volume of concrete needed to make the water pipe?

A 0.06 m^3 **B** 0.09 m^3
C 0.70 m^3 **D** 0.99 m^3 *(1 mark)*

(Q22, **2017 HSC**) **Medium**

21 In the circle, centre O, the area of the quadrant is 100 cm^2.

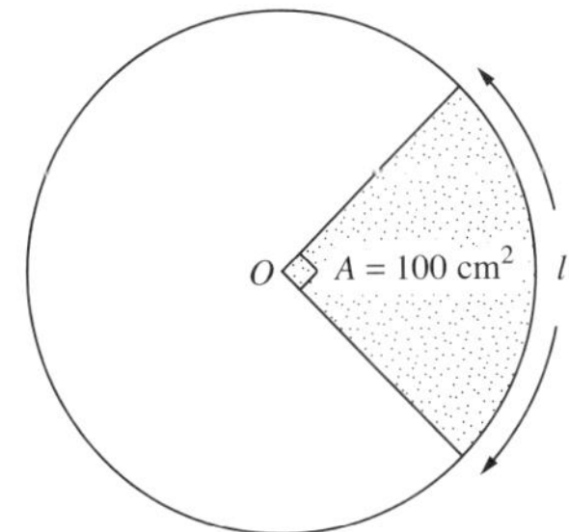

What is the arc length l, correct to one decimal place?

A 8.9 cm **B** 11.3 cm
C 17.7 cm **D** 25.1 cm *(1 mark)*

(Q25, **2017 HSC**) **Medium**

22 A solid is made up of a sphere sitting partially inside a cone.

The sphere, centre O, has a radius of 4 cm and sits 2 cm inside the cone. The solid has a total height of 15 cm. The solid and its cross-section are shown.

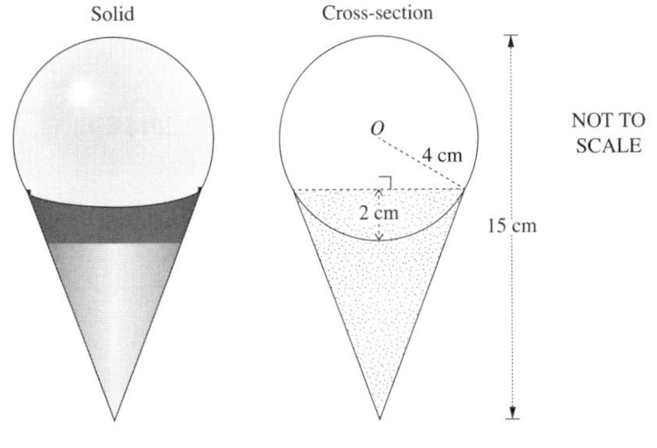

What is the volume of the cone, correct to the nearest cm^3? *(3 marks)*

(Q30e, **2017 HSC**) **Hard**

23 A container is in the shape of a triangular prism which has a capacity of 12 litres. The area of the base is 240 cm^2.

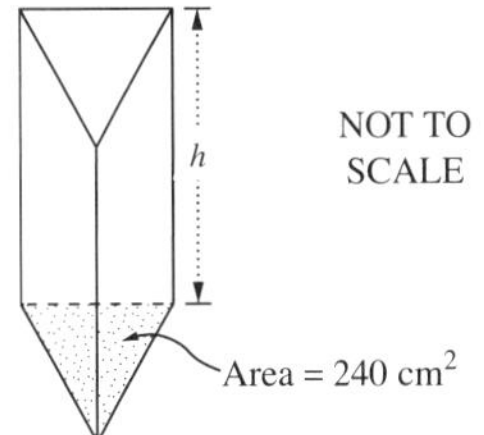

What is the distance, h, between the two triangular ends of the container?

A 5 cm **B** 20 cm
C 25 cm **D** 50 cm *(1 mark)*

(Q12, **2016 HSC**) **Hard**

24 The width (W) of a river can be calculated using two similar triangles, as shown in the diagram.

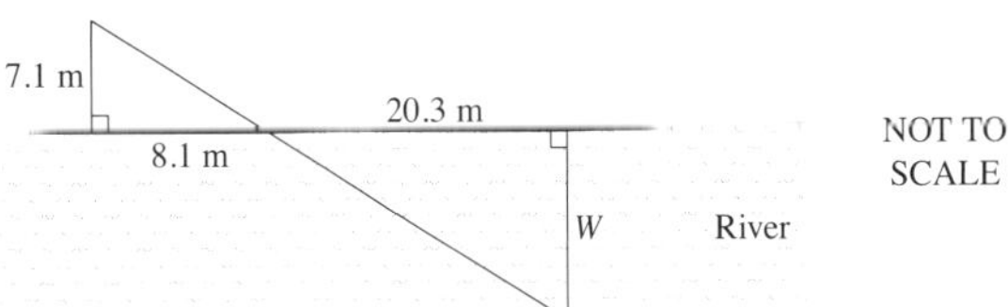

What is the approximate width of the river?

A 17.8 m **B** 19.3 m
C 23.2 m **D** 24.9 m *(1 mark)*

(Q16, **2016 HSC**) **Easy**

25 Calculate the surface area of a sphere with a radius of 5 cm, correct to the nearest whole number.

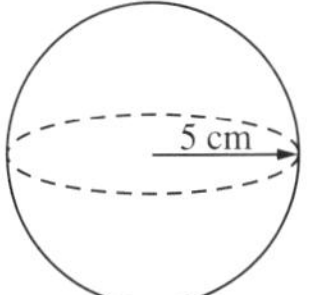

(1 mark)

(Q26a, **2016 HSC**) **Easy**

26 The field diagram shows a block of land $ABCD$ that has been surveyed. All measurements are in metres.

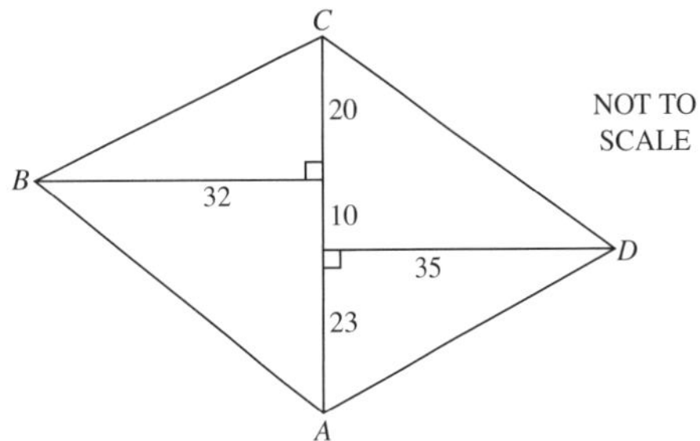

Calculate the length of AB, correct to the nearest metre. *(2 marks)*

(Q26d, **2016 HSC**) Easy

27 A company makes large marshmallows. They are in the shape of a cylinder with diameter 5 cm and height 3 cm, as shown in the diagram.

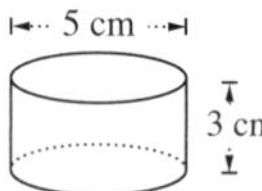

i Find the volume of one of these large marshmallows, correct to one decimal place. *(2 marks)* Easy

ii A cake is to be made by stacking 24 of these large marshmallows and filling the gaps between them with chocolate. The diagrams show the cake and its top view. The shading shows the gaps to be filled with chocolate.

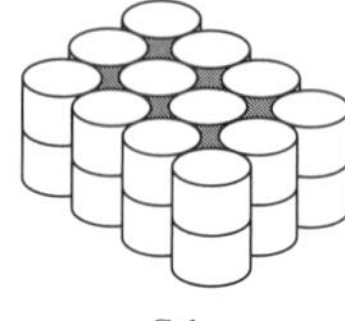
Cake

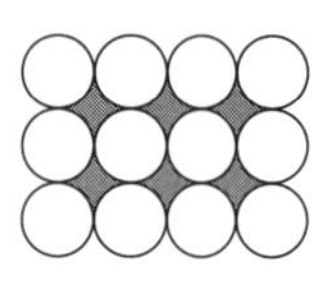
Top view

What volume of chocolate will be required? Give your answer correct to the nearest whole number. *(3 marks)* Hard

(Q28e, **2016 HSC**)

28 The area of a roof is 30 m^2. Any rain that falls on the roof flows directly onto a garden.

Calculate how many litres of water flow onto the garden when 20 mm of rain falls on the roof. *(2 marks)*

(Q30a, **2016 HSC**) Hard

29 The Louvre Pyramid in Paris has a square base with side length 35 m and a perpendicular height of 22 m.

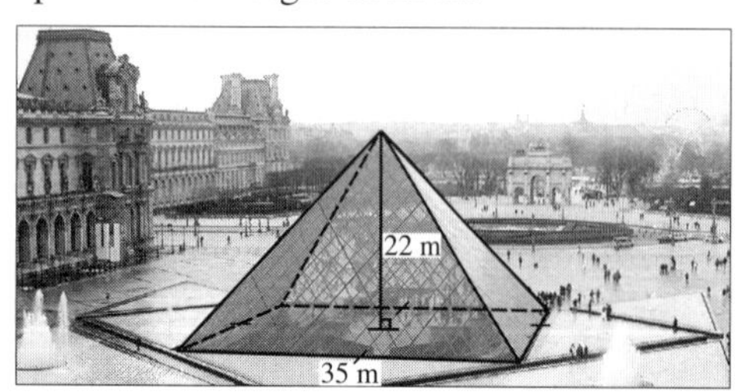

What is the volume of this pyramid, to the nearest m^3?

A 257 m^3 **B** 1027 m^3
C 8983 m^3 **D** 26 950 m^3 *(1 mark)*

(Q8, **2015 HSC**) Easy

30 Approximately 71% of Earth's surface is covered by water. Assume Earth is a sphere.

Calculate the number of square kilometres covered by water. *(2 marks)*

[radius of earth = 6400 km]

(Q26f, **2015 HSC**) Medium

31 At a particular time during the day, a tower of height 19.2 metres casts a shadow.

At the same time, a person who is 1.65 metres tall casts a shadow 5 metres long.

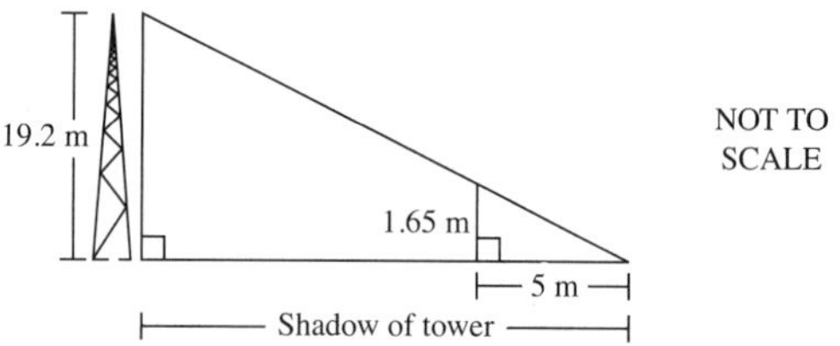

What is the length of the shadow cast by the tower at that time? *(2 marks)*

(Q27a, **2015 HSC**) Medium

32 The diagram shows an annulus.

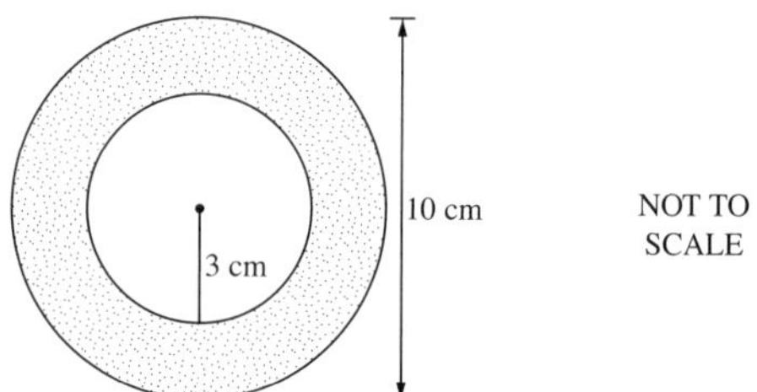

Calculate the area of the annulus. *(1 mark)*

(Q28a, **2015 HSC**) Easy

33 A path 1.5 metres wide surrounds a circular lawn of radius 3 metres.

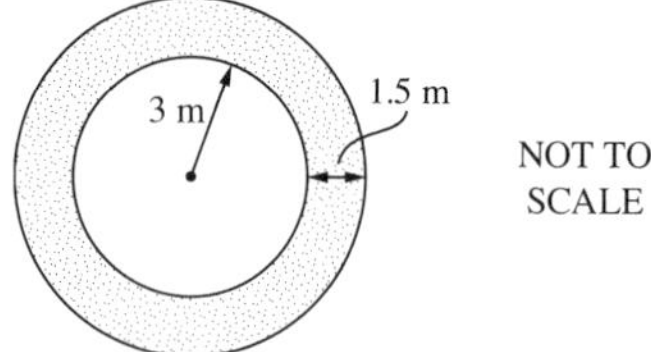

What is the approximate area of the path?

A 7.1 m^2 **B** 21.2 m^2
C 35.3 m^2 **D** 56.5 m^2 *(1 mark)*

(Q12, **2014 HSC**) Easy

34 A grain silo is made up of a cylinder with a hemisphere (half a sphere) on top. The outside of the silo is to be painted.

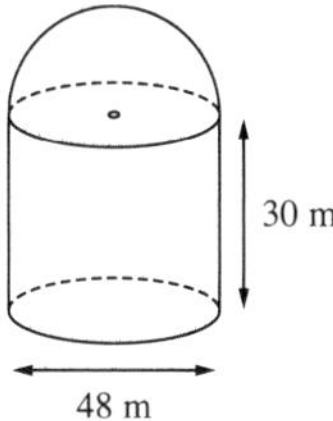

What is the area to be painted?

A 8143 m^2 **B** $11\,762 \text{ m}^2$
C $12\,667 \text{ m}^2$ **D** $23\,524 \text{ m}^2$ *(1 mark)*

(Q25, **2014 HSC**) **Medium**

35 The base of a water tank is in the shape of a rectangle with a semicircle at each end, as shown.

The tank is 1400 mm long, 560 mm wide, and has a height of 810 mm.

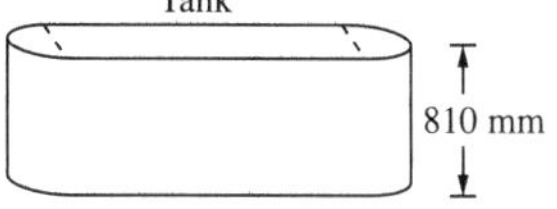

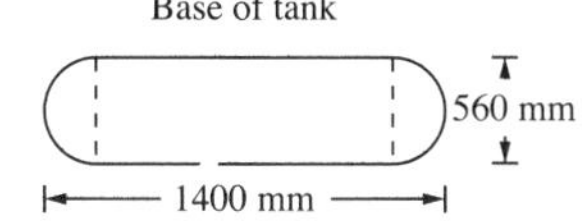

NOT TO SCALE

What is the capacity of the tank, to the nearest litre? *(4 marks)*

(Q27c, **2014 HSC**) **Hard**

36 A square pyramid fits exactly on top of a cube to form a solid.

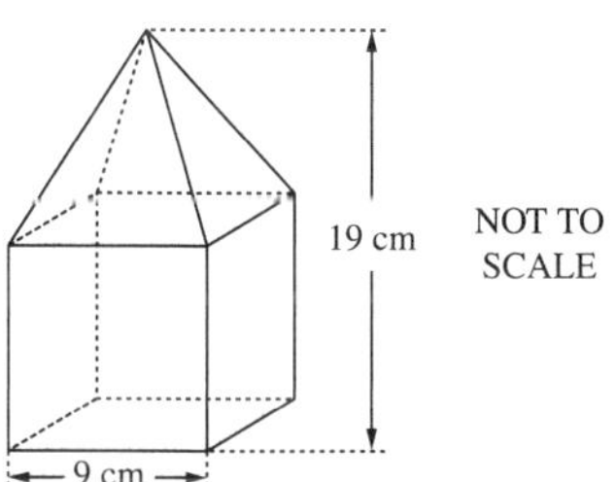

What is the volume of the solid?

A 513 cm^3 **B** 999 cm^3
C 1242 cm^3 **D** 1539 cm^3 *(1 mark)*

(Q12, **2013 HSC**) **Medium**

37 The shaded region shows a quadrant with a rectangle removed.

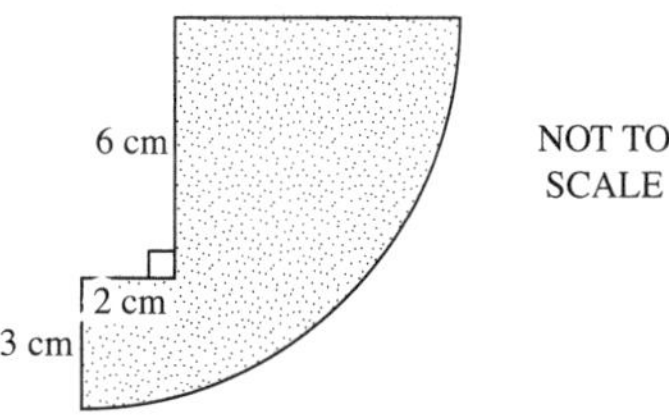

What is the area of the shaded region, to the nearest cm^2?

A 38 cm^2 **B** 52 cm^2
C 61 cm^2 **D** 70 cm^2 *(1 mark)*

(Q16, **2013 HSC**) **Easy**

38 Triangles ABC and DEF are similar.

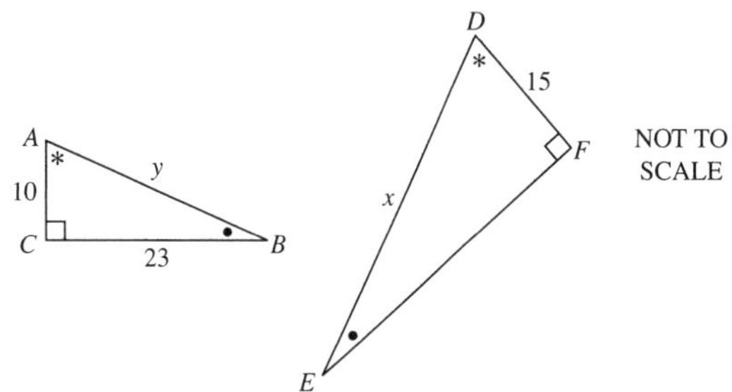

Which expression could be used to find the value of x?

A $y \times \frac{10}{15}$ **B** $y \times \frac{10}{23}$
C $y \times \frac{15}{10}$ **D** $y \times \frac{23}{15}$ *(1 mark)*

(Q17, **2013 HSC**) **Medium**

39 A logo is designed using half of an annulus.

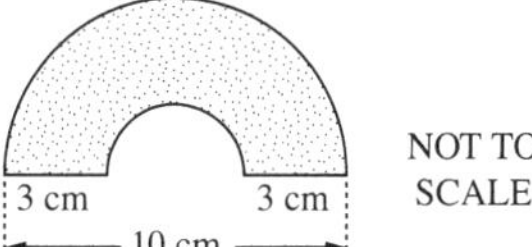

What is the area of the logo, to the nearest cm^2?

A 25 cm^2 **B** 33 cm^2
C 132 cm^2 **D** 143 cm^2 *(1 mark)*

(Q19, **2013 HSC**) **Medium**

40 A net is made using four rectangles and two trapeziums. It is folded to form a solid.

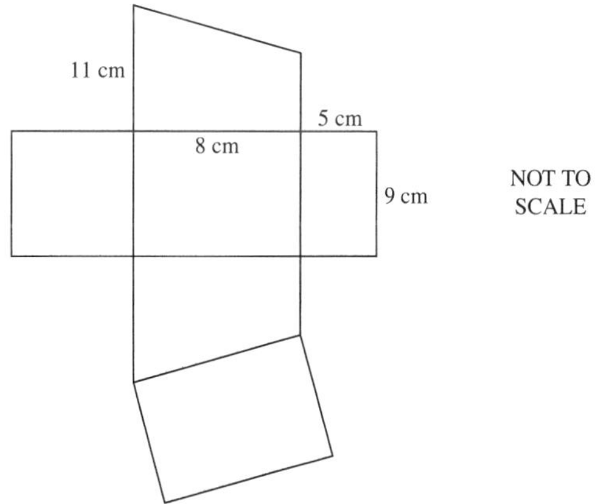

What is the volume of the solid, in cm^3?

A 360 cm^3 **B** 434 cm^3
C 440 cm^3 **D** 576 cm^3 *(1 mark)*

(Q25, **2013 HSC**) **Medium**

41 What is the volume of this rectangular prism in cubic centimetres?

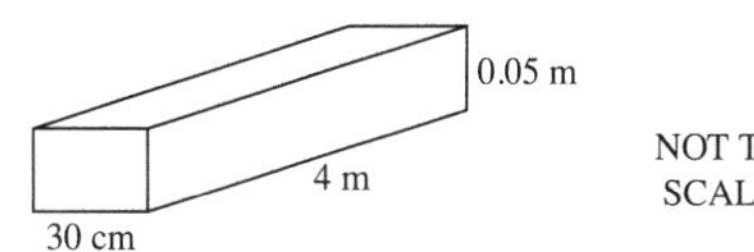

A 6 cm^3 **B** 600 cm^3
C 60000 cm^3 **D** 6000000 cm^3

(1 mark)

(Q6, **2012 HSC**) Medium

42 The solid shown is made of a cylinder with a hemisphere (half a sphere) on top.

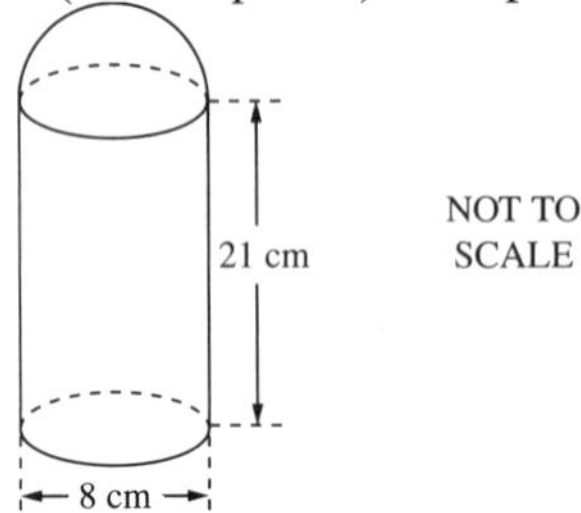

What is the total surface area of the solid, to the nearest square centimetre?

A 628 cm^2 **B** 679 cm^2
C 729 cm^2 **D** 829 cm^2 *(1 mark)*

(Q25, **2012 HSC**) Hard

43 The sector shown has a radius of 13 cm and an angle of 230°.

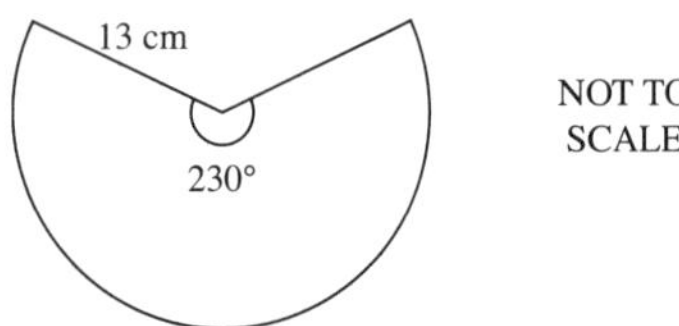

What is the perimeter of the sector to the nearest centimetre? *(2 marks)*

(Q27b, **2012 HSC**) Medium

44 Jacques and a flagpole both cast shadows on the ground. The difference between the lengths of their shadows is 3 metres.

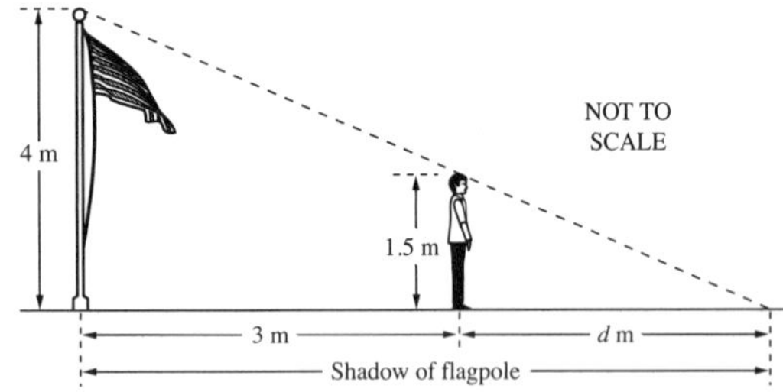

What is the value of d, the length of Jacques' shadow? *(3 marks)*

(Q28c, **2012 HSC**) Medium

45 A field diagram has been drawn from an offset survey.

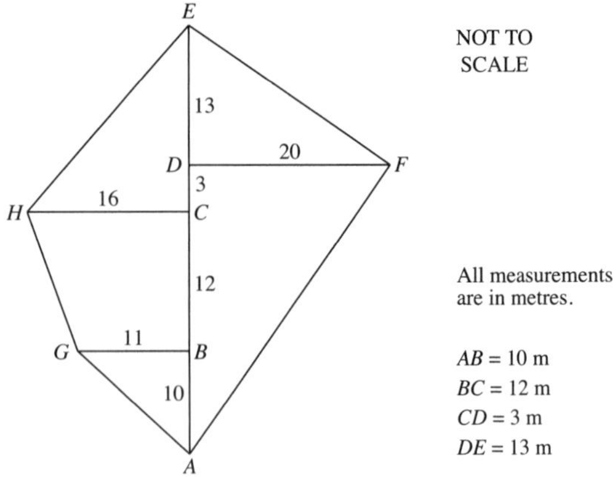

What is the distance from G to H, correct to the nearest metre?

A 11 **B** 13
C 16 **D** 20 *(1 mark)*

(Q3, **2010 HSC**) Medium

46 What is the area of the shaded part of this quadrant, to the nearest square centimetre?

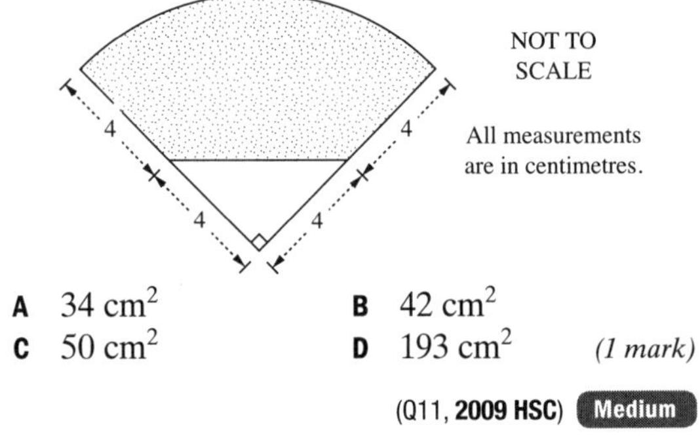

A 34 cm^2 **B** 42 cm^2
C 50 cm^2 **D** 193 cm^2 *(1 mark)*

(Q11, **2009 HSC**) Medium

47 How many square centimetres are in 0.0075 square metres?

A 0.75 **B** 7.5
C 75 **D** 7500 *(1 mark)*

(Q12, **2009 HSC**) Medium

48 Two identical spheres fit exactly inside a cylindrical container, as shown.

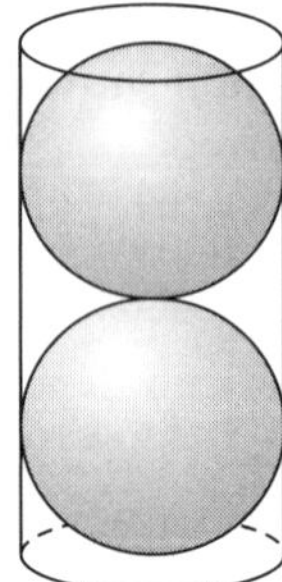

The diameter of each sphere is 12 cm.

What is the volume of the cylindrical container, to the nearest cubic centimetre?

A 1357 cm^3 **B** 2714 cm^3
C 5429 cm^3 **D** 10857 cm^3 *(1 mark)*

(Q19, **2009 HSC**) Medium

49 The diagram shows the shape and dimensions of a terrace which is to be tiled.

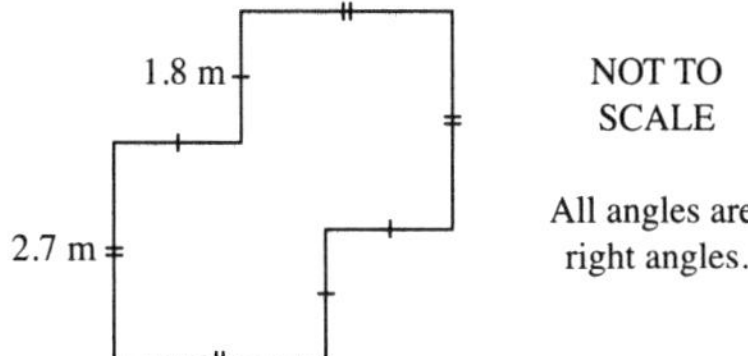

i Find the area of the terrace. *(2 marks)* Medium

ii Tiles are sold in boxes. Each box holds one square metre of tiles and costs $55. When buying the tiles, 10% more tiles are needed, due to cutting and wastage.

Find the total cost of the boxes of tiles required for the terrace. *(2 marks)* Medium

(Q23c, **2009 HSC**)

50 What is the surface area of the open box?

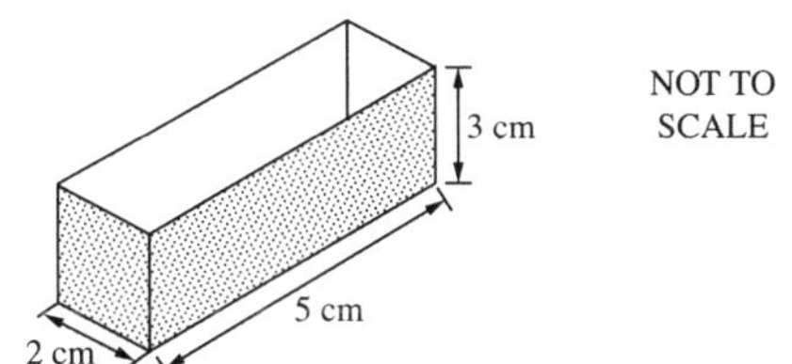

A 10 cm^2 **B** 30 cm^2
C 52 cm^2 **D** 62 cm^2 *(1 mark)*

(Q2, **2008 HSC**) Medium

51 The diagram shows the floor of a shower. The drain in the floor is a circle with a diameter of 10 cm.

What is the area of the shower floor, excluding the drain?

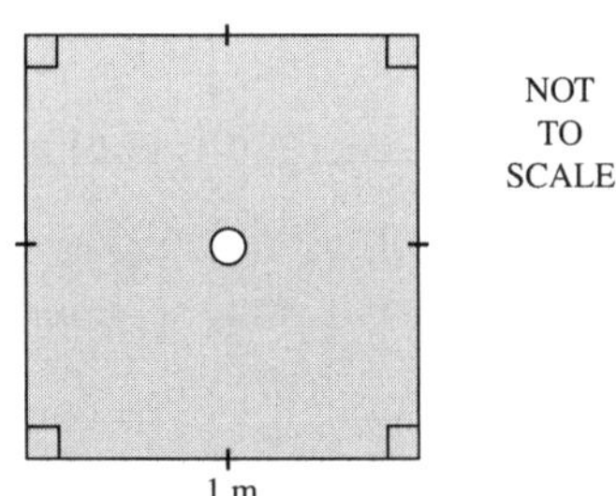

A 9686 cm^2 **B** 9921 cm^2
C 9969 cm^2 **D** 10 000 cm^2 *(1 mark)*

(Q11, **2008 HSC**) Medium

52 A point P lies between a tree, 2 metres high, and a tower, 8 metres high. P is 3 metres away from the base of the tree.

From P, the angles of elevation to the top of the tree and to the top of the tower are equal.

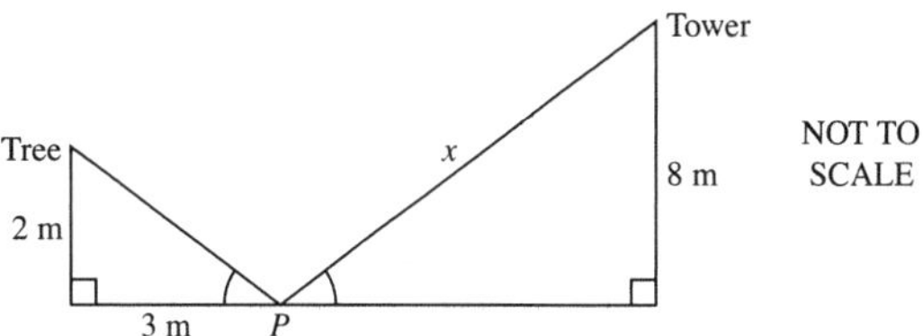

What is the distance, x, from P to the top of the tower?

A 9 m **B** 9.61 m
C 12.04 m **D** 14.42 m *(1 mark)*

(Q20, **2008 HSC**) Medium

53 What scale factor has been used to transform Triangle A to Triangle B?

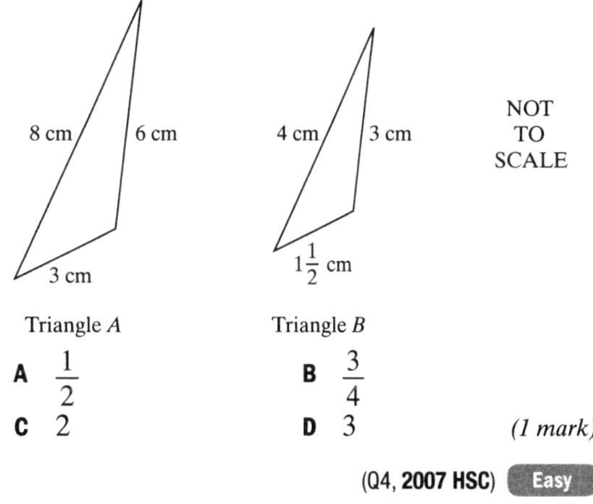

A $\frac{1}{2}$ **B** $\frac{3}{4}$
C 2 **D** 3 *(1 mark)*

(Q4, **2007 HSC**) Easy

54 The radius of a sphere is increased by 10%.

What is the percentage increase in its surface area?

A 10% **B** 20%
C 21% **D** 33% *(1 mark)*

(Q20, **2006 HSC**) Hard

55 The roof of this greenhouse is a square pyramid with identical triangular faces. The sides of the greenhouse are rectangles and there is no floor. The dimensions of the greenhouse are shown on the diagram.

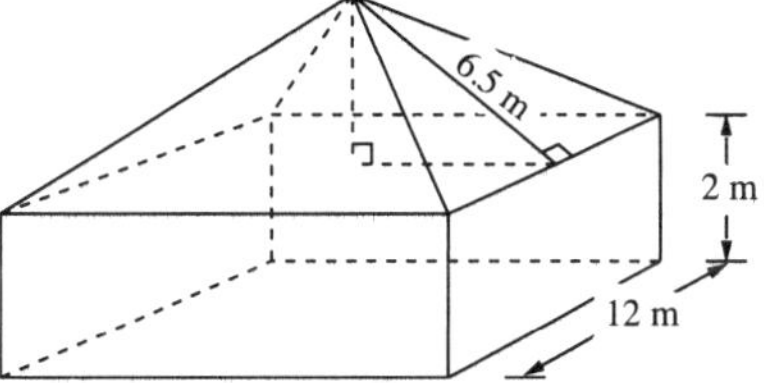

i Sketch a possible net of the greenhouse. *(1 mark)* Medium

ii Calculate the surface area of the greenhouse. *(2 marks)* Medium

(Q26b, **2006 HSC**)

56 A model yacht has two triangular sails. These triangles are similar to each other. Some dimensions of the sails, in centimetres, are shown on the diagram.

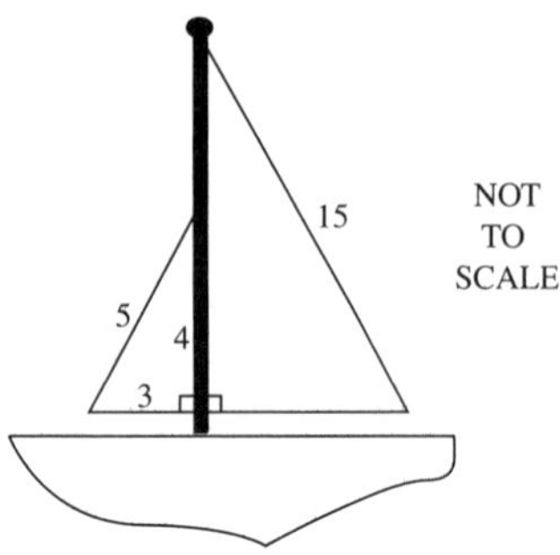

What is the total area of both sails?

A 24 cm^2 **B** 27 cm^2
C 60 cm^2 **D** 97 cm^2 *(1 mark)*

(Q18, **2005 HSC**) **Medium**

---

57 A clay brick is made in the shape of a rectangular prism with dimensions as shown.

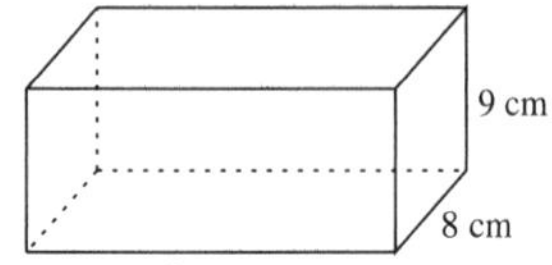

i Calculate the volume of the clay brick. *(1 mark)* **Easy**

Three identical cylindrical holes are made through the brick as shown. Each hole has a radius of 1.4 cm.

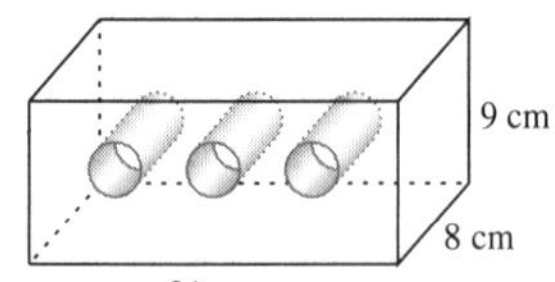

ii What is the volume of clay remaining in the brick after the holes have been made? (Give your answer to the nearest cubic centimetre.) *(3 marks)* **Hard**

iii What percentage of clay is removed by making the holes through the brick? (Give your answer correct to one decimal place.) *(1 mark)* **Medium**

(Q23b, **2005 HSC**)

---

58 The diagram shows the shape of Carmel's garden bed. All measurements are in metres.

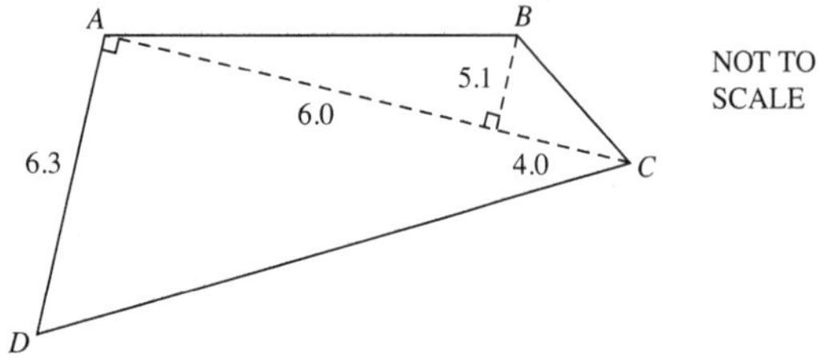

i Show that the area of the garden bed is 57 square metres. *(2 marks)* **Easy**

ii Carmel decides to add a 5 cm layer of straw to the garden bed. Calculate the volume of straw required. Give your answer in cubic metres. *(2 marks)* **Easy**

iii Each bag holds 0.25 cubic metres of straw. How many bags does she need to buy? *(2 marks)* **Easy**

iv A straight fence is to be constructed joining point A to point B. Find the length of this fence to the nearest metre. *(2 marks)* **Easy**

(Q23a, **2004 HSC**)

---

59 Calculate the height (h metres) of the tree in the diagram. All measurements are in metres.

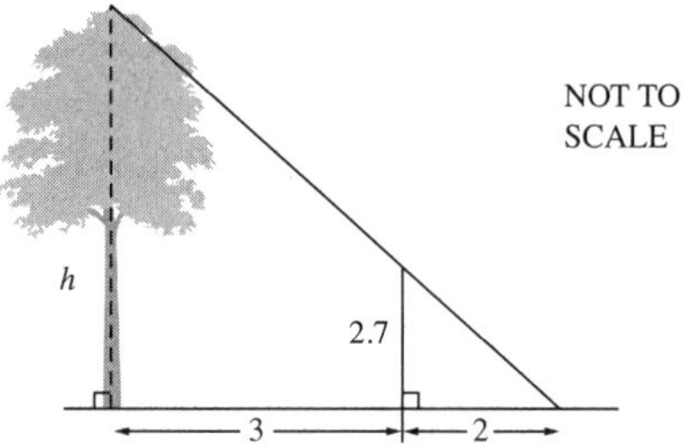

(2 marks)

(Q23c, **2004 HSC**) **Easy**

---

60 A swimming pool has a length of 6 m and a width of 5 m. The depth of the pool is 1 m at one end and 3.5 m at the other end, as shown in the diagram.

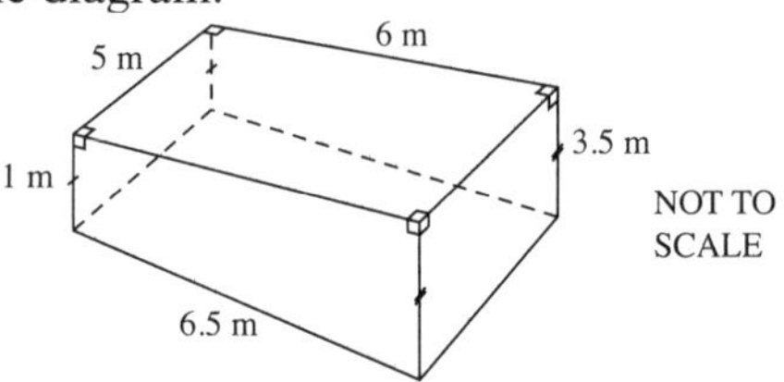

What is the volume of this pool in cubic metres?

A 67.5 **B** 105
C 109.375 **D** 113.75 *(1 mark)*

(Q9, **2003 HSC**) **Medium**

---

61 The roof of the Sydney Opera House is covered with 1.056 million tiles. If each tile covers 175 cm^2, what area is covered by the tiles?

A 184.8 m^2 **B** 18 480 m^2
C 184 800 m^2 **D** 1 848 000 m^2 *(1 mark)*

(Q19, **2003 HSC**) **Medium**

---

62 In her garden, Keryn has a birdbath in the shape of a hemisphere (half a sphere).

The internal diameter is 45 cm.

NOT TO SCALE

What is the internal surface area of this birdbath? (Give your answer to the nearest square centimetre.) *(2 marks)*

(Q23b, **2003 HSC**) **Medium**

63 Peta is designing an eight-cylinder racing engine. Each cylinder has a bore (diameter) of 10.0 cm and a stroke (height) of 7.8 cm, as shown below.

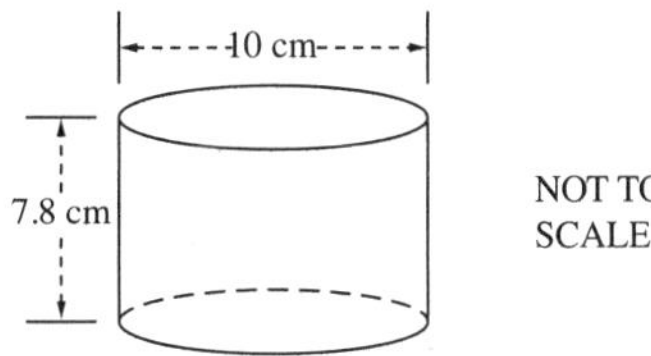

NOT TO SCALE

i Calculate the volume of each cylinder, correct to the nearest cubic centimetre. *(2 marks)* **Easy**

ii The capacity of the engine is the sum of the capacities of the eight cylinders. Does Peta's engine meet the racing requirement that the capacity should be under 5 litres? Justify your answer with a mathematical calculation. *(3 marks)* **Medium**

(Q23d, **2003 HSC**)

64 The Great Pyramid of Egypt has a square base of side 230 m. Its perpendicular height is 135 m.

What is the volume of the pyramid?

A $10\,350$ m^3 **B** $1\,397\,250$ m^3
C $2\,380\,500$ m^3 **D** $7\,141\,500$ m^3 *(1 mark)*

(Q3, **2002 HSC**) **Easy**

65 The game of Beach Quidditch is played with a large hollow spherical ball made from gold vinyl. The diameter of the ball is 1.2 metres.

If the vinyl costs \$32 per square metre, which of the following is closest to the cost of the vinyl for one ball?

A \$29 **B** \$145
C \$232 **D** \$579 *(1 mark)*

(Q10, **2002 HSC**) **Medium**

66 The sheets of paper Jenny uses in her photocopier are 21 cm by 30 cm. The paper is 80 gsm, which means that one square metre of this paper has a mass of 80 grams. Jenny has a pile of this paper weighing 25.2 kg.

How many sheets of paper are in the pile?

A 500 **B** 2000
C 2500 **D** 5000 *(1 mark)*

(Q21, **2002 HSC**) **Medium**

67 A shelf 20 cm wide is attached to a wall, under a light.

i The diagram shows the end view, ED, of the shelf attached to a wall AC.

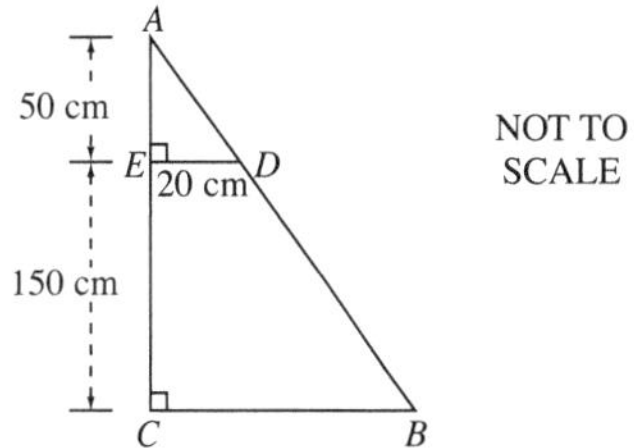

NOT TO SCALE

When the wall light at A is turned on, the shelf casts a shadow CB on the floor.

1 Name a pair of similar figures in the diagram. *(1 mark)* **Medium**

2 Calculate the enlargement factor between these two similar figures. *(1 mark)* **Medium**

3 What is the length of the shadow CB? *(1 mark)* **Medium**

ii The shelf is moved to a new position d cm below the light. The length of the shadow is now x cm.

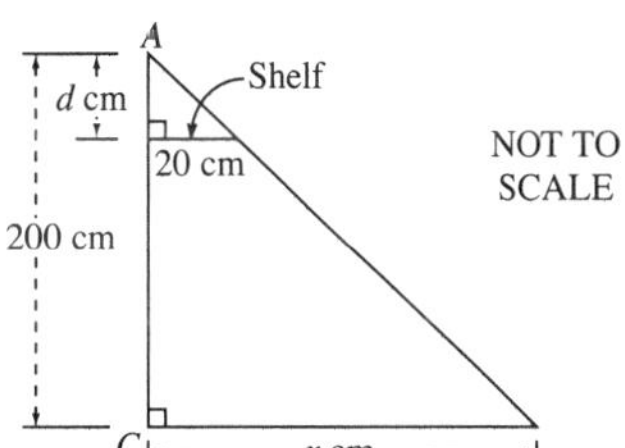

NOT TO SCALE

Write down an equation relating d and x. *(1 mark)* **Medium**

(Q25a, **2002 HSC**)

68 **i** The orbits of Earth and Venus around the Sun are almost circular, and in the same plane.

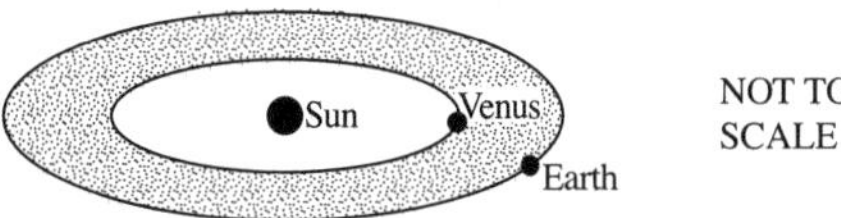

NOT TO SCALE

Earth is 1.496×10^8 km from the Sun.

Venus is 1.082×10^8 km from the Sun.

Treating the space between the orbits as an annulus, calculate its area. Write your answer in scientific notation correct to two significant figures. *(2 marks)* **Medium**

ii Rearrange the formula for the area of an annulus, $A = \pi(R^2 - r^2)$, to make R the subject. *(2 marks)* **Medium**

iii A small metal washer is to be made in the shape of an annulus with inner radius 0.75 mm.

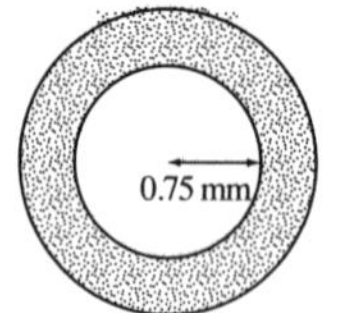

NOT TO SCALE

The area of the face of the washer (shaded on the diagram) is to be 6.79 mm^2. Calculate the outer radius correct to two decimal places. *(2 marks)* **Medium**

(Q25c, **2002 HSC**)

---

69 This is a sketch of a sector of a circle.

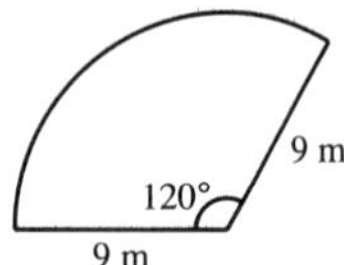

Calculate the area of this sector (correct to one decimal place).

A 9.4 m^2 **B** 18.8 m^2
C 36.8 m^2 **D** 84.8 m^2 *(1 mark)*

(Q3, **2001 HSC**) 

---

70 A sphere has a volume of 360 cm^3.

What is its radius (correct to one decimal place)?

A 1.7 cm **B** 4.4 cm
C 8.1 cm **D** 9.3 cm *(1 mark)*

$V = \frac{4}{3}\pi r^3$

(Q18, **2001 HSC**) **Medium**

Year 11 Perimeter, area and volume—Worked answers

1 A has perimeter
4×8 cm = 32 cm.
B has perimeter
$2 \times (11 + 3)$ cm = 28 cm.
C has perimeter
3×10 cm = 30 cm.
D has perimeter
$(9 + 3 + 2 \times 4)$ cm = 20 cm
The largest perimeter is 32 cm.
Answer A

2 Trapezoidal rule: $A = \frac{h}{2}(d_f + d_l)$
Two applications:
$A = \frac{6}{2}(1.2 + 2) + \frac{6}{2}(2 + 1.4)$
$= 19.8$
Answer B

3 Volume of half a sphere:
$V = \frac{2}{3}\pi r^3$ ✓
$= \frac{2}{3} \times \pi \times 2^3$
$= 16.7551608\ldots$
$= 16.8 \text{ m}^3$ (1 d.p.) ✓
(2 marks)

4 Bottom part has 5 square faces.
Area $= 5 \times 8^2$
$= 320 \text{ cm}^2$ ✓
Top part:
Two triangular sections:
Area $= 2 \times \frac{1}{2} \times 8 \times 3$
$= 24 \text{ cm}^2$ ✓
Two rectangular faces:
Area $= 2 \times 8 \times 5$
$= 80 \text{ cm}^2$
Total area $= (320 + 24 + 80) \text{ cm}^2$
$= 424 \text{ cm}^2$
The surface area of the composite solid is 424 cm^2. ✓
(3 marks)

5 Rectangle:
$P = 2 \times (4 + 2)$ cm
$= 12$ cm

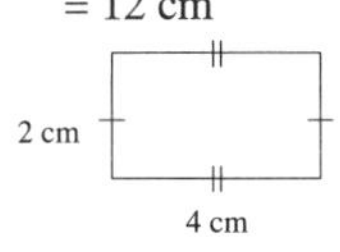

Answer A

6 The diameter is 16 cm so $r = 8$.
$V = \frac{1}{2} \times \frac{4}{3}\pi r^3$ ✓
$= \frac{2}{3} \times \pi \times 8^3$
$= 1072.33029\ldots$
$= 1072 \text{ cm}^3$ (nearest cubic centimetre) ✓
(2 marks)

7 **a** Area $\approx \frac{120}{2}[70 + 80] + \frac{120}{2}[80 + 140]$ ✓
$= 22\,200$
∴ the area is approximately 22 200 m^2. ✓ *(2 marks)*

b Area of lake
$= 240 \times 200 - (22200 + 13800)$
$= 12\,000$
∴ the area of the lake is 12 000 m^2. ✓
Volume $= 12\,000 \times 3.6$
$= 43\,200$
∴ the volume of the lake is 43 200 m^3.
As 43 200 m^3
= 43 200 kL
= 43.2 ML, or 43 ML (to nearest whole)
∴ the lake contains 43 ML. ✓
(2 marks)

8 **a** $h = 18.6 \div 3$
$= 6.2$ ✓
Area $\approx \frac{6.2}{2}[0 + 9.4] + \frac{6.2}{2}[9.4 + 16.4] + \frac{6.2}{2}[16.4 + 21.3]$
$= 225.99$
∴ the area is approximately 225.99 cm^2. ✓ *(2 marks)*

b Surface area
$= 2 \times 225.99 + 128.7 + 21.3 \times 4 + 18.6 \times 4$ ✓
$= 740.28$
= 740 (nearest whole)
∴ the surface area is 740 cm^2. ✓ *(2 marks)*

9 Area $\approx \frac{20}{2}[43 + 45] + \frac{20}{2}[45 + 34] + \frac{20}{2}[34 + 38]$ ✓
$= 2390$
∴ the area is approximately 2390 m^2. ✓ *(2 marks)*

10 1.215 hectares = 12 150 m^2
$A = \frac{45}{2}(110 + a) + \frac{45}{2}(a + 154)$
$= 12\,150$ ✓
$22.5(110 + 2a + 154) = 12\,150$
$2a + 264 = 540$
$2a = 276$
$a = 138$ ✓
(2 marks)

11 As diameter = 8 cm, then radius = 4 cm.
$h = 24 - 4$
$= 20$ ✓
Volume = volume of cylinder + volume of hemisphere
$= \pi r^2 h + \frac{1}{2} \times \frac{4}{3}\pi r^3$
$= \pi \times 4^2 \times 20 + \frac{1}{2} \times \frac{4}{3} \times \pi \times 4^3$ ✓
$= 1139.350936\ldots$
= 1140 (3 sig. figs)
∴ the volume is 1140 cm^3. ✓
Surface area = SA of cylinder + SA of hemisphere
$= \pi r^2 + 2\pi rh + \frac{1}{2} \times 4\pi r^2$
$= \pi \times 4^2 + 2 \times \pi \times 4 \times 20 + \frac{1}{2} \times 4 \times \pi \times 4^2$
$= 653.4512719\ldots$
= 653 (3 sig. figs)
∴ the surface area is 653 cm^2. ✓
(4 marks)

12 **a** Area $\approx 24 \times 55 - [\frac{12}{2}(35 + 22) + \frac{12}{2}(22 + 30) + \frac{12}{2}(20 + 5) + \frac{12}{2}(5 + 10)]$ ✓
$= 426$
∴ the area is approximately 426 m^2. ✓ *(2 marks)*

b Volume $= 426 \times 0.6$
$= 255.6$
∴ the volume is 255.6 m^3. ✓
∴ the lake contains 255 600 L.
Number of times
$= 255\,600 \div 4$
$= 63\,900$
∴ it would take Bozo 63 900 times. ✓ *(2 marks)*

13 Using $r = 14$ and $h = 24$:
Volume of cylinder + volume of hemisphere
$= \pi r^2 h + \frac{1}{2} \times \frac{4}{3}\pi r^3$ ✓
$= \pi \times 14^2 \times 24 + \frac{1}{2} \times \frac{4}{3} \times \pi \times 14^3$
$= 20\,525.072\ldots$
= 20 525 (nearest whole)
∴ the volume is 20 525 cm^3. ✓
(2 marks)

14 The diagram shows the base of a pool with a consistent depth of 1.8 m.
a Area $\approx \frac{5}{2}(5 + 7.2) + \frac{5}{2}(7.2 + 12)$
$= 78.5$
∴ the area is approximately 78.5 m^2. ✓

Volume = 78.5 × 1.8
= 141.3
∴ the volume is approximately 141.3 m^3. ✓

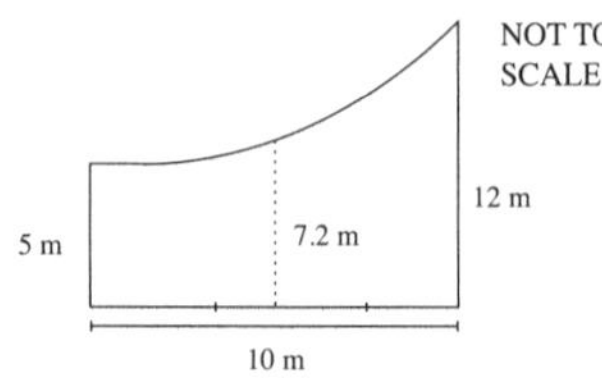

(2 marks)

b 16 mm = 0.016 m
Volume = 78.5 × 0.016
= 1.256 ✓
∴ the amount of water is approximately 1.256 m^3, or 1256 L.
Water rate = 10L/8 seconds
= 1.25L/s ✓
Time = 1256 ÷ 1.25
= 1004.8 seconds
= 16.7466666... min
= 17 (nearest whole)
∴ it would take 17 minutes. ✓

(3 marks)

15 $V = \frac{1}{3}Ah$
$= \frac{1}{3} \times 4x \times 3x \times 2x$
$= 8x^3$
The volume, in cubic metres, is $8x^3$.
Answer A

16 Let r cm be the radius.
By Pythagoras' theorem:
$r^2 + 6^2 = 10^2$
$r^2 = 64$
$r = 8 \quad (r > 0)$
Arc length of quadrant
$= \frac{1}{4} \times 2\pi r$
$= \frac{1}{4} \times 2 \times \pi \times 8$
= 12.5663706...
= 12.6 cm (1 d.p.)
Total perimeter
= (12.6 + 10 + 6 + 8) cm
= 36.6 cm
Answer B

17 $r = 3.8 \div 2$
= 1.9
$A = \frac{1}{2} \times 2\pi rh$ ✓
$= \pi \times 1.9 \times 10$
= 59.6902604... m^2
= 60 m^2 (nearest m^2) ✓

(2 marks)

18 1.26 ML = 1 260 000 L ✓
= 1260 m^3
$V = \pi r^2 h$
$1260 = \pi \times 9^2 \times h$
$h = \frac{1260}{\pi \times 9^2}$ ✓
= 4.951 487 11...
= 4.95 (2 d.p.)
The height of the water tank is 4.95 m to two decimal places. ✓

(3 marks)

19 $V = Ah$
$= \frac{1.2}{2} \times (3.6 + 2.4) \times 1.5$
= 5.4 m^3
Answer A

20 $A = \pi(R^2 - r^2)$
$R = 0.35 + 0.1 = 0.45, \quad r = 0.35$
$A = \pi \times (0.45^2 - 0.35^2)$
= 0.251 3274... m^2
$V = Ah$
= 0.251 3274... × 2.8
= 0.703 716 754...
= 0.70 m^3 (2 d.p.)
Answer C

21 $A = \frac{1}{4}\pi r^2$
So $\frac{1}{4}\pi r^2 = 100$
$\pi r^2 = 400$
$r^2 = 127.323954...$
$r = 11.283791... \quad (r > 0)$
Now l
$= \frac{\theta}{360} 2\pi r$
$= \frac{1}{4} \times 2 \times \pi \times 11.283791...$
= 17.7245385...
= 17.7 cm (1 d.p.)
Answer C

22 The circular cross-section of the sphere has radius 4 cm.
So the diameter is 8 cm.
If the circle sits 2 cm into the cone then 6 cm remains above the cone.
Let h cm be the height of the cone.
$h = 15 - 6$
= 9 ✓
AB is a diameter of the cone.
Let the radius of the cone be r cm.
By Pythagoras' theorem:
$r^2 + 2^2 = 4^2$
$r^2 + 4 = 16$
$r^2 = 12$ ✓

Volume of cone:
$V = \frac{1}{3}\pi r^2 h$
$= \frac{1}{3} \times \pi \times 12 \times 9$
= 113.097 335...
= 113 (nearest unit)
The volume of the cone is 113 cm^3 to the nearest cubic centimetre. ✓

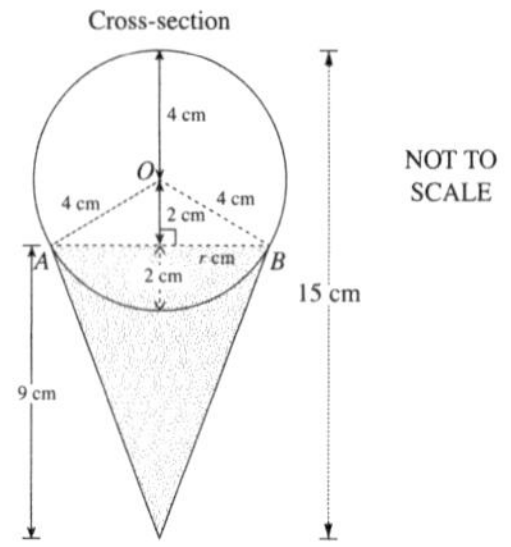

(3 marks)

23 Capacity = 12 L = 12 000 mL
So volume is 12 000 cm^3.
$V = Ah$
$12\,000 = 240 \times h$
$h = 12\,000 \div 240$
= 50

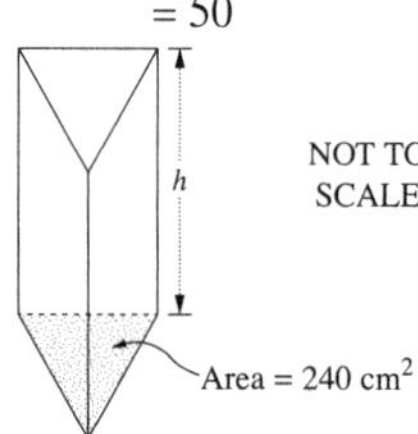

The distance between the ends is 50 cm.
Answer D

24 Corresponding sides of similar triangles are in proportion.

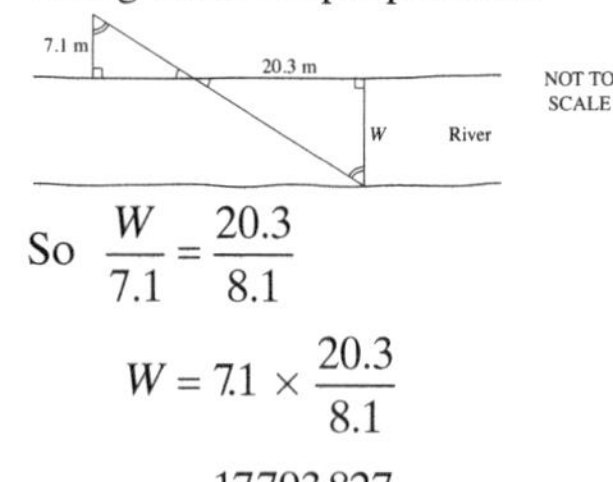

So $\frac{W}{7.1} = \frac{20.3}{8.1}$
$W = 7.1 \times \frac{20.3}{8.1}$
= 17.793 827....
= 17.8 (1 d.p.)
The approximate width of the river is 17.8 m.
Answer A

25 $A = 4\pi r^2$
$= 4 \times \pi \times 5^2$
= 314.159 265...
= 314 cm^2 (nearest square centimetre)

(1 mark)

26

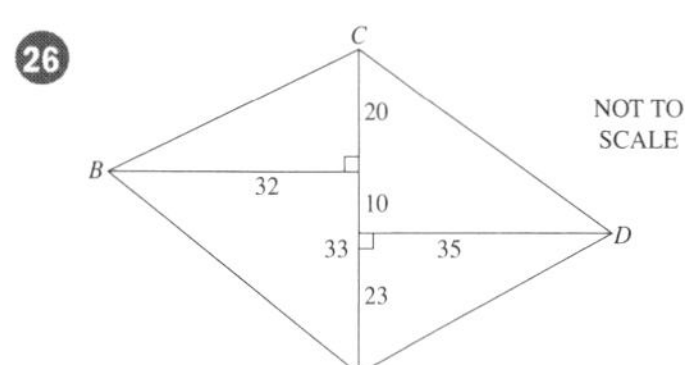

$AB^2 = 32^2 + 33^2$
$= 2113$ ✓
$AB = 45.9673797\ldots$
$= 46$ m (nearest metre) ✓

(2 marks)

27 **i** $V = \pi r^2 h$
$= \pi \times 2.5^2 \times 3$ ✓
$= 58.9048622\ldots$
$= 58.9\text{ cm}^3$ (1 d.p.) ✓

(2 marks)

ii The volume of chocolate required between 4 marshmallows is the same as that in a square-based prism of height 6 cm surrounding a cylinder (made up of 2 marshmallows). ✓

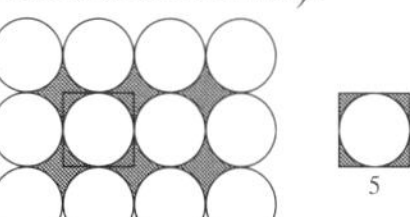

There are 6 of those spaces. ✓

So $V = 6 \times (5 \times 5 \times 6 - 2 \times 58.9048622\ldots)$
$= 193.14165\ldots$
$= 193\text{ cm}^3$ (nearest cubic centimetre) ✓

Or: Consider the rectangle whose vertices are the centres of the 4 corner marshmallows. ✓

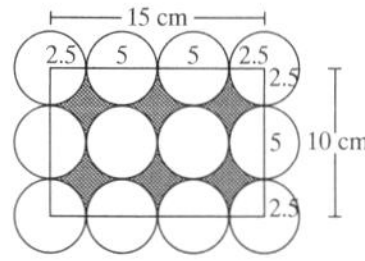

The shaded area is the area of the rectangle minus the area of 6 circles (2 full circles plus 6 semicircles plus 4 quadrants). ✓

So $V = (15 \times 10 - 6 \times \pi \times 2.5^2) \times 6$
$= 193.141652\ldots$

So the volume of chocolate is 193 cm^3 to the nearest cubic centimetre. ✓ *(3 marks)*

28 20 mm = 0.02 m
$V = Ah$
$= 30 \times 0.02$
$= 0.6\text{ m}^3$ ✓

Water on garden = 0.6×1000 L
= 600 L ✓

(2 marks)

29 $V = \frac{1}{3}Ah$
$= \frac{1}{3} \times 35^2 \times 22$
$= 8983.3333\ldots\text{ m}^3$
$= 8983\text{ m}^3$ (nearest m^3)

Answer C

30 $A = 4\pi r^2$
$= 4 \times \pi \times 6400^2$
$= 514\,718\,540.4\ldots\text{ km}^2$ ✓

Area of water
$= 0.71 \times 514\,718\,540.4\ldots$
$= 365\,450\,163.7\ldots\text{ km}^2$

So around 365.5 million square kilometres are covered by water. ✓

(2 marks)

31 The two triangles have a common angle and both have a right angle so they are equiangular and similar triangles.

Let x m be the length of the shadow.

$\frac{x}{5} = \frac{19.2}{1.65}$ ✓

$x = 5 \times \frac{19.2}{1.65}$
$= 58.181818\ldots$
$= 58.18$ (2 d.p.)

The length of the shadow is 58.18 m correct to two decimal places. ✓

(2 marks)

32 $A = \pi(R^2 - r^2)$
$= \pi \times (5^2 - 3^2)$
$= 50.2654824\ldots$
$= 50.3\text{ cm}^2$ (1 d.p.)

(1 mark)

33 $A = \pi(R^2 - r^2)$
$= \pi \times (4.5^2 - 3^2)$
$= 35.3429\ldots$
$\approx 35.3\text{ m}^2$

Answer C

34 $r = 24, h = 30$

$A = 2\pi rh + 2\pi r^2$
$= 2 \times \pi \times 24 \times 30 + 2 \times \pi \times 24^2$
$= 8143.008\,15\ldots$
$= 8143\text{ m}^2$ (nearest square metre)

Answer A

35 Total length of tank = 1400 mm
= 1.4 m

Radius of each semicircle
= 280 mm
= 0.28 m ✓

Length of rectangle
= (1.4 − 0.56) m
= 0.84 m

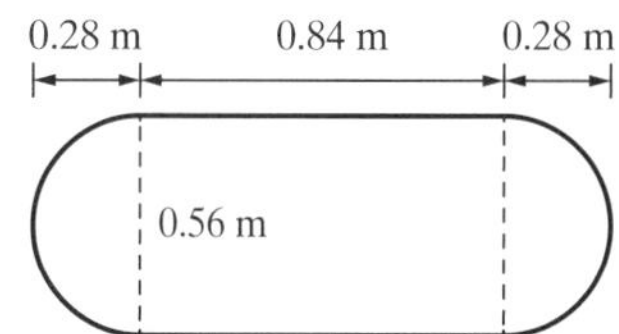

$A = 0.84 \times 0.56 + \pi \times 0.28^2$
$= 0.7167008\ldots\text{ m}^2$ ✓

Height of tank = 810 mm
= 0.81 m

$V = Ah$
$= 0.7167008\ldots \times 0.81$
$= 0.5805276\ldots\text{ m}^3$ ✓

Capacity = $0.5805276\ldots \times 1000$ L
= 580.5276... L
= 581 L (nearest litre) ✓

(4 marks)

36 Volume of cube = 9^3 cm^3
$= 729\text{ cm}^3$

Height of pyramid = (19 − 9) cm
= 10 cm

Volume of pyramid:

$V = \frac{1}{3}Ah$
$= \frac{1}{3} \times 9^2 \times 10$
$= 270$

The volume of the pyramid is 270 cm^3.

Total volume = (729 + 270) cm^3
= 999 cm^3

Answer B

37 Radius of quadrant = (6 + 3) cm
= 9 cm

Area of quadrant:

$A = \frac{1}{4}\pi r^2$
$= \frac{1}{4} \times \pi \times 9^2$
$= 63.617\,25\ldots$

The area of the quadrant is 64 cm^2 to the nearest cm^2

Rectangle: Area = $(6 \times 2)\text{ cm}^2$
$= 12\text{ cm}^2$

Shaded area $= (64 - 12)\text{ cm}^2$
$= 52\text{ cm}^2$

Answer B

38 The triangles are similar so the corresponding sides are in proportion.

So $\dfrac{x}{y} = \dfrac{15}{10}$

$x = y \times \dfrac{15}{10}$

Answer C

39 Diameter of outer circle is 10 cm so the radius is 5 cm. $R = 5$

Diameter of inner circle is 4 cm so $r = 2$

$A = \dfrac{1}{2}\pi(R^2 - r^2)$

$= \dfrac{1}{2} \times \pi \times (5^2 - 2^2)$

$= 32.986\,72\ldots$

So the area is 33 cm^2 to the nearest cm^2.

Answer B

40 The solid is a trapezoidal prism.

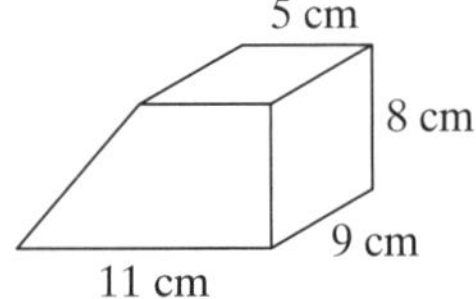

Area of trapezium:

$A = \dfrac{1}{2}h(a + b)$

$= \dfrac{1}{2} \times 8 \times (5 + 11)$

$= 64$

$V = AH$

$= 64 \times 9$

$= 576$

The volume of the solid is 576 cm^3.

Answer D

41 $4\text{ m} = 400\text{ cm}$

$0.05\text{ m} = 5\text{ cm}$

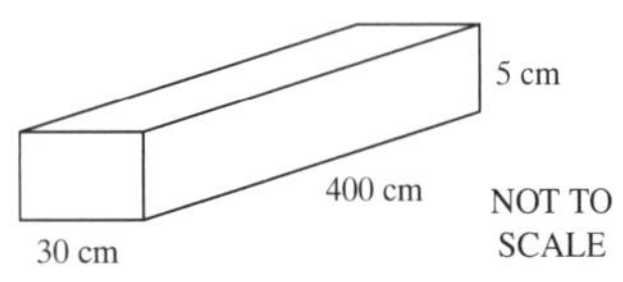

$V = lbh$

$= 400 \times 30 \times 5$

$= 60\,000$

The volume is 60 000 cm^3.

Answer C

42 Hemisphere: $A = 2\pi r^2$

Cylinder open at one end:

$A = 2\pi rh + \pi r^2$

Total surface area:

$A = 2\pi rh + 3\pi r^2$

$= 2 \times \pi \times 4 \times 21 + 3 \times \pi \times 4^2$

$= 678.584\,0132\ldots$

$= 679\text{ cm}^2$

[nearest square centimetre]

Answer B

43

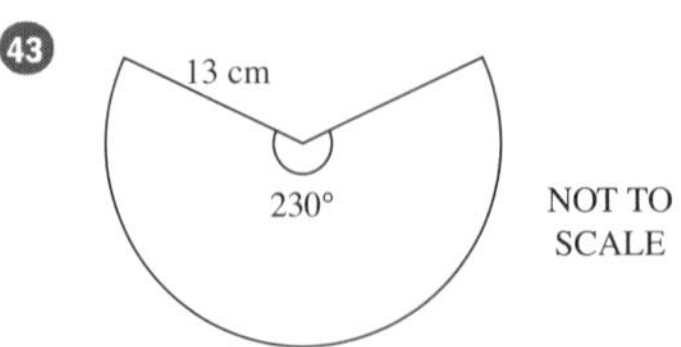

Arc length: $l = \dfrac{\theta}{360}2\pi r$

$= \dfrac{230}{360} \times 2 \times \pi \times 13$

$= 52.185\,3446\ldots$ ✓

$= 52\text{ cm}$ [nearest cm] ✓

Perimeter $= 52\text{ cm} + 2 \times 13\text{ cm}$
$= 78\text{ cm}$

The perimeter of the sector is 78 cm, to the nearest centimetre. ✓

(2 marks)

44

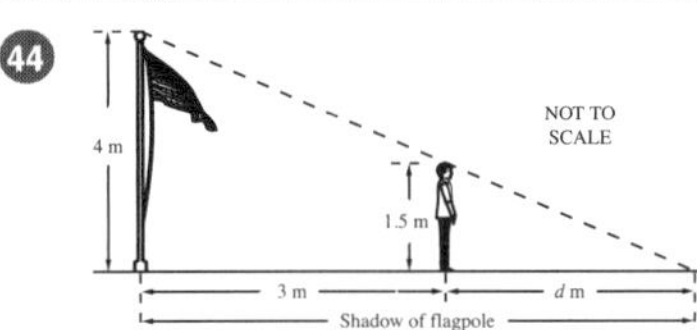

The triangle formed by the flagpole and its shadow and the triangle formed by Jacques and his shadow are equiangular, so they are similar.

So $\dfrac{d}{1.5} = \dfrac{3 + d}{4}$ ✓

$4d = 4.5 + 1.5d$ ✓

$2.5d = 4.5$

$d = 1.8$ ✓

(3 marks)

45 Consider trapezium $HCBG$.

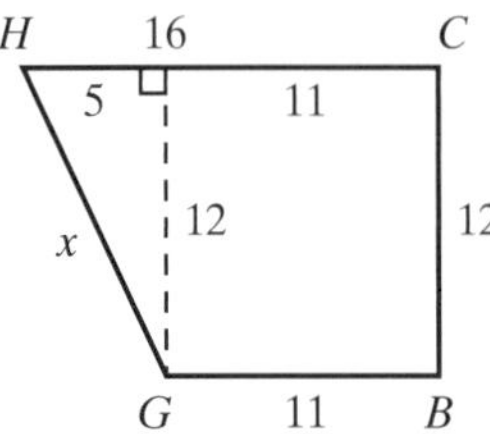

Let x m be the distance from G to H.

By Pythagoras' theorem:

$x^2 = 5^2 + 12^2$

$= 25 + 144$

$= 169$

$x = \sqrt{169}\quad (x > 0)$

$= 13$

The distance from G to H is 13 metres.

Answer B

46 Area of sector

$= \dfrac{\theta}{360}\pi r^2$

$= \dfrac{90}{360} \times \pi \times 8^2$

$= 50.2654\ldots\text{ cm}^2$

Area of triangle

$= \dfrac{1}{2} \times 4 \times 4$

$= 8\text{ cm}^2$

$\therefore$ Shaded area

$= 50.2654\ldots - 8$

$= 42.2654\ldots$

$= 42\text{ cm}^2$ to nearest cm^2

Answer B

47 $1\text{ m}^2 = 100\text{ cm} \times 100\text{ cm}$

$= 10\,000\text{ cm}^2$

$\therefore 0.0075\text{ m}^2 = 0.0075 \times 10\,000$

$= 75\text{ cm}^2$

Answer C

48 $V = \pi r^2 h$

$= \pi \times 6^2 \times 24$

$= 2714.3360\ldots$

$= 2714\text{ cm}^3$ to nearest cm^3

Answer B

49 **i**

1.8 m, 1.8 m, 2.7 m, 1.8 m, 2.7 m, 1.8 m

Length of large square

$= 2.7 + 1.8$

$= 4.5\text{ m}$ ✓

Area of terrace

= Area of large square

− area of 2 smaller squares

$= (4.5 \times 4.5) - (1.8 \times 1.8 \times 2)$

$= 13.77\text{ m}^2$ ✓

(2 marks)

ii Area of tiles required
$= 110\% \times 13.77$ ✓
$= 15.147 \text{ m}^2$
$\therefore$ 16 boxes will be needed.

$\therefore$ Cost of boxes of tiles
$= 16 \times \$55$
$= \$880$ ✓
(2 marks)

50 Surface area
$= (2 \times 3) \times 2 + (5 \times 3) \times 2 + (2 \times 5)$
$= 12 + 30 + 10$
$= 52 \text{ cm}^2$

Answer C

51 Area of shower floor
$= 1 \text{ m} \times 1 \text{ m}$
$= 100 \text{ cm} \times 100 \text{ cm}$
$= 10\,000 \text{ cm}^2$
Area of drain
$= \pi \times 5^2$
$= 78.5398 \ldots \text{ cm}^2$
$\therefore$ Required area
$= 10\,000 - 78.5398 \ldots$
$= 9921.4601 \ldots$
$\doteqdot 9921 \text{ cm}^2$
Answer B

52 The two triangles are similar as they have equal angles.

Let y be the distance from P to the top of the tree.

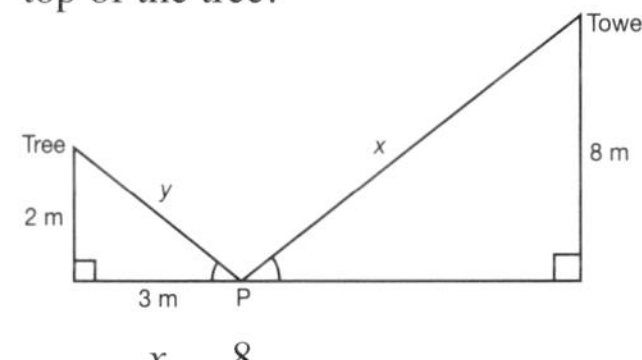

$\therefore \frac{x}{y} = \frac{8}{2}$

Also, $y^2 = 2^2 + 3^2$
$= 4 + 9$
$\therefore y = \sqrt{13}$
$\therefore \frac{x}{\sqrt{13}} = \frac{8}{2} = 4$
$\therefore x = 4\sqrt{13}$
$= 14.4222 \ldots$
$\doteqdot 14.42 \text{ m}$

Answer D

53 The side-lengths of triangle B are half of the side-lengths of triangle A.

$\therefore$ The scale factor is $\frac{1}{2}$.

Note that each side length of triangle A is multiplied by $\frac{1}{2}$ to obtain the lengths of the sides of B.

Answer A

54 $A = 4\pi r^2$
Let $r = 10$ cm
$\therefore A = 4 \times \pi \times 10^2$
$= 1256.63 \ldots$

New radius $= 10 + 10\% \times 10$
$= 11$ cm

$\therefore A = 4 \times \pi \times 11^2$
$= 1520.53 \ldots$

$\therefore$ Percentage increase
$= \frac{1520.53\ldots - 1256.63\ldots}{1256.63\ldots} \times 100$
$= 21\%$

Answer C

55 **i**

(1 mark)

ii Area of rectangle $= 12 \times 2$
$= 24 \text{ m}^2$

Area of triangle $= \frac{1}{2} \times 12 \times 6.5$
$= 39 \text{ m}^2$ ✓
$\therefore$ Surface area of greenhouse
$= 4 \times 24 + 4 \times 39$
$= 96 + 156$
$= 252 \text{ m}^2$ ✓
(2 marks)

56 Since the triangles are similar, and since $5 \times 3 = 15$, the other sides must be multiplied by 3 to find their lengths.

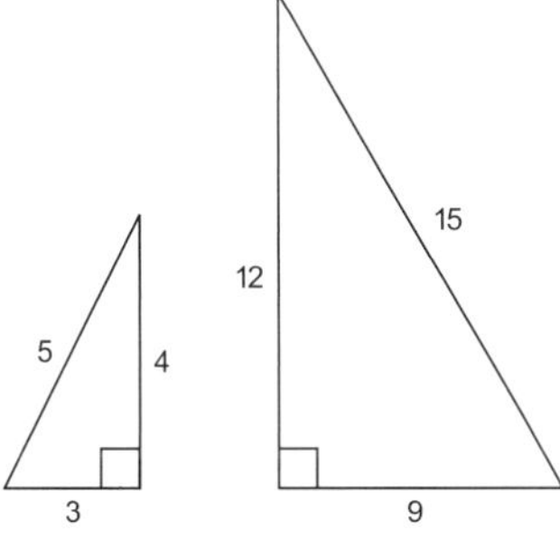

Area of both sails
$= \frac{3 \times 4}{2} + \frac{9 \times 12}{2}$
$= 6 + 54$
$= 60 \text{ cm}^2$
Answer C

57 **i** $V = 21 \times 9 \times 8$
$= 1512 \text{ cm}^3$
(1 mark)

ii Each hole is a cylinder of radius 1.4 cm and height 8 cm.

Volume of each cylinder
$= \pi r^2 h$
$= \pi \times 1.4^2 \times 8$
$= 49.2601 \ldots \text{ cm}^3$ ✓

Volume of 3 cylinders
$= 3 \times 49.2601 \ldots \text{ cm}^3$
$= 147.7805 \ldots \text{ cm}^3$ ✓

$\therefore$ Volume of clay remaining
$= 1512 - 147.7805 \ldots$
$= 1364.2194 \ldots$ ✓
$= 1364 \text{ cm}^3$ to the nearest cm^3
(3 marks)

iii Percentage of clay removed
$= \frac{147.7805\ldots}{1512} \times 100$
$= 9.7738 \ldots$
$= 9.8\%$ to one decimal place
(1 mark)

58 **i** Area of ΔABC
$= \frac{1}{2} \times 10 \times 5.1$
$= 25.5 \text{ m}^2$ ✓

Area of ΔACD
$= \frac{1}{2} \times 10 \times 6.3$
$= 31.5 \text{ m}^2$
$\therefore$ Total area of the garden bed
$= 25.5 + 31.5$
$= 57 \text{ m}^2$ ✓
(2 marks)

ii $V = \text{area} \times \text{height}$
$= 57 \times 0.05$ (5 cm = 0.05 m) ✓
$= 2.85 \text{ m}^3$ ✓
(2 marks)

iii Number of bags required
$= 2.85 \div 0.25$ ✓
$= 11.4$ ✓
$\therefore$ She will need to buy 12 bags.
(2 marks)

iv $AB^2 = 6.0^2 + 5.1^2$ (Pythagoras)
$= 36 + 26.01$
$= 62.01$ ✓

$\therefore AB = \sqrt{62.01}$
$= 7.874 \ldots$
$= 8$ metres to nearest metre. ✓
(2 marks)

59 $\frac{h}{2.7} = \frac{5}{2}$ (using similar triangles)
$\therefore h = \frac{5 \times 2.7}{2}$ ✓
$= 6.75$ m ✓
(2 marks)

60 Volume = area of trapezium × width

$= \frac{1}{2} h(a + b) \times 5$

$= \frac{1}{2} \times 6 \times (1 + 3.5) \times 5$

$= 67.5 \text{ m}^3$

Answer A

61 Area covered by tiles

$= 1.056 \times 10^6 \times 175 \text{ cm}^2$

$= 184\,800\,000 \text{ cm}^2$ (by calc.)

$= 18\,480 \text{ m}^2$

(NB: $100 \times 100 \text{ cm}^2 = 1 \text{ m}^2$

$10\,000 \text{ cm}^2 = 1 \text{ m}^2$)

Answer B

62 Surface area of a sphere

$= 4\pi r^2$

∴ Internal surface area of birdbath

$= \frac{1}{2} \times 4\pi r^2$ ✓

$= \frac{1}{2} \times 4\pi \times \left(\frac{45}{2}\right)^2$

$= 3180.8625\ldots \text{ cm}^2$

$= 3181 \text{ cm}^2$ to nearest square cm ✓

(2 marks)

63 **i** $V = \pi r^2 h$

$= \pi \times 5^2 \times 7.8$ ✓

$= 612.61\ldots \text{ cm}^3$

$= 613 \text{ cm}^3$ to nearest cm^3 ✓

(2 marks)

ii Total volume

$= 8 \times 612.61\ldots \text{ cm}^3$ ✓

$= 4900.88\ldots \text{ cm}^3$ ✓

$= 4900.88\ldots$ mL ($1 \text{ cm}^3 = 1$ mL)

$= 4.9008\ldots$ L

∴ Peta's engine does meet the racing requirements as it is under 5 litres. ✓

(3 marks)

64 Volume

$= \frac{1}{3} \times$ area of base × height

$= \frac{1}{3} \times (230 \times 230) \times 135$

$= 2\,380\,500 \text{ m}^3$

Answer C

65 Surface area of ball

$= 4\pi r^2$

$= 4\pi \times (0.6)^2$

$= 4.523\ldots \text{ m}^2$

∴ Cost of vinyl

$= 4.523\ldots \times \$32$

$= \$144.76$

$\doteqdot \$145$

Answer B

66 Weight of paper = 25.2 kg

= 25 200 g

Square metres of paper

$= 25\,200 \div 80$

$= 315$

Area of one sheet $= 0.21 \times 0.3$

$= 0.063 \text{ m}^2$

∴ Number of sheets $= 315 \div 0.063$

$= 5000$

Answer D

67 **i** **1** ΔEAD and ΔCAB are similar triangles. (They each have a right angle at E and C, and they each have angle A in common.)

(1 mark)

2 $AE = 50$ cm

$AC = 150 + 50$

$= 200$ cm

∴ Enlargement factor is 4. ($200 \div 50$)

(1 mark)

3 $CB = 4 \times 20$

$= 80$ cm

(1 mark)

ii Since ΔEAD and ΔCAB are similar,

$\frac{d}{200} = \frac{20}{x}$

$\therefore d = \frac{20}{x} \times 200$

$\therefore d = \frac{4000}{x}$

or $x = \frac{4000}{d}$

(1 mark)

68 **i** $A = \pi(R^2 - r^2)$

$= \pi\big((1.496 \times 10^8)^2 - (1.082 \times 10^8)^2\big)$ ✓

$= \pi(1.067\,292 \times 10^{16})$

$= 3.3529\ldots \times 10^{16}$ km

$= 3.4 \times 10^{16}$ km to 2 sig. figures. ✓

(2 marks)

ii $A = \pi(R^2 - r^2)$

$\therefore \frac{A}{\pi} = R^2 - r^2$ ✓

$\therefore R^2 = \frac{A}{\pi} + r^2$

$\therefore R = \sqrt{\frac{A}{\pi} + r^2}$ ✓

(2 marks)

iii Using (ii) above

$R = \sqrt{\frac{6.79}{\pi} + 0.75^2}$ ✓

$= \sqrt{2.7238\ldots}$

$= 1.6504\ldots$ mm

$= 1.65$ mm to 2 decimal places ✓

(2 marks)

69 $A = \frac{\theta}{360} \times \pi r^2$

$= \frac{120}{360} \times \pi \times 9^2$

$= 84.8230\ldots$

$\doteqdot 84.8 \text{ m}^2$ to 1 decimal place

Answer D

70 $V = \frac{4}{3}\pi r^3$

$360 = \frac{4}{3}\pi r^3$

$360 \div \frac{4}{3} = \pi r^3$

$\pi r^3 = 270$

$r^3 = 270 \div \pi$

$= 85.943\ldots$

$\therefore r = \sqrt[3]{85.943\ldots}$

$\doteqdot 4.4$ cm

Answer B

1 The price and the power consumption of two different brands of television are shown.

Television A	*Television B*
Price: $900 Power: 176 W	Price: $921.90 Power: 160 W

The average cost for electricity is 25c/kWh. A particular family watches an average of 3 hours of television per day.

a The annual cost of electricity for Television *A* for this family is $48.18. For this family, what is the difference in the annual cost of electricity between Television *A* and Television *B*? *(2 marks)* Medium

b For this family, how many years will it take for the total cost of buying and using Television *A* to be equal to the total cost of buying and using Television *B*? *(2 marks)* Hard

(Q27, **2021 HSC**)

2 Amanda uses 80 kilocalories of energy per kilometre while she is running.

She eats a burger that contains 2180 kilojoules of energy. How many kilometres will she need to run to use up all the energy from the burger? Give your answer correct to one decimal place.
(1 kilocalorie = 4.184 kilojoules) *(2 marks)*

(Q24, **2019 HSC**) Easy

3 The table shows an estimated number of kilojoules burned per kilogram of body mass per 30 minutes in different activities.

Activity	Energy used in 30 minutes
Walking: 6 km/h	9.22 kJ/kg
Cycling: 20 km/h	18.43 kJ/kg
Swimming: 50 m/min	20.73 kJ/kg
Jogging: 10 km/h	21.19 kJ/kg
Running: 16 km/h	32.35 kJ/kg

a Emma has a mass of 58 kg and rides her bike at an average speed of 20 km/h for 90 minutes. What is the amount of energy she has used, to the nearest kilojoule? *(1 mark)* Medium

b James, who weighs 70 kg, drinks a 600-mL bottle of cola which contains 1080 kJ. For how long must James run at 16 km/h to burn off the energy contained in the bottle, to the nearest minute? *(2 marks)* Medium

Bonus question (see page iv)

4 The table shows the number of Calories per 100 grams for three macronutrients.

Source of energy	Number of Calories per 100 grams
Carbohydrate	400
Protein	400
Fat	900

Use the conversion 1 Calorie = 4.184 kJ to find the number of kilojoules in 12.4 g of carbohydrates, to the nearest kilojoule? *(2 marks)*

Bonus question Medium

5 Every day, a 1200-watt microwave oven is used for 45 minutes at 40% power. Electricity is charged at $0.25 per kWh. What is the cost of running this microwave oven for 180 days? *(3 marks)*

(Q28c, **2018 HSC**) Medium

6 Electricity costs $0.27 per kWh.

How much does 20 kWh cost? *(1 mark)*

(Q26a, **2017 HSC**) Easy

7 The cost of buying a new heater is $990. It uses energy according to the following energy label.

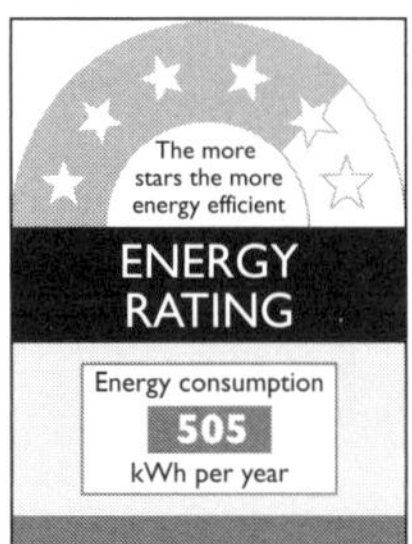

Energy is charged at the rate of $0.35/ kWh.

How much will it cost in total to purchase and then run this heater for five years? *(2 marks)*

(Q28b, **2016 HSC**) Easy

8 The energy consumption of a computer in standby mode is 21 watts. The cost of electricity is 31 cents per kWh.

A school computer room has 20 computers.

How much will the school save by switching off all 20 computers during 11 weeks of school holidays? *(2 marks)*

(Q30a, **2015 HSC**) **Medium**

9 In a household of 4, each member uses an average of 13 minutes of hot water per day.

The household uses a 9 kW hot water unit.

Electricity is charged at 11.97c/kWh when the hot water unit is being used.

What is the electricity cost for the hot water used by this household in one week?

A \$1.63 **B** \$6.54

C \$392.14 **D** \$653.56 *(1 mark)*

(Q20, **2014 HSC**) **Medium**

Year 11 Units of energy and mass—Worked answers

1 **a** Cost for 176 W = \$48.18
Cost for 160 W
= \$48.18 ÷ 176 × 160
= \$43.80 ✓
Difference = \$48.18 – \$43.80
= \$4.38 ✓
(2 marks)

b Difference in price
= \$921.90 – \$900
= \$21.90 ✓
Now, \$21.90 ÷ \$4.38 = 5.
So it will take 5 years for the total cost of both televisions to be the same. ✓
(2 marks)

2 80 kilocalories
= 80 × 4.184 kilojoules
= 334.72 kJ ✓
Distance = (2180 ÷ 334.72) km
= 6.512 9063… km
= 6.5 km (1 d.p.) ✓
(2 marks)

3 **a** Energy used
= 58 × 18.43 × 3
= 3206.82
= 3207 (nearest whole)
∴ Emma will use 3207 kJ of energy. *(1 mark)*

b Total energy per half hour
= 70 × 32.35
= 2264.5 ✓
Number of minutes
= 1080 ÷ 2264.5 × 30
= 14.307 794 22…
= 14 (nearest whole)
∴ it will take 14 minutes. ✓
(2 marks)

4 Number
= 12.4 ÷ 100 × 400 × 4.184 ✓
= 207.5264…
= 208
∴ 208 kJ ✓ *(2 marks)*

5 Power used per day
$= 0.4 \times 1200\text{ W} \times \frac{3}{4}\text{ h}$ ✓
= 360 Wh
Total power use = 180 × 360 Wh
= 64 800 Wh
= 64.8 kWh ✓
Cost = 64.8 × \$0.25
= \$16.20 ✓ *(3 marks)*

6 Cost = 20 × \$0.27
= \$5.40 *(1 mark)*

7 Energy used in 5 years
= 5 × 505 kWh
= 2525 kWh ✓
Cost of this energy
= 2525 × \$0.35
= \$883.75
Total cost = \$990 + \$883.75
= \$1873.75 ✓
(2 marks)

8 11 weeks = 11 × 7 × 24 hours
= 1848 h
Energy usage
= 20 × 21 W × 1848 h
= 776 160 Wh
= 776.16 kWh ✓
Cost = 776.16 × \$0.31
= \$240.6096
= \$241 (nearest dollar)
So, the school would save around \$241 by switching off the computers. ✓
(2 marks)

9 Total time = 7 × 4 × 13 minutes
= 364 minutes
= 6.066 66… h

Power used
= 9 kW × 6.066 66… h
= 54.6 kWh

Cost = 54.6 × \$0.1197
= \$6.535 62
= \$6.54 (nearest cent)

Answer B

1 City A is in Sweden and is located at (58°N, 16°E). Sydney, in Australia, is located at (33°S, 151°E).

Robert lives in Sydney and needs to give an online presentation to his colleagues in City A starting at 5:00 pm Thursday, local time in Sweden.

What time and day, in Sydney, should Robert start his presentation?

It is given that 15° = 1 hour time difference. Ignore daylight saving. *(3 marks)*

(Q20, **2021 HSC**) Medium

2 The Coordinated Universal Time (UTC) of Auckland is +12 hours and the UTC of Chicago is –5 hours.

When the time in Chicago is 2 pm, Thursday, what is the time in Auckland?

A 9 pm, Wednesday **B** 7 am, Thursday
C 9 pm, Thursday **D** 7 am, Friday *(1 mark)*

(Q5, **2019 HSC**) Medium

3 A plane leaves Miami, Florida (UTC–4) at 2:40 pm on Monday and arrives in Copenhagen, Denmark (UTC+2) at 6:55 am on Tuesday. If the flight distance is 7840 km what is the average speed of the aeroplane, to the nearest km/h? *(2 marks)*

Bonus question (see page iv) Hard

4 Holly flies from Auckland (UTC+12) across the International Date Line to Honolulu (UTC–10). The plane departs Auckland at 9:20 pm Friday and flies 7155 kilometres at an average speed of 810 km/h.

What is the local time when the plane lands in Honolulu? *(3 marks)*

Bonus question Hard

5 A plane left Dubai (UTC+4) at 1:30 pm on Thursday. It flew non-stop to Dallas (UTC–5) at an average speed of 810 km/h, arriving in Dallas at 8:45 pm on Thursday. What was the distance the plane flew? *(2 marks)*

Bonus question Hard

6 The time in Brisbane is $4\frac{1}{2}$ hours ahead of the time in New Delhi. John flew from New Delhi to Brisbane via Singapore. His plane left New Delhi at 11.30 am (New Delhi time), stopped for 3 hours in Singapore, and arrived in Brisbane at 9.00 am the following day (Brisbane time).

What was the plane's total flying time? *(3 marks)*

(Q29a, **2018 HSC**) Medium

7 Island A and island B are both on the equator. Island B is west of island A. The longitude of island A is 5°E and the angle at the centre of Earth (O), between A and B, is 30°.

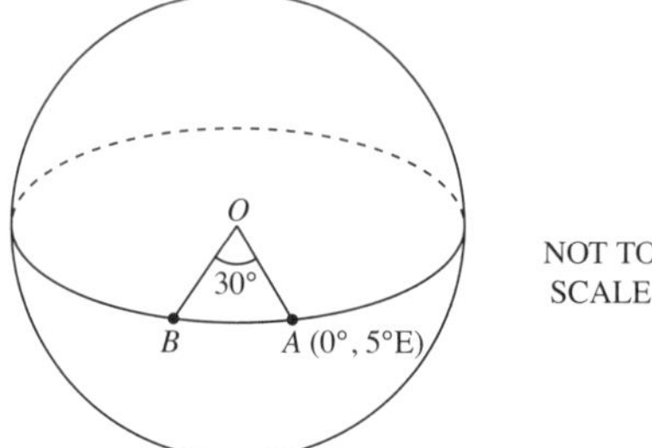

i What is the longitude of island B? *(1 mark)* Hard

ii What time is it on island B when it is 10 am on island A? *(1 mark)* Medium

(Q27d, **2017 HSC**)

8 Melbourne is located at (38°S, 145°E) and Dubai is located at (24°N, 55°E).

i Calculate the difference in longitude between Melbourne and Dubai. *(1 mark)* Easy

ii Show that the time difference between Melbourne and Dubai is 6 hours. *(1 mark)* Easy

iii A plane leaves Melbourne on Friday at 11.30 pm. The flight time to Dubai is 15 hours.
What will be the time and the day in Dubai when the plane is due to land? *(2 marks)* Easy

(Q27e, **2016 HSC**)

9 Stockholm is located at 59°N 18°E and Darwin is located at 13°S 131°E.

What is the time difference between Stockholm and Darwin? (Ignore time zones and daylight saving.)

A 184 minutes **B** 288 minutes
C 452 minutes **D** 596 minutes *(1 mark)*

(Q14, **2015 HSC**) Easy

10 Singapore is located at 1°N 104°E and Sydney is located at 34°S 151°E.

What is the time difference between Singapore and Sydney? (Ignore daylight saving.) *(2 marks)*

(Q26g, **2014 HSC**) **Medium**

11 Karin is in Athens, which is two hours ahead of Greenwich Mean Time. Marco is in New York, which is five hours behind Greenwich Mean Time.

i Karin is going to ring Marco at 10 pm on Tuesday, Athens time.
What day and time will it be in New York when she rings? *(1 mark)* **Medium**

ii Marco is going to fly from New York to Athens. His flight will leave on Wednesday at 9 am, New York time, and will take 11 hours. What day and time will it be in Athens when he arrives? *(2 marks)* **Easy**

(Q27e, **2013 HSC**)

12 Perth in Western Australia is 8 hours ahead of Greenwich in England. Cape Town in South Africa is 2 hours ahead of Greenwich.

What is the time in Cape Town when it is 1 pm in Perth?

A 3 am **B** 7 am
C 7 pm **D** 11 pm *(1 mark)*

(Q3, **2011 HSC**) **Easy**

13 In this diagram of the Earth, O represents the centre and B lies on both the Equator and the Greenwich Meridian.

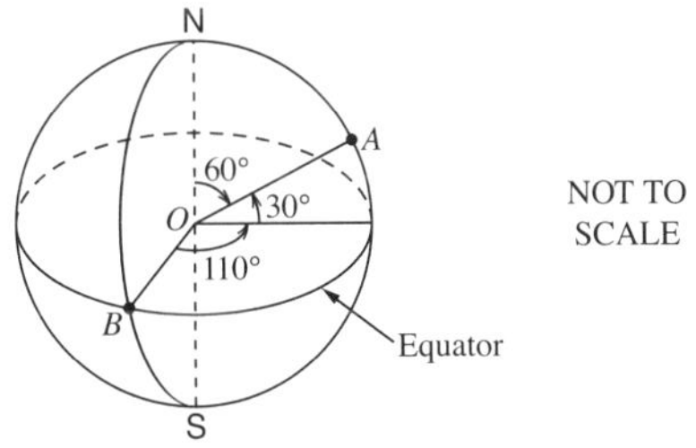

What is the latitude and longitude of point A?

A 30°N 110°E **B** 30°N 110°W
C 60°N 110°E **D** 60°N 110°W *(1 mark)*

(Q15, **2010 HSC**) **Easy**

14 Lee wants to call home to Sydney (34°S, 151°E) from Papeete, Tahiti (17°S, 149°W) when it is 7 pm on Friday in Sydney.

Give the time and day in Papeete when she should make the call. (Ignore time zones.) *(2 marks)* **Medium**

(Q25c ii, **2010 HSC**)

15 Osaka is at 34°N, 135°E, and Denver is at 40°N, 105°W.

i Show that there is a 16-hour time difference between the two cities. (Ignore time zones.) *(2 marks)* **Medium**

ii John lives in Denver and wants to ring a friend in Osaka. In Denver it is 9 pm Monday.
What time and day is it in Osaka then? *(1 mark)* **Easy**

iii John's friend in Osaka sent him a text message which happened to take 14 hours to reach him. It was sent at 10 am Thursday, Osaka time.
What was the time and day in Denver when John received the text? *(2 marks)* **Hard**

(Q26b, **2009 HSC**)

16 Kim lives in Perth (32°S, 115°E). He wants to watch an ice hockey game being played in Toronto (44°N, 80°W) starting at 10.00 pm on Wednesday.

What is the time in Perth when the game starts?

A 9.00 am on Wednesday
B 7.40 pm on Wednesday
C 12.20 am on Thursday
D 11.00 am on Thursday *(1 mark)*

(Q20, **2007 HSC**) **Medium**

17 Cassie flew from London (52°N, 0°E) to Manila (15°N, 120°E).

Her plane left London at 9.30 am Monday (London time), stopped for 5 hours in Singapore and arrived in Manila at 4.00 pm Tuesday (Manila time).

What was the total flying time? (Ignore time zones.) *(3 marks)*

(Q26d, **2006 HSC**) **Medium**

18 The location of Town A is 25°N 45°E. The location of Town B is 10°N 105°E.

Which of the following is true? (Ignore time zones.)

A Town A is four hours behind Town B.
B Town A is four hours ahead of Town B.
C Town A is one hour behind Town B.
D Town A is one hour ahead of Town B.
(1 mark)

(Q19, **2005 HSC**) **Easy**

19 Kathmandu is 30° west of Perth. Using the longitude difference, what is the time in Kathmandu when it is noon in Perth?

A 10:00 am **B** 11:30 am
C 12:30 pm **D** 2:00 pm *(1 mark)*

(Q10, **2003 HSC**) **Easy**

20 The table shows the approximate coordinates for two cities.

City	*Latitude*	*Longitude*
Buenos Aires	35° S	60° W
Adelaide	35° S	140° E

i What is the time difference between Adelaide and Buenos Aires? (Ignore time zones.) *(1 mark)* **Medium**

ii Roy lives in Adelaide and his cousin Juan lives in Buenos Aires. Roy wants to telephone Juan at 7 pm on a Friday night, Buenos Aires time. At what time, and on what day, should Roy make the call? *(2 marks)* **Medium**

(Q25b, **2002 HSC**)

21 In this diagram of the Earth, O represents the centre and G represents Greenwich. The point A lies on the equator.

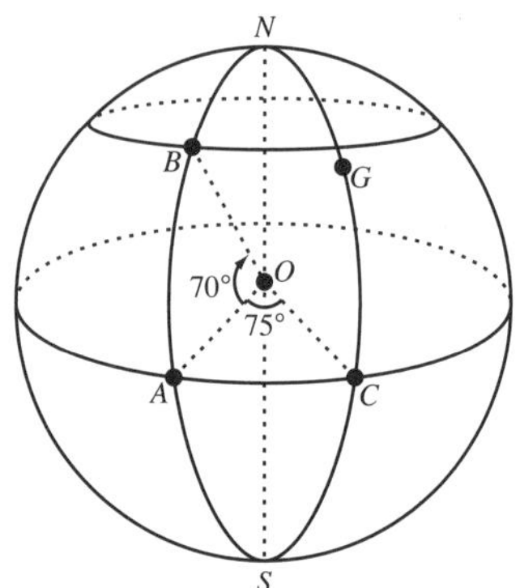

NOT TO SCALE

i What is the time difference between Greenwich and point A? (Ignore time zones.) *(1 mark)* **Medium**

ii What is the latitude of point B? *(1 mark)* **Easy**

(Q24b, **2001 HSC**)

Year 11 Working with time—Worked answers

1 Difference in longitude
$= 151° - 16°$
$= 135°$ ✓
Time difference $= (135 \div 15)$ h
$= 9$ h ✓
Now Sydney is 9 hours ahead of A.
So the time in Sydney is 9 h after 5 pm Thursday.
The time is 2:00 am Friday. ✓
(3 marks)

2 5 hours after 2 pm is 7 pm.
12 hours after 7 pm is 7 am the next day.
The time in Auckland is 7 am Friday.
Answer D

3 $4 + 2 = 6$
Copenhagen is 6 hours ahead of Miami.
Plane lands at 6:55 am minus 6 h = 12:55 am Miami time.
Time of flight
= 2:40 pm Mon. to 12:55 am Tues.
= 10 h 15 min
= 10.25 h ✓
Speed $= \dfrac{7840}{10.25}$
$= 764.8780488\ldots$
$= 765$ (nearest whole) ✓
$\therefore$ the average speed is 765 km/h.
(2 marks)

4 Time of flight $= 7155 \div 810$
$= 8\frac{5}{6}$ h
= 8h 50 min ✓
Honolulu is 22 hours behind Auckland.
9:20 pm plus 8 h 50 min is 6:10 am Saturday (Auckland time). ✓
Subtracting 22 hours gives 8:10 am Friday. ✓ *(3 marks)*

5 $4 + 5 = 9$
$\therefore$ time difference = Dallas is 9 hours behind Dubai.
Dubai 1:30 pm Thurs. = Dallas 4:30 am Thurs.
Flight time = 4:30 am to 8:45 pm
= 16 h 15 min
= 16.25 h ✓
Distance = Speed × Time
$= 810 \times 16.25$
$= 13\,162.5$
$\therefore$ the distance was 13 162.5 km. ✓ *(2 marks)*

6 $4\frac{1}{2}$ h ahead of 11:30 am is 4 pm. ✓
So the plane left New Delhi at 4 pm Brisbane time.
From 4.00 pm until 9.00 am the following day is 17 hours. ✓
Flying time $= (17 - 3)$ h
$= 14$ h ✓
(3 marks)

7 **i** B is 30° west of A.
The longitude of A is 5° E, so 5° west of A is the prime meridian.
Now $30 - 5 = 25$ so B is 25° west of the prime meridian.
So the longitude of B is 25° W.
(1 mark)

ii Angular difference = 30°
Time difference
$= (30 \div 15)$ h
$= 2$ h
B is west of A so it is 2 h behind A.
The time on island B would be 8 am.
(1 mark)

8 **i** Difference in longitude
$= 145° - 55°$
$= 90°$ *(1 mark)*

ii Every 15° difference in longitude is 1 hour difference in time.
So time difference
$= (90 : 15)$ h
$= 6$ h *(1 mark)*

iii The plane leaves at 11:30 pm Melbourne time.
Dubai is 6 h behind Melbourne time so the plane leaves at 5:30 pm Dubai time. ✓
It lands 15 hours later.
15 hours after 5:30 pm is 8:30 am the next day.
So the plane is due to land at 8.30 am Saturday Dubai time. ✓ *(2 marks)*

9 Difference in longitude
$= 131° - 18°$
$= 113°$
Time difference
$= 113 \times 4$ minutes
$= 452$ minutes
Answer C *(1 mark)*

10 Singapore (1 °N, 104 °E),
Sydney (34 °S, 151 °E)
Angular difference in longitude
$= 151° - 104°$
$= 47°$ ✓
Time difference $= 47 \times 4$ min
$= 188$ min
$= 3$ h 8 min
(Ignoring time zones) ✓
(2 marks)

11 **i** New York is 7 hours behind Athens time.
So 10 pm Athens time is 3 pm New York time.
In New York it will be 3 pm on Tuesday when Karin rings.
(1 mark)

ii At 9 am Wednesday in New York it will be 4 pm Wednesday in Athens. ✓
The flight takes 11 hours.
So 11 hours after 4 pm will be 3 am.
Marco will arrive in Athens at 3 am on Thursday. ✓
(2 marks)

12 When it is 1 pm in Perth it is 5 am in Greenwich. So it will be 7 am in Cape Town.
Answer B

13 A is 30° north of the Equator and 110° east of the Greenwich meridian.
The coordinates of A are 30°N 110°E.
Answer A

14 Difference in longitude
$= 151° + 149°$
$= 300°$
Time difference
$= (300 \div 15)$ hours
$= 20$ hours ✓
Papeete is 20 hours behind Sydney time.
Lee should call home when it is 11 pm Thursday in Papeete. ✓
(2 marks)

15 **i** Longitude difference
$= 135° + 105°$
$= 240°$ ✓
Time difference
$= 240° \div 15$
$= 16$ hours ✓
(2 marks)

ii Denver is 16 hours behind Osaka as it is further west.

∴ 9 pm Monday in Denver is 9 pm Monday + 16 hours in Osaka, i.e. 1 pm Tuesday in Osaka.

(1 mark)

iii Osaka is 16 hours ahead of Denver. ✓

∴ John received the text, 10 am Thursday – 16 hours + 14 hours ✓

i.e. 8 am Thursday in Denver.

(2 marks)

16 Longitude difference
$= 115° + 80°$
$= 195°$

Time difference
$= 195° \div 15°$
$= 13$ hours

∴ Perth is 13 hours ahead of Toronto.

∴ The time will be 11:00 am on Thursday.

Answer D

17 London (52°N, 0°E)
Manila (15°N, 120°E)

Longitude difference = 120°
∴ Time difference = 120 ÷ 15
= 8 hours ✓

Now 4 pm Tuesday in Manila is 8 am Tuesday in London.

∴ Total time taken is $22\frac{1}{2}$ hours (from Monday 9:30 am to Tuesday 8 am London time), including 5 hours stopping time. ✓

∴ Total flying time
$= 22\frac{1}{2} - 5$
$= 17\frac{1}{2}$ hours ✓

(3 marks)

18

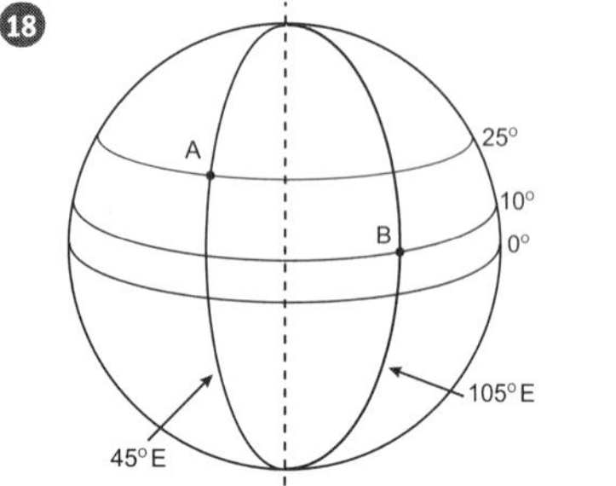

Since 15° = 1 hour,
time difference
= longitude difference ÷ 15
= (105 – 45) ÷ 15
= 60 ÷ 15
= 4 hours

B is ahead 4 hours as it is further east.

Answer A

19 Time difference = 30° × 4 min
= 120 min
= 2 h

∴ Kathmandu is 2 h behind Perth, i.e. 10:00 am.

Answer A

20 **i** Longitude difference
$= 60° + 140°$
$= 200°$

Time difference
$= 200 \div 15$
$= 13.\dot{3}$ h
$= 13$ h 20 min

(1 mark)

ii Adelaide is 13 h 20 min ahead of Buenos Aires as it is to the east. ✓

∴ 7 pm Fri, Buenos Aires time is (7 pm Fri + 13 h 20 min) Adelaide time, i.e. 8:20 am Saturday, Adelaide time. ✓

(2 marks)

21 **i** Time difference = 75° ÷ 15
= 5 hours

(1 mark)

ii 70° N

(1 mark)

1 Adam purchased some office furniture five years ago. It depreciated by $2300 each year based on the straight-line method of depreciation. The salvage value of the furniture is now $7500.

Find the initial value of the office furniture. *(2 marks)*

(Q19, **2021 HSC**) Easy

2 What is the interest earned, in dollars, if $800 is invested for x months at a simple interest rate of 3% per annum?

A $2x$
B $24x$
C $200x$
D $2400x$ *(1 mark)*

(Q9, **2019 HSC**) Hard

3 Part of a supermarket receipt is shown.

SUPERMARKET	
RECEIPT	
Date: 22/09/2019	
Description	$
*Chocolates 300 g	*A*
Tomatoes 1 kg	5.00
Natural almonds 400 g	*B*
Cheese slices 500 g	8.50
Milk 2 L	3.20
Bananas 570 g	2.85
Total for 6 items	**36.25**
GST included in total	0.70

*GST of 10% is included in the price of item.

Determine the missing values, A and B, to complete the receipt. *(2 marks)*

(Q29, **2019 HSC**) Medium

4 Jodie takes out a loan of $40 000 and is charged a reducible interest rate of 5.2% p.a. At the end of each month she is charged interest and then she makes a repayment of $650. Jodie uses a spreadsheet to calculate the balance owing after each monthly repayment.

Loan table				
	Amount:	**$40 000.00**		
	Annual interest rate:	**5.20%**		
	Monthly repayment (R):	**$650.00**		
Month	Principal (P)	Interest (I)	$P+I$	$P+I-R$
1	$40 000.00	$173.33	$40 173.33	$39 523.33
2	$39 523.33	$171.27	$39 694.60	**A**
3	$39 044.60	$169.19	$39 213.79	$38 563.79
4	$38 563.79	$167.11	$38 730.90	$38 080.90
5	$38 080.90	$165.02	$38 245.92	$37 595.92
6	$37 595.92	$162.92	$37 758.84	$37 108.84
7	$37 108.84	$160.80	$37 269.64	$36 619.64
8	$36 619.64	**B**	$36 778.33	$36 128.33
9	$36 128.33	$156.56	$36 284.88	$35 634.88
10	$35 634.88	$154.42	$35 789.30	$35 139.30
11	$35 139.30	$152.27	$35 291.57	$34 641.57
12	$34 641.57	$150.11	$34 791.68	$34 141.68

a Calculate the values in the following cells:
i A *(1 mark)* Easy
ii B *(1 mark)* Medium
b What is the total interest charged by the financial institution in the first 12 months? *(1 mark)* Medium

Bonus question (see page iv)

5 The table below shows the monthly repayments per $1000 on a bank loan for various annual interest rates.

Term	5.5%	6.0%	6.5%	7.0%	7.5%	8.0%
20 years	$6.8684	$7.1643	$7.4581	$7.7506	$8.0560	$8.3669
30 years	$5.6754	$5.9955	$6.3233	$6.6503	$6.9921	$7.3404

a To the nearest dollar, what is the total amount of interest paid using monthly repayments over the period of the loan of $465 000 for 20 years at 7.5% p.a.? *(2 marks)* Medium

b Many years ago Kate borrowed $240 000 for a term of 30 years at an interest rate of 6.5% p.a. Today, Kate still owes $173 000 and decides to repay the balance with a new loan from a different financial institution which offers her 5.5% p.a. interest for 20 years. What is the difference in the two monthly repayments? *(2 marks)* **Hard**

Bonus question

6 Colin's April credit card statement shows an opening balance of $1460 and a purchase of $1135 on 17 April. The credit card has no interest-free period.

Compound interest is charged at the rate of 16.2425% per annum, calculated daily, including the date of purchase and the date the account is paid. The minimum payment is 3.5% of the closing balance.

a What is the closing balance on this credit card for April? *(2 marks)* **Hard**

b What is the total amount of interest charged for April? *(1 mark)* **Medium**

c If Colin only pays the minimum payment on 30 April, what will be the opening balance for May? *(1 mark)* **Medium**

Bonus question

7 Yanika opens a new credit card account, with interest and fees as shown.

> Interest
> - Flat rate of 12.3% per annum
> - No interest-free period
>
> Fees
> - $0 for online repayments
> - $3 for repayments in cash
>
> (fee added to balance immediately after repayment)

Yanika makes a single purchase of $849 with the credit card.

i Show that the balance owing on the credit card 24 days after making the purchase is $855.87. *(2 marks)* **Medium**

ii Yanika makes her first repayment 24 days after making the purchase. She makes a cash repayment of $450. What is the balance owing on the credit card immediately after her repayment is made and the repayment fee has been charged? *(1 mark)* **Medium**

(Q28d, **2018 HSC**)

8 Rachel bought a motorcycle advertised for $7990. She paid a $500 deposit and took out a flat-rate loan to repay the balance. Simple interest was charged at a rate of 7% per annum on the amount borrowed. She repaid the loan over 2 years, making equal weekly repayments.

Calculate the weekly repayment. *(3 marks)*

(Q26g, **2017 HSC**) **Medium**

9 Marge borrowed $19000 to buy a used car. Interest on the loan was charged at 4.8% pa at the end of each month. She made a repayment of $436 at the end of every month. The table below sets out her monthly repayment schedule for the first four months of the loan.

Month	*Amount owing at start of month*	*Interest charged*	*Repayment*	*Amount owing at end of month*
1	*A*	$76.00	$436.00	$18 640.00
2	$18 640.00	*X*	$436.00	$18 278.56
3	$18 278.56	$73.11	$436.00	$17 915.67
4	$17 915.67	$71.66	$436.00	*B*

i Some values in the table are missing. Write down the values for *A* and *B*. *(2 marks)* **Easy**

ii Calculate the value of *X*. *(2 marks)* **Easy**

iii Marge repaid this loan over four years. What is the total amount that Marge repaid? *(1 mark)* **Medium**

(Q27d, **2016 HSC**)

10 A camera costs $449, including 12% GST. What is the price of the camera without GST, correct to the nearest dollar?

A $395 **B** $401
C $437 **D** $503 *(1 mark)*

(Q15, **2015 HSC**) **Medium**

11 On 20 August, tickets were purchased for $425 using a credit card. No other purchases were made using this card in August. Simple interest was charged at a rate of 18.4% per annum. There was no interest-free period. The period for which interest was charged included the date of purchase and the date of payment.

What amount was paid when the account was paid in full on 31 August? *(2 marks)*

(Q29a, **2015 HSC**) **Medium**

12 Lynne invests $1000 for a term of 15 months. Interest is paid at a flat rate of 3.75% per annum.

How much will Lynne's investment be worth at the end of the term?

A $1046.88 **B** $1047.09
C $1296.88 **D** $1468.75 *(1 mark)*

(Q9, **2013 HSC**) **Medium**

13 Polly borrowed $11 000. She repaid the loan in full at the end of two years with a lump sum of $12 000.

What annual simple interest rate was she charged?

A 4.17% **B** 4.55%
C 8.33% **D** 9.09% *(1 mark)*

(Q13, **2013 HSC**) Easy

14 Heather used her credit card to purchase a plane ticket valued at $1990 on 28 January 2011. She made no other purchases on her credit card account in January. She paid the January account in full on 19 February 2011.

The credit card account has no interest free period. Simple interest is charged daily at the rate of 20% per annum, including the date of purchase and the date the account is paid.

How much interest did she pay, to the nearest cent? *(2 marks)*

(Q26c, **2012 HSC**) Hard

15 Matthew bought a laptop priced at $2800. He paid a 10% deposit and made monthly repayments of $95.20 for 3 years.

What annual flat rate of interest was Matthew charged? Justify your answer with suitable calculations. *(4 marks)*

(Q28e, **2012 HSC**) Hard

16 A television was purchased for $2100 on 12 April 2011 using a credit card. Simple interest was charged at a rate of 19.74% per annum for purchases on this credit card. There were no other purchases on this credit card account.

There was no interest-free period. The period for which interest was charged included the date of purchase and the date of payment.

What amount was paid when the account was paid in full on 20 May 2011?

A $2143.16 **B** $2143.59
C $2144.29 **D** $2144.74 *(1 mark)*

(Q10, **2011 HSC**) Hard

17 Ying borrowed $250 000 to buy a house. The interest rate and monthly repayment for her loan are shown in the spreadsheet.

	A	B	C	D	E
1	**Home Loan Table**			*This table assumes the same number of days in each month, ie*	
2		Amount =	$250 000		
3		Annual Interest Rate =	7.65%	Interest = Rate/12 × Principal	
4		Monthly Repayment (*R*) =	$1871.94		
5					
6	**Month**	**Principal (*P*)**	**Interest (*I*)**	$P + I$	$P + I - R$
7	1	$250 000.00	$1593.75	$251 593.75	$249 721.81
8	2	$249 721.81	$1591.98	$251 313.79	$249 441.85
9	3	$249 441.85	$1590.19	$251 032.04	
10	4				

Sheet 1 / Sheet 2

What is the total interest charged for the first four months of this loan?

A $6364.32 **B** $6366.11
C $6369.67 **D** $6376.25 *(1 mark)*

(Q22, **2011 HSC**) Medium

18 Furniture priced at $20 000 is purchased. A deposit of 15% is paid.

The balance is borrowed using a flat-rate loan at 19% per annum interest, to be repaid in equal monthly instalments over five years.

What will be the amount of each monthly instalment? Justify your answer with suitable calculations. *(4 marks)*

(Q26c, **2011 HSC**) Medium

19 A new phone was purchased for $725 which included 10% GST.

What was the price of the phone without GST, correct to the nearest cent?

A $65.91 **B** $72.50
C $652.50 **D** $659.09 *(1 mark)*

(Q2, **2010 HSC**) Medium

20 Minjy invests $2000 for 1 year and 5 months. The simple interest is calculated at a rate of 6% per annum.

What is the total value of the investment at the end of this period?

A $2170 **B** $2180
C $3003 **D** $3700 *(1 mark)*

(Q5, **2010 HSC**) Medium

21 Which of the following graphs shows the lowest rate of depreciation over the given time period?

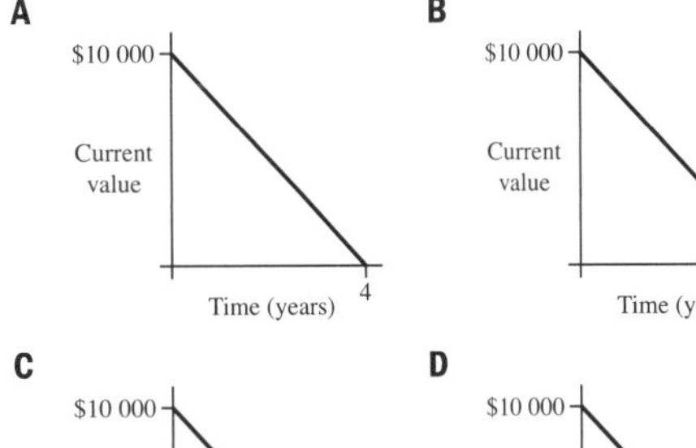

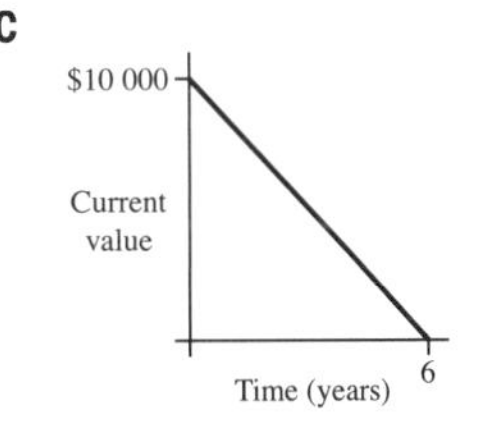

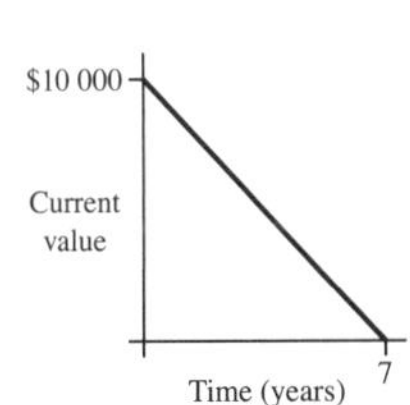

(Q11, **2010 HSC**) Easy

22 In July, Ms Alott received a statement for her credit card account. The account has no interest free period. Simple interest is calculated and charged to her account on the statement date.

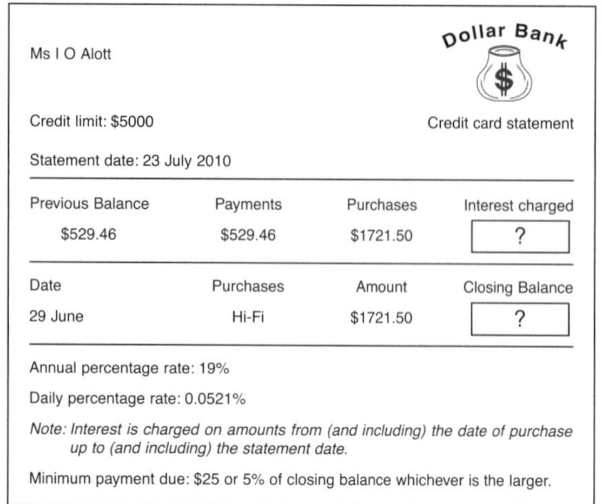

Ms I O Alott

Dollar Bank

Credit limit: $5000 — Credit card statement

Statement date: 23 July 2010

Previous Balance	Payments	Purchases	Interest charged
$529.46	$529.46	$1721.50	?

Date	Purchases	Amount	Closing Balance
29 June	Hi-Fi	$1721.50	?

Annual percentage rate: 19%

Daily percentage rate: 0.0521%

Note: Interest is charged on amounts from (and including) the date of purchase up to (and including) the statement date.

Minimum payment due: $25 or 5% of closing balance whichever is the larger.

What is the minimum payment due on this account?

A $22.42 **B** $25.00

C $86.08 **D** $87.20 *(1 mark)*

(Q22, **2010 HSC**) **Hard**

23 William wants to buy a car. He takes out a loan for $28 000 at 7% per annum interest for four years.

Monthly repayments for loans at different interest rates are shown in the spreadsheet.

	A	B	C	D	E
1		Monthly repayments			
2		Term of loan (in months)		48	
3					
4	Amount	Interest rate p.a.			
5	borrowed	6%	7%	8%	9%
6	$27 000	$634.10	$646.55	$659.15	$671.90
7	$27 500	$645.84	$658.52	$671.36	$684.34
8	$28 000	$657.58	$670.49	$683.56	$696.78
9	$28 500	$669.32	$682.47	$695.77	$709.22
10	$29 000	$681.07	$694.44	$707.97	$721.67
11	$29 500	$692.81	$706.41	$720.18	$734.11
12	$30 000	$704.55	$718.39	$732.39	$746.55

Sheet 1 / Sheet 2

How much interest does William pay over the term of this loan? *(2 marks)*

(Q25b, **2010 HSC**) **Medium**

24 Lou bought a plasma TV which was priced at $3499. He paid $1000 deposit and got a loan for the balance that was paid off by 24 monthly instalments of $135.36.

What simple interest rate per annum, to the nearest percent, was charged on his loan?

A 11% **B** 15%

C 30% **D** 46% *(1 mark)*

(Q20, **2009 HSC**) **Hard**

25 Ali is buying a speedboat at Betty's Boats.

Betty's Boats

Cash price
$16 000

OR

Terms
15% deposit plus
$320 per month for
5 years

What is the amount of interest Ali will have to pay if he chooses to buy the boat on terms?

A $3200 **B** $5600

C $19 200 **D** $21 600 *(1 mark)*

(Q15, **2008 HSC**) **Medium**

26 The price of a CD is $22.00, which includes 10% GST.

What is the amount of GST included in this price?

A $2.00 **B** $2.20

C $19.80 **D** $20.00 *(1 mark)*

(Q6, **2007 HSC**) **Easy**

27 In June, Ms Bigspender received a statement for her credit card account.

The account has no interest-free period. Simple interest is calculated and charged to her account on the statement date.

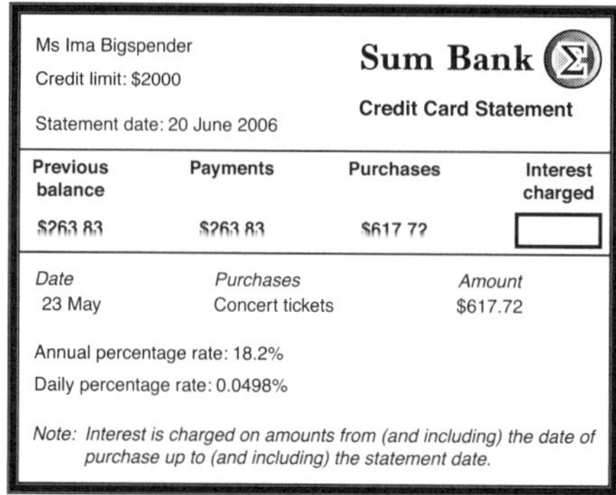

Ms Ima Bigspender

Credit limit: $2000

Statement date: 20 June 2006

Sum Bank

Credit Card Statement

Previous balance	Payments	Purchases	Interest charged
$263.83	$263.83	$617.72	

Date	*Purchases*	*Amount*
23 May	Concert tickets	$617.72

Annual percentage rate: 18.2%

Daily percentage rate: 0.0498%

Note: Interest is charged on amounts from (and including) the date of purchase up to (and including) the statement date.

i For how many days is she charged interest on her purchase? *(1 mark)* **Easy**

ii Calculate the interest charged to her account. *(2 marks)* **Medium**

(Q25b, **2006 HSC**)

28 Rita purchased a camera for $880 while on holidays in Australia. This price included 10% GST. When she left Australia she received a refund of the GST.

What was Rita's refund?

A $80 **B** $88

C $792 **D** $800 *(1 mark)*

(Q17, **2004 HSC**) **Medium**

29 Iliana buys several items at the supermarket. The docket for her purchases is shown.

XYZ Supermarket

27/04/03 12.53

MILK 1L	$1.19
* PEPSI 1.25L	$1.29
* DISINFECTANT	$7.23
* TEA TREE OIL	$4.13
SPINACH	$1.64
SOUP	$1.57
TOTAL	$17.05

10% GST INCLUDED
IN COST OF TAXABLE ITEMS

* = TAXABLE ITEMS

What is the amount of GST included in the total?

A $1.15 **B** $1.27
C $1.55 **D** $1.71 *(1 mark)*

(Q20, **2003 HSC**) **Medium**

30 Zoë plans to borrow money to buy a car and considers the following repayment guide:

Fortnightly car loan repayment guide

Amount borrowed	*Length of loan* 1 year	2 years	3 years
($)	($)	($)	($)
10 000	410	217	153
10 500	430	228	161
11 000	451	239	168
11 500	471	249	176
12 000	492	260	183
12 500	512	271	191
13 000	532	282	199
13 500	553	293	206
14 000	573	303	214
14 500	594	314	221
15 000	614	325	229
15 500	635	336	237
16 000	655	347	244

Zoë wishes to borrow $15 500 and pay back the loan in fortnightly instalments over two years.

What is the flat rate of interest per annum on this loan? *(3 marks)*

(Q24c, **2003 HSC**) **Medium**

31 In one year, the population of a city increased by 20%. The next year, it decreased by 10%.

What was the percentage increase in the population over the two years?

A 8% **B** 10%
C 15% **D** 30% *(1 mark)*

(Q19, **2002 HSC**) **Medium**

32 A computer was purchased for $2500 and depreciated over six years, as shown in the graph.

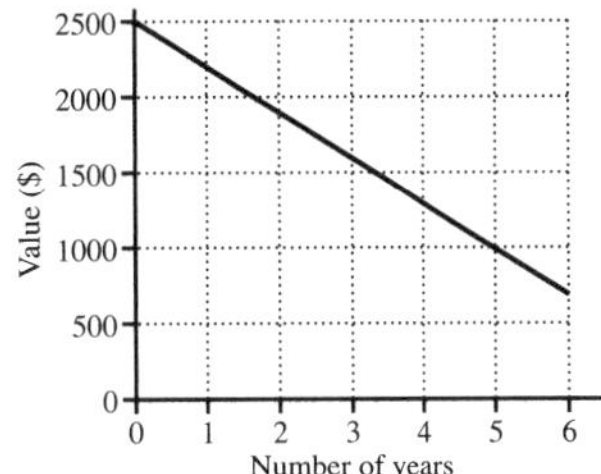

By how much did the computer depreciate each year?

A $200 **B** $250
C $300 **D** $350 *(1 mark)*

(Q9, **2001 HSC**) **Medium**

33 George buys a television for $574.20, including 10% GST.
What is the value of the GST component? *(1 mark)*

(Q27a, **2001 HSC**) **Easy**

34 A car is purchased for $42 000.
Use the declining balance method to calculate the salvage value of the car after 4 years at a depreciation rate of 15% per annum. *(2 marks)*

(Q27b, **2001 HSC**) **Easy**

Year 11 Interest and depreciation—Worked answers

1 $V_0 = \$7500 + 5 \times \$2300$ ✓
$= \$19000$
The initial value of the furniture was \$19000. ✓
(2 marks)

2 $P = 800,\ r = 0.03,\ n = \frac{x}{12}$
$I = Prn$
$= 800 \times 0.03 \times \frac{x}{12}$
$= 2x$
Answer A

3 Chocolates are the only item with GST included in the price.
GST = \$0.70
So the price of the chocolates without GST is \$7.00 and the price on the docket will be \$7.70. ✓
Total for 6 items = \$36.25
Total for 5 items
= \$7.70 + \$5.00 + \$8.50 + \$3.20 + \$2.85
= \$27.25
Cost of almonds
= \$36.25 – \$27.25 = \$9.00
So $A = 7.70$ and $B = 9.00$ ✓
(2 marks)

4 **a** **i** \$39 694.60 – \$650
= \$39 044.60 *(1 mark)*
ii \$36 619.64 × 0.052 ÷ 12
= \$158.69 *(1 mark)*
b Total interest
= 12 × \$650 – (\$40 000 – \$34 141.68)
= \$1941.68
∴ total interest of \$1941.68
(1 mark)

5 **a** Total interest
= 465 × 8.056 × 12 × 20 – \$465 000 ✓
= 434 049.60
= 434 050 (nearest whole)
∴ the total interest is \$434 050. ✓ *(2 marks)*
b Repayment 1
= 240 × 6.3233
= 1517.59 (2 dec. pl.) ✓
Repayment 2
= 173 × 6.8684
= 1188.23 (2 dec. pl.)
Difference
= 1517.59 – 1188.23
= 329.36
∴ the difference is \$329.36. ✓ *(2 marks)*

6 **a** 16.2425% per annum
= 0.0445% per day ✓
Closing balance
$= 1460 \times 1.000445^{30} + 1135 \times 1.000445^{14}$
= 2621.708 829...
= 2621.71 (2 dec. pl.)
∴ Colin's closing balance is \$2621.71. ✓ *(2 marks)*
b Interest
= 2621.71 – (1460 + 1135)
= 26.71
∴ interest is \$26.71. *(1 mark)*
c Minimum payment for April
= 0.035 × 2621.71
= 91.759 85 ...
= 91.76
Opening balance for May
= 2621.71 – 91.76
= 2529.95
∴ opening balance of \$2529.95. *(1 mark)*

7 **i** $P = \$849,\ r = 0.123,\ n = \frac{24}{365}$
$I = Prn$
$= \$849 \times 0.123 \times \frac{24}{365}$ ✓
= \$6.866 432 87...
= \$6.87 (nearest cent)
Balance owing
= \$849 + \$6.87
= \$855.87 ✓ *(2 marks)*
ii Balance owing
= \$855.87 – \$450 + \$3.00
= \$408.87 *(1 mark)*

8 Price = \$7990, Deposit = \$500
Balance owing = \$7990 – \$500
= \$7490
Interest rate = 7%
$P = \$7490,\ r = 0.07,\ n = 2$
$I = Prn$
= \$7490 × 0.07 × 2
= \$1048.60 ✓
Total to repay
= \$7490 + \$1048.60
= \$8538.60 ✓
Now 2 years = 2 × 52 weeks
= 104 weeks
Weekly repayments
= \$8538.60 ÷ 104
= \$82.101 923....
= \$82.10 (nearest cent) ✓
(3 marks)

9 **i** $A = \$19000$ ✓
$B = \$17915.67 + \$71.66 - \$436.00$
= \$17 551.33 ✓ *(2 marks)*
ii $X = \frac{0.048}{12} \times \$18640.00$ ✓
= \$74.56 ✓ *(2 marks)*
iii Total repaid = 4 × 12 × \$436
= \$20 928 *(1 mark)*

10 Original price + GST
= 100% + 12%
= 112%
So 112% = \$449
1% = \$449 ÷ 112
100% = \$449 ÷ 112 × 100
= \$400.892 85...
= \$401 (nearest dollar)
Answer B

11 From, and including, 20 August until, and including, 31 August is 12 days.
$I = Prn$
$= \$425 \times \frac{0.184}{365} \times 12$ ✓
= \$2.570 9589...
= \$2.57 (nearest cent)
Amount paid = \$425 + \$2.57
= \$427.57 ✓
(2 marks)

12 $I = Prn$
$= \$1000 \times 0.0375 \times \frac{15}{12}$
= \$46.875
= \$46.88 [nearest cent]
Value of investment
= \$1000 + \$46.88
= \$1046.88
Answer A

13 Interest = \$12 000 – \$11 000
= \$1000
$I = Prn$
$1000 = 11\,000 \times r \times 2$
$1000 = 22\,000r$
$r = \frac{1000}{22\,000}$
= 0.045 454 545...
The interest rate
= 4.545 4545... %
= 4.55% [2 d.p.]
Answer B

14 Number of days = 4 + 19
= 23
$P = \$1990\quad r = 0.2\quad n = \frac{23}{365}$ ✓
$I = Prn$
$= \$1990 \times 0.2 \times \frac{23}{365}$
= \$25.079 452...
= \$25.08 [nearest cent]
Heather paid \$25.08 interest, to the nearest cent. ✓
(2 marks)

15 Deposit = 0.1 × $2800
= $280 ✓

Balance owing
= $2800 − $280
= $2520

Total repayments
= $95.20 × 3 × 12
= $3427.20

Interest
= $3427.20 − $2520
= $907.20 ✓

I = $907.20 P = $2520 n = 3

$I = Prn$

$907.20 = $2520 × r × 3
= 7560r$ ✓

r = $907.20 ÷ $7560
= 0.12

The annual rate of interest was 12%. ✓

(4 marks)

16 From, and including, 12 April until, and including, 20 May is 39 days.

$P = \$2100, r = 0.1974, n = \frac{39}{365}$

$I = Prn$

$= \$2100 \times 0.1974 \times \frac{39}{365}$

= $44.293 315...

= $44.29 [nearest cent]

Total amount = $2100 + $44.29
= $2144.29

Answer C

17 For month 3:

$P + I - R$
= $251 032.04 − $1871.94
= $249 160.10

For month 4:

P = $249 160.10

$I = \frac{0.0765}{12} \times \$249\,160.10$

= $1588.395 638...
= $1588.40 [nearest cent]

Total interest
= $1593.75 + $1591.98 + $1590.19 + $1588.40
= $6364.32

Answer A

18 Deposit = 15% of $20 000
= 0.15 × $20 000
= $3000 ✓

Balance owing
= $20 000 − $3000
= $17 000

Interest = $17 000 × 0.19 × 5
= $16 150 ✓

Total to be repaid
= $17 000 + $16 150
= $33 150 ✓

Number of payments
= 5 × 12
= 60

Amount of each payment
= $33 150 ÷ 60
= $552.50 ✓

(4 marks)

19 Let the price without GST be 100%.
The price with GST will be (100 + 10)% or 110%.

So 110% = $725.00

10% = $65.909 090...
(after dividing by 11)

100% = $659.090 909...
(after multiplying by 10)

So the price without GST is $659.09.

Answer D

20 $P = \$2000, \; r = 0.06, \; n = 1\frac{5}{12}$

$I = Prn$

$= \$2000 \times 0.06 \times 1\frac{5}{12}$

= $170

Value of investment
= $2000 + $170
= $2170

Answer A

21 At the lowest rate of depreciation, an item will retain some value for a longer period of time.

Answer D

22 Number of days from 29 June to 23 July is 25.

So n = 25

Daily interest rate = 0.0521%
= 0.000 521

So r = 0.000 521

P = $1721.50

$I = Prn$

= $1721.50 × 0.000 521 × 25
= $22.422 537...
= $22.42 [nearest cent]

Closing balance
= $1721.50 + $22.42
= $1743.92

5% of closing balance
= 0.05 × $1743.92
= $87.196
= $87.20 [nearest cent]

This is more than $25.

So the minimum payment is $87.20.

Answer D

23 On a $28 000 loan at 7% p.a. over 4 years the repayments are $670.49 per month.

Number of repayments = 48

Total repaid = 48 × $670.49
= $32 183.52 ✓

Interest = $32 183.52 − $28 000
= $4183.52 ✓

(2 marks)

24 Price paid = $1000 + 24 × $135.36
= $4284.64

Interest = $4284.64 − $3499
= $749.64

$I = Prn$

where n = 24 months = 2 years

749.64 = 2499 × r × 2

749.64 = 4998r

$r = \frac{749.64}{4998}$

= 0.1499 . . .
= 0.1499 . . . × 100
= 15% to nearest per cent

Answer B

25 Deposit = 0.15 × $16 000
= $2400

Balance = $16 000 − $2400
= $13 600

Repayments = $320 × 12 × 5
= $19 200

Interest paid
= Total price − cash price
= $2400 + $19 200 − $16 000
= $5600

Answer B

26 110% of price = $22.00
∴ 10% of price = $22.00 ÷ 11
= $2.00

Answer A

27 **i** 29 days
(9 days in May, 20 days in June)

(1 mark)

ii $I = Prn$

$= 617.72 \times \frac{0.0498}{100} \times 29$ ✓

$= \$8.92$ ✓

(2 marks)

28 110% of price = \$880

∴ 10% of price = \$880 ÷ 11

= \$80

Answer A

29 Total of taxable items (incl. GST)

= \$1.29 + \$7.23 + \$4.13

= \$12.65

∴ 110% of total = \$12.65

∴ 10% of total = \$12.65 ÷ 11

= \$1.15

Answer A

30 From the table, Zoe's fortnightly repayments are \$336.

Total repayments

= \$336 × 26 × 2

= \$17 472 ✓

Interest charged

= \$17 472 – \$15 500

= \$1972 ✓

Interest rate per annum

$= \frac{1972}{15\,500} \times 100 \div 2$

= 6.361. . .%

≑ 6.4% p.a. ✓

(3 marks)

31 Let the initial population be 100.

After 1 year,

population = 1.2 × 100

= 120

After 2 years,

population = 0.9 × 120

= 108

∴ Increase in population over 2 years is 8 people.

∴ Percentage increase

$= \frac{8}{100} \times 100$

= 8%

Answer A

32 When $n = 0$, value = \$2500

When $n = 5$, value = \$1000

∴ Depreciation $= \frac{\$1500}{5}$

= \$300 per year

Answer C

33 110% of price = \$574.20

10% of price = \$574.20 ÷ 11

= \$52.20

∴ GST component = \$52.20

(1 mark)

34 $S = V_0(1 - r)^n$

$= 42\,000(1 - 0.15)^4$ ✓

$= \$21\,924.26$ ✓

(2 marks)

1 The table shows the income tax rates for the 2020–2021 financial year.

Taxable income	Tax payable on this income
0 – \$18 200	Nil
\$18 201 – \$45 000	19 cents for each \$1 over \$18 200
\$45 001 – \$120 000	\$5092 plus 32.5 cents for each \$1 over \$45 000
\$120 001 – \$180 000	\$29 467 plus 37 cents for each \$1 over \$120 000
\$180 001 and over	\$51 667 plus 45 cents for each \$1 over \$180 000

William has a gross annual salary of \$84 000. He has allowable tax deductions of \$900 for home-office equipment and \$474 for union fees. William must also pay a Medicare Levy of 2% of his taxable income.

Calculate the total tax payable by William including the Medicare Levy. *(3 marks)*

(Q22, **2021 HSC**) Medium

2 The table shows the income tax rates for the 2019–2020 financial year.

Taxable income	Tax payable on this income
\$0 to \$18 200	Nil
\$18 201 to \$37 000	19c for each \$1 over \$18 200
\$37 001 to \$90 000	\$3572 plus 32.5c for each \$1 over \$37 000
\$90 001 to \$180 000	\$20 797 plus 37c for each \$1 over \$90 000
\$180 001 and over	\$54 097 plus 45c for each \$1 over \$180 000

For the 2019–2020 financial year, Wally had a taxable income of \$122 680. During the year, he paid \$3000 per month in Pay As You Go (PAYG) tax.

Calculate Wally's tax refund, ignoring the Medicare levy. *(3 marks)*

(Q20, **2020 HSC**) Medium

3 Julia earns \$28 per hour. Her hourly pay rate increases by 2%.How much will she earn for a 4-hour shift with this increase?

A \$2.24 **B** \$28.56
C 112 **D** \$114.24 *(1 mark)*

(Q7, **2019 HSC**) Easy

4 The table shows the income tax rates for the 2018–2019 financial year.

Taxable income	Tax payable on this income
\$0 to \$18 200	Nil
\$18 201 to \$37 000	19c for each \$1 over \$18 200
\$37 001 to \$90 000	\$3572 plus 32.5c for each \$1 over \$37 000
\$90 001 to \$180 000	\$20 797 plus 37c for each \$1 over \$90 000
\$180 001 and over	\$54 097 plus 45c for each \$1 over \$180 000

The Medicare levy is calculated as 2% of taxable income.

For the 2018–2019 financial year, Charlie pays a Medicare levy of \$1934.80.

Calculate the tax payable on Charlie's taxable income. *(3 marks)*

(Q32, **2019 HSC**) Medium

5 A nanny charges \$15 per hour, or part thereof, for looking after a child.

What does the nanny charge for looking after a child from 8 am until 3.20 pm on a particular day?

A \$105 **B** \$108
C \$110 **D** \$120 *(1 mark)*

(Q8, **2018 HSC**) Medium

6 To determine the retail price of an item, a shop owner increases its cost price by 30%. In a sale, the retail price is reduced by 30% to give the sale price.

How does the sale price compare to the cost price?

A The sale price is less than the cost price.
B The sale price is the same as the cost price.
C The sale price is more than the cost price.
D It is impossible to compare without knowing the cost price. *(1 mark)*

(Q14, **2018 HSC**) Medium

7 Last year, Luke's taxable income was \$87 000 and the tax payable on this income was \$19 822. This year, Luke's taxable income has increased by \$16 800.

i Use the table to calculate the tax payable by Luke this year. *(2 marks)* Easy

Taxable income (\$)	Tax payable
\$0 – \$18 200	Nil
\$18 201 – \$37 000	19c for each \$1 over \$18 200
\$37 001 – \$87 000	\$3572 plus 32.5c for each \$1 over \$37 000
\$87 001 – \$180 000	\$19 822 plus 37c for each \$1 over \$87 000
\$180 001 and over	\$54 232 plus 45c for each \$1 over \$180 000

ii How much extra money will Luke have this year, after paying tax, as a result of the increase in his taxable income? Ignore the Medicare levy. *(2 marks)* Easy

(Q30b, **2018 HSC**)

8 Tom earns a weekly wage of \$1025. He also receives an additional allowance of \$87.50 per day when handling toxic substances.

What is Tom's income in a fortnight in which he handles toxic substances on 5 separate days?

A $1112.50 **B** $1462.50
C $2225.00 **D** $2487.50 *(1 mark)*

(Q6, **2017 HSC**) Easy

9 Sabrina's taxable income is $86 725 in a particular year.

The table below is used to calculate her tax payable. In addition, she pays the Medicare levy, which is 2% of her taxable income.

Taxable income ($)	*Tax payable*
$0 – $18 200	Nil
$18 201 – $37 000	19c for each $1 over $18 200
$37 001 – $87 000	$3572 plus 32.5c for each $1 over $37 000
$87 001 – $180 000	$19 822 plus 37c for each $1 over $87 000
$180 001 and over	$54 232 plus 45c for each $1 over $180 000

Calculate Sabrina's net income in that year. *(3 marks)*

(Q29b, **2017 HSC**) Medium

10 Isabella works a 35-hour week and is paid at an hourly rate of $18. Any overtime hours worked are paid at time-and-a-half. In a particular week, she earned $1008.

How many hours in total did Isabella work in this week to earn this amount?

A 37.3 **B** 42
C 49 **D** 56 *(1 mark)*

(Q20, **2016 HSC**) Easy

11 Jenny earns a yearly salary of $63 752. Her annual leave loading is $17\frac{1}{2}$% of four weeks pay.

Calculate her total pay for her four weeks of annual leave. *(3 marks)*

(Q26e, **2016 HSC**) Easy

12 Theo is completing his tax return. He has a gross salary of $82 521 and income from a rental property totalling $10 920. He is claiming $13 420 in allowable deductions.

i Determine Theo's taxable income. *(1 mark)* Easy

ii Using the tax table below, calculate Theo's tax payable. *(2 marks)* Easy

Taxable income	*Tax on this income*
$0 – $18 200	Nil
$18 201 – $37 000	19c for each $1 over $18 200
$37 001 – $80 000	$3572 plus 32.5c for each $1 over $37 000
$80 001 – $180 000	$17 547 plus 37c for each $1 over $80 000
$180 001 and over	$54 547 plus 45c for each $1 over $180 000

iii In addition to the above tax, Theo must also pay a Medicare levy of $1600.42.

Theo has already paid $20 525 as Pay As You Go (PAYG) tax.

Should Theo receive a tax refund or will he owe more tax? Justify your answer with calculations. *(2 marks)* Medium

(Q26f, **2016 HSC**)

13 Gayle's gross pay each week is $952.25.

The following deductions are taken from her gross pay each week:

- tax $180.93
- superannuation $85.70
- union membership $21.40
- health fund $38.15.

What is Gayle's net pay each week?

A $326.18 **B** $626.07
C $771.32 **D** $952.25 *(1 mark)*

(Q3, **2015 HSC**) Easy

14 Jane sells jewellery. Her commission is based on a sliding scale of 6% on the first $2000 of her sales, 3.5% on the next $1000, and 2% thereafter.

What is Jane's commission when her total sales are $5670?

A $188.40 **B** $208.40
C $321.85 **D** $652.05 *(1 mark)*

(Q13, **2014 HSC**) Easy

15 Luke's normal rate of pay is $24.80 per hour. In one week he worked 14 hours at the normal rate, 4 hours at time-and-a-half, and $3\frac{1}{2}$ hours at double time. He was also paid a wet weather allowance of $50 for the week.

What was his pay for the week?

A $583.20 **B** $620.40
C $669.60 **D** $719.60 *(1 mark)*

(Q3, **2013 HSC**) Easy

16 An enterprise agreement has the following annual salary arrangements:

Base Salary	
Step 1	$35 000
Step 2	$40 000
Step 3	$45 000

Leadership Allowance	
Leader 1	$5000
Leader 2	$7500
Leader 3	$10 000

George's employer pays 6% more than the enterprise agreement. He is on Step 3 and receives an allowance for Leader 2.

What is George's gross monthly pay?

A $4375.00 **B** $4412.50
C $4600.00 **D** $4637.50 *(1 mark)*

(Q11, **2013 HSC**) **Medium**

---

17 The table shows the tax payable to the Australian Taxation Office for different taxable incomes.

Taxable income	*Tax on this income*
\$0 – \$18 200	Nil
\$18 201 – \$37 000	19c for each \$1 over \$18 200
\$37 001 – \$80 000	\$3572 plus 32.5c for each \$1 over \$37 000
\$80 001 – \$180 000	\$17 547 plus 37c for each \$1 over \$80 000
\$180 001 and over	\$54 547 plus 45c for each \$1 over \$180 000

Peta has a gross annual salary of $84 000. She has tax deductions of $1000 for work-related travel and $500 for stationery. The Medicare levy that she pays is calculated at 1.5% of her taxable income.

Peta has already paid $18 500 in tax.

Will Peta receive a tax refund or will she owe money to the Australian Taxation Office? Justify your answer by calculating the refund or amount owed. *(4 marks)*

(Q27b, **2013 HSC**) 

---

18 Jo qualifies for both Rent Assistance and Youth Allowance and receives a fortnightly payment from the government.

Rent Assistance is $119.40 per fortnight.

The maximum Youth Allowance is $402.70 per fortnight. It is reduced by 50 cents in the dollar for any income earned over $236 per fortnight.

Jo earns $300 per fortnight from a part-time job.

What is the total payment Jo receives each fortnight from the government?

A $370.70 **B** $372.10
C $458.60 **D** $490.10 *(1 mark)*

(Q18, **2012 HSC**) 

---

19 Tai earns a gross weekly wage of $1024. Each week her deductions are:

- tax instalment of $296.40
- health fund contribution of $24.50
- union fees of $15.80.

She also pays $3640 over the year as her share of the household expenses.

What percentage of her net wage does Tai pay for household expenses? *(3 marks)*

(Q27a, **2012 HSC**) **Easy**

---

20 Simon is a mechanic who receives a normal rate of pay of $22.35 per hour for a 40-hour week.

When he is needed for emergency call-outs he is paid a special allowance of $150 for that week. Additionally, every time he is called out to an emergency he is paid for a minimum of 4 hours at double time.

In the week beginning 2 February, 2011 Simon worked 40 hours normal time and was needed for emergency call-outs. His emergency call-out log book for the week is shown.

Employee: Simon Week: 2/2/11 to 8/2/11	
Date	*Duration of call-out*
3/2/11	5 hours
5/2/11	1.5 hours

What was Simon's total pay for that week?

A $1189.28 **B** $1296.30
C $1334.55 **D** $1446.30 *(1 mark)*

(Q19, **2011 HSC**) 

---

21 Sri has a gross salary of $56 350. She has tax deductions of $350 for union fees, $2000 in work-related expenses and $250 in donations to charities.

The Medicare levy is 1.5% of her taxable income.

Calculate Sri's Medicare levy. *(3 marks)*

(Q23a, **2011 HSC**) **Medium**

---

22 This advertisement appeared in a newspaper.

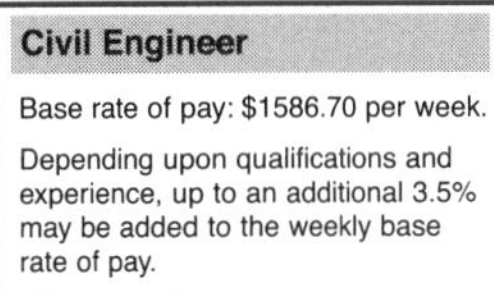

What is the maximum possible salary per annum for this civil engineer, correct to the nearest dollar? *(2 marks)*

(Q23a, **2010 HSC**) **Medium**

---

23 Billy worked for 35 hours at the normal hourly rate of pay and for five hours at double time. He earned $561.60 in total for this work.

What was the normal hourly rate of pay?

A $7.02 **B** $12.48
C $14.04 **D** $16.05 *(1 mark)*

(Q10, **2009 HSC**) **Easy**

24 Using the tax table, what is the tax payable on $43 561?

Taxable income	*Tax payable*
$0 – $12 000	Nil
$12 001 – $30 000	Nil plus 30 cents for each $1 over $12 000
$30 001 – $45 000	$5400 plus 40 cents for each $1 over $30 000
$45 001 – $60 000	$11 400 plus 50 cents for each $1 over $45 000
over $60 000	$18 900 plus 55 cents for each $1 over $60 000

A $5424.40 **B** $10 824.40
C $16 224.40 **D** $17 424.40 *(1 mark)*

(Q6, **2008 HSC**) Easy

25 Luke's normal rate of pay is $15 per hour. Last week he was paid for 12 hours, at time-and-a-half.
How many hours would Luke need to work this week, at double time, to earn the same amount?

A 4 **B** 6
C 8 **D** 9 *(1 mark)*

(Q7, **2008 HSC**) Easy

26 Bob is employed as a salesman. He is offered two methods of calculating his income.

Method 1:	Commission only of 13% on all sales
Method 2:	$350 per week plus a commission of 4.5% on all sales

Bob's research determines that the average sales total per employee per month is $15 670.

i Based on his research, how much could Bob expect to earn in a year if he were to choose Method 1? *(2 marks)* Easy

ii If Bob were to choose a method of payment based on the average sales figures, state which method he should choose in order to earn the greater income. Justify your answer with appropriate calculations. *(3 marks)* Medium

(Q24a, **2008 HSC**)

27 Joe is about to go on holidays for four weeks. His weekly salary is $280 and his holiday loading is $17\frac{1}{2}$% of four weeks pay.
What is Joe's total pay for the four weeks holiday?

A $196 **B** $329
C $1169 **D** $1316 *(1 mark)*

(Q3, **2007 HSC**) Easy

28 Myles is in his third year as an apprentice film editor.

i Myles purchased film-editing equipment for $5000. After 3 years it has depreciated to $3635 using the straight-line method.
Calculate the rate of depreciation per year as a percentage. *(2 marks)* Medium

ii Myles earns $800 per week. Calculate his taxable income for this year if the only allowable deduction is the amount of depreciation of his film-editing equipment in the third year of use. *(1 mark)* Medium

iii Use this tax table to calculate Myles's tax payable. *(2 marks)* Medium

Taxable income ($)	*Tax payable*
$0 – $10 000	Nil
$10 001 – $28 000	Nil plus 25 cents for each $1 over $10 000
$28 001 – $50 000	$4500 plus 30 cents for each $1 over $28 000
$50 001 – $100 000	$11 100 plus 40 cents for each $1 over $50 000
over $100 000	$31 100 plus 60 cents for each $1 over $100 000

(Q26b, **2007 HSC**)

29 A salesman earns $200 per week plus $40 commission for each item he sells.

How many items does he need to sell to earn a total of $2640 in two weeks?

A 33 **B** 56
C 61 **D** 66 *(1 mark)*

(Q5, **2006 HSC**) Medium

30 This income tax table is used to calculate Evelyn's tax payable.

Taxable income	*Tax payable*
$0 – $20 000	Nil
$20 001 – $45 000	Nil plus 10 cents for each $1 over $20 000
$45 001 – $70 000	$2500 plus 35 cents for each $1 over $45 000
$70 001 and above	$11 250 plus 52 cents for each $1 over $70 000

Evelyn's taxable income increases from $50 000 to $80 000.

What percentage of her increase will she pay in additional tax?

A 15.25% **B** 40.7%
C 43.5% **D** 52% *(1 mark)*

(Q22, **2006 HSC**) Medium

31 Two groups of people were surveyed about their weekly wages. The results are shown in the box-and-whisker plots.

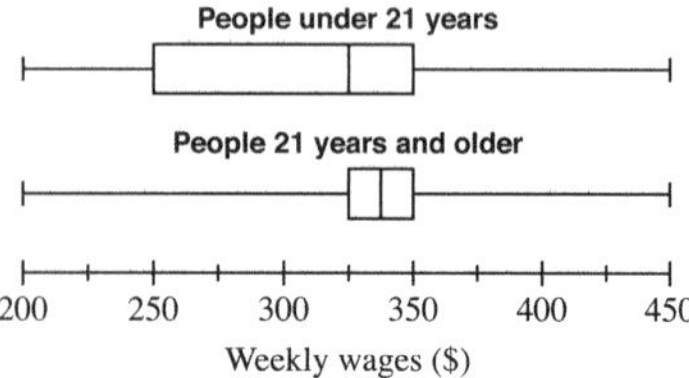

Which of the following statements is true for the people surveyed?

A The same percentage of people in each group earned more than \$325 per week.
B Approximately 75% of people under 21 years earned less than \$350 per week.
C Approximately 75% of people 21 years and older earned more than \$350 per week.
D Approximately 50% of people in each group earned between \$325 and \$350 per week. *(1 mark)*

(Q22, **2005 HSC**) Easy

32 Using the tax table, determine the tax payable on a taxable income of \$47 000.

Taxable income	*Tax on this income*
\$0 – \$6 000	NIL
\$6 001 – \$22 000	16 cents for each \$1 over \$6 000
\$22 001 – \$45 000	\$2 560 plus 25 cents for each \$1 over \$22 000
\$45 001 – \$60 000	\$8 310 plus 40 cents for each \$1 over \$45 000
\$60 001 and over	\$14 310 plus 48 cents for each \$1 over \$60 000

A \$8310.40 **B** \$9109.60
C \$9110.00 **D** \$10 310.40 *(1 mark)*

(Q10, **2004 HSC**) Easy

33 Stan worked for 24 hours as shown on his pay slip.

What was his hourly rate of pay?
A \$11.20 **B** \$12.13
C \$14.56 **D** \$18.20 *(1 mark)*

(Q20, **2004 HSC**) Easy

34 David is paid at these rates:

Weekday rate	\$18.00 per hour
Saturday rate	Time-and-a-half
Sunday rate	Double time

His time sheet for last week is:

	Start	*Finish*	*Unpaid break*
Friday	9.00 am	1.30 pm	30 minutes
Saturday	9.00 am	4.00 pm	1 hour
Sunday	8.00 am	2.00 pm	1 hour

i Calculate David's gross pay for last week. *(3 marks)* Easy

ii David decides not to work on Saturdays. He wants to keep his weekly gross pay the same. How many extra hours at the weekday rate must he work? *(1 mark)* Medium

(Q27b, **2004 HSC**)

35 Dora works for \$9.60 per hour for eight hours each day on Thursday and Friday. On Saturday she works for six hours at time-and-a-half.

How much does Dora earn in total for Thursday, Friday and Saturday?
A \$192.00 **B** \$211.20
C \$240.00 **D** \$316.80 *(1 mark)*

(Q3, **2003 HSC**) Easy

36 Vicki earns a taxable income of \$58 624 from her job with an insurance company. She pays \$14 410.80 tax on this income.

i Vicki has a second job which pays \$900 gross income per month.
What is Vicki's total annual taxable income from both jobs, assuming that she has no allowable tax deductions? *(1 mark)* Easy

ii Use the tax table below to calculate the total tax payable on her income from both jobs. *(2 marks)* Easy

Taxable income	*Tax payable*
\$0–\$6 000	NIL
\$6 001–\$22 000	18 cents for each \$1 over \$6 000
\$22 001–\$55 000	\$2 880 plus 30 cents for each \$1 over \$22 000
\$55 001–\$66 000	\$12 780 plus 45 cents for each \$1 over \$55 000
\$66 001 and over	\$17 730 plus 48 cents for each \$1 over \$66 000

iii Show that Vicki's monthly net income from her second job is \$486.44. *(3 marks)* Medium

(Q24b, **2003 HSC**)

37 The graph shows the tax payable for taxable incomes up to \$60 000 in a proposed tax system.

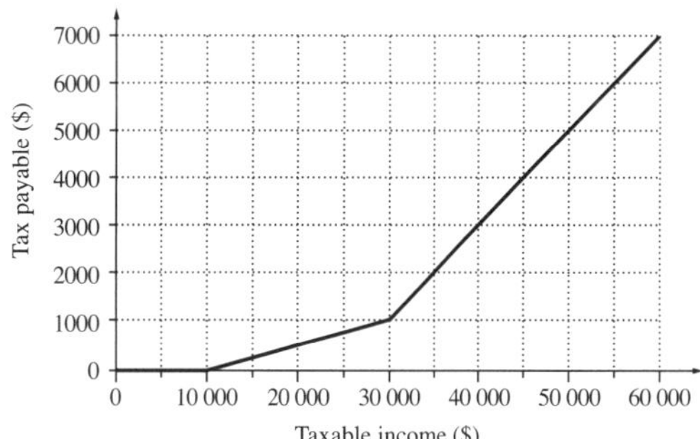

How much of each dollar earned over \$30 000 is payable in tax?

A 10 cents **B** 12 cents
C 20 cents **D** 23 cents *(1 mark)*

(Q22, **2002 HSC**) **Medium**

---

38 Jordan's gross pay is \$1500 per fortnight.

i Fortnightly deductions from Jordan's gross pay are:
- \$269.17 for tax;
- \$7.88 for union fees;
- \$16.25 for private health insurance.

Calculate his fortnightly net pay. *(1 mark)* 

ii Jordan is paid an annual leave loading of $17\frac{1}{2}\%$ of 4 weeks' gross pay. Calculate his annual leave loading. *(1 mark)* **Easy**

(Q23a, **2002 HSC**)

---

39 Alex works in a shop where the normal weekday rate of pay is \$12 per hour. On Saturdays she is paid time-and-a-half.

How much did Alex earn in a week in which she worked for seven hours on Thursday and three hours on Saturday?

A \$84 **B** \$120
C \$138 **D** \$180 *(1 mark)*

(Q1, **2001 HSC**) **Easy**

---

40 The table shows personal income tax rates.

Taxable income	*Tax on this income*
\$0 – \$6 000	Nil
\$6 001 – \$20 000	17 cents for each \$1 over \$6 000
\$20 001 – \$50 000	\$2 380 plus 30 cents for each \$1 over \$20 000
\$50 001 – \$60 000	\$11 380 plus 42 cents for each \$1 over \$50 000
\$60 001 and over	\$15 580 plus 47 cents for each \$1 over \$60 000

Sandra has a gross income of \$60 780 and deductions that total \$2420.

What is the tax payable on Sandra's taxable income?

A \$13 526.60 **B** \$14 891.20
C \$15 946.60 **D** \$17 084.00 *(1 mark)*

(Q10, **2001 HSC**) **Easy**

Year 11 Earning and managing money—Worked answers

1 Taxable income
= \$84 000 – \$900 – \$474
= \$82 626 ✓
Tax payable on that income
= \$5092 + \$0.325 × (82 626 – 45 000)
= \$17 320.45 ✓
Medicare levy = 2% of \$82 626
= 0.02 × \$82 626
= \$1652.52
Total tax payable
= \$17 320.45 + \$1652.52
= \$18 972.97 ✓
(3 marks)

2 Taxable income = \$122 680
Tax on that income
= \$20 797 + \$0.37 × (122 680 – 90 000)
= \$32 888.60 ✓
PAYG tax paid
= 12 × \$3000
= \$36 000 ✓
Tax refund
= \$36 000 – \$32 888.60
= \$3111.40 ✓
(3 marks)

3 2% of \$28 = \$0.56
New hourly pay rate = \$28.56
Julia's earnings = 4 × \$28.56
= \$114.24
Answer D

4 \$1934.80 is 2% or $\frac{1}{50}$ of Charlie's taxable income. ✓
Taxable income = 50 × \$1934.80
= \$96 740 ✓
Tax payable = \$20 797 + \$0.37 × (96 740 – 90 000)
= \$23 290.80 ✓
(3 marks)

5 From 8 am until 3 pm is 7 hours.
From 3 pm until 3.20 pm is another part of an hour so 8 hours will be charged altogether.
Total charge = 8 × \$15
= \$120
Answer D

6 Retail price is 130% of cost price.
Sale price is 70% of retail price.
So sale price is 70% of 130% of cost price or 91% of cost price.
The sale price is less than the cost price.
Answer A

7 **i** Tax payable
= \$19 822 + \$0.37 × 16 800 ✓
= \$19 822 + \$6216
= \$26 038 ✓ *(2 marks)*

ii Extra tax paid = \$6216 ✓
Extra money
= \$16 800 – \$6216
= \$10 584
Luke will have \$10 584 extra this year. ✓ *(2 marks)*

8 Tom's fortnightly income
= 2 × \$1025 + 5 × \$87.50
= \$2487.50
Answer D

9 Taxable income = \$86 725
Tax payable
= \$3572 + \$0.325 × (86 725 – 37 000)
= \$19 732.625
= \$19 732.63 (nearest cent) ✓
Medicare levy = 0.02 × \$86 725
= \$1734.50
Total tax = \$19 732.63 + \$1734.50
= \$21 467.13 ✓
Net income
= \$86 725 – \$21 467.13
= \$65 257.87 ✓
(3 marks)

10 Normal earnings = 35 × \$18
= \$630
Total overtime pay
= \$1008 – \$630
= \$378
Overtime rate
= 1.5 × \$18
= \$27 per hour
Overtime hours
= \$378 ÷ \$27 = 14
Total hours worked = 35 + 14
= 49
Answer C

11 Weekly pay = \$63 752 ÷ 52
= \$1226 ✓
So 4 weeks pay = 4 × \$1226
= \$4904
Loading = 0.175 × \$4904
= \$858.20 ✓
Total holiday pay
= \$4904 + \$858.20
= \$5762.20 ✓
(3 marks)

12 **i** Taxable income
= \$82 521 + \$10 920 – \$13 420
= \$80 021
(1 mark)

ii Tax payable
= \$17 547 + \$0.37 × (80 021 – 80 000) ✓
= \$17 554.77 ✓
(2 marks)

iii Total to be paid
= \$17 554.77 + \$1600.42
= \$19 155.19 ✓
But Theo has paid \$20 525.
So Theo will receive a tax refund because he has already paid more than is necessary.
Refund
= \$20 525 – \$19 155.19
= \$1369.81 ✓
(2 marks)

13 Total deductions
= \$180.93 + \$85.70 + \$21.40 + \$38.15
= \$326.18
Net pay = \$952.25 – \$326.18
= \$626.07
Answer B

14 6% of \$2000 = \$120
3.5% of \$1000 = \$35
2% of \$2670 = \$53.40
Total = \$208.40
Answer B

15 Normal pay = 14 × \$24.80
= \$347.20
Time-and-a-half overtime pay
= 4 × 1.5 × \$24.80
= \$148.80
Double-time pay
= 3.5 × 2 × \$24.80
= \$173.60
Total pay
= \$347.20 + \$148.80 + \$173.60 + \$50
= \$719.60
Answer D

16 On Step 3 with allowance for Leader 2 the enterprise agreement pay
= \$45 000 + \$7500
= \$52 500
6% more than this amount
= 1.06 × \$52 500
= \$55 650
Monthly pay = \$55 650 ÷ 12
= \$4637.50
Answer D

17 Total deductions = $1000 + $500
= $1500

Taxable income = $84 000 – $1500
= $82 500 ✓

From the tax table:

tax = $17 547 + $0.37 × 2500
= $18 472 ✓

Medicare levy = 1.5% × $82 500
= $1237.50

Total tax payable
= $18 472 + $1237.50
= $19 709.50 ✓

So, Peta owes money to the Australian Taxation Office because the tax payable is greater than the tax she has already paid.

Difference
= $19 709.50 – $18 500
= $1209.50

Peta must pay an additional $1209.50 ✓

(4 marks)

18 Income above $236
= $300 – $236
= $64

Reduction = 0.5 × $64
= $32

Jo's youth allowance
= $402.70 – $32
= $370.70

Total payment
= $370.70 + $119.40
= $490.10

Answer D

19 Total deductions
= $296.40 + $24.50 + $15.80
= $336.70 ✓

Net wage = $1024 – $336.70
= $687.30

Weekly share of household expenses
= $3640 ÷ 52
= $70 ✓

Percentage of net wage

$= \frac{70}{687.30} \times 100\%$

= 10.184 78... %
= 10.2% [1 decimal place] ✓

(3 marks)

20 Normal weekly pay = 40 × $22.35
= $894.00

For the first call-out Simon is paid for 5 hours at double time.

For the second call-out Simon is paid for 4 hours at double time.

Pay for call-outs = 9 × 2 × $22.35
= $402.30

Total pay
= $894 + $150 + $402.30
= $1446.30

Answer D

21 Total tax deductions
= $350 + $2000 + $250
= $2600 ✓

Taxable income
= $56 350 – $2600
= $53 750 ✓

Medicare levy = 1.5% of $53 750
= 0.015 × $53 750
= $806.25 ✓

(3 marks)

22 3.5% of weekly base rate
= 0.035 × $1586.70
= $55.5345
= $55.53 [nearest cent] ✓

Maximum weekly pay
= $1586.70 + $55.53
= $1642.23

Maximum annual salary
= 52 × $1642.23
= $85 395.96
= $85 396 [nearest dollar] ✓

(2 marks)

23 Hours worked at normal rate
= 35 + 2 × 5
= 45

∴ Hourly rate of pay
= $561.60 ÷ 45
= $12.48

Answer B

24 Tax payable
= $5400 + 0.40 × $(43 561 – 30 000)
= $5400 + $5424.40
= $10 824.40

Answer B

25 12 hours pay at time-and-a-half
= 12 × 1.5 × $15
= $270

Double-time pay
= 2 × $15
= $30 per hour

Number of hours worked
= $270 ÷ $30
= 9

Answer D

26 **i** Expected income (Method 1)

$= \frac{13}{100} \times \$15\,670 \times 12$ ✓

= $24 445.20 ✓

(2 marks)

ii Expected income (Method 2)

$= \$350 \times 52 + \frac{4.5}{100} \times \$15\,670 \times 12$ ✓

= $18 200 + $8461.80
= $26 661.80 ✓

Based on the calculations above and in **i**, Bob should choose Method 2. ✓

(3 marks)

27 Total pay
= 4 weeks pay + loading

$= 4 \times \$280 + 17\frac{1}{2}\% \times 4 \times \$280$

= $1120 + $196
= $1316

Answer D

28 **i** $S = V_0 - Dn$

$\therefore 3635 = 5000 - D \times 3$
$3D = 5000 - 3635$
$= 1365$
$\therefore D = \$455.$ ✓

∴ Rate of depreciation

$= \frac{455}{5000} \times 100$

= 9.1% ✓

(2 marks)

ii Taxable income
= Income – deductions
= ($800 × 52) – $455
= $41 600 – $455
= $41 145

(1 mark)

iii Tax payable
= $4500 + 0.30 × ($41 145 – $28 000) ✓
= $4500 + 0.3 × $13 145
= $4500 + $3943.50
= $8443.50 ✓

(2 marks)

29 Pay for 2 weeks = $400

Commission from sales
= $2640 – $400
= $2240

No. of items sold = $2240 ÷ $40
= 56

Answer B

30 Tax on \$50 000
= \$2500 + 0.35 × (\$50 000 − \$45 000)
= \$2500 + \$1750
= \$4250

Tax on \$80 000
= \$11 250 + 0.52 × \$10 000
= \$16 450 (by calc.)

Additional tax = \$16 450 − \$4250
= \$12 200

Percentage of increase

$= \frac{12\,200}{(80\,000 - 50\,000)} \times 100$

$= \frac{12\,200}{30\,000} \times 100$

= 40.6666...%
≑ 40.7%

Answer B

31 A is not true because 50% of people under 21 earned more than \$325, but 75% of the others earned more than \$325. (∴ C is not true either.)
D is not true because only 25% of people under 21 earned between \$325 and \$350 per week.
B is true. 75% of people under 21 earned less than \$350 per week.

Answer B

32 Tax payable
= \$8310 + 0.40 × (\$47 000 − \$45 000)
= \$8310 + 0.40 × \$2000
= \$9110

Answer C

33 Normal hours worked

$= 20 + 4 \times 1\frac{1}{2}$

= 26

∴ Hourly rate
= \$291.20 ÷ 26
= \$11.20

Answer A

34 **i** Friday: 4 × \$18 = \$72 ✓

Saturday: $6 \times 1\frac{1}{2} \times \$18 = \$162$

Sunday: 5 × 2 × \$18 = \$180 ✓

∴ Total gross pay
= \$72 + \$162 + \$180
= \$414 ✓

(3 marks)

ii Saturday's pay was the equivalent of $6 \times 1\frac{1}{2} = 9$ hours.

(1 mark)

35 Thursday earnings
= 8 × \$9.60 = \$76.80

Friday earnings
= 8 × \$9.60 = \$76.80

Saturday earnings

$= 6 \times 1\frac{1}{2} \times \$9.60 = \$86.40$

Total earnings
= \$76.80 + \$76.80 + \$86.40
= \$240

Answer C

36 **i** Taxable income from Job 1
= \$58 624

Taxable income from Job 2
= \$900 × 12
= \$10 800

∴ Total taxable income
= \$10 800 + \$58 624
= \$69 424

(1 mark)

ii Total tax payable
= \$17 730 ✓
+ 0.48(\$69 424 − \$66 000)
= \$17 730 + 0.48 × \$3424
= \$17 730 + \$1643.52
= \$19 373.52 ✓

(2 marks)

iii Tax on second job
= Tax payable − Tax paid
= \$19 373.52 − \$14 410.80
= \$4962.72 ✓

∴ Monthly tax on second job
= \$4962.72 ÷ 12
= \$413.56 ✓

∴ Net monthly income from second job
= \$900 − \$413.56
= \$486.44 ✓

(3 marks)

37 Tax on income of \$30 000
= \$1000 (from graph)

Tax on income of \$40 000
= \$3000 (from graph)

∴ Additional tax per extra \$10 000 earned = \$2000

∴ Amount per dollar payable in tax.

$= \frac{\$2000}{\$10\,000} = \$0.20$

= 20 cents

Answer C

38 **i** Fortnightly net pay
= \$1500 − \$269.17 − \$7.88 − \$16.25
= \$1206.70

(1 mark)

ii Annual leave loading

$= \frac{17\frac{1}{2}}{100} \times 2 \times \$1500$

= \$525.00

(1 mark)

39 Amount earned

$= 7 \times \$12 + 3 \times 1\frac{1}{2} \times \$12$

= \$84 + \$54
= \$138

Answer C

40 Taxable income
= \$60 780 − \$2420
= \$58 360

Tax payable
= \$11 380 + 0.42 × (\$58 360 − \$50 000)
= \$11 380 + \$3511.20
= \$14 891.20

Answer B

1 Mary is 18 years old and has just purchased comprehensive motor vehicle insurance. The following excesses apply to claims for at-fault motor vehicle accidents.

- Basic excess of $850 for each claim
- An additional age excess of $1600 for drivers under 25 years of age
- An additional age excess of $400 for drivers 25 years of age or over with no more than 2 years driving experience

How much would Mary be required to pay as excess if she made a claim as the driver at fault in a car accident?

A $1600 **B** $850 + $400
C $850 + $1600 **D** $850 + $1600 + $400 *(1 mark)*

(Q6, **2019 HSC**) Easy

2 The driving distance from Alex's home to his work is 20 km. He drives to and from work five times each week. His car uses fuel at the rate of 8 L /100 km.

How much fuel does he use driving to and from work each week?

A 16 L **B** 20 L
C 25 L **D** 40 L *(1 mark)*

(Q5, **2018 HSC**) Easy

3 Sam is the driver at fault in a car accident.

Which of the following is covered by Sam's compulsory third-party (CTP) insurance?

A Repairs to Sam's car
B Injury to the other driver
C Damage to the other driver's car
D Cost of repairing a building damaged in the accident *(1 mark)*

(Q15, **2018 HSC**) Medium

4 David earns a gross income of $890 per week. Each week, 25% of this income is deducted in taxation. David budgets to save 20% of his net income.

How much does he budget to save each week?

A $44.50 **B** $133.50
C $489.50 **D** $534.00 *(1 mark)*

(Q21, **2018 HSC**) Easy

5 Kate is comparing two different models of car. Car A uses fuel at the rate of 9 L /100 km. Car B uses 3.5 L /100 km.

Suppose Kate plans on driving 8000 km in the next year.

How much less fuel will she use driving car B instead of car A?

A 280 L **B** 440 L
C 720 L **D** 1000 L *(1 mark)*

(Q14, **2017 HSC**) Easy

6 An old washing machine uses 130 L of water per load. A new washing machine uses 50 L per load.

How much water is saved each year if two loads of washing are done each week using the new machine?

A 2600 L **B** 4160 L
C 5200 L **D** 8320 L *(1 mark)*

(Q9, **2016 HSC**) Easy

7 Alice intends to buy a car and insure it.

Briefly describe what each of these types of insurance covers:

- Compulsory third-party insurance (CTP)
- Non-compulsory third-party property insurance. *(2 marks)*

(Q27a, **2016 HSC**) Medium

8 An insurance company offers customers the following discounts on the basic annual premium for car insurance.

Type of discount	*Discount*
Multi-policy discount (Owner has more than 2 insurance policies with the company)	15%
No-claim bonus (Owner has had at least 5 years without an insurance claim)	20%
Combined CTP and comprehensive insurance (Owner has both insurances with the company)	$50

If a customer is eligible for more than one discount, subsequent discounts are applied to the already discounted premium. The combined compulsory third party (CTP) and comprehensive insurance discount is always applied last.

Jamie has three insurance policies, including combined CTP and comprehensive insurance, with this company. He has used this company for 8 years and he has never made a claim.

The basic annual premium for his car insurance is $870.

How much will Jamie need to pay after the discounts are applied?

A \$482.44 **B** \$515.50
C \$541.60 **D** \$557.60 *(1 mark)*

(Q25, **2015 HSC**) Medium

9 Alex is buying a used car which has a sale price of \$13 380. In addition to the sale price there are the following costs:

Transfer of registration \$30
Stamp Duty ....

i Stamp Duty for this car is calculated at \$3 for every \$100, or part thereof, of the sale price.
Calculate the Stamp Duty payable. *(1 mark)* Medium

ii Alex borrows the total amount to be paid for the car including Stamp Duty and transfer of registration. Interest on the loan is charged at a flat rate of 7.5% per annum. The loan is to be repaid in equal monthly instalments over 3 years.
Calculate Alex's monthly repayments. *(4 marks)* Medium

iii Alex wishes to take out comprehensive insurance for the car for 12 months.
The cost of comprehensive insurance is calculated using the following:

Base rate	**\$845**
Fire Service Levy (FSL)	1% of base rate
Stamp Duty	5.5% of the total of base rate and FSL
GST	10% of the total of base rate and FSL.

Find the total amount that Alex will need to pay for comprehensive insurance. *(3 marks)* Easy

iv Alex has decided he will take out the comprehensive car insurance rather than the less expensive non-compulsory third-party car insurance. What extra cover is provided by the comprehensive car insurance? *(1 mark)* Medium

(Q27a, **2014 HSC**)

10 Xuso is comparing the costs of two different ways of travelling to university.

Xuso's motorcycle uses one litre of fuel for every 17 km travelled. The cost of fuel is \$1.67/L and the distance from her home to the university car park is 34 km. The cost of travelling by bus is \$36.40 for 10 single trips.

Which way of travelling is cheaper and by how much? Support your answer with calculations. *(2 marks)*

(Q27b, **2014 HSC**) Easy

11 A section of Jim's electricity bill is shown.

Energy Used and Costs

METER ID	THIS READING	LAST READING	ENERGY USED	RATE (per kWh)	COST
Peak Energy Charge					
TMV04221/01	531.2	274.8	256.4 kWh	47.7700c	\$122.48
Shoulder Energy Charge					
TMV04221/02	A	560.9	523.5 kWh	19.4000c	\$101.56
Off-peak (Night) Energy Charge					
TMV04221/03	242.5	0.0	242.5 kWh	9.6000c	\$23.28

i What is the value of A? *(1 mark)* Medium

ii How much will Jim save if he uses 154 kWh of energy at the Off-peak rate rather than at the Peak rate? *(2 marks)* Medium

(Q26d, **2013 HSC**)

12 Warrick has a net income of \$590 per week. He has created a budget to help manage his money.

	A	B	C
1	**Warrick's Weekly Budget**		
2			
3	Rent	\$175	
4	Petrol	\$45	
5	Car registration	\$10	
6	Car maintenance	\$15	
7	Food	\$90	
8	Electricity	X	
9	Telephone and internet	\$40	
10	Insurances	\$30	
11	Entertainment	\$70	
12	Clothes and gifts	\$50	
13	Savings	\$40	
14	**Total**	**\$590**	

Sheet 1 Sheet 2

i Find the value of **X**, the amount that Warrick allocates towards electricity each week. *(1 mark)* Easy

ii Warrick has an unexpectedly high telephone and internet bill. For the last three weeks, he has put aside his savings as well as his telephone and internet money to pay the bill.
How much money has he put aside altogether to pay the bill? *(1 mark)* Easy

iii The bill for the telephone and internet is \$620. It is due in two weeks time. Warrick realises he has not put aside enough money to pay the bill.

How could Warrick reallocate non-essential funds in his budget so he has enough money to pay the bill? Justify your answer with suitable reasons and calculations. *(3 marks)* **Medium**

(Q23d, **2010 HSC**)

13 Margaret has a weekly income of $900 and allocates her money according to the budget shown in the sector graph.

How long will it take Margaret to save $3600?

A 4 weeks **B** 5 weeks
C 16 weeks **D** 18 weeks *(1 mark)*

(Q7, **2007 HSC**) **Medium**

14 Reece is preparing his annual budget for 2006. His expected income is:

- $90 every week as a swimming coach
- Interest earned from an investment of $5000 at a rate of 4% per annum.

His planned expenses are:

- $30 every week on transport
- $12 every week on lunches
- $48 every month on entertainment.

Reece will save his remaining income. He uses the spreadsheet below for his budget.

	A	B	C	D	E	F	G	H
1		REECE'S ANNUAL BUDGET FOR 2006						
2								
3		INCOME				EXPENSES		
4								
5	Wages			$4,680		Transport		$(Y)
6	Interest on investment			$(X)		Lunches		$624
7						Entertainment		$(Z)
8								
9								

Sheet 1 / Sheet 2

i Determine the values of (X), (Y) and (Z). (Assume there are exactly 52 weeks in a year.) *(3 marks)* **Medium**

ii At the beginning of 2006, Reece starts saving.
Will Reece have saved enough money during 2006 for a deposit of $2100 on a car if he keeps to his budget? Justify your answer with suitable calculations. *(2 marks)* **Medium**

(Q25a, **2005 HSC**)

Year 11 Budgeting and household expenses—Worked answers

1 Mary must pay the basic excess, plus the age excess because she is under 25 years.
She must pay \$850 + \$1600.
Answer C

2 Total distance = 5 × 2 × 20 km
= 200 km
At 8 L/100 km 16 L of fuel will be used.
Answer A

3 Injury to the other driver is covered by CTP insurance.
Answer B

4 Net income = 75% of \$890
= \$667.50
Budgeted savings
= 20% of \$667.50
= \$133.50
Answer B

5 8000 km = 80 × 100 km
Car A uses 9 L/100 km.
Fuel used by car A = 80 × 9 L
= 720 L
Car B uses 3.5 L/100 km.
Fuel used by car B = 80 × 3.5 L
= 280 L
Difference = (720 – 280) L
= 440 L
Answer B

6 Saving per load = (130 – 50) L
= 80 L
Yearly savings = 2 × 52 × 80 L
= 8320 L
Answer D

7 CTP insurance covers injuries to passengers and other people in an accident. ✓
Third-party property insurance covers damage to other people's cars and property. ✓
(2 marks)

8 Multi-policy discount
= 15% of \$870
= \$130.50
Price after discount
= \$870 – \$130.50
= \$739.50
No-claim bonus discount
= 20% of \$739.50
= \$147.90
Price after this discount
= \$739.50 – \$147.90
= \$591.60
Combined insurance discount = \$50
Final price
= \$591.60 – \$50
= \$541.60
Answer C

9 **i** \$13 380 = 133 × \$100 + \$80
So there will be 134 lots of \$3 charged.
Stamp duty = 134 × \$3
= \$402
(1 mark)

ii Total borrowed
= \$13 380 + \$402 + \$30
= \$13 812 ✓
$I = Prn$
= \$13 812 × 0.075 × 3
= \$3107.70 ✓
Total to repay
= \$13 812 + \$3107.70
= \$16 919.70 ✓
Now 3 years = 3 × 12 months
= 36 months
Each repayment
= \$16 919.70 ÷ 36
= \$469.991 666...
= \$469.99 (nearest cent) ✓
(4 marks)

iii Base rate = \$845
FSL = 1% of \$845
= \$8.45 ✓
Base rate + FSL
= \$845 + \$8.45
= \$853.45
Stamp duty = 5.5% of \$853.45
= 0.055 × \$853.45
= \$46.939 75
= \$46.94 ✓ (nearest cent)
GST = 10% of \$853.45
= \$85.345
= \$85.35 (nearest cent)
Total = \$853.45 + \$46.94 + \$85.35
= \$985.74 ✓
(3 marks)

iv Comprehensive car insurance covers damage or theft of the car of the person taking out the insurance whereas (noncompulsory) third party car insurance only covers damage done by that vehicle to other people's cars and property.
(1 mark)

10 Fuel used in single trip
= (34 ÷ 17) L
= 2 L
Cost of fuel = 2 × \$1.67
= \$3.34 ✓
Cost of single bus trip
= \$36.40 ÷ 10
= \$3.64
Difference = \$3.64 + \$3.34
= \$0.30
So, based on fuel costs alone, it is 30 cents cheaper for each single trip to travel by motorcycle. ✓
[But this ignores all the other costs associated with owning and operating the motorcycle, so is not a true picture of the difference in costs.]
(2 marks)

11 **i** A = 560.9 + 523.5
= 1084.4
(1 mark)

ii Difference in rates
= (47.77 – 9.6) c/kWh
= 38.17 c/kWh ✓
For 154 kW:
Savings = 154 × 38.17c
= 5878.18c
= \$58.78 ✓
[nearest cent]
(2 marks)

12 **i** Total shown in spreadsheet
= \$(175 + 45 + 10 + 15 + 90 + 40 + 30 + 70 + 50 + 40)
= \$565
X = \$590 – \$565
= \$25
(1 mark)

ii Amount put aside each week
= \$40 + \$40
= \$80
Total for 3 weeks = 3 × \$80
= \$240
(1 mark)

iii Amount still required
= \$620 − \$240
= \$380

Amount from 2 more weeks of savings and phone and internet funds
= 2 × \$80
= \$160

Extra amount required
= \$380 − \$160
= \$220 ✓

Extra amount required per week
= \$220 ÷ 2
= \$110

Non-essential items are entertainment and clothing and gifts.

Total of these = \$70 + \$50
= \$120 ✓

Possible answers could show how Warrick could find the extra amount needed from entertainment and clothing and gifts funds. ✓

(3 marks)

13 Angle size of savings = 80°
(by measurement)

∴ Amount saved per week

$= \frac{80}{360} \times \$900$

= \$200

Time taken to save \$3600
= \$3600 ÷ \$200
= 18 weeks

Answer D

14 **i** X = interest on investment

$= \frac{4}{100} \times \$5000$

= \$200 ✓

Y = annual transport expenses
= 52 × \$30
= \$1560 ✓

Z = annual entertainment expenses
= 12 × \$48
= \$576 ✓

(3 marks)

ii Total income
= \$4680 + \$200
= \$4880

Total expenses
= \$1560 + \$624 + \$576
= \$2760

Savings
= \$4880 − \$2760
= \$2120 ✓

Yes. Reece will be able to save the \$2100 deposit for his car. ✓

(2 marks)

1 A school collected data related to the reasons given by students for arriving late. The Pareto chart shows the data collected.

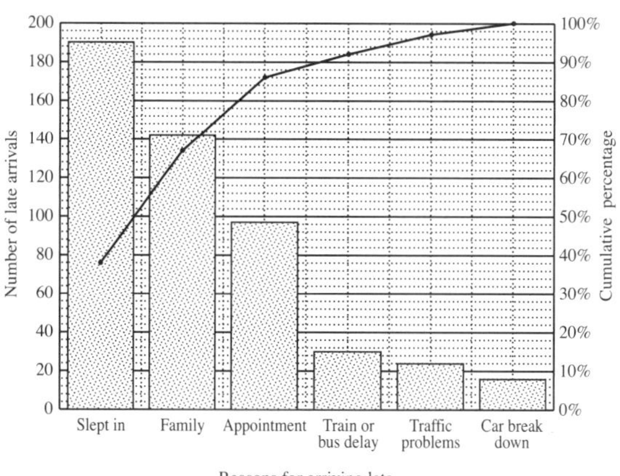

What percentage of students gave the reason 'Train or bus delay'?

A 6% **B** 15%
C 30% **D** 92% *(1 mark)*

(Q10, **2019 HSC**) **Hard**

---

2 A survey asked the following question.

'How many brothers do you have?'

How would the responses be classified?

A Categorical, ordinal
B Categorical, nominal
C Quantitative, discrete
D Quantitative, continuous *(1 mark)*

(Q3, **2018 HSC**) **Easy**

---

3 A factory's quality control department has tested every 50th item produced for possible defects.

What type of sampling has been used?

A Random **B** Stratified
C Systematic **D** Quantitative *(1 mark)*

(Q4, **2017 HSC**) **Easy**

---

4 Which set of data is classified as categorical and nominal?

A blue, green, yellow
B small, medium, large
C 5.2 cm, 6 cm, 7.21 cm
D 4 people, 5 people, 9 people *(1 mark)*

(Q7, **2016 HSC**) **Medium**

---

5 On a school report, a student's record of completing homework is graded using the following codes.

C = consistently
U = usually
S = sometimes
R = rarely
N = never

What type of data is this?

A Categorical, ordinal
B Categorical, nominal
C Quantitative, continuous
D Quantitative, discrete *(1 mark)*

(Q4, **2015 HSC**) **Easy**

---

6 A survey was conducted where people were asked which of two brands of smartphones they preferred. The results were:

- 48% preferred Brand X
- 52% preferred Brand Y.

A graph displaying the data is to be included in a magazine article. The editor of the magazine wishes to ensure that the graph is not misleading in any way.

Which graph should the editor choose to include in the article?

A

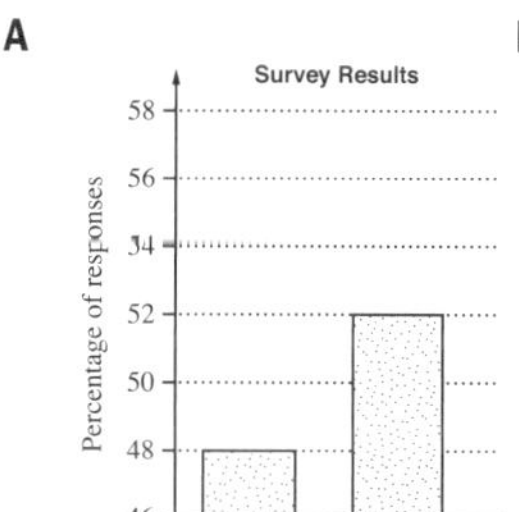

B

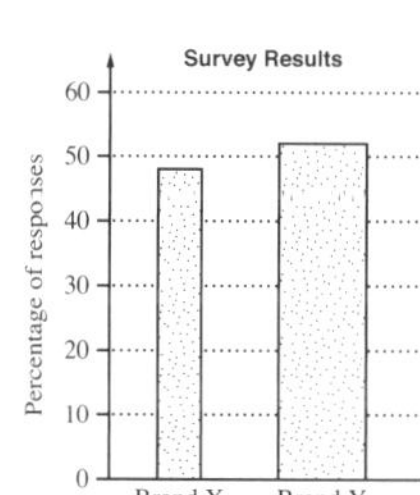

C

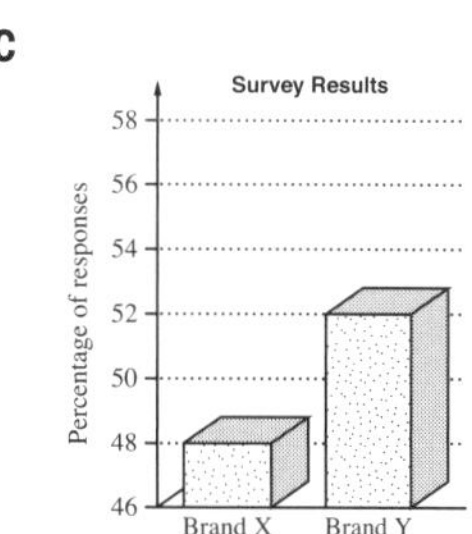

D

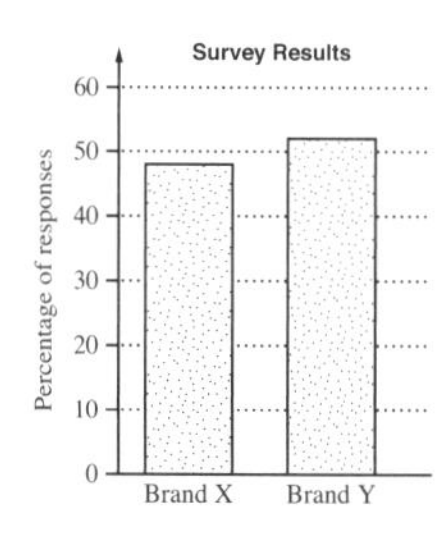

(1 mark)

(Q6, **2013 HSC**) **Easy**

---

7 A high school has 100 students in each year group, Year 7 to Year 12. A survey is to be conducted to determine the average number of text messages sent per month by students at the school.

Which of the following would provide the most representative sample for this survey?

A All Year 7 students
B All Physics students in Years 11 and 12
C 20 students chosen at random from each year group
D 120 students chosen at random from the school roll *(1 mark)*

(Q8, **2013 HSC**) Easy

---

8 Handmade chocolates are checked for size and shape. Every 30th chocolate is sampled.

Which term best describes this type of sampling?

A Census **B** Random
C Stratified **D** Systematic *(1 mark)*

(Q2, **2012 HSC**) Easy

---

9 Greg needs to conduct a statistical inquiry into how much time people aged 18–25 years have spent accessing social media websites in the last two weeks. He has decided to survey a sample of students from his university.

The process of statistical inquiry includes the following steps, which are NOT in order.

A Writing a report
B Posing questions
C Organising data
D Analysing data and drawing conclusions
E Collecting data
F Summarising and displaying data

i Using the letters A, B, C, D, E and F, list the steps in the most appropriate order for Greg to conduct his statistical inquiry. *(2 marks)* Easy

ii Greg conducts his statistical inquiry. At which step in the process would he have drawn this graph?

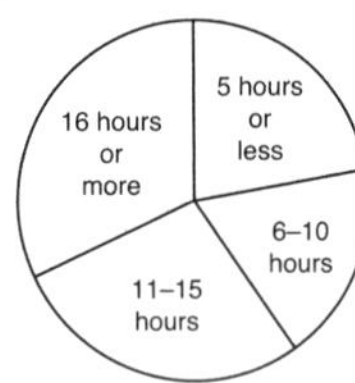

(1 mark) Easy

(Q26d, **2012 HSC**)

---

10 A study on the mobile phone usage of NSW high school students is to be conducted. Data is to be gathered using a questionnaire.

The questionnaire begins with the three questions shown.

Q1: Do you own a mobile phone?
Yes ☐ No ☐

Q2: Which phone company do you use?
....

Q3: Do you use pre-paid or a plan?
Pre-paid ☐ Plan ☐

i Classify the type of data that will be collected in Q2 of the questionnaire. *(1 mark)* Easy

ii Write a suitable question for this questionnaire that would provide quantitative data. *(1 mark)* Easy

iii An initial study is to be conducted using a stratified sample. Describe a method that could be used to obtain a representative stratified sample. *(1 mark)* Medium

(Q25a, **2011 HSC**)

---

11 The graph below displays data collected at a school on the number of students in each Year group, who own a mobile phone.

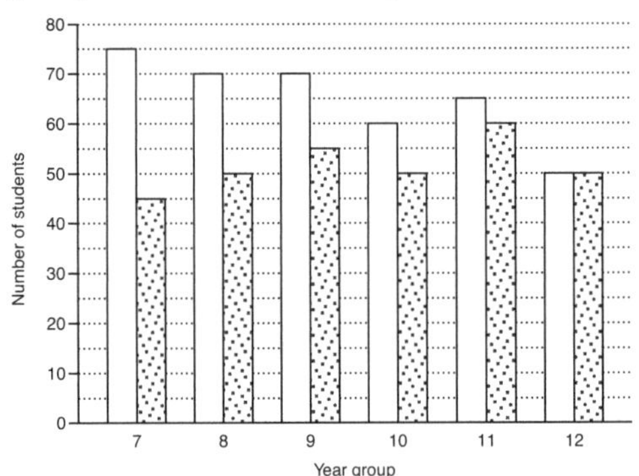

i Which Year group has the highest percentage of students with mobile phones? *(1 mark)* Easy

ii Two students are chosen at random, one from Year 9 and one from Year 10. Which student is more likely to own a mobile phone? Justify your answer with suitable calculations. *(2 marks)* Easy

iii Identify a trend in the data shown in the graph. *(1 mark)* Easy

(Q25b, **2011 HSC**)

---

12 The eye colours of a sample of children were recorded.

When analysing this data, which of the following could be found?

A Mean **B** Median
C Mode **D** Range *(1 mark)*

(Q3, **2009 HSC**) 

13 Jamie wants to know how many songs were downloaded legally from the internet in the last 12 months by people aged 18–25 years. He has decided to conduct a statistical inquiry.

After he collects the data, which of the following shows the best order for the steps he should take with the data to complete his inquiry?

A Display, organise, conclude, analyse
B Organise, display, conclude, analyse
C Display, organise, analyse, conclude
D Organise, display, analyse, conclude *(1 mark)*

(Q5, **2009 HSC**) Easy

14 The Australian Bureau of Statistics provides the NSW government with data on the age of residents living in different areas across the state. After analysing this data, the government makes decisions relating to the provision of services or facilities.

Give an example of a possible decision the government might make and describe how the data might justify this decision. *(2 marks)*

(Q24c, **2009 HSC**) Easy

15 List TWO ways in which this graph is misleading.

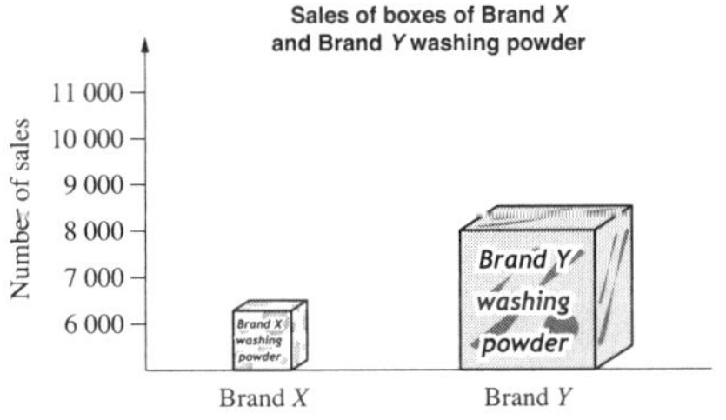

(2 marks)

(Q24a, **2006 HSC**) Easy

Year 11 Classifying and representing data—Worked answers

1 Cumulative percentage for train or bus delay = 92%
Previous cumulative percentage = 86%
So percentage for train or bus delay
= (92 – 86)%
= 6%
Answer A

2 The responses are quantitative because they are numbers. The data is discrete because the responses are whole numbers.
Answer C

3 Testing every 50th item is systematic sampling.
Answer C

4 Blue, green, yellow
[Data in types that cannot be ordered]
Answer A

5 Categorical, ordinal data
[Data divided into groups which can be ranked.]
Answer A

6 The graph that should be included is D.
[The graph in option A doesn't begin at 0 on the vertical axis. The columns in graph B are not the same width and graph C uses 3D columns.]
Answer D

7 The most representative sample would have equal numbers of students from each year group. Choosing 20 students at random from each year group will provide the most representative sample.
Answer C

8 It is an example of systematic sampling.
Answer D

9 **i** B, E, C, F, D, A ✓✓ *(2 marks)*
ii A graph is a data display. This would be done in step F. *(1 mark)*

10 **i** The type of data will be categorical nominal. *(1 mark)*
ii (Any question that requires an answer that involves counting. Some possible answers are:)
'About how many phone calls do you make each day?'
'How many text messages did you send last week?' *(1 mark)*
iii (Any study that surveys students in proportion to the numbers in a particular group.)
A study might determine the number of boys and girls in each year level in urban, regional and remote areas and then select 1% of each of these groups to take part in a survey. *(1 mark)*

11 **i** Year 12 has the highest percentage of students with mobile phones (100%). *(1 mark)*
ii In Year 9 there are 70 students and 55 have a mobile phone.

Percentage with mobile
$= \frac{55}{70} \times 100\%$
$= 78\frac{4}{7}\%$ ✓

In Year 10 there are 60 students and 50 have a mobile phone.
Percentage with mobile
$= \frac{50}{60} \times 100\%$
$= 83\frac{1}{3}\%$ ✓

So a Year 10 student is more likely to have a mobile phone. *(2 marks)*
iii (One possible answer:)
As the year level increases so does the percentage of students with mobile phones. *(1 mark)*

12 Only the mode can be found for categorical data.
Answer C

13 Answer D

14 For example, data might show that in a particular area, there is a large number of young families, that is, a large number of pre-school age children. The government might then decide to provide more childcare services in that area. ✓✓ *(2 marks)*

15 **1** By showing the sales of Brands X and Y as cubes (volume), the actual difference between the two brands has been exaggerated. ✓
2 The vertical axis representing sales does not start at zero. If it did, Brand X would not appear that much less than Brand Y. ✓ *(2 marks)*

1 The stem-and-leaf plot shows the number of goals scored by a team in each of ten netball games.

0	6	8		
1	2	4	5	
2	1	5	5	9
3	5			

What is the mode of this dataset?

A 5 **B** 18

C 25 **D** 29 *(1 mark)*

(Q3, **2021 HSC**) Easy

2 The number of downloads of a song on each of twenty consecutive days is shown in the following graph.

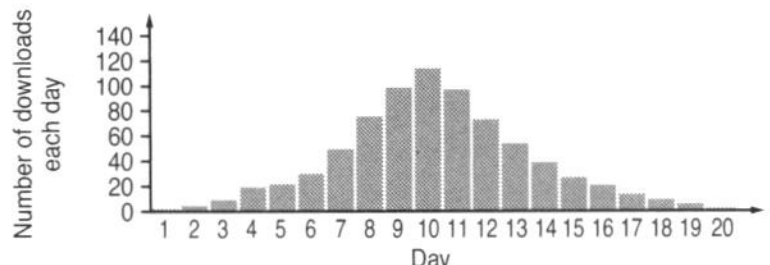

Which of the following graphs best shows the cumulative number of downloads up to and including each day?

A

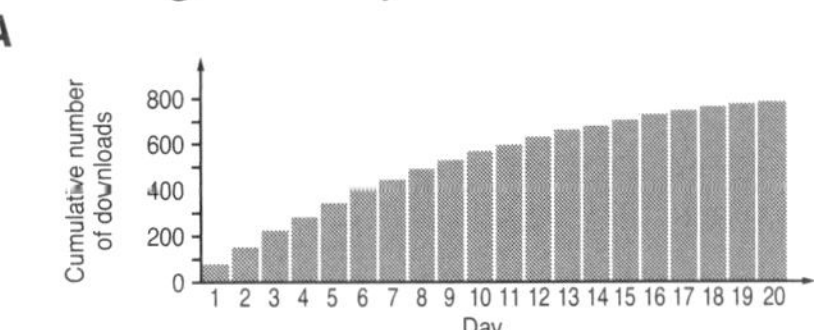

B

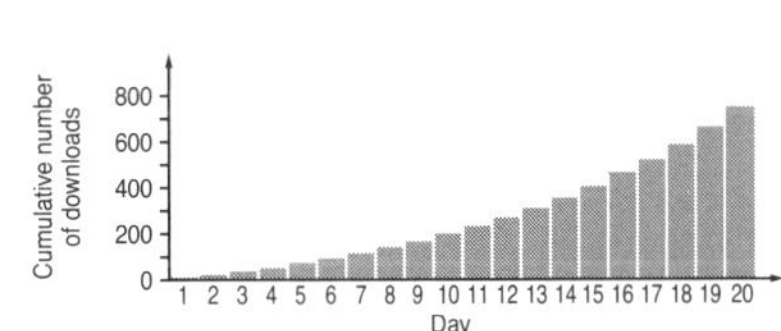

C

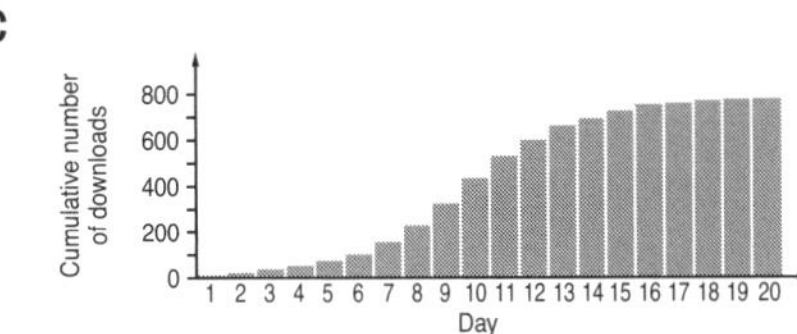

D

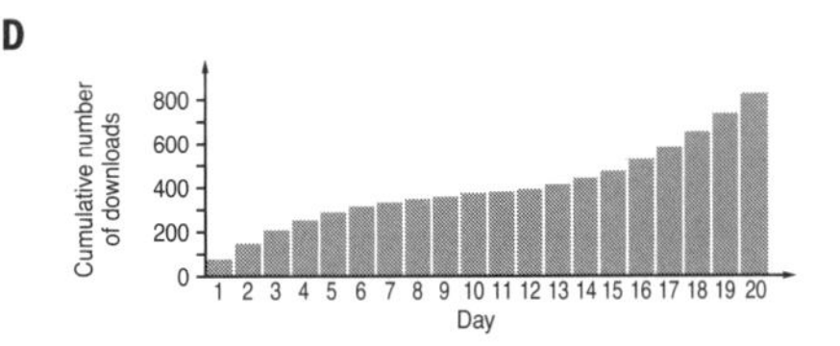

(1 mark)

CQ (Q7, **2021 HSC**) Medium

3 The five-number summary of a dataset is given.

Lowest score = 1
Lower quartile (Q_1) = 4
Median (Q_2) = 7
Upper quartile (Q_3) = 10
Highest score = 20

Is 20 an outlier? Justify your answer with calculations. *(2 marks)*

(Q17, **2021 HSC**) Medium

4 Which histogram best represents a dataset that is positively skewed?

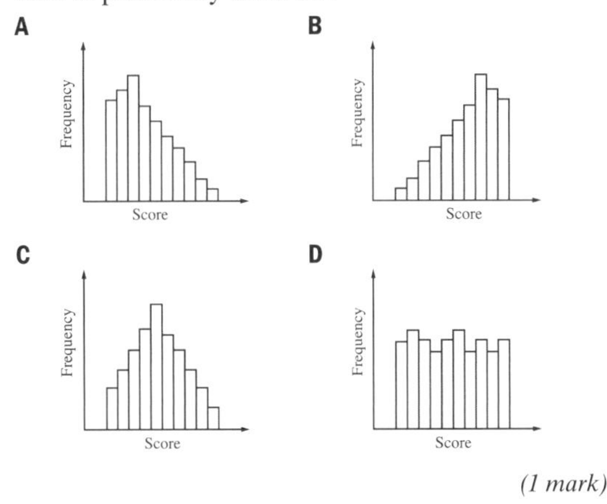

(1 mark)

(Q7, **2020 HSC**) Hard

5 Consider the following dataset.

1 5 9 10 15

Suppose a new value, x, is added to this dataset, giving the following.

1 5 9 10 15 x

It is known that x is greater than 15. It is also known that the difference between the means of the two datasets is equal to ten times the difference between the medians of the two datasets.

Calculate the value of x. *(4 marks)*

(Q28, **2020 HSC**) Medium

6 The heights, in centimetres, of 10 players on a basketball team are shown.

170, 180, 185, 188, 192, 193, 193, 194, 196, 202

Is the height of the shortest player on the team considered an outlier? Justify your answer with calculations. *(3 marks)*

(Q19, **2019 HSC**) Medium

7 Two netball teams, Team A and Team B, each played 15 games in a tournament. For each team, the number of goals scored in each game was recorded.

The frequency table shows the data for Team A.

Number of goals	*Frequency*
19	1
20	0
21	1
22	1
23	1
24	3
25	0
26	4
27	3
28	1

The data for Team B was analysed to create the box-plot shown.

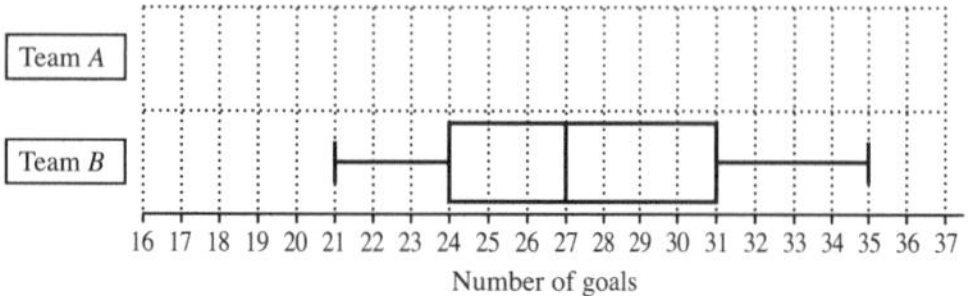

Compare the distributions of the number of goals scored by the two teams. Support your answer with the construction of a box-plot for the data for Team A. *(5 marks)*

(Q39, **2019 HSC**) **Hard**

8 The results of a quiz marked out of 40 are represented in the box plot.

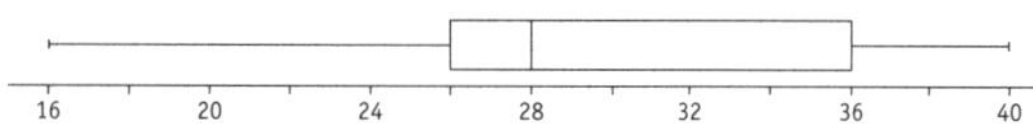

a A student is chosen at random. What is the probability that their score is at least 90%? *(1 mark)* **Medium**

b Determine whether a score of 16 is an outlier. Use calculations to support your argument. *(2 marks)* **Medium**

Bonus question (see page iv)

9 A set of scores has the following five-number summary.

lower extreme = 2
lower quartile = 5
median = 6
upper quartile = 8
upper extreme = 9

What is the range?

A 2 **B** 3
C 6 **D** 7 *(1 mark)*

(Q1, **2018 HSC**) **Easy**

10 A set of data is displayed in this dot plot.

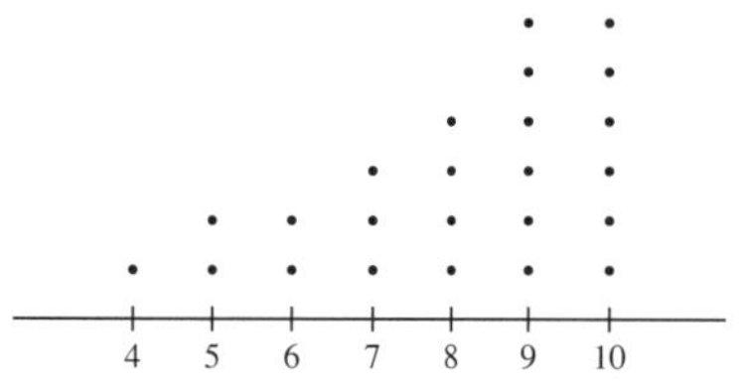

Which of the following best describes this set of data?

A Symmetrical
B Positively skewed
C Negatively skewed
D Normally distributed *(1 mark)*

(Q6, **2018 HSC**) **Medium**

11 A set of data is summarised in this frequency distribution table.

Score	*Frequency*
3	1
4	2
5	6
6	7
7	9
8	5
	Total = 30

Which of the following is true about the data?

A Mode = 7, median = 5.5
B Mode = 7, median = 6
C Mode = 9, median = 5.5
D Mode = 9, median = 6 *(1 mark)*

(Q11, **2018 HSC**) **Medium**

12 The graph displays the mean monthly rainfall in Sydney and Perth.

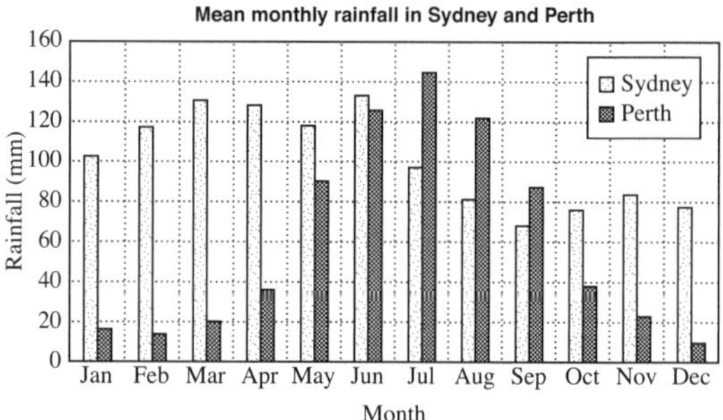

i For how many months is the mean monthly rainfall higher in Perth than in Sydney? *(1 mark)* **Easy**

ii For which of the two cities is the standard deviation of the mean monthly rainfall smaller? Justify your answer WITHOUT calculations. *(1 mark)* **Medium**

(Q26d, **2018 HSC**)

13 A cumulative frequency table for a data set is shown.

Score	*Cumulative frequency*
1	5
2	9
3	16
4	20
5	34
6	42

What is the interquartile range of this data set? *(2 marks)*

(Q26e, **2018 HSC**) **Hard**

14 The box-and-whisker plot for a set of data is shown.

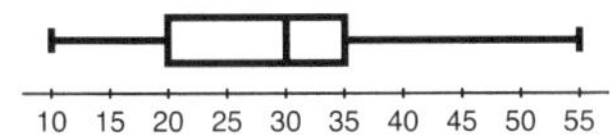

What is the median of this set of data?

A 15 **B** 20
C 30 **D** 35 *(1 mark)*

(Q1, **2017 HSC**) **Easy**

15 Jamal surveyed eight households in his street. He asked them how many kilolitres (kL) of water they used in the last year. Here are the results.

220, 105, 101, 450, 37, 338, 151, 205

i Calculate the mean of this set of data. *(1 mark)* **Easy**

ii What is the standard deviation of this set of data, correct to one decimal place? *(1 mark)* **Medium**

(Q27a, **2017 HSC**)

16 A set of data has a lower quartile (Q_L) of 10 and an upper quartile (Q_U) of 16.

What is the maximum possible range for this set of data if there are no outliers? *(2 marks)*

(Q30a, **2017 HSC**) **Medium**

17 A soccer referee wrote down the number of goals scored in 9 different games during the season.

2, 3, 3, 3, 5, 5, 8, 9, ☐

The last number has been omitted. The range of the data is 10.

What is the five-number summary for this data set?

A 2, 3, 5, 8.5, 12 **B** 2, 3, 5, 8.5, 10
C 2, 3, 5, 8, 12 **D** 2, 3, 5, 8, 10 *(1 mark)*

(Q19, **2016 HSC**) **Medium**

18 A grouped data frequency table is shown.

Class interval	*Frequency*
1–5	3
6–10	6
11–15	8
16–20	9

What is the mean for this set of data?

A 6.5 **B** 10.5
C 11.9 **D** 12.4 *(1 mark)*

(Q21, **2016 HSC**) **Medium**

19 The box-and-whisker plots show the results of a History test and a Geography test.

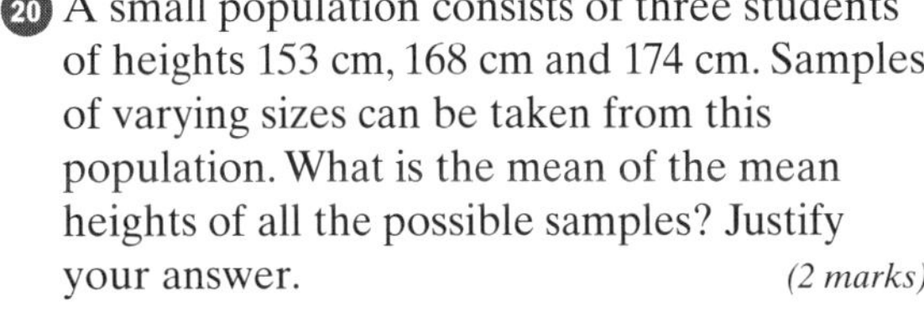

In History, 112 students completed the test. The number of students who scored above 30 marks was the same for the History test and the Geography test.

How many students completed the Geography test?

A 8 **B** 50
C 56 **D** 112 *(1 mark)*

(Q22, **2016 HSC**) **Easy**

20 A small population consists of three students of heights 153 cm, 168 cm and 174 cm. Samples of varying sizes can be taken from this population. What is the mean of the mean heights of all the possible samples? Justify your answer. *(2 marks)*

(Q27b, **2016 HSC**) **Medium**

21 The heights of 400 students were measured. The results are displayed in this cumulative frequency polygon.

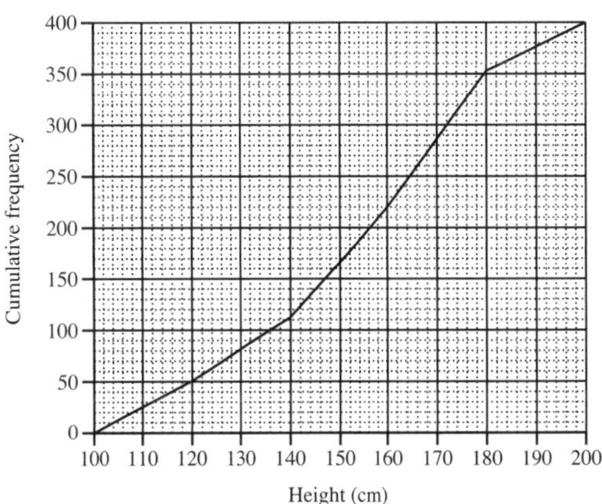

Use the polygon to estimate the interquartile range. *(2 marks)*

(Q27c, **2016 HSC**) **Medium**

22 The ages of members of a dance class are shown in the back-to-back stem-and-leaf plot.

Women		Men
2	3	4 6
4 2	4	2 2 5 6 8
8 8 5 4 0 0	5	3
9 4 3 3	6	3

Pat claims that the women who attend the dance class are generally older than the men.

Is Pat correct? Justify your answer by referring to the median and skewness of the two sets of data. *(3 marks)*

(Q29c, **2016 HSC**) **Medium**

23 The times, in minutes, that a large group of students spend on exercise per day are presented in the box-and-whisker plot.

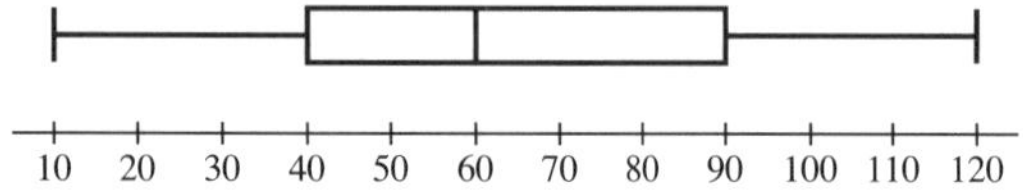

What percentage of these students spend between 40 minutes and 60 minutes per day on exercise?

A 17% **B** 20%
C 25% **D** 50% *(1 mark)*

(Q6, **2015 HSC**) **Medium**

24 In a small business, the seven employees earn the following wages per week:

$300, $490, $520, $590, $660, $680, $970.

i Is the wage of $970 an outlier for this set of data? Justify your answer with calculations. *(3 marks)* **Medium**

ii Each employee receives a $20 pay increase. What effect will this have on the standard deviation? *(1 mark)* **Medium**

(Q27d, **2015 HSC**)

25 Data from 200 recent house sales are grouped into class intervals and a cumulative frequency histogram is drawn.

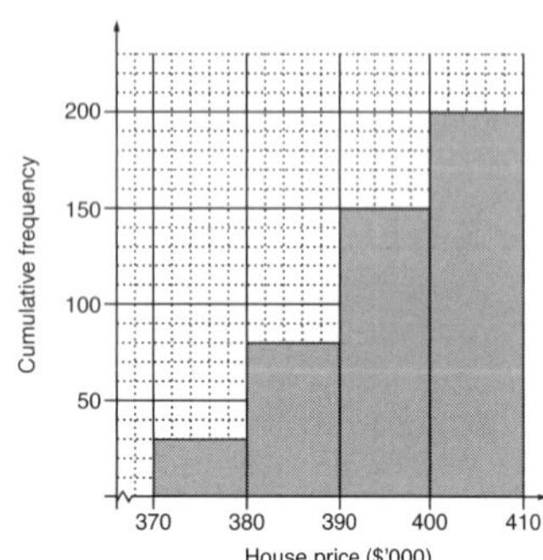

i Use the graph to estimate the median house price. *(1 mark)* **Hard**

ii By completing the table, calculate the mean house price. *(3 marks)* **Hard**

Class Centre ($'000)	*Frequency*

(Q29d, **2015 HSC**)

26 Twenty Year 12 students were surveyed. These students were asked how many hours of sport they play per week, to the nearest hour.

The results are shown in the frequency table.

Hours per week	*Frequency*
0–2	5
3–5	10
6–8	3
9–11	2

What is the mean number of hours of sport played by the students per week?

A 3.3 **B** 4.3
C 5.0 **D** 5.3 *(1 mark)*

(Q14, **2014 HSC**) **Medium**

27 The times taken for 160 music downloads were recorded, grouped into classes and then displayed using the cumulative frequency histogram shown.

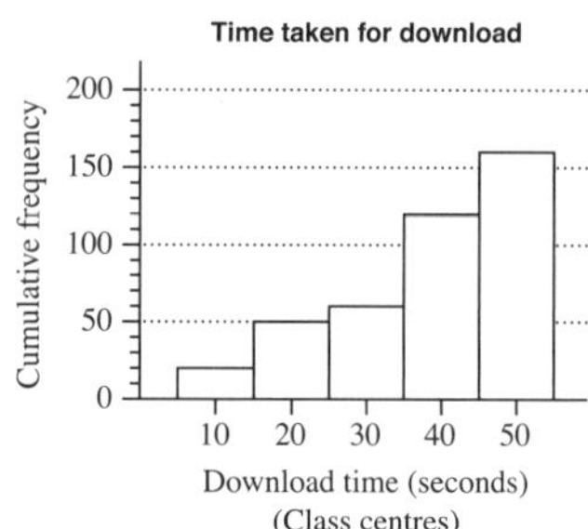

On the diagram, draw the lines that are needed to find the median download time. *(2 marks)*

(Q26e, **2014 HSC**) **Medium**

28 Terry and Kim each sat twenty class tests. Terry's results on the tests are displayed in the box-and-whisker plot shown in part **i**.

i Kim's 5-number summary for the tests is 67, 69, 71, 73, 75.
Draw a box-and-whisker plot to display Kim's results below that of Terry's results.

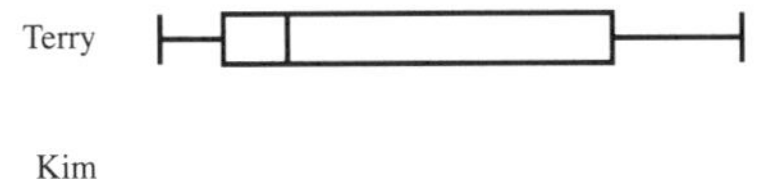

(1 mark) Easy

ii What percentage of Terry's results were below 69? *(1 mark)* Medium

iii Terry claims that his results were better than Kim's. Is he correct? Justify your answer by referring to the summary statistics and the skewness of the distributions. *(4 marks)* Hard

(Q29c, **2014 HSC**)

29 The July sale prices for properties in a suburb were:

\$552 000, \$595 000, \$607 000, \$607 000, \$682 000 and \$685 000

On 1 August, another property in the same suburb was sold for over one million dollars.

If this property had been sold in July, what effect would it have had on the mean and median sale prices for July?

A Both the mean and the median would have changed.

B Neither the mean nor the median would have changed.

C The mean would have changed and the median would have stayed the same.

D The mean would have stayed the same and the median would have changed. *(1 mark)*

(Q14, **2013 HSC**) Easy

30 The frequency histogram shows the number of goals scored by a football team in each game in a season.

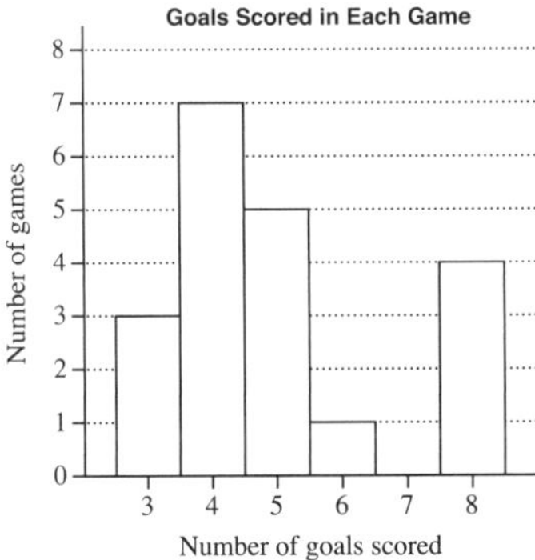

What was the mean number of goals scored per game by this team?

A 4 **B** 4.5

C 5 **D** 5.5 *(1 mark)*

(Q15, **2013 HSC**) Medium

31 Write down a set of six data values that has a range of 12, a mode of 12 and a minimum value of 12. *(2 marks)*

(Q26b, **2013 HSC**) Medium

32 Jason travels to work by car on all five days of his working week, leaving home at 7 am each day. He compares his travel times using roads without tolls and roads with tolls over a period of 12 working weeks.

He records his travel times (in minutes) in a back-to-back stem-and-leaf plot.

Travel time (minutes)

Without tolls		*With tolls*
9	3	5 8 9 9
9 9 8 7 7 6 5 5 4 4 3 2 0	4	0 1 2 6 7 7 8 8 8 9
9 8 7 5 4 3 3 3 2 2 2 2 1 1 0	5	2 4 4 5 6 8 9
1	6	1 3 5 7
	7	0 2 8
	8	2
	9	0

i What is the modal travel time when he uses roads without tolls? *(1 mark)* Easy

ii What is the median travel time when he uses roads without tolls? *(1 mark)* Medium

iii Describe how the two data sets differ in terms of the spread and skewness of their distributions. *(2 marks)* Medium

(Q26f, **2013 HSC**)

33 A retailer has collected data on the number of televisions that he sold each week in 2012.

He grouped the data into classes and displayed the data using a cumulative frequency histogram and polygon (ogive).

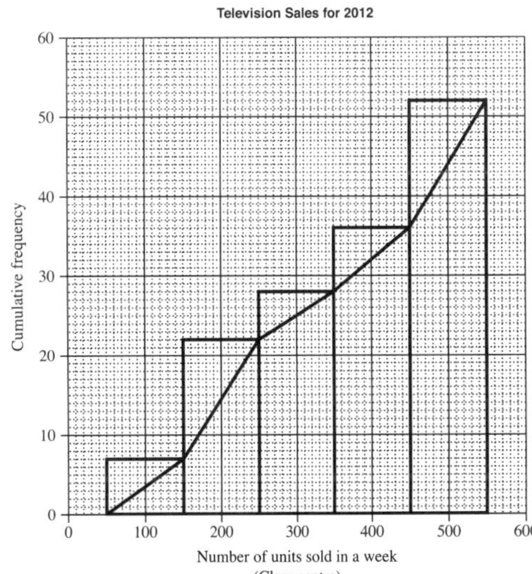

i Use the cumulative frequency polygon to determine the interquartile range. *(2 marks)* Hard

ii Oscar said that the retailer sold 300 televisions in 6 of the weeks in 2012. Is he correct? Give a reason for your answer. *(1 mark)* Hard

(Q27c, **2013 HSC**)

34 A set of 15 scores is displayed in a stem-and-leaf plot.

Stem	Leaf
5	3 4
6	2 6 7
7	7 7 8 9
8	2 4
9	1 3 5 7

What is the median of these scores?

A 7 **B** 8
C 77 **D** 78 *(1 mark)*

(Q1, **2012 HSC**) **Easy**

35 The test results in English and Mathematics for a class were recorded and displayed in the box-and-whisker plots.

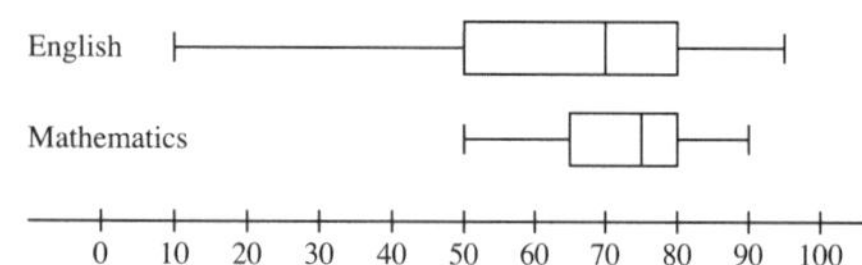

i What is the interquartile range for English? *(1 mark)* **Easy**

ii Compare and contrast the two data sets by referring to the skewness of the distributions and the measures of location and spread. *(3 marks)* **Medium**

(Q28d, **2012 HSC**)

36 A set of data is displayed in this box-and-whisker plot.

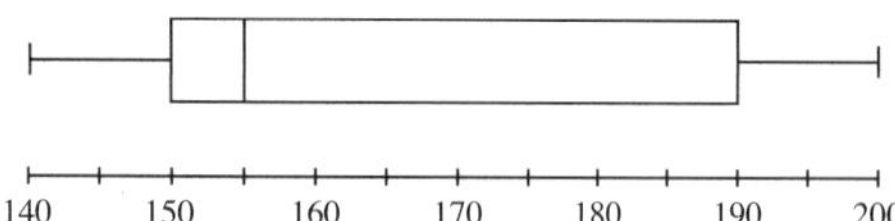

Which of the following best describes this set of data?

A Symmetrical
B Positively skewed
C Negatively skewed
D Normally distributed *(1 mark)*

(Q7, **2011 HSC**) **Easy**

37 The sets of data, X and Y, are displayed in the histograms.

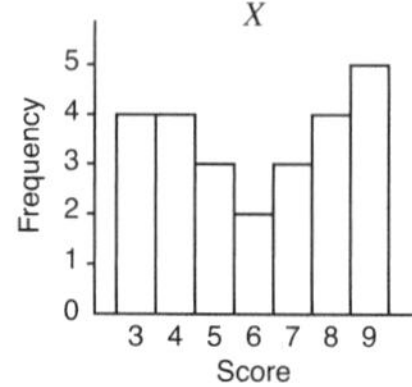

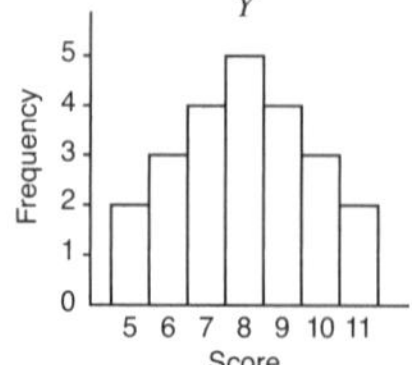

Which of these statements is true?

A X has a larger mode and Y has a larger range.
B X has a larger mode and the ranges are the same.
C The modes are the same and Y has a larger range.
D The modes are the same and the ranges are the same. *(1 mark)*

(Q11, **2011 HSC**) **Medium**

38 A data set of nine scores has a median of 7.

The scores 6, 6, 12 and 17 are added to this data set.

What is the median of the data set now?

A 6 **B** 7
C 8 **D** 9 *(1 mark)*

(Q14, **2011 HSC**) **Medium**

39 The heights of the players in a basketball team were recorded as 1.8 m, 1.83 m, 1.84 m, 1.86 m and 1.92 m. When a sixth player joined the team, the average height of the players increased by 1 centimetre.

What was the height of the sixth player?

A 1.85 m **B** 1.86 m
C 1.91 m **D** 1.93 m *(1 mark)*

(Q17, **2011 HSC**) **Medium**

40 Data was collected from 30 students on the number of text messages they had sent in the previous 24 hours. The set of data collected is displayed.

Male		*Female*
9 9 8 7 6 5 5 4 2 1	0	8 9
1 1 0 0	1	1 1 2 5 6 8 8 8
0	2	0 1 7
	3	4
	4	
	5	
	6	
1	7	

i What is the outlier for this set of data? *(1 mark)* **Medium**

ii What is the interquartile range of the data collected from the female students? *(1 mark)* **Medium**

(Q25d, **2011 HSC**)

41 The results of a survey are displayed in the dot plot.

What is the range of this data?

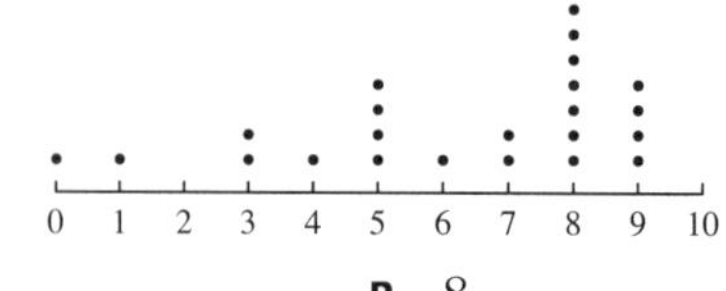

A 7 **B** 8
C 9 **D** 10 *(1 mark)*

(Q1, **2010 HSC**) **Easy**

42 This back-to-back stem-and-leaf plot displays the test results for a class of 26 students.

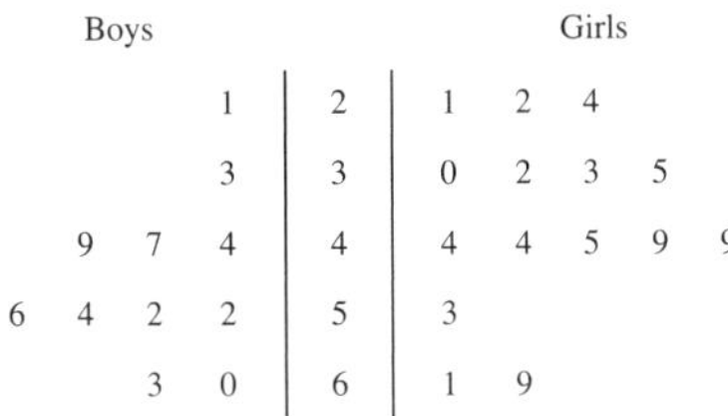

Boys		Girls
1	2	1 2 4
3	3	0 2 3 5
9 7 4	4	4 4 5 9 9
6 4 2 2	5	3
3 0	6	1 9

What is the median test result for the class?

A 44   **B** 46

C 48   **D** 49   *(1 mark)*

(Q16, **2010 HSC**) **Medium**

43 A new shopping centre has opened near a primary school. A survey is conducted to determine the number of motor vehicles that pass the school each afternoon between 2.30 pm and 4.00 pm.

The results for 60 days have been recorded in the table and are displayed in the cumulative frequency histogram.

Score	*Class centre*	*Frequency*	*Cumulative frequency*
100–124	112	10	10
125–149	137	X	25
150–174	162	20	45
175–199	187	15	60

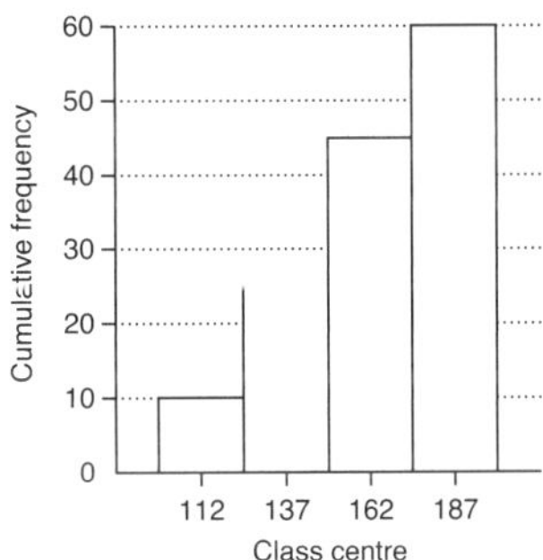

i Find the value of X in the table. *(1 mark)* **Easy**

ii Carefully copy the cumulative frequency histogram into your writing booklet.
On the cumulative frequency histogram you have copied draw a cumulative frequency polygon (ogive) for this data. *(1 mark)* **Hard**

iii Use your graph to determine the median. Show, by drawing lines on your graph, how you arrived at your answer. *(1 mark)* **Hard**

iv Prior to the opening of the new shopping centre, the median number of motor vehicles passing the school between 2.30 pm and 4.00 pm was 57 vehicles per day.
What problem could arise from the change in the median number of motor vehicles passing the school before and after the opening of the new shopping centre?
Briefly recommend a solution to this problem. *(2 marks)* **Medium**

(Q26b, **2010 HSC**)

44 The graphs show the distribution of the ages of children in Numbertown in 2000 and 2010.

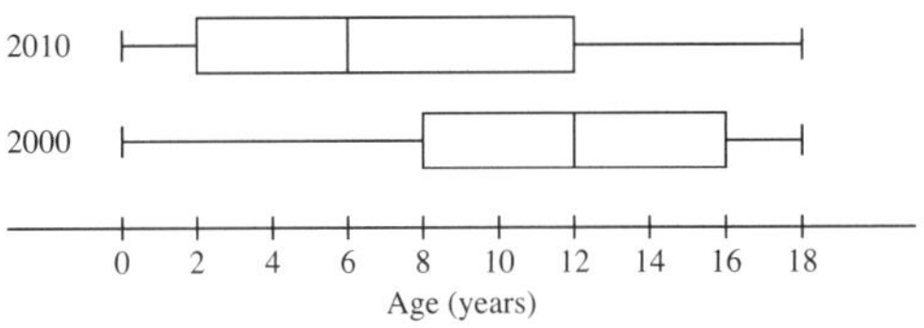

i In 2000 there were 1750 children aged 0–18 years.
How many children were aged 12–18 years in 2000? *(1 mark)* **Medium**

ii The number of children aged 12–18 years is the same in both 2000 and 2010.
How many children aged 0–18 years are there in 2010? *(1 mark)* **Hard**

iii Identify TWO changes in the distribution of ages between 2000 and 2010. In your answer, refer to measures of location or spread or the shape of the distributions. *(2 marks)* **Hard**

iv What would be ONE possible implication for government planning, as a consequence of this change in the distribution of ages? *(1 mark)* **Medium**

(Q27b, **2010 HSC**)

45 The mean of a set of ten scores is 14. Another two scores are included and the new mean is 16.

What is the mean of the two additional scores?

A 4   **B** 16

C 18   **D** 26   *(1 mark)*

(Q21, **2009 HSC**) **Hard**

46 The diagram below shows a stem-and-leaf plot for 22 scores.

Stem	Leaf
2	3 5 9
3	1 4 7 9
4	2 4 4 5 7
5	1 2 4
6	2 3 7
7	5 8 8 8

i What is the mode for this data? *(1 mark)* **Easy**

ii What is the median for this data? *(1 mark)* **Easy**

(Q24a, **2009 HSC**)

47 In a school, boys and girls were surveyed about the time they usually spend on the internet over a weekend. These results were displayed in box-and-whisker plots, as shown below.

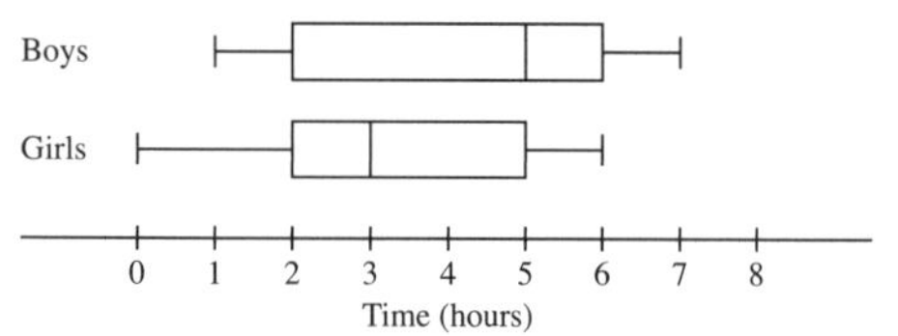

i Find the interquartile range for boys. *(1 mark)* **Easy**

ii What percentage of girls usually spend 5 or less hours on the internet over a weekend? *(1 mark)* **Easy**

iii Jenny said that the graph shows that the same number of boys as girls usually spend between 5 and 6 hours on the internet over a weekend.
Under what circumstances would this statement be true? *(1 mark)* **Hard**

(Q26a, **2009 HSC**)

48 The stem-and-leaf plot represents the daily sales of soft drink from a vending machine. If the range of sales is 43, what is the value of (*N*)?

Stem	Leaf
2	(*N*) 5 5
3	4 7 7 9
4	0 5 8
5	2
6	0 7

A 4 **B** 5
C 24 **D** 25 *(1 mark)*

(Q3, **2008 HSC**) **Easy**

49 What is the median of the following set of scores?

Score	*Frequency*
12	13
14	6
16	2
18	12
Total	33

A 12 **B** 13
C 14 **D** 15 *(1 mark)*

(Q8, **2008 HSC**) **Easy**

50 The marks for a Science test and a Mathematics test are presented in box-and-whisker plots.

Science test

Mathematics test

Which measure must be the same for both tests?

A Mean **B** Range
C Median **D** Interquartile range *(1 mark)*

(Q10, **2008 HSC**) **Medium**

51 The height of each student in a class was measured and it was found that the mean height was 160 cm.

Two students were absent. When their heights were included in the data for the class, the mean height did not change.

Which of the following heights are possible for the two absent students?

A 155 cm and 162 cm
B 152 cm and 167 cm
C 149 cm and 171 cm
D 143 cm and 178 cm *(1 mark)*

(Q13, **2008 HSC**) **Medium**

52 Christina has completed three Mathematics tests. Her mean mark is 72%. What mark (out of 100) does she have to get in her next test to increase her mean mark to 73%? *(2 marks)*

(Q23f, **2008 HSC**) **Medium**

53 Leanne copied a two-way table into her book.

	Male	*Female*	*Totals*
Full-time work	279	356	635
Part-time work	187	439	716
Totals	466	885	1351

Leanne made an error in copying one of the values in the shaded section of the table.

Which value has been incorrectly copied?

A The number of males in full-time work
B The number of males in part-time work
C The number of females in full-time work
D The number of females in part-time work *(1 mark)*

(Q16, **2007 HSC**) **Easy**

54 Ms Wigginson decided to survey a sample of 10% of the students at her school.
The school enrolment is shown in the table.

Year	7	8	9	10	11	12	*Total*
Number of students	225	232	233	230	150	130	1200

She surveyed the same number of students in each year group.

How would the numbers of students surveyed in Year 10 and Year 11 have changed if Ms Wigginson had chosen to use a stratified sample based on year groups?

A Increased in both Year 10 and Year 11
B Decreased in both Year 10 and Year 11
C Increased in Year 10 and decreased in Year 11
D Decreased in Year 10 and increased in Year 11 *(1 mark)*

(Q17, **2007 HSC**) **Medium**

55 This set of data is arranged in order from smallest to largest.

$$5, 6, 11, x, 13, 18, 25$$

The range is six less than twice the value of x.

Which one of the following is true?

A The median is 12 and the interquartile range is 7.
B The median is 12 and the interquartile range is 12.
C The median is 13 and the interquartile range is 7.
D The median is 13 and the interquartile range is 12. *(1 mark)*

(Q21, **2007 HSC**) **Hard**

56 A set of examination results is displayed in a cumulative frequency histogram and polygon (ogive).

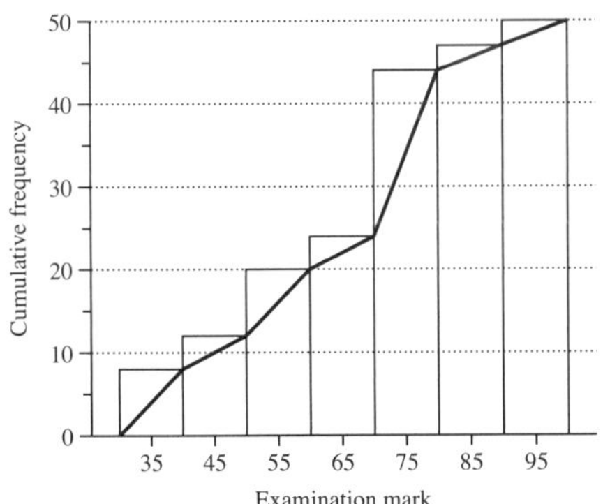

Sanath knows that his examination mark is in the 4th decile.

Which of the following could have been Sanath's examination mark?

A 37 **B** 57
C 67 **D** 77 *(1 mark)*

(Q22, **2007 HSC**) **Hard**

57 Consider the following set of scores:

$$3, 5, 5, 6, 8, 8, 9, 10, 10, 50.$$

i Calculate the mean of the set of scores. *(1 mark)* **Easy**

ii What is the effect on the mean and on the median of removing the outlier? *(2 marks)* **Medium**

(Q24a, **2007 HSC**)

58 Barry constructed a back-to-back stem-and-leaf plot to compare the ages of his students.

Ages of students attending Barry's Ballroom Dancing Studio

Females		*Males*
9	1	1 2 3
7	2	0 2 2 2 4 5
5	3	0 0 1 7
5 2	4	6 7
3 2 0	5	2
4 4 2 1	6	4 4

i Write a brief statement that compares the distribution of the ages of males and females from this set of data. *(1 mark)* **Easy**

ii What is the mode of this set of data? *(1 mark)* **Easy**

iii Liam decided to use a grouped frequency distribution table to calculate the mean age of the students at Barry's Ballroom Dancing Studio.
For the age group 30–39 years, what is the value of the product of the class centre and the frequency? *(2 marks)* **Hard**

iv Liam correctly calculated the mean from the grouped frequency distribution table to be 39.5.
Caitlyn correctly used the original data in the back-to-back stem-and-leaf plot and calculated the mean to be 38.2.
What is the reason for the difference in the two answers? *(1 mark)* **Medium**

(Q24d, **2007 HSC**)

59 A set of scores is displayed in a stem-and-leaf plot.

1	2 2 3
2	5 8
3	8 9
4	1 3 9

What is the median of this set of scores?

A 28 **B** 30
C 33 **D** 47 *(1 mark)*

(Q4, **2006 HSC**) **Easy**

60 Which of these graphs best represents positively skewed data with the smaller standard deviation?

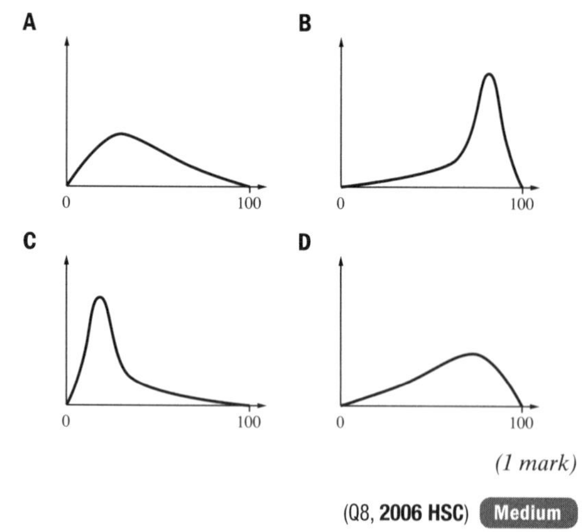

(1 mark)

(Q8, **2006 HSC**) **Medium**

61 The mean of a set of 5 scores is 62.

What is the new mean of the set of scores after a score of 14 is added?

A 38 **B** 54
C 62 **D** 76 *(1 mark)*

(Q12, **2006 HSC**) **Medium**

62 Vicki wants to investigate the number of hours spent on homework by students at her high school.

i Briefly describe a valid method of randomly selecting 200 students for a sample. *(1 mark)* **Medium**

ii Vicki chooses her sample and asks each student how many hours (to the nearest hour) they usually spend on homework during one week. The responses are shown in the frequency table.

Number of hours spent on homework in a week	*Frequency*
0 to 4	69
5 to 9	72
10 to 14	38
15 to 19	21

What is the mean amount of time spent on homework? *(2 marks)* **Hard**

(Q23c, **2006 HSC**)

63 The heights of the 60 members of a choir were recorded. These results were grouped and then displayed as a cumulative frequency histogram and polygon.

The shortest person in the choir is 140 cm and the tallest is 190 cm.

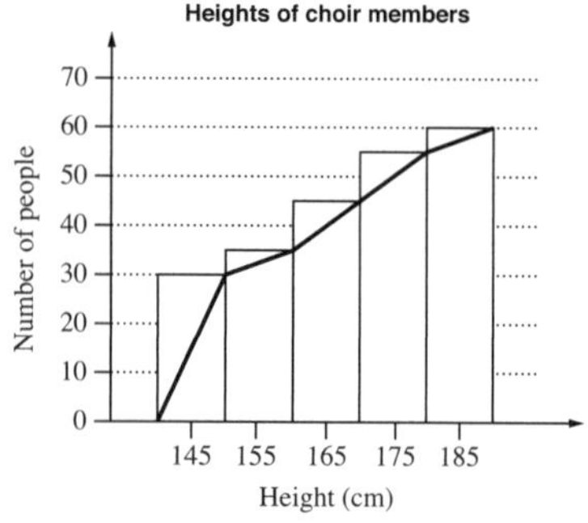

Draw an accurate box-and-whisker plot to represent the data. *(3 marks)*

(Q24c, **2006 HSC**) **Hard**

64 What is the mean of the set of scores?

$$3, 4, 5, 6, 6, 8, 8, 8, 15$$

A 6 **B** 7
C 8 **D** 9 *(1 mark)*

(Q1, **2005 HSC**) **Easy**

65 A set of data is represented by the cumulative frequency histogram and ogive.

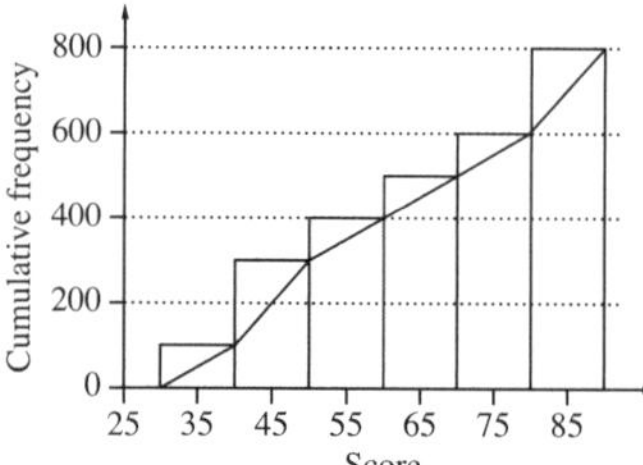

What is the best approximation for the interquartile range for this set of data?

A 25 **B** 30
C 35 **D** 40 *(1 mark)*

(Q9, **2005 HSC**) **Medium**

66 Two groups of people were surveyed about their weekly wages. The results are shown in the box-and-whisker plots.

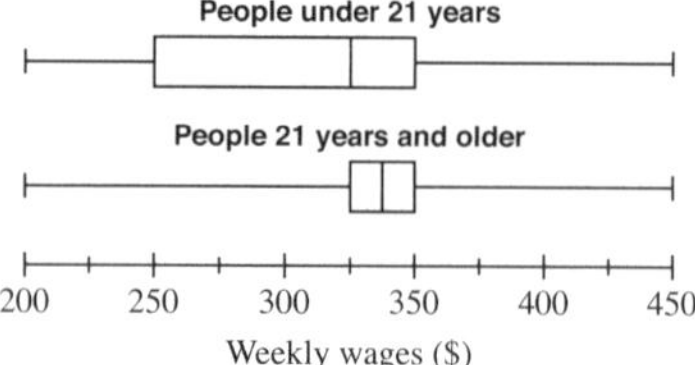

Which of the following statements is true for the people surveyed?

A The same percentage of people in each group earned more than $325 per week.
B Approximately 75% of people under 21 years earned less than $350 per week.
C Approximately 75% of people 21 years and older earned more than $350 per week.

D Approximately 50% of people in each group earned between \$325 and \$350 per week. *(1 mark)*

(Q22, **2005 HSC**) **Medium**

67 **i** Draw a stem-and-leaf plot for the following set of scores.

21 45 29 27 19 35 23 58 34 27

(2 marks) **Easy**

ii What is the median of the set of scores? *(1 mark)* **Easy**

iii Comment on the skewness of the set of scores. *(1 mark)* **Medium**

(Q24a, **2005 HSC**)

Use the set of scores 1, 3, 3, 3, 4, 5, 7, 7, 12 to answer Questions 68 and 69.

68 What is the range of the set of scores?

A 6 **B** 9
C 11 **D** 12 *(1 mark)*

(Q6, **2004 HSC**) **Easy**

69 What are the median and the mode of the set of scores?

A Median 3, mode 5
B Median 3, mode 3
C Median 4, mode 5
D Median 4, mode 3 *(1 mark)*

(Q7, **2004 HSC**) **Easy**

70 This box-and-whisker plot represents a set of scores.

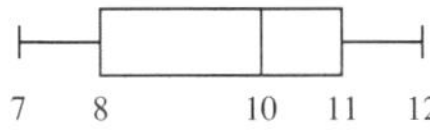

What is the interquartile range of this set of scores?

A 1 **B** 2
C 3 **D** 5 *(1 mark)*

(Q12, **2004 HSC**) **Easy**

71 Joy asked the students in her class how many brothers they had. The answers were recorded in a frequency table as follows.

Number of brothers	*Frequency*
0	5
1	10
2	3
3	1
4	1

One of the students is chosen at random.

What is the probability that this student has at least two brothers?

A 0.10 **B** 0.15
C 0.25 **D** 0.75 *(1 mark)*

(Q13, **2003 HSC**) **Easy**

72 The results of a geography test are displayed in a stem-and-leaf plot.

Stem	Leaf
2	3 3 4 5
3	5 6 6 6 7 7
4	1 2
5	0 0 4

What is the range of the data?

A 15 **B** 27
C 29 **D** 31 *(1 mark)*

(Q1, **2002 HSC**) **Easy**

73 Students were surveyed about the number of movies they had watched in the last week. The results are shown in this *cumulative* frequency histogram.

How many students said they watched four movies last week?

A 5 **B** 10
C 25 **D** 35 *(1 mark)*

(Q17, **2002 HSC**) **Medium**

74 Jane and Sam are in a Geography class of 12 students. The class is going on a three-day excursion by bus.

The students are asked to each pack one bag for the trip. The bags are weighed, and the weights (in kg) are listed in order as follows:

8, 9, 10, 10, 15, 18, 22, 25, 29, 35, 38, 41

i A bag is selected at random. What is the probability that the chosen bag weighs more than 30 kg? *(1 mark)* **Easy**

ii While Sam waits for the bus to be ready, he works out the five number summary for the weights of the bags:

8, 10, 20, 32, 41

Using this five number summary, construct an accurate box-and-whisker plot to display the distribution of the weights of the bags. *(2 marks)* **Easy**

iii Calculate the interquartile range of the weights. *(1 mark)* **Easy**

(Q24a, **2002 HSC**)

75 After three small quizzes, Vicki has an average mark of 5. She wants to increase her average to 6.

What mark must she score in the next quiz for her average mark to be exactly 6? *(2 marks)*

(Q26a, **2002 HSC**) **Medium**

76 Roxy selected 30 students at random from Year 12 at her high school, and asked each of them how many text messages they had sent from a mobile phone within the last day. The results are summarised in the following table.

Number of text messages sent	*Frequency*
0	3
1	3
2	4
3	4
4	9
5	7

- **i** Calculate the mean number of text messages sent. (Give your answer correct to two decimal places.) *(1 mark)* **Easy**
- **ii** Calculate the sample standard deviation. (Give your answer correct to two decimal places.) *(1 mark)* **Easy**
- **iii** Determine the median number of text messages sent. *(1 mark)* **Easy**
- **iv** Describe the skewness of the data. *(1 mark)* **Medium**
- **v** There are 150 students in Year 12 at Roxy's school. Use the sample data in the table to estimate how many of these Year 12 students would have sent more than three text messages within the last day. *(1 mark)* **Medium**

(Q26b, **2002 HSC**)

77 The following frequency table shows Ravdeep's scores on a number of quizzes.

Score	*Frequency*
1	2
2	3
3	5
4	2
5	1

Which expression gives Ravdeep's mean score?

A $\dfrac{2+6+15+8+5}{13}$ **B** $\dfrac{2+6+15+8+5}{5}$

C $\dfrac{1+2+3+4+5}{13}$ **D** $\dfrac{1+2+3+4+5}{5}$

(1 mark)

(Q8, **2001 HSC**) **Medium**

Use the back-to-back stem-and-leaf plot to answer Questions 78 and 79.

SCORES ON A CLASS TEST

Boys		*Girls*
9 8 8	0	6
5 4 4 2 2	1	2 2 5 8
9 3 1 1	2	1 3 3 4 5 5 6
5 4 2	3	2 2 4

78 What is the range of scores in this class test?

A 27 **B** 28
C 29 **D** 35 *(1 mark)*

(Q15, **2001 HSC**) **Easy**

79 Find the median score for the boys in this class test.

A 12 **B** 15
C 19 **D** 21 *(1 mark)*

(Q16, **2001 HSC**) **Medium**

80 Andy and her biology class went to two large city parks and measured the heights of the trees in metres.

In Central Park there were 25 trees. In East Park there were 27 trees. The data sets were displayed in two box-and-whisker plots.

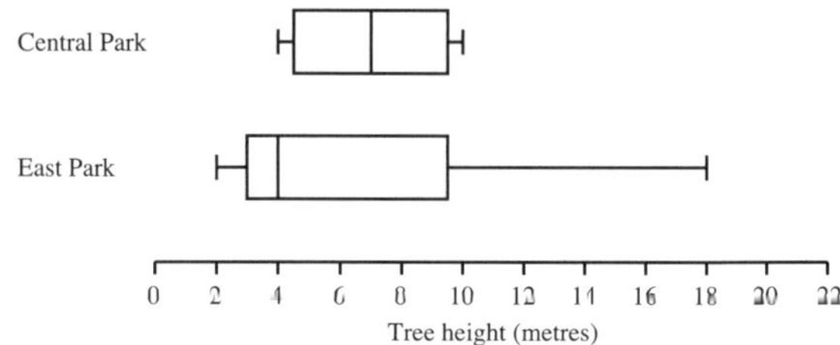

- **i** In which park is the tallest tree, and how high is it? *(2 marks)* **Easy**
- **ii** What is the median height of trees in Central Park? *(1 mark)* **Easy**
- **iii** Compare and contrast the two data sets by examining the shape and skewness of the distributions, and the measures of location and spread. *(3 marks)* **Medium**

(Q23c, **2001 HSC**)

Year 11 Summary statistics—Worked answers

1 There are two scores of 25 and only one of all the others.
So the mode is 25.
Answer C

2 The cumulative number is the total up to and including each day. The cumulative value after one day is the same as the value for that day, so A and D can be eliminated. The cumulative value after 20 days cannot be much more than the cumulative value after 19 days, so B can be eliminated.
Answer C

3 $IQR = Q_3 - Q_1$
$= 10 - 4$
$= 6$ ✓
Now $Q_3 + 1.5 \times IQR$
$= 10 + 1.5 \times 6$
$= 19$
So 20 is an outlier because it is more than $Q_3 + 1.5 \times IQR$. ✓
(2 marks)

4 Positively skewed data has a 'tail' on the right.

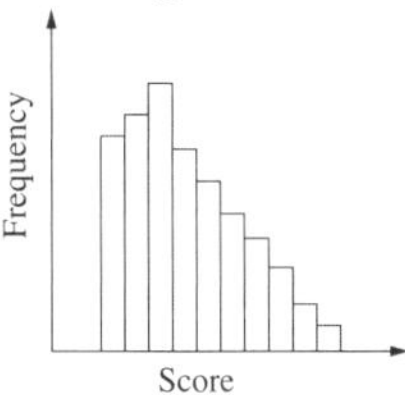

Answer A

5 1 5 9 10 15

$$\text{Mean} = \frac{1+5+9+10+15}{5}$$
$= 8$ ✓
Median = 9
1 5 9 10 15 x
Median = 9.5 ✓
The difference between the medians is 0.5.
So the difference between the means is 10×0.5 or 5.
As $x > 15$ the mean will have increased by 5.
New mean = 13 ✓
So

$$\frac{1+5+9+10+15+x}{6} = 13$$
$40 + x = 13 \times 6$
$40 + x = 78$
$x = 38$ ✓
(4 marks)

6 170, 180, 185, 188, 192, 193, 193, 194, 196, 202
Median = 192.5 ✓
$Q_1 = 185$
$Q_3 = 194$
$IQR = 194 - 185$
$= 9$ ✓
Now $Q_1 - 1.5 \times IQR$
$= 185 - 1.5 \times 9$
$= 171.5$
Now $170 < 171.5$
So the height of the shortest player could be considered an outlier. ✓
(3 marks)

7 Team A: lower extreme is 19 and upper extreme 28.
Median = 26
$Q_1 = 23$ and $Q_3 = 27$ ✓

The spread is much greater for Team B than for Team A. Both the range and interquartile range are greater for Team B. ✓
In general, Team B scored more goals per game than Team A. The upper extreme for Team B is 7 goals higher than for Team A and the lower extreme, median and both quartiles are all higher for Team B than for Team A. In 50% of games Team B scored 27 goals or more whereas in only 25% of games Team A had 27 or more goals. ✓✓
The data for Team B is fairly symmetrical, perhaps slightly positively skewed, while that for Team A is negatively skewed. ✓
(5 marks)

8 **a** $90\% = \frac{36}{40}$
36 in the upper quartile (Q3).
$\therefore$ the probability is $\frac{1}{4}$, or 25%.
(1 mark)

b $Q1 = 26$ and
$IQR = 36 - 26 = 10$. ✓
Consider 16:
Is $16 < Q1 - 1.5 \times IQR$?
$16 < 26 - 1.5 \times 10$?
$16 < 26 - 15$?
No.
$\therefore$ 16 is not an outlier. ✓
(2 marks)

9 Range = upper extreme − lower extreme
$= 9 - 2$
$= 7$
Answer D

10 The data is negatively skewed.
Answer C

11 The greatest frequency is 9 for a score of 7, so the mode is 7.
Total frequency is 30 so the median is the average of the 15th and 16th scores.
But both the 15th and 16th scores are 6. So the median is 6.
Answer B

12 **i** Rainfall is higher in Perth in July, August and September. So the mean monthly rainfall is higher in Perth than in Sydney for 3 months.
(1 mark)

ii The standard deviation of the mean monthly rainfall will be smaller in Sydney because the monthly rainfall is more consistent in Sydney and there is a greater spread in Perth.
(1 mark)

13 The median is the average of the 21st and 22nd scores.
Lower quartile is the 11th score.
So $Q_L = 3$ ✓
Upper quartile is the 32nd score.
So $Q_U = 5$
$IQR = Q_U - Q_L$
$= 5 - 3$
$= 2$ ✓
(2 marks)

14 median is 30

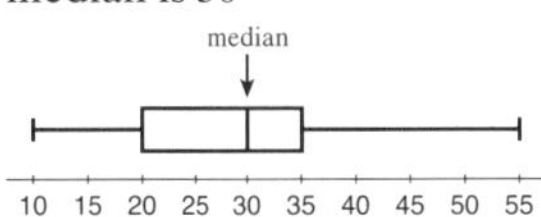

Answer C

15 220, 105, 101, 450, 37, 338, 151, 205
i mean = 200.875 (by calculator)
(1 mark)

ii standard deviation
$= 127.357\,211\ldots$
$= 127.4$ (1 d.p.)
(1 mark)

16 $Q_L = 10$, $Q_U = 16$
$IQR = Q_U - Q_L$
$= 16 - 10$
$= 6$
$Q_L - 1.5 \times IQR = 10 - 1.5 \times 6$
$= 1$ ✓
$Q_U + 1.5 \times IQR = 16 + 1.5 \times 6$
$= 25$

If there are no outliers there are no scores less than 1 or greater than 25.

Maximum range = 25 – 1

= 24 ✓

(2 marks)

17 First number = 2 and range = 10

So last number = 2 + 10 = 12

2, 3, 3, 3, 5, 5, 8, 9, 12

Median = 5

Lower quartile = 3

Upper quartile $= \dfrac{8+9}{2} = 8.5$

So five-number summary is 2, 3, 5, 8.5, 12.

Answer A

18

Class interval	*Class centre*	*Frequency*
1–5	3	3
6–10	8	6
11–15	13	8
16–20	18	9

$\bar{x} = 12.423076\ldots$

$= 12.4$ (1 d.p.)

Answer D

19 For the History test, 30 is the upper quartile, so $\frac{1}{4}$ of 112 or 28 scored above 30.

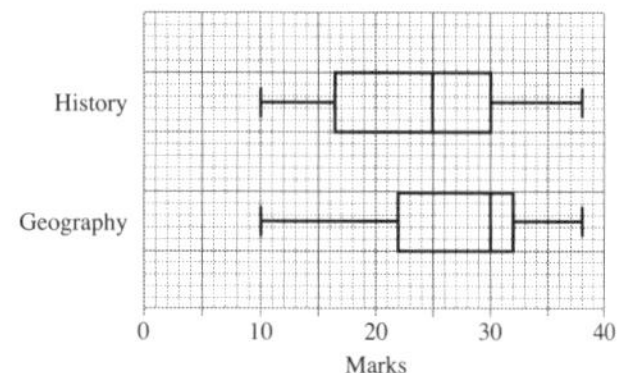

For the Geography test, 30 is the median. So 28 students is half the number of those who did the test.

Number completing the Geography test

= 2 × 28 = 56

Answer C

20 153 cm, 168 cm, 174 cm

$\bar{x} = \dfrac{153+168+174}{3}$ cm

= 165 cm ✓

So the mean of the mean heights of all the possible samples will be 165 cm because the mean of all possible sample means is equal to the mean of the population. ✓

(2 marks)

21 Upper quartile ≈ 172 ✓

Lower quartile ≈ 136

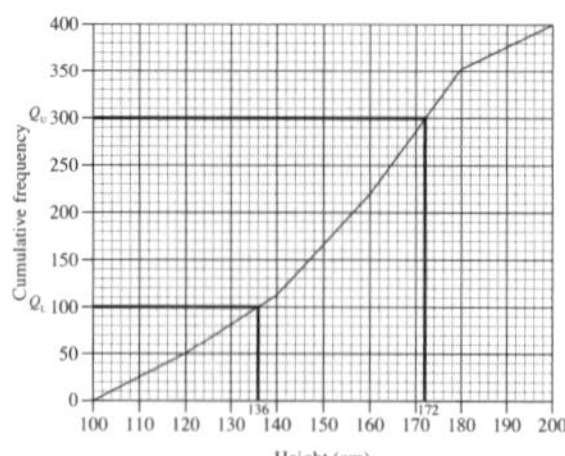

Interquartile range = 172 – 136

= 36 ✓

(2 marks)

22 Median for women = 55 ✓

Median for men = 45

Yes, Pat is correct. The median age of the women is 10 years older than the median age of the men. Only one man is older than the median age of the women. ✓

The data is also skewed. The data for the women is negatively skewed, meaning that there are more ages towards the top end of the range, whereas the data ✓ for the men is positively skewed.

(3 marks)

23 40 min is the lower quartile and 60 min is the median. So 25% of students will spend between 40 and 60 minutes exercising.

Answer C

24 \$300, \$490, \$520, \$590, \$660, \$680, \$970

i Median = \$590

Q_U = \$680

Q_L = \$490

$IQR = Q_U - Q_L$

= \$680 – \$490

= \$190 ✓

$Q_U + 1.5 \times IQR$

= \$680 + 1.5 × \$190

= \$965 ✓

Now \$970 > \$965

So \$970 is an outlier for this set of data. ✓

(3 marks)

ii The standard deviation will not change because the spread of marks from the mean will not change.

(1 mark)

25 **i**

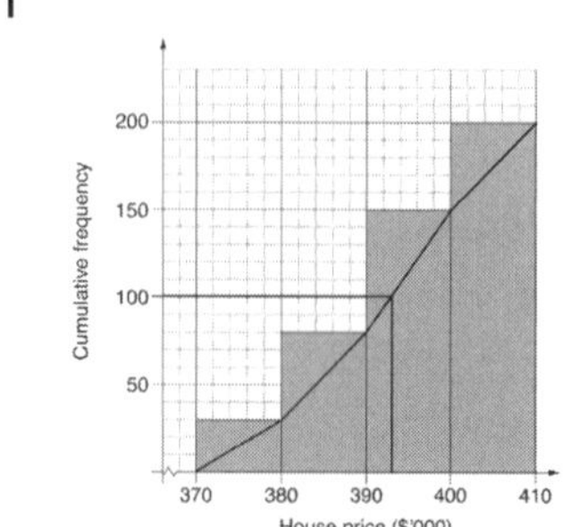

Using the graph the median house price is about \$393 000.

(1 mark)

ii

Class Centre (\$'000)	Frequency
375	30
385	50
395	70
405	50

By calculator the mean price is \$392 000. ✓✓✓

(3 marks)

26

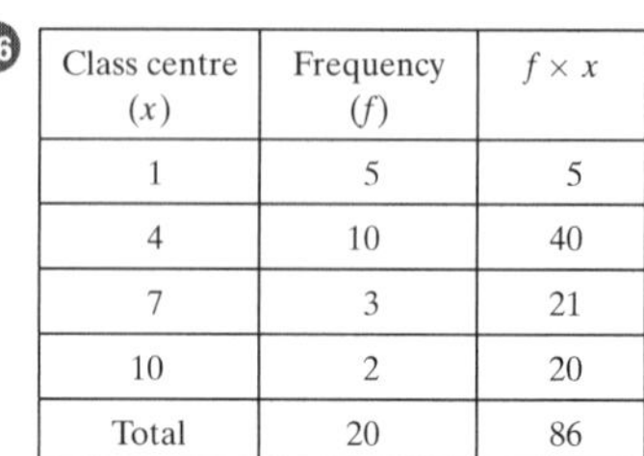

Class centre (x)	Frequency (f)	$f \times x$
1	5	5
4	10	40
7	3	21
10	2	20
Total	20	86

Mean = 86 ÷ 20

= 4.3

Answer B

27

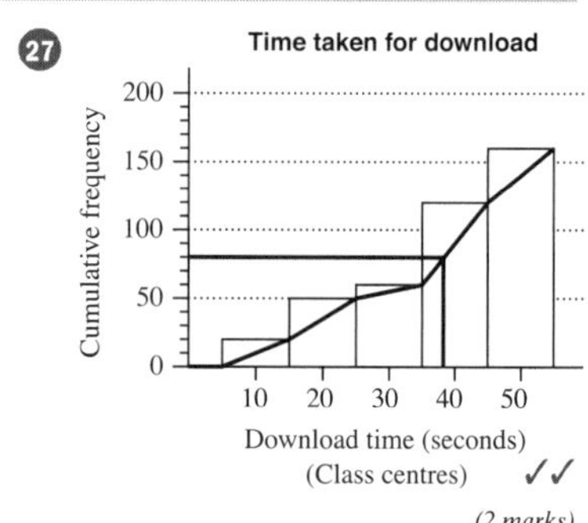

✓✓

(2 marks)

28 **i**

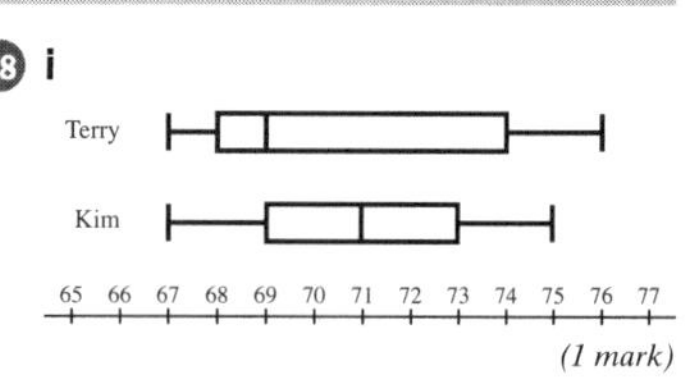

(1 mark)

ii 69 is the median for Terry. So 50% of Terry's results were below 69.

(1 mark)

iii No, Terry is not correct. ✓ Although he has the highest score, his box plot is positively skewed and only 50% of his scores are above 69 whereas 75% of Kim's scores are above that mark. ✓ Kim's box plot is symmetrical and both the range and interquartile range are smaller than those of Terry, meaning that Kim's scores are more consistent. ✓ The upper quartile for Terry was 1 mark higher than that of Kim, as is the upper extreme, so Terry did perform well with the top

25% of scores. ✓

Terry and Kim had the same lowest score.

(4 marks)

29 The mean would have increased because the amount being added is greater than any of the July prices.

The median would still be \$607 000 (because the two middle prices were the same).

So the mean would have changed but the median would have stayed the same.

Answer C

30 Number of games

$= 3 + 7 + 5 + 1 + 4$

$= 20$

Number of goals

$= 3 \times 3 + 7 \times 4 + 5 \times 5 + 1 \times 6 + 4 \times 8$

$= 100$

Mean number of goals

$= 100 \div 20$

$= 5$

Answer C

31 minimum value = 12, range = 12

maximum value = 12 + 12

= 24 ✓

A possible set is 12, 12, 12, 12, 12, 24 ✓

[or any set of 6 scores with lowest score 12, highest score 24 and more scores of 12 than of any other value].

(2 marks)

32

Travel time (minutes)

Without tolls		With tolls
9	3	5 8 9 9
9 9 8 7 7 6 5 5 4 4 3 2 0	4	0 1 2 6 7 7 8 8 8 9
9 8 7 5 4 3 3 3 2 2 2 2 1 1 0	5	2 4 4 5 6 8 9
1	6	1 3 5 7
	7	0 2 8
	8	2
	9	0

i Without tolls, there are 4 travel times of 52 minutes. No other travel time occurs as often.

So the modal travel time on roads without tolls is 52 minutes.

(1 mark)

ii There are 30 travel times shown for roads without tolls. So the median is the average of the 15th and 16th times,

$$\text{Median} = \frac{50 + 51}{2}$$

$$= 50.5$$

The median travel time on roads without tolls is 50.5 minutes.

(1 mark)

iii The data set for travel times without tolls is fairly symmetrical while that for the times with tolls is positively skewed. ✓

The spread of times is much greater for the travel times with tolls. The range for the travel times with tolls is $2\frac{1}{2}$ times larger than for those without tolls. ✓

(2 marks)

33 i Maximum cumulative frequency is 52.

Now $52 \div 4 = 13$

and $13 \times 3 = 39$

So the lower quartile will be found at a cumulative frequency of 13 and the upper quartile at 39.

From the graph the lower quartile is 190 and the upper quartile is 470. ✓

Interquartile range

$= 470 - 190 = 280$ ✓

(2 marks)

ii No, Oscar is not correct. The retailer has grouped the data, so Oscar should have said that he sold between 250 and 350 televisions in 6 of the weeks.

(1 mark)

34 The median of 15 scores is the 8th score.

5	3 4
6	2 6 7
7	7 7 (8) 9
8	2 4
9	1 3 5 7

The 8th score is 78.

Answer D

35

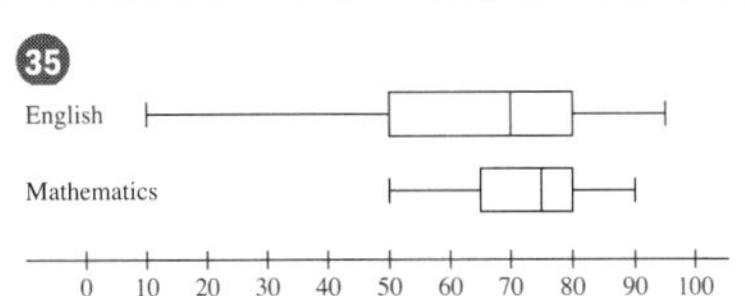

i English: lower quartile = 50

upper quartile = 80

interquartile range

$= 80 - 50$

$= 30$

(1 mark)

ii [Different answers are possible.]

Both data sets are negatively skewed. ✓

The spread of marks for English is much greater than for Mathematics. It is more than double. (The range for English is 85 while the range for Maths is 40.)

The interquartile range for English (30) is also double that for Maths (15). ✓

The highest mark in English (95) was higher than the highest mark in Maths (90), but the upper quartiles are the same (80). The median (70 for English, 75 for Maths), lower quartile (50 for English, 65 for Maths) and lower extreme (10 for English, 50 for Maths) are all higher for Maths than English. ✓

25% of students scored marks in English that were less than any marks scored in Maths.

(3 marks)

36

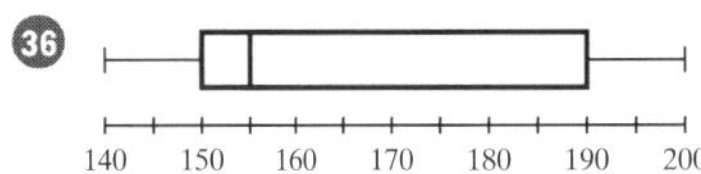

The median of this data is much closer to the lower extreme and lower quartile than it is to the upper extreme and upper quartile so the data is positively skewed.

Answer B

37 For X:

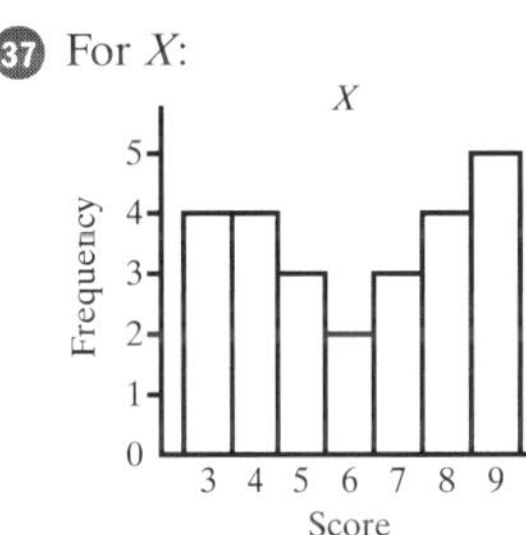

mode = 9

range = 9 − 3

= 6

For Y:

mode = 8

range = 11 − 5

= 6

So X has a larger mode and the ranges are the same.

Answer B

38 Two of the new scores are less than the original median and two are greater than the original median. So the median will not change.

The median will be 7.

Answer B

39 Total height of first 5 players
$= (1.8 + 1.83 + 1.84 + 1.86 + 1.92)$ m
$= 9.25$ m

Average height $= (9.25 \div 5)$ m
$= 1.85$ m

The average height increases by 1 cm.

New average height
$= (1.85 + 0.01)$ m
$= 1.86$ m

Total height of 6 players
$= 6 \times 1.86$ m
$= 11.16$ m

Height of sixth player
$= (11.16 - 9.25)$ m
$= 1.91$ m

Answer C

40 **i** Outlier = 71

(1 mark)

ii For female students:
Median = 17
Lower quartile = 11
Upper quartile = 20
Interquartile range = 20 – 11
= 9

(1 mark)

41 Highest score = 9
Lowest score = 0
Range = 9 – 0
= 9

Answer C

42 There are 26 students. The median is the average of the 13th and 14th scores (when the scores are arranged in order).

All students

2	1 1 2 4
3	0 2 3 3 5
4	4 4 4 5 7 9 9 9
5	2 2 3 4 6
6	0 1 3 9

The 13th score is 45 and the 14th score is 47.

Median $= \dfrac{45 + 47}{2}$
$= 46$

Answer B

43 **i** $X = 25 - 10$
$= 15$

(1 mark)

ii

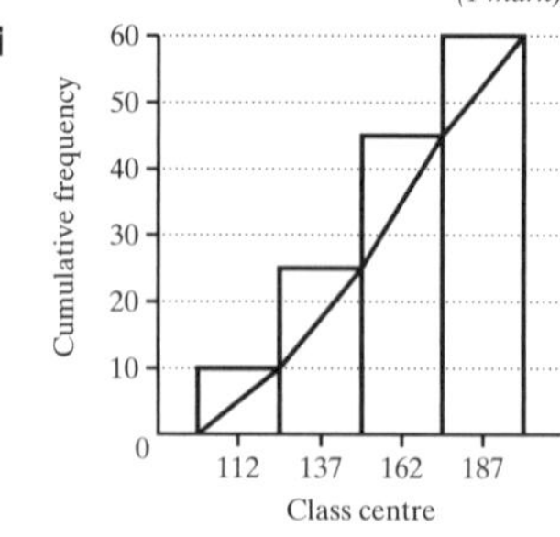

(1 mark)

iii median ≈ 155

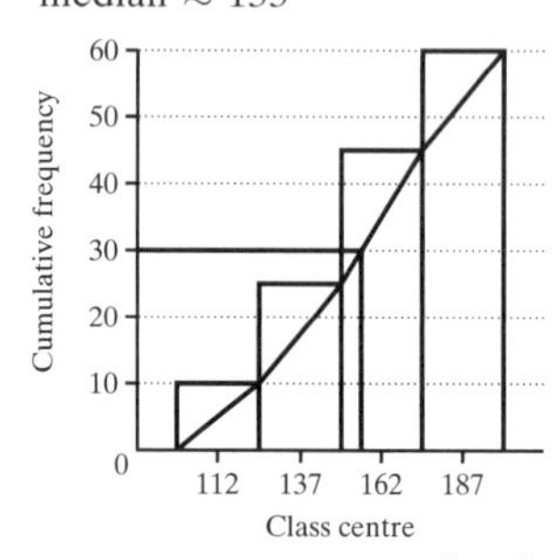

(1 mark)

iv Possible answers for the problem:
- increased traffic and congestion around the school
- safety issues for the school children. ✓

Possible answers for the solution to the problem:
- create a bypass around the school
- lower the speed limit around the school
- install flashing lights to increase awareness of motorists of the danger. ✓

(2 marks)

44

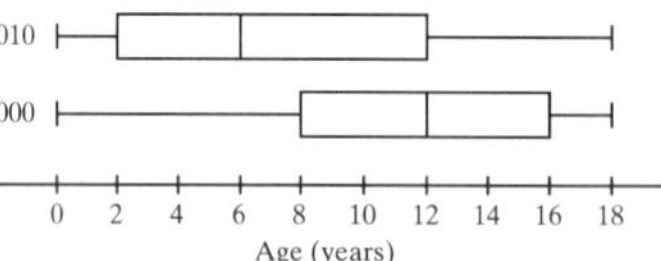

i In 2000, the median is 12.
So 50% or $\frac{1}{2}$ of the children were aged 12–18 years.
Number of children
$= \frac{1}{2} \times 1750$
$= 875$

(1 mark)

ii In 2010, 875 children are aged from 12–18.
Now 12 is the upper quartile.
So 875 is 25% or $\frac{1}{4}$ of the total children.
Number of children
$= 875 \times 4$
$= 3500$

(1 mark)

iii Some possible answers are:
- The median age has decreased from 12 to 6. ✓
- In 2000 the ages were negatively skewed with more than 50% of the children aged 10 or over. In 2010 the ages were positively skewed with more than 50% of the children aged 8 or less.
- Although the range has not changed the interquartile range has increased from 8 to 10. ✓

(2 marks)

iv Some possible answers are:
- need to plan for more schools
- need to provide more child-care places
- need to improve the level of services for families with children.

(1 mark)

45 Total of ten scores $= 14 \times 10$
$= 140$

$\therefore \dfrac{140 + \text{two new scores}}{12} = 16$

$140 + \text{two new scores} = 192$
$\text{two new scores} = 52$

$\therefore$ Mean of two new scores $= \dfrac{52}{2}$
$= 26$

Answer D

46 **i** Mode is 78 (occurs 3 times).

(1 mark)

ii Median
= average of 11th and 12th scores
$= \dfrac{45 + 47}{2}$
$= 46$

(1 mark)

47 **i** IQR = UQ – LQ
= 6 – 2
= 4

(1 mark)

ii Each section of the box-and-whisker plot represents 25%, therefore 75% of girls spend less than 5 hours on the internet.

(1 mark)

iii 25% of boys and girls spend between 5 and 6 hours on the internet but this is only the same number of boys and girls if there are an equal number of boys and girls in the survey.

(1 mark)

48 Range
= highest score – lowest score
∴ 43 = 67 – lowest score
∴ Lowest score = 67 – 43
= 24
∴ The value of N is 4.
Answer A

49 Since there are 33 scores, the middle score is the $\left(\frac{33+1}{2}\right)$th score, i.e. the 17th score.

There are 13 scores of 12, and 6 scores of 14, so the 17th score is 14.

Answer C

50 Interquartile range is the same, as the size of the rectangles is the same.
∴ The upper quartile – the lower quartile will have the same value.
Answer D

51 If the mean did not change, then the mean of the heights of the 2 absent students must be 160 cm.

(C): $\frac{149+171}{2} = 160$

Answer C

52 Total marks from 3 tests
= 72 × 3
= 216 ✓

New mean $= \frac{216 + \text{new mark}}{4}$

$73 = \frac{216 + \text{new mark}}{4}$

292 = 216 + new mark
new mark = 292 – 216
= 76 ✓

(2 marks)

53 In line 1,
279 + 356 = 635
∴ 279 and 356 are correct.
In column 1,
279 + 187 = 466
∴ 279 and 187 are correct
∴ 439 is incorrect which is females in part-time work.
Answer D

54 No. of students surveyed
= 10% of 1200
= 120

No. of students surveyed from each year
= 120 ÷ 6
= 20

In a stratified sample:

Year 10: $\frac{230}{1200} \times 120 = 23$

∴ She would sample 3 more.

Year 11: $\frac{150}{1200} \times 120 = 15$

∴ She would sample 5 less.

Answer C

55 Range = 25 – 5
= 20

∴ 20 = $2x$ – 6
$2x$ = 26
x = 13

∴ The set of data is
5, 6, 11, 13, 13, 18, 25

∴ The median is 13.

∴ The interquartile range
= upper quartile – lower quartile
= 18 – 6
= 12

Answer D

56 The 4th decile is the 40th percentile.
40% of 50 = 20
∴ Scores between 55 and 60.
Answer B

57 **i** Mean

$= \frac{\text{sum of scores}}{\text{number of scores}}$

$= \frac{3 + 2(5) + 6 + 2(8) + 9 + 2(10) + 50}{10}$

$= \frac{114}{10}$

= 11.4

(1 mark)

ii Removing the outlier of 50 does not affect the median which will still be 8. ✓

However, the mean will now be $\frac{64}{9} = 7.\dot{1}$, which is significantly lower. ✓

(2 marks)

58 **i** There are more males attending the studio and the distribution of their ages is positively skewed. There are more younger males than females. The distribution of ages of females is negatively skewed.

(1 mark)

ii Mode for the whole group is 64.

(1 mark)

iii For the group 30–39 years, the class centre $= \frac{30 + 39}{2}$
= 34.5 ✓

∴ Class centre × frequency
∴ = 34.5 × 5
∴ = 172.5 ✓

(2 marks)

iv The grouped frequency distribution table uses the class centres to calculate the mean, while Caitlyn used the raw data making her mean more accurate.

(1 mark)

59 Counting in to the middle scores, these are 28 and 38.

∴ Median $= \frac{28 + 38}{2}$
= 33

Answer C

60 A and C are positively skewed, but C is less spread out (smaller standard deviation).
Answer C

61 $\bar{x}$ = 62 for 5 scores

$\bar{x} = \frac{\text{sum of scores}}{\text{no. of scores}}$

∴ $62 = \frac{\text{sum of scores}}{5}$

∴ Sum of scores = 62 × 5
= 310

New mean $= \frac{310 + 14}{6}$
= 54

Answer B

62 **i** A stratified random sample.

Vicki could take a sample of students from each year in the proportion of their representation in the school.

She could then choose these students by systematically going through class lists.

(1 mark)

ii $\bar{x} = \frac{\text{sum of scores}}{\text{no. of scores}}$

$= \frac{(2 \times 69) + (7 \times 72) + (12 \times 38) + (17 \times 21)}{200}$ ✓

$= \frac{1455}{200}$

$= 7.275$ hours ✓

(N.B. For grouped data, you must use the class centre as the score.)

(2 marks)

63 5 summary statistics:
Lowest score = 140 cm
Lower quartile = 145 cm
Median = 150 cm ✓
Upper quartile = 170 cm
Maximum score = 190 cm ✓

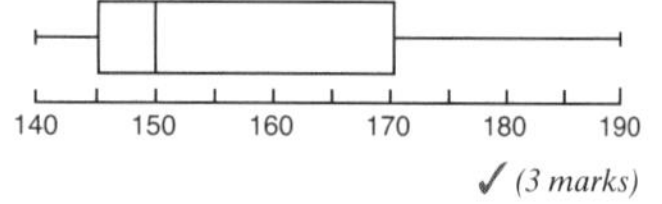

✓ *(3 marks)*

64 Mean $= \frac{3+4+5+6+6+8+8+8+15}{9}$

$= 7$

Answer B

65

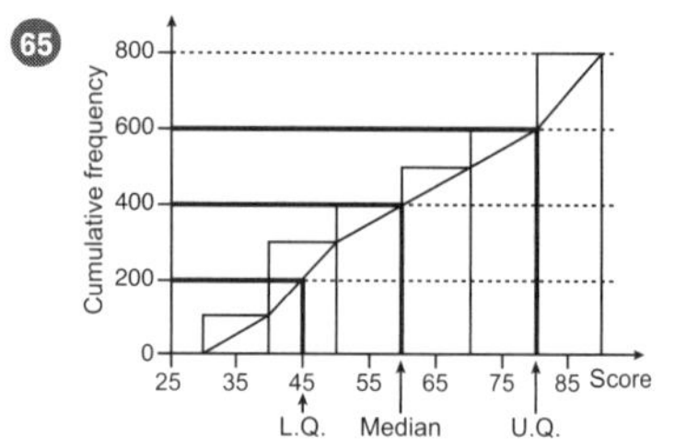

From graph,

Median = 60
Upper quartile = 80
Lower quartile = 45

∴ Interquartile range
= 80 – 45
= 35

Answer C

66 A is not true because 50% of people under 21 earned more than \$325, but 75% of the others earned more than \$325. (∴ C is not true either.)
D is not true because only 25% of people under 21 earned between \$325 and \$350 per week.
B is true. 75% of people under 21 earned less than \$350 per week.
Answer B

67 i

Stem	Leaf
1	9
2	1 3 7 7 9
3	4 5
4	5
5	8

✓✓ *(2 marks)*

ii Median = 28 *(1 mark)*

iii Slightly positively skewed. *(1 mark)*

68 Range
= highest score – lowest score
= 12 – 1
= 11
Answer C

69 Median = middle score
= 4
Mode = most frequent score
= 3
Answer D

70 Interquartile range
= upper quartile – lower quartile
= 11 – 8
= 3
Answer C

71
P(At least 2 brothers)

$= \frac{\text{No. of students with 2, 3 or 4 brothers}}{20}$

$= \frac{5}{20}$

$= 0.25$

Answer C

72 Range = 54 – 23
= 31
Answer D

73 No. of students who watched 4 or less movies = 35
No. of students who watched 3 or less movies = 25
∴ 10 students watched 4 movies.
Answer B

74 i $P(\text{More than 30 kg}) = \frac{3}{12}$

$= \frac{1}{4}$

(1 mark)

ii 0 5 10 15 20 25 30 35 40 45

✓✓ *(2 marks)*

iii $IQR = 32 - 10$
$= 22$ kg *(1 mark)*

75 $\bar{x} = \frac{\Sigma x}{n}$

$\therefore 6 = \frac{(5 \times 3) + \text{new score}}{4}$ ✓

∴ 24 = 15 + new score
∴ new score = 24 – 15
= 9 ✓ *(2 marks)*

76 i $\bar{x} = 3.133...$
= 3.13 to 2 decimal places *(1 mark)*

ii σ_{n-1}
= 1.655...
= 1.66 to 2 decimal places *(1 mark)*

iii Median = 4
(The average of the 15th and 16th scores.) *(1 mark)*

iv The data is negatively skewed (seen by the lower frequency of the lower scores). *(1 mark)*

v Estimate $= \frac{16}{30} \times 150$
= 80 students *(1 mark)*

77

$\bar{x} = \frac{(1 \times 2) + (2 \times 3) + (3 \times 5) + (4 \times 2) + (5 \times 1)}{\text{Total of Frequency}}$

$= \frac{2 + 6 + 15 + 8 + 5}{13}$

Answer A

78 Range = Highest – Lowest
= 35 – 6
= 29
Answer C

79 There are 15 scores. The median is the 8th score. By counting, the median is 15.
Answer B

80 i In East Park, the tallest tree is 18 m. ✓✓ *(2 marks)*

ii Median = 7 m *(1 mark)*

iii

Central Park	East Park
• shape is symmetrical	• very positively skewed ✓
• median = 7 m	• median = 4 m ✓
• range of 6 m with heights from 4 m to 10 m	• range of 16 m with heights from 2 m to 18 m ✓
• $UQ = 9.5$ m, $LQ \doteqdot 4.5$ m ∴ $IQR = 5$ m	• $UQ = 9.5$ m, $LQ = 3$ m ∴ $IQR = 6.5$ m

(3 marks)

1 There are 8 chocolates in a box. Three have peppermint centres (P) and five have caramel centres (C).

Kim randomly chooses a chocolate from the box and eats it. Sam then randomly chooses and eats one of the remaining chocolates.

A partially completed probability tree is shown.

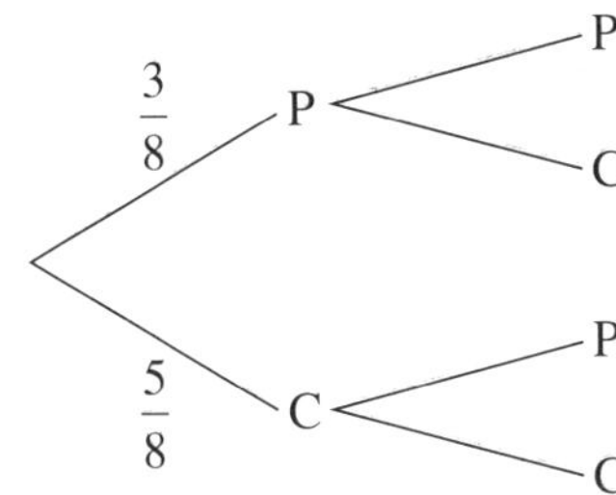

What is the probability that Kim and Sam choose chocolates with different centres?

A $\frac{15}{64}$ **B** $\frac{15}{56}$

C $\frac{15}{32}$ **D** $\frac{15}{28}$ *(1 mark)*

(Q11, **2021 HSC**) 

2 The top of a rectangular table is divided into 8 equal sections as shown.

1	2	3	4
5	6	7	8

A standard die with faces labelled 1 to 6 is rolled onto the table. The die is equally likely to land in any of the 8 sections of the table. If the die does not land entirely in one section of the table, it is rolled again.

A score is calculated by multiplying the value shown on the top face of the die by the number shown in the section of the table where the die lands.

What is the probability of getting a score of 6?

A $\frac{1}{48}$ **B** $\frac{1}{12}$

C $\frac{1}{8}$ **D** $\frac{1}{6}$ *(1 mark)*

(Q15, **2020 HSC**) Medium

3 A roulette wheel has the numbers $0, 1, 2, \ldots, 36$ where each of the 37 numbers is equally likely to be spun.

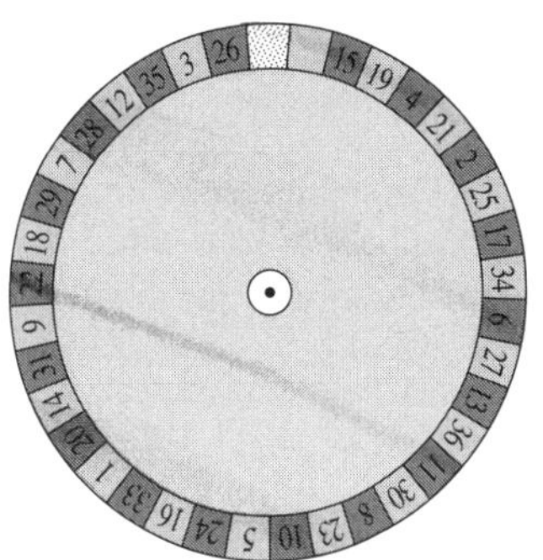

If the wheel is spun 18 500 times, calculate the expected frequency of spinning the number 8. *(2 marks)*

(Q20, **2019 HSC**) Easy

4 A bowl of fruit contains 17 apples of which 9 are red and 8 are green.

Dennis takes one apple at random and eats it. Margaret also takes an apple at random and eats it.

By drawing a probability tree diagram, or otherwise, find the probability that Dennis and Margaret eat apples of the same colour. *(3 marks)*

(Q25, **2019 HSC**) Hard

5 An experiment has three distinct outcomes, A, B and C. Outcome A occurs 50% of the time. Outcome B occurs 23% of the time.

What is the expected number of times outcome C would occur if the experiment is conducted 500 times?

A 115 **B** 135

C 250 **D** 365 *(1 mark)*

(Q9, **2018 HSC**) Easy

6 During a year, the maximum temperature each day was recorded. The results are shown in the table.

	Number of days with maximum temperature less than 25°C	*Number of days with maximum temperature greater than or equal to 25°C*	*Total number of days*
Summer	42	48	90
Autumn	57	35	92
Winter	71	21	92
Spring	53	38	91
Total number of days	223	142	365

From the days with a maximum temperature less than 25°C, one day is selected at random.

What is the probability, to the nearest percentage, that the selected day occurred during winter?

A 19% **B** 25%
C 32% **D** 77% *(1 mark)*

(Q20, **2018 HSC**) Easy

7 Jeremy rolled a biased 6-sided die a number of times. He recorded the results in a table.

Number	1	2	3	4	5	6
Frequency	23	19	48	20	21	19

What is the relative frequency of rolling a 3? *(1 mark)*

(Q26a, **2018 HSC**) Medium

8 A game consists of two tokens being drawn at random from a barrel containing 20 tokens. There are 17 tokens labelled 10 cents and 3 tokens labelled $2. The player wins the total value of the two tokens drawn.

Complete the probability tree by writing the missing probabilities in the boxes.

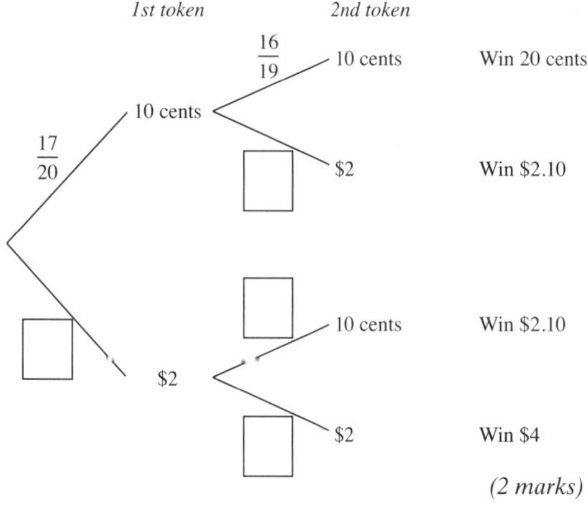

(2 marks)

(Q30d i, **2018 HSC**) Easy

9 In a survey of 200 randomly selected Year 12 students it was found that 180 use social media.

Based on this survey, approximately how many of 75 000 Year 12 students would be expected to use social media?

A 60 000 **B** 67 500
C 74 980 **D** 75 000 *(1 mark)*

(Q5, **2017 HSC**) Easy

10 The faces on a twenty-sided die are labelled $0.05, $0.10, $0.15, … , $1.00.

The die is rolled once.

What is the probability that the amount showing on the upper face is more than 50 cents but less than 80 cents?

A $\frac{1}{4}$ **B** $\frac{3}{10}$

C $\frac{7}{20}$ **D** $\frac{1}{2}$ *(1 mark)*

(Q15, **2017 HSC**) Medium

11 A deck of 52 playing cards contains 12 picture cards. Two cards from the deck are drawn at random and placed on a table.

What is the probability, correct to four decimal places, that exactly one picture card is on the table?

A 0.0498 **B** 0.1810
C 0.3550 **D** 0.3620 *(1 mark)*

(Q24, **2017 HSC**) Hard

12 A group of Year 12 students was surveyed. The students were asked whether they live in the city or the country. They were also asked if they have ever waterskied.

The results are recorded in the table.

	Have waterskied	Have never waterskied
Live in the city	150	2500
Live in the country	70	800

i A person is selected at random from the group surveyed.
Calculate the probability that the person lives in the city and has never waterskied. *(2 marks)* Easy

ii A newspaper article claimed that Year 12 students who live in the country are more likely to have waterskied than those who live in the city.
Is this true, based on the survey results? Justify your answer with relevant calculations. *(2 marks)* Medium

(Q29c, **2017 HSC**)

13 A group of 485 people was surveyed. The people were asked whether or not they smoke. The results are recorded in the table.

	Smokers	*Non-smokers*	*Total*
Male	88	176	264
Female	68	153	221
	156	329	485

A person is selected at random from the group.

What is the approximate probability that the person selected is a smoker OR is male?

A 33% **B** 18%
C 68% **D** 87% *(1 mark)*

(Q23, **2016 HSC**) **Hard**

14 A cricket team is about to play two matches. The probability of the team having a win, a loss or a draw is 0.7, 0.1 and 0.2 respectively in each match. The possible results in the two matches are displayed in the probability tree diagram.

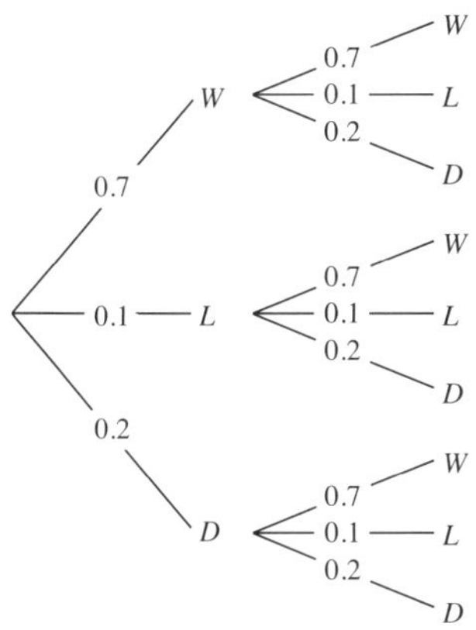

i What is the probability of the team having a win and a draw, in any order? *(2 marks)* **Medium**

ii Paul claims that 1.4 is the probability of the team winning both matches. Give one reason why this is NOT correct. *(1 mark)* **Medium**

(Q28c, **2016 HSC**)

15 The probability of winning a game is $\frac{7}{10}$.
Which expression represents the probability of winning two consecutive games?

A $\frac{7}{10} \times \frac{6}{9}$ **B** $\frac{7}{10} \times \frac{6}{10}$

C $\frac{7}{10} \times \frac{7}{9}$ **D** $\frac{7}{10} \times \frac{7}{10}$ *(1 mark)*

(Q16, **2015 HSC**) **Easy**

16 The table shows the relative frequency of selecting each of the different coloured jelly beans from packets containing green, yellow, black, red and white jelly beans.

Colour	*Relative frequency*
Green	0.32
Yellow	0.13
Black	0.14
Red	
White	0.24

i What is the relative frequency of selecting a red jelly bean? *(1 mark)* **Easy**

ii Based on this table of relative frequencies, what is the probability of NOT selecting a black jelly bean? *(1 mark)* **Easy**

(Q26e, **2015 HSC**)

17 On a tray there are 12 hard-centred chocolates (H) and 8 soft-centred chocolates (S). Two chocolates are selected at random. A partially completed probability tree is shown.

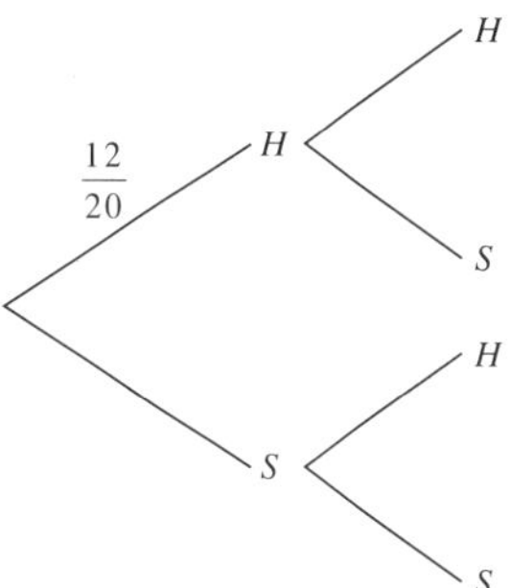

What is the probability of selecting one of each type of chocolate? *(3 marks)*

(Q30b, **2015 HSC**) **Medium**

18 A group of 150 people was surveyed and the results recorded.

Survey results

	Owns a mobile	*Does not own a mobile*	*Total*
Male	42	28	70
Female	63	17	80
	105	45	150

A person is selected at random from the surveyed group.

What is the probability that the person selected is a male who does not own a mobile?

A $\frac{28}{150}$ **B** $\frac{45}{150}$

C $\frac{28}{70}$ **D** $\frac{45}{70}$ *(1 mark)*

(Q8, **2014 HSC**) **Easy**

19 In Mathsville, there are on average eight rainy days in October.

Which expression could be used to find a value for the probability that it will rain on two consecutive days in October in Mathsville?

A $\frac{8}{31} \times \frac{7}{30}$ **B** $\frac{8}{31} \times \frac{7}{31}$

C $\frac{8}{31} \times \frac{8}{30}$ **D** $\frac{8}{31} \times \frac{8}{31}$ *(1 mark)*

(Q16, **2014 HSC**) **Hard**

20 Jaz has 2 bags of apples.

Bag A contains 4 red apples and 3 green apples.
Bag B contains 3 red apples and 1 green apple.

Jaz chooses an apple from one of the bags.

Which tree diagram could be used to determine the probability that Jaz chooses a red apple?

A

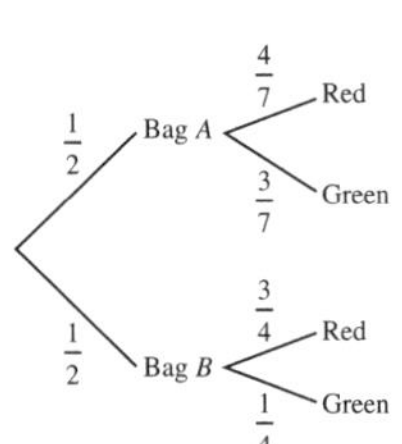

B

$\frac{1}{2}$ Bag A
$\frac{7}{11}$ Red
$\frac{4}{11}$ Green
$\frac{1}{2}$ Bag B
$\frac{7}{11}$ Red
$\frac{4}{11}$ Green

C

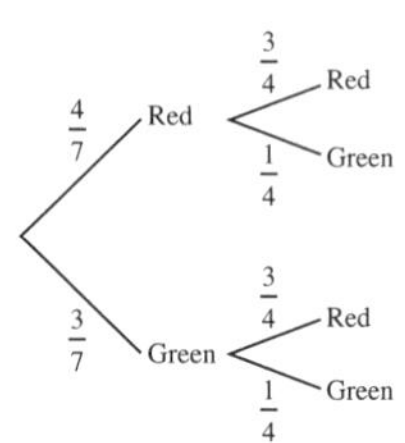

D

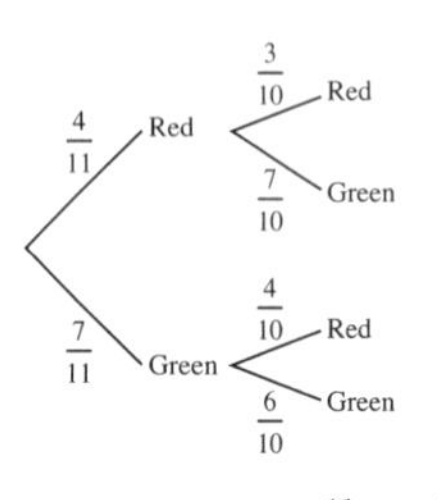

(1 mark)

(Q19, **2014 HSC**) Easy

21 A fair coin is tossed three times. Using a tree diagram, or otherwise, calculate the probability of obtaining two heads and a tail in any order. *(2 marks)*

(Q28c, **2014 HSC**) Medium

22 Which of the following events would be LEAST likely to occur?

A Tossing a fair coin and obtaining a head
B Rolling a standard six-sided die and obtaining a 3
C Randomly selecting the letter 'G' from the 26 letters of the alphabet
D Winning first prize in a raffle of 100 tickets in which you have 4 tickets *(1 mark)*

(Q1, **2013 HSC**) Easy

23 In an experiment, a standard six-sided die was rolled 72 times. The results are shown in the table.

Number on die	*Frequency*
1	6
2	12
3	10
4	20
5	9
6	15

Which number on the die was obtained the expected number of times?

A 1 **B** 2
C 3 **D** 6 *(1 mark)*

(Q7, **2013 HSC**) Easy

24 Students studying vocational education courses were surveyed about their living arrangements.

	Females	*Males*	*Totals*
Living with parent(s)	46	155	201
Not living with parent(s)	182	122	304
Totals	228	277	505

One of these students is selected at random.

What is the probability that this student is male and living with his parent(s)?

A 31% **B** 40%
C 56% **D** 77% *(1 mark)*

(Q10, **2013 HSC**) Easy

25 Two unbiased dice, each with faces numbered 1, 2, 3, 4, 5 and 6, are rolled. What is the probability of obtaining a sum of 6?

A $\frac{1}{6}$ **B** $\frac{1}{12}$
C $\frac{5}{12}$ **D** $\frac{5}{36}$ *(1 mark)*

(Q18, **2013 HSC**) Hard

26 The probability that Michael will score more than 100 points in a game of bowling is $\frac{31}{40}$.

i A commentator states that the probability that Michael will score less than 100 points in a game of bowling is $\frac{9}{40}$.
Is the commentator correct? Give a reason for your answer. *(1 mark)* Hard

ii Michael plays two games of bowling. What is the probability that he scores more than 100 points in the first game and then again in the second game? *(1 mark)* Hard

(Q26c, **2013 HSC**)

27 In a class there are 15 girls (G) and 7 boys (B). Two students are chosen at random to be class representatives.

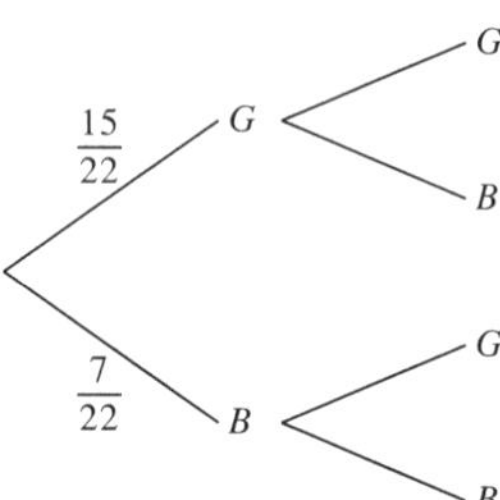

i Copy and complete the tree diagram in your answer booklet. *(2 marks)* Medium

ii What is the probability that the two students chosen are of the same gender? *(2 marks)* Medium

(Q30b, **2013 HSC**)

28 Two unbiased dice, each with faces numbered 1, 2, 3, 4, 5 and 6, are rolled.

What is the probability of a 6 appearing on at least one of the dice?

A $\frac{1}{6}$ **B** $\frac{11}{36}$

C $\frac{25}{36}$ **D** $\frac{5}{6}$ *(1 mark)*

(Q12, **2012 HSC**) **Hard**

29 A spinner with different coloured sectors is spun 40 times. The results are recorded in the table.

Colour obtained	*Frequency*
Red	2
Yellow	4
Blue	6
Orange	☐
Green	10
Purple	12

What is the relative frequency of obtaining the colour orange?

A $\frac{3}{20}$ **B** $\frac{1}{5}$

C 6 **D** 8 *(1 mark)*

(Q17, **2012 HSC**) **Medium**

30 The dot plot shows the number of push-ups that 13 members of a fitness class can do in one minute.

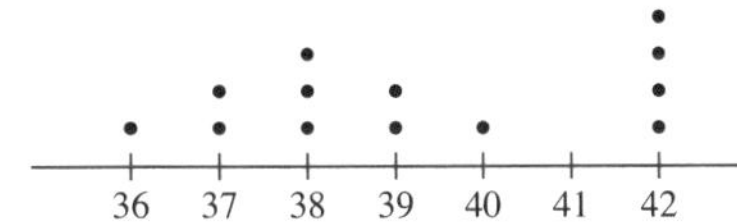

i What is the probability that a member selected at random from the class can do more than 38 push-ups in one minute? *(1 mark)* **Easy**

ii A new member who can do 32 push-ups in one minute joins the class. Does the addition of this new member to the class change the probability calculated in part **i**? Justify your answer. *(1 mark)* **Easy**

(Q26e, **2012 HSC**)

31 A box contains 33 scarves made from two different fabrics. There are 14 scarves made from silk (S) and 19 made from wool (W).

Two girls each select, at random, a scarf to wear from the box.

i Copy and complete the probability tree diagram in your answer booklet. *(2 marks)* **Medium**

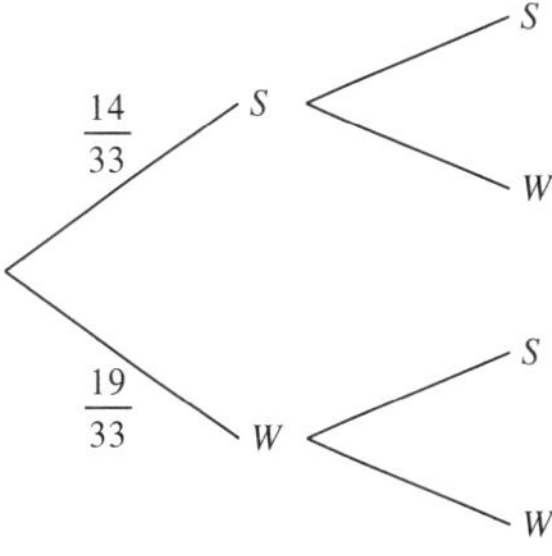

ii Calculate the probability that the two scarves selected are made from silk. *(1 mark)* **Medium**

iii Calculate the probability that the two scarves selected are made from different fabrics. *(2 marks)* **Medium**

(Q27e, **2012 HSC**)

32 Which of the following could be the probability of an event occurring?

A 1 **B** $\frac{6}{5}$

C 1.27 **D** 145% *(1 mark)*

(Q2, **2011 HSC**) **Easy**

33 An unbiased coin is tossed 10 times.

A tail is obtained on each of the first 9 tosses.

What is the probability that a tail is obtained on the 10th toss?

A $\frac{1}{2^{10}}$ **B** $\frac{1}{2}$

C $\frac{1}{10}$ **D** $\frac{9}{10}$ *(1 mark)*

(Q15, **2011 HSC**) **Easy**

34 A die was rolled 72 times. The results for this experiment are shown in the table.

Number obtained	*Frequency*
1	16
2	11
3	***A***
4	8
5	12
6	15

i Find the value of ***A***. *(1 mark)* **Easy**

ii What was the relative frequency of obtaining a 4? *(1 mark)* **Medium**

iii If the die was unbiased, which number was obtained the expected number of times? *(1 mark)* **Easy**

(Q24b, **2011 HSC**)

35 At another school, students who use mobile phones were surveyed. The set of data is shown in the table.

	Pre-paid	*Plan*	TOTAL
Female students	172	147	319
Male students	158	103	261
TOTAL	330	250	

i How many students were surveyed at this school? *(1 mark)* **Easy**

ii Of the female students surveyed, one is chosen at random. What is the probability that she uses pre-paid? *(1 mark)* **Easy**

iii Ten new male students are surveyed and all ten are on a plan. The set of data is updated to include this information. What percentage of the male students surveyed are now on a plan? Give your answer to the nearest per cent. *(1 mark)* **Medium**

(25c, **2011 HSC**)

36 The two spinners shown are used in a game.

Spinner A

Spinner B

Each arrow is spun once. The score is the total of the two numbers shown by the arrows.

A table is drawn up to show all scores that can be obtained in this game.

		Spinner B			
		1	1	2	3
	1	2	2	3	4
Spinner A	1	2	2	3	4
	3	4	4	X	6

i What is the value of X in the table? *(1 mark)* **Easy**

ii What is the probability of obtaining a score less than 4? *(1 mark)* **Easy**

iii On Spinner B, a 2 is obtained. What is the probability of obtaining a score of 3? *(1 mark)* **Medium**

(Q26a, **2011 HSC**)

37 A bag contains red, green, yellow and blue balls.

Colour	*Probability*
Red	$\frac{1}{3}$
Green	$\frac{1}{4}$
Yellow	?
Blue	$\frac{1}{6}$

The table shows the probability of choosing a red, green or blue ball from the bag.

If there are 12 yellow balls in the bag, how many balls are in the bag altogether?

A 16 **B** 36

C 48 **D** 60 *(1 mark)*

(Q8, **2010 HSC**) **Medium**

38 A group of 347 people was tested for flu and the results were recorded. The flu test results are not always accurate.

	Test results		
	Test indicated flu	*Test did not indicate flu*	*Total*
People with flu	72	3	75
People without flu	16	256	272
	88	259	347

A person is selected at random from the tested group.

What is the probability that their test result is accurate, to the nearest per cent?

A 21% **B** 22%

C 95% **D** 96% *(1 mark)*

(Q12, **2010 HSC**) **Easy**

39 Lou and Ali are on a fitness program for one month. The probability that Lou will finish the program successfully is 0.7 while the probability that Ali will finish successfully is 0.6. The probability tree shows this information.

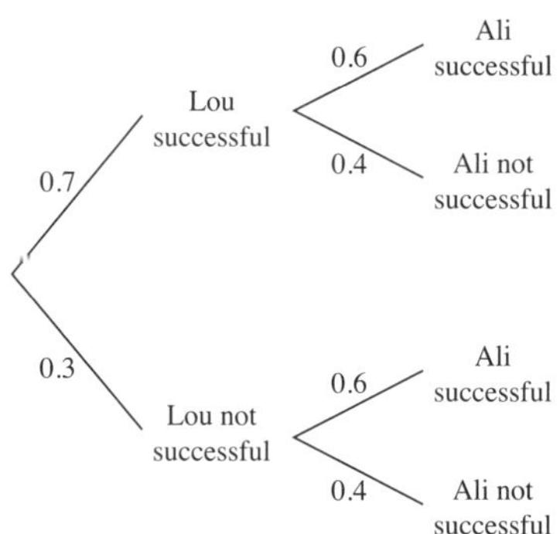

What is the probability that only one of them will be successful?

A 0.18 **B** 0.28

C 0.42 **D** 0.46 *(1 mark)*

(Q20, **2010 HSC**) **Medium**

40 On Saturday, Jonty recorded the colour of T-shirts worn by the people at his gym. The results are shown in the graph.

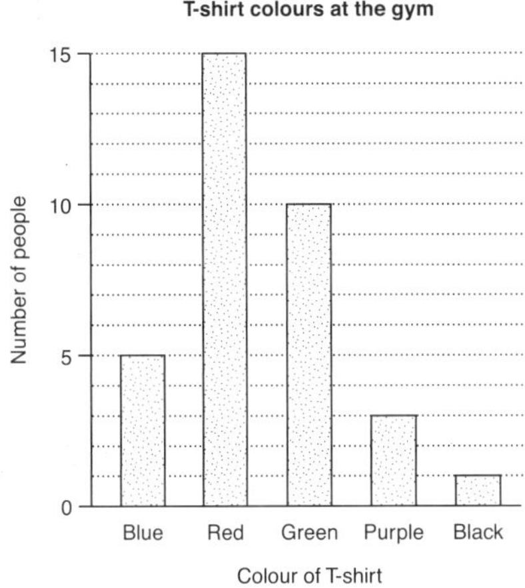

i How many people were at the gym on Saturday? (Assume everyone was wearing a T-shirt.) *(1 mark)* Easy

ii What is the probability that a person selected at random at the gym on Saturday, would be wearing either a blue or green T-shirt? *(1 mark)* Easy

(Q23c, **2010 HSC**)

41 A newspaper states: 'It will most probably rain tomorrow.'

Which of the following best represents the probability of an event that will most probably occur?

A $33\frac{1}{3}\%$ **B** 50%

C 80% **D** 100% *(1 mark)*

(Q1, **2009 HSC**) Easy

42 A wheel has the numbers 1 to 20 on it, as shown in the diagram. Each time the wheel is spun, it stops with the marker on one of the numbers.

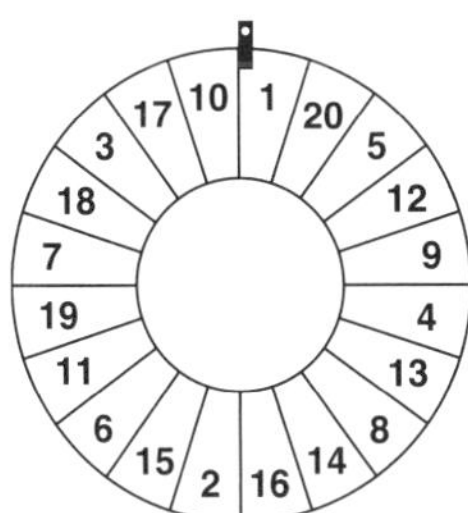

The wheel is spun 120 times.

How many times would you expect a number less than 6 to be obtained?

A 20 **B** 24

C 30 **D** 36 *(1 mark)*

(Q9, **2009 HSC**) Medium

43 In each of three raffles, 100 tickets are sold and one prize is awarded.

Mary buys two tickets in one raffle. Jane buys one ticket in each of the other two raffles.

Determine who has the better chance of winning at least one prize. Justify your response using probability calculations. *(4 marks)*

(Q27c, **2009 HSC**) Hard

44 In an experiment, two unbiased dice, with faces numbered 1, 2, 3, 4, 5, 6, are rolled 18 times. The difference between the numbers on their uppermost faces is recorded each time. Juan performs this experiment twice and his results are shown in the tables.

Experiment 1

Difference	*Frequency*
0	3
1	3
2	2
3	4
4	3
5	3

Experiment 2

Difference	*Frequency*
0	4
1	4
2	3
3	3
4	2
5	2

Juan states that Experiment 2 has given results that are closer to what he expected than the results given by Experiment 1.

Is he correct? Explain your answer by finding the sample space for the dice differences and using theoretical probability. *(4 marks)*

(Q28d, **2009 HSC**) Hard

45 A bag contains some marbles. The probability of selecting a blue marble at random from this bag is $\frac{3}{8}$.

Which of the following could describe the marbles that are in the bag?

A 3 blue, 8 red

B 6 blue, 11 red

C 3 blue, 4 red, 4 green

D 6 blue, 5 red, 5 green *(1 mark)*

(Q16, **2008 HSC**) Medium

46 A die has faces numbered 1 to 6. The die is biased so that the number 6 will appear more often than each of the other numbers. The numbers 1 to 5 are equally likely to occur.

The die was rolled 1200 times and it was noted that the 6 appeared 450 times.

Which statement is correct?

A The probability of rolling the number 5 is expected to be $\frac{1}{7}$.

B The number 6 is expected to appear 2 times as often as any other number.

C The number 6 is expected to appear 3 times as often as any other number.

D The probability of rolling an even number is expected to be equal to the probability of rolling an odd number. *(1 mark)*

(Q22, **2008 HSC**) **Medium**

47 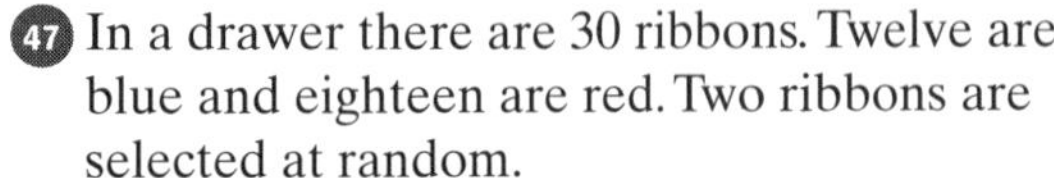

In a drawer there are 30 ribbons. Twelve are blue and eighteen are red. Two ribbons are selected at random.

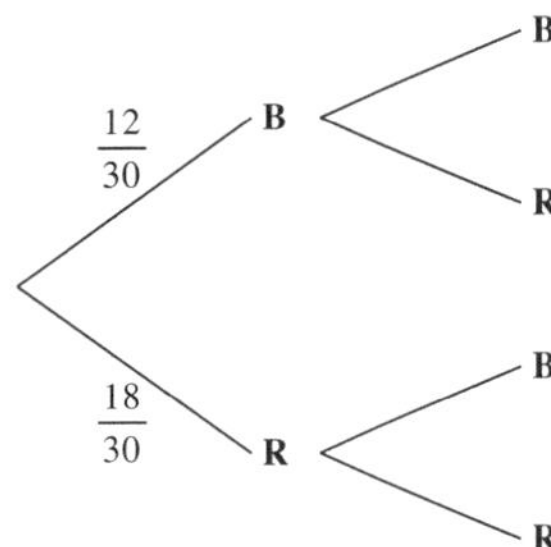

i Copy and complete the probability tree diagram. *(1 mark)* **Medium**

ii What is the probability of selecting a pair of ribbons which are the same colour? *(2 marks)* **Medium**

(Q25b, **2008 HSC**)

48 Cecil invited 175 movie critics to preview his new movie. After seeing the movie, he conducted a survey. Cecil has almost completed the two-way table.

	Aged <40	*Aged ≥40*	*Totals*
Movie critics who liked the movie	65		102
Movie critics who did not like the movie		31	
Totals		(A)	175

i Determine the value of (A). *(1 mark)* **Easy**

ii A movie critic is selected at random. What is the probability that the critic was less than 40 years old and did not like the movie? *(2 marks)* **Easy**

iii Cecil believes that his movie will be a box office success if 65% of the critics who were surveyed liked the movie. Will this movie be considered a box office success? Justify your answer. *(1 mark)* **Medium**

(Q26a, **2008 HSC**)

49 The retirement ages of two million people are displayed in a table.

Retirement age	*Number of people* (thousands)
36–40	5
41–45	10
46–50	20
51–55	35
56–60	180
61–65	700
66–70	500
71–75	400
76–80	150

i What is the relative frequency of the 51–55 year retirement age? *(1 mark)* **Easy**

ii Describe the distribution. *(1 mark)* **Medium**

(Q26b, **2008 HSC**)

50 Each student in a class is given a packet of lollies. The teacher records the number of red lollies in each packet using a frequency table.

Number of red lollies in each packet	*Frequency*
0	2
1	4
2	2
3	7
4	3
5	1

What is the relative frequency of a packet of lollies containing more than three red lollies?

A $\frac{4}{19}$ **B** $\frac{4}{15}$

C $\frac{11}{19}$ **D** $\frac{11}{15}$ *(1 mark)*

(Q2, **2007 HSC**) **Medium**

51 Each time she throws a dart, the probability that Mary hits the dartboard is $\frac{2}{7}$.

She throws two darts, one after the other.

What is the probability that she hits the dartboard with both darts?

A $\frac{1}{21}$ **B** $\frac{4}{49}$

C $\frac{2}{7}$ **D** $\frac{4}{7}$ *(1 mark)*

(Q10, **2007 HSC**) **Medium**

52 Give an example of an event that has a probability of exactly $\frac{3}{4}$. *(1 mark)*

(Q25a, **2007 HSC**) **Easy**

53 In a stack of 10 DVDs, there are 5 rated **PG**, 3 rated **G** and 2 rated **M**.

i A DVD is selected at random. What is the probability that it is rated **M**? *(1 mark)* **Easy**

ii Grant chooses two DVDs at random from the stack. Copy or trace the tree diagram into your writing booklet. Complete your tree diagram by writing the correct probability on each branch. *(2 marks)* **Medium**

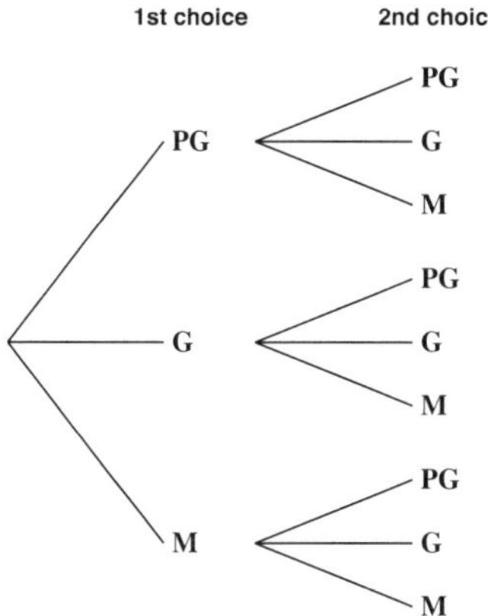

iii Calculate the probability that Grant chooses two DVDs with the same rating. *(2 marks)* **Medium**

(Q25c, **2007 HSC**)

54 Two unbiased dice, A and B, with faces numbered 1, 2, 3, 4, 5 and 6 are rolled. The numbers on the uppermost faces are noted. This table shows all the possible outcomes.

Die *A*

Die *B*	*1*	*2*	*3*	*4*	*5*	*6*
1	1,1	1,2	1,3	1,4	1,5	1,6
2	2,1	2,2	2,3	2,4	2,5	2,6
3	3,1	3,2	3,3	3,4	3,5	3,6
4	4,1	4,2	4,3	4,4	4,5	4,6
5	5,1	5,2	5,3	5,4	5,5	5,6
6	6,1	6,2	6,3	6,4	6,5	6,6

A game is played where the difference between the highest number showing and the lowest number showing on the uppermost faces is calculated.

What is the probability that the difference between the numbers showing on the uppermost faces of the two dice is one? *(1 mark)* **Medium**

(Q28a i, **2007 HSC**)

55 The probability of an event occurring is $\frac{9}{10}$.

Which statement best describes the probability of this event occurring?

A The event is likely to occur.
B The event is certain to occur.
C The event is unlikely to occur.
D The event has an even chance of occurring. *(1 mark)*

(Q1, **2006 HSC**) **Easy**

56 Marcella is planning to roll a standard six-sided die 60 times.

How many times would she expect to roll the number 4?

A 6 **B** 10
C 15 **D** 20 *(1 mark)*

(Q6, **2006 HSC**) **Easy**

57 Kay randomly selected a marble from a bag of marbles, recorded its colour and returned it to the bag. She repeated this process a number of times.

Colour	*Tally*	*Frequency*
Red	𝍸 ||	7
Blue	|||	3
Yellow	||	2
Green	||||	4
Purple	𝍸 |||	8

Based on these results, what is the best estimate of the probability that Kay will choose a green marble on her next selection?

A $\frac{5}{24}$ **B** $\frac{1}{24}$
C $\frac{1}{6}$ **D** $\frac{1}{5}$ *(1 mark)*

(Q10, **2006 HSC**) **Easy**

58 Sonia buys three raffle tickets.

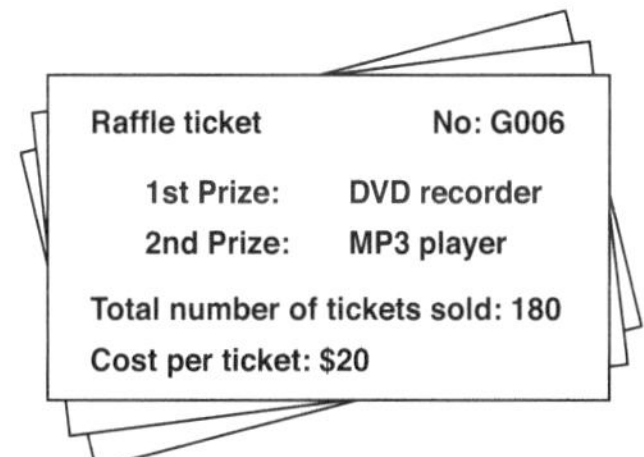
Raffle ticket No: G006
1st Prize: DVD recorder
2nd Prize: MP3 player
Total number of tickets sold: 180
Cost per ticket: $20

i What is the probability that Sonia wins first prize? *(1 mark)* **Easy**
ii What is the probability that she wins both prizes? *(2 marks)* **Medium**

(Q25c, **2006 HSC**)

59 A new test has been developed for determining whether or not people are carriers of the Gaussian virus.

Two hundred people are tested. A two-way table is being used to record the results.

	Positive	*Negative*
Carrier	74	12
Not a carrier	16	***A***

i What is the value of A? *(1 mark)* **Easy**
ii A person selected from the tested group is a carrier of the virus.

What is the probability that the test results would show this? *(2 marks)* **Medium**

iii For how many of the people tested were their test results inaccurate? *(1 mark)* **Medium**

(Q26c, **2006 HSC**)

60 On a bridge, the toll of \$2.50 is paid in coins collected by a machine. The machine only accepts two-dollar coins, one-dollar coins and fifty-cent coins.

i List the different combinations of coins that could be used to pay the \$2.50 toll. *(1 mark)* **Easy**

ii Jill has three two-dollar coins, six one-dollar coins and two fifty-cent coins. She selects two coins at random. What is the probability that she selects exactly \$2.50? *(3 marks)* **Hard**

(Q28a, **2006 HSC**)

61 Four radio stations reported the probability of rain as shown in the table.

Radio station	*Probability of rain*
2AT	0.53
2BW	17%
2CZ	$\frac{13}{25}$
2DL	0.6

Which radio station reported the highest probability of rain?

A 2AT **B** 2BW

C 2CZ **D** 2DL *(1 mark)*

(Q3, **2005 HSC**) **Easy**

62 The diagram shows a spinner.

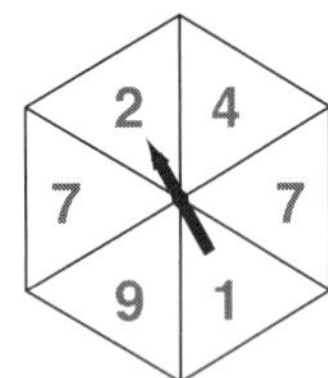

The arrow is spun and will stop in one of the six sections.

What is the probability that the arrow will stop in a section containing a number greater than 4?

A $\frac{2}{5}$ **B** $\frac{2}{3}$

C $\frac{1}{3}$ **D** $\frac{1}{2}$ *(1 mark)*

(Q11, **2005 HSC**) **Easy**

63 On a television game show, viewers voted for their favourite contestant. The results were recorded in the two-way table.

	Male viewers	*Female viewers*
Contestant 1	1372	3915
Contestant 2	2054	3269

One male viewer was selected at random from all of the male viewers.

What is the probability that he voted for Contestant 1?

A $\frac{1372}{10\,610}$ **B** $\frac{1372}{5287}$

C $\frac{1372}{3426}$ **D** $\frac{1372}{2054}$ *(1 mark)*

(Q16, **2005 HSC**) **Medium**

64 There are 100 tickets sold in a raffle. Justine sold all 100 tickets to five of her friends. The number of tickets she sold to each friend is shown in the table.

Friend	*Number of tickets*
Danielle	45
Khalid	5
Nancy	10
Shani	14
Herman	26
Total	100

i Justine claims that each of her friends is equally likely to win first prize.
Give a reason why Justine's statement is NOT correct. *(1 mark)* **Easy**

ii What is the probability that first prize is NOT won by Khalid or Herman? *(2 marks)* **Medium**

(Q23a, **2005 HSC**)

65 Moheb owns five red and seven blue ties. He chooses a tie at random for himself and puts it on. He then chooses another tie at random, from the remaining ties, and gives it to his brother.

i What is the probability that Moheb chooses a red tie for himself? *(1 mark)* **Easy**

ii Copy the tree diagram into your writing booklet.
Complete your tree diagram by writing the correct probability on each branch.

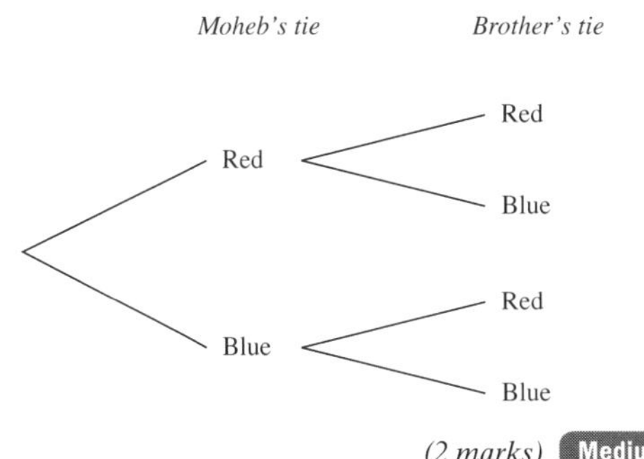

(2 marks) **Medium**

iii Calculate the probability that both of the ties are the same colour. *(2 marks)* **Medium**

(Q23c, **2005 HSC**)

66 Which fraction is equal to a probability of 25%?

A $\frac{1}{25}$ **B** $\frac{1}{4}$

C $\frac{1}{3}$ **D** $\frac{1}{2}$ *(1 mark)*

(Q1, **2004 HSC**) **Easy**

67 Two dice are rolled. What is the probability that only one of the dice shows a six?

A $\frac{5}{36}$ **B** $\frac{1}{6}$

C $\frac{5}{18}$ **D** $\frac{11}{36}$ *(1 mark)*

(Q18, **2004 HSC**) **Hard**

68 Lie detector tests are not always accurate. A lie detector test was administered to 200 people.
The results were:

- 50 people lied. Of these, the test indicated that 40 had lied;
- 150 people did NOT lie. Of these, the test indicated that 20 had lied.

i Copy the table into your writing booklet and complete it using the information above.

	Test indicated a lie	*Test did not indicate a lie*	*Total*
People who lied			50
People who did NOT lie			150

(2 marks) **Easy**

ii For how many of the people tested was the lie detector test accurate? *(1 mark)* **Easy**

iii For what percentage of the people tested was the test accurate? *(1 mark)* **Easy**

iv What is the probability that the test indicated a lie for a person who did NOT lie? *(1 mark)* **Medium**

(Q25c, **2004 HSC**)

69 From 5 boys and 7 girls, two children will be chosen at random to work together on a project.

Which of the following probability trees could be used to determine the probability of choosing a boy and a girl?

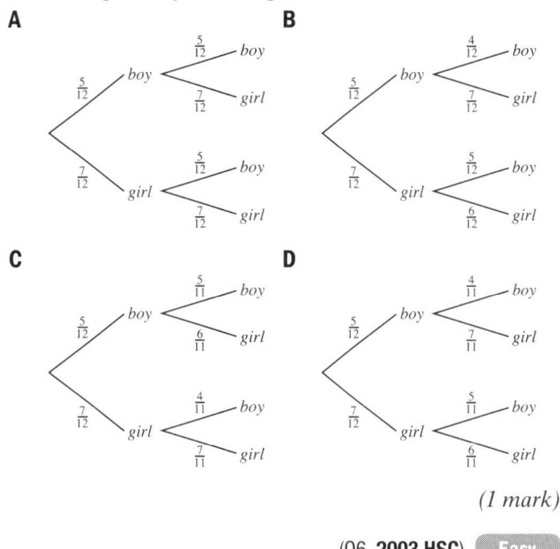

(1 mark)

(Q6, **2003 HSC**) **Easy**

70 Joy asked the students in her class how many brothers they had. The answers were recorded in a frequency table as follows.

Number of brothers	*Frequency*
0	5
1	10
2	3
3	1
4	1

One of the students is chosen at random.
What is the probability that this student has at least two brothers?

A 0.10 **B** 0.15

C 0.25 **D** 0.75 *(1 mark)*

(Q13, **2003 HSC**) **Easy**

71 A celebrity mathematician, Karl, arrives in Sydney for one of his frequent visits. Karl is known to stay at one of three Sydney hotels.

Hotel X is his favourite, and he stays there on 50% of his visits to Sydney. When he does not stay at Hotel X, he is equally likely to stay at Hotels Y or Z.

i What is the probability that he will stay at Hotel Z? *(2 marks)* **Easy**

ii On his first morning in Sydney, Karl always flips a coin to decide if he will have a cold breakfast or a hot breakfast. If the coin comes up heads he has a cold breakfast. If the coin comes up tails he has a hot breakfast.

1 List all the possible combinations of hotel and breakfast choices. *(2 marks)* **Medium**

2 Give a brief reason why these combinations are not all equally likely. *(1 mark)* **Medium**

3 Calculate the probability that Karl stays at Hotel Z and has a cold breakfast. *(1 mark)* **Medium**

(Q27a, **2003 HSC**)

72 Sarah has two packets of jelly beans. Each packet contains one black and five yellow jelly beans. Sarah takes one jelly bean from each packet without looking.

What is the probability that both of the jelly beans are black?

A $\frac{1}{36}$ **B** $\frac{1}{12}$

C $\frac{1}{6}$ **D** $\frac{1}{3}$ *(1 mark)*

(Q5, **2002 HSC**) **Medium**

73 Rob, Alex and Tan plan a swimming race against each other. Rob and Alex are each twice as likely as Tan to win the race. What is the probability that Tan will win the race?

A $\frac{1}{6}$ **B** $\frac{1}{5}$

C $\frac{1}{4}$ **D** $\frac{1}{3}$ *(1 mark)*

(Q20, **2002 HSC**) **Hard**

74 There is one seat at the back of the excursion bus that is very popular among the students.

Before the excursion, a draw is conducted to determine who will sit in the popular seat. The names of the 12 students are placed in a hat and 3 names are drawn without replacement. The first name drawn determines who will sit in the seat on the first day. The second name drawn determines who will sit in the seat on the second day. The third name drawn determines who will sit in the seat on the third day.

i What is the probability that Jane's name is the first drawn? *(1 mark)* **Medium**

ii What is the probability that Jane's name is the second drawn? *(1 mark)* **Medium**

iii What is the probability that Jane's name will NOT be one of the three names drawn from the hat? *(2 marks)* **Hard**

(Q24c, **2002 HSC**)

75 The arrow is spun and will point to one of the four colours when it stops.

If the arrow is spun twice, what is the probability that it points to the same colour both times it stops?

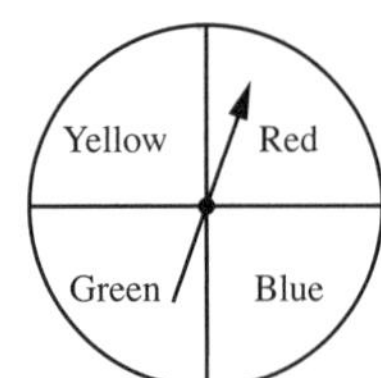

A $\frac{1}{16}$ **B** $\frac{1}{8}$

C $\frac{1}{4}$ **D** $\frac{1}{2}$ *(1 mark)*

(Q11, **2001 HSC**) **Medium**

76 Sonia has written letters to four of her friends and sealed the letters in envelopes. Now she does not know which envelope contains which letter.

If Sonia addresses the envelopes to her four friends at random, what is the probability that each envelope contains the correct letter?

A $\frac{1}{256}$ **B** $\frac{1}{24}$

C $\frac{1}{16}$ **D** $\frac{1}{4}$ *(1 mark)*

(Q22, **2001 HSC**) **Hard**

77 Five men and three women are living on an island, but not all will be able to stay.

i If one person is selected at random, what is the probability that this person is female? *(1 mark)* **Easy**

ii Two people are to be randomly selected to leave the island.

1 Copy the tree diagram into your writing booklet, and complete the diagram by writing the probabilities on all the branches.

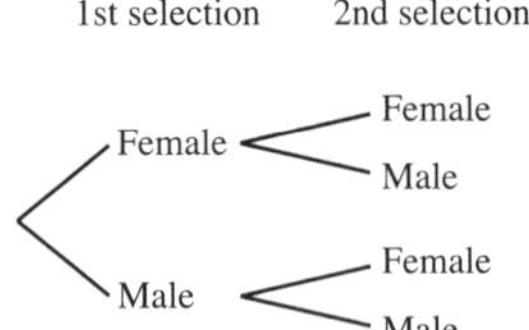

(2 marks) **Medium**

2 Calculate the probability that the selection includes exactly one female. *(2 marks)* **Medium**

iii Antoinette is one of the women on the island. Before the two people are randomly selected to leave, Antoinette calculates her chance of remaining on the island. She concludes that she has a good chance of remaining.

Do you agree? Justify your answer. *(2 marks)* **Hard**

(Q25a, **2001 HSC**)

Year 11 Relative frequency and probability—Worked answers

1 Once Kim has eaten a chocolate only 7 remain.

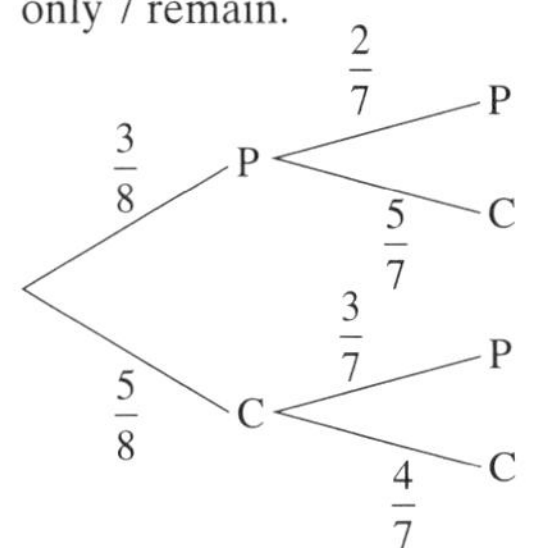

Probability of different centres
$= P(PC) + P(CP)$
$= \frac{3}{8} \times \frac{5}{7} + \frac{5}{8} \times \frac{3}{7}$
$= \frac{15}{28}$
Answer D

2 Ways of getting a score of 6:

- rolling 6 in section 1
- rolling 3 in section 2
- rolling 2 in section 3
- rolling 1 in section 6.

So required probability
$= 4 \times \frac{1}{6} \times \frac{1}{8}$
$= \frac{1}{12}$
Answer B

3 $P(\text{spinning } 8) = \frac{1}{37}$ ✓
Expected frequency of 8
$= \frac{1}{37} \times 18\,500$
$= 500$ ✓
(2 marks)

4 $P(\text{same colour})$
$= P(RR) + P(GG)$ ✓
$= \frac{9}{17} \times \frac{8}{16} + \frac{8}{17} \times \frac{7}{16}$
$= \frac{8}{17}$ ✓

Dennis, Margaret
$\frac{9}{17}$ R: $\frac{8}{16}$ R, $\frac{8}{16}$ G
$\frac{8}{17}$ G: $\frac{9}{16}$ R, $\frac{7}{16}$ G ✓
(3 marks)

5 Outcome C occurs $(100 - (50 + 23))\%$ or 27% of the time.
Expected number $= 27\%$ of 500
$= 135$
Answer B

6 Required probability
$= \frac{71}{223} \times 100\%$
$= 31.838\,5650\ldots\%$
$= 32\%$ (nearest per cent)
Answer C

7 Total frequency
$= 23 + 19 + 48 + 20 + 21 + 19$
$= 150$
Relative frequency of 3
$= \frac{48}{150}$
$= \frac{8}{25}$ (or 0.32)
(1 mark)

8 i

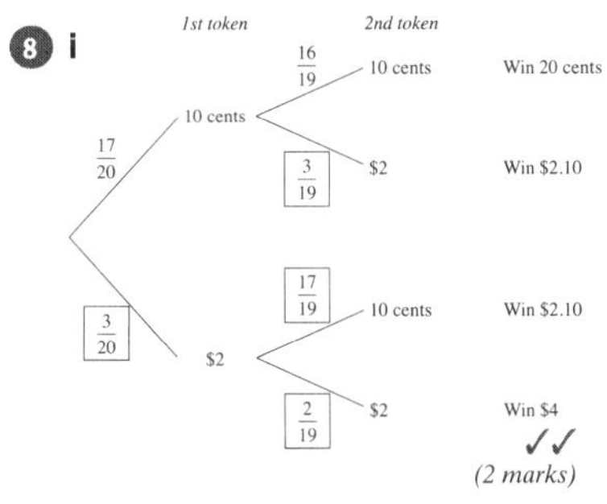

✓✓
(2 marks)

9 Expected number
$= \frac{180}{200} \times 75\,000$
$= 67\,500$
Answer B

10 The amounts more than 50c but less than 80c are \$0.55, \$0.60, \$0.65, \$0.70 and \$0.75.
$P(\text{between 50c and 80c}) = \frac{5}{20}$
$= \frac{1}{4}$
Answer A

11 $P(\text{exactly one picture card})$
$= P(\text{picture, not picture}) + P(\text{not picture, picture})$
$= \frac{12}{52} \times \frac{40}{51} + \frac{40}{52} \times \frac{12}{51}$
$= \frac{80}{221}$
$= 0.361\,990\,95\ldots$
$= 0.3620$ (4 d.p.)

$\frac{12}{52}$ picture card: $\frac{11}{51}$ picture card, $\frac{40}{51}$ not a picture card
$\frac{40}{52}$ not a picture card: $\frac{12}{51}$ picture card, $\frac{39}{51}$ not a picture card

Answer D

12 i Total surveyed
$= 150 + 2500 + 70 + 800$
$= 3520$ ✓
$P(\text{lives in city and never waterskied})$
$= \frac{2500}{3520}$
$= \frac{125}{176}$ ✓
(2 marks)

ii Number living in the city
$= 150 + 2500$
$= 2650$
$P(\text{city waterskier})$
$= \frac{150}{2650}$
$= 5.660\,3773\ldots\%$ ✓
Number living in the country
$= 70 + 800$
$= 870$
$P(\text{country waterskier})$
$= \frac{70}{870}$
$= 8.045\,977\ldots\%$
So, based on these figures, a person living in the country is slightly more likely to have waterskied than someone living in the city. ✓
(2 marks)

13 $P(\text{smoker or male})$
$= 1 - P(\text{female non-smoker})$
$= 1 - \frac{153}{485}$
$= \frac{332}{485} \times 100\%$
$= 68.4536\ldots\%$
$= 68\%$ (nearest per cent)
Answer C

14 i $P(\text{win, draw})$
$= P(WD) + P(DW)$ ✓
$= 0.7 \times 0.2 + 0.2 \times 0.7$
$= 0.28$ ✓
(2 marks)

ii The probability of an event cannot be greater than 1.
[The actual probability is $0.7 \times 0.7 = 0.49$.]
(1 mark)

15 Probability $= \frac{7}{10} \times \frac{7}{10}$
Answer D

16 **i** Total of given frequencies
$= 0.32 + 0.13 + 0.14 + 0.24$
$= 0.83$
Relative frequency of red
$= 1 - 0.83$
$= 0.17$
(1 mark)

ii $P(\text{black jelly bean}) = 0.14$
$P(\text{not choosing black})$
$= 1 - 0.14$
$= 0.86$
(1 mark)

17

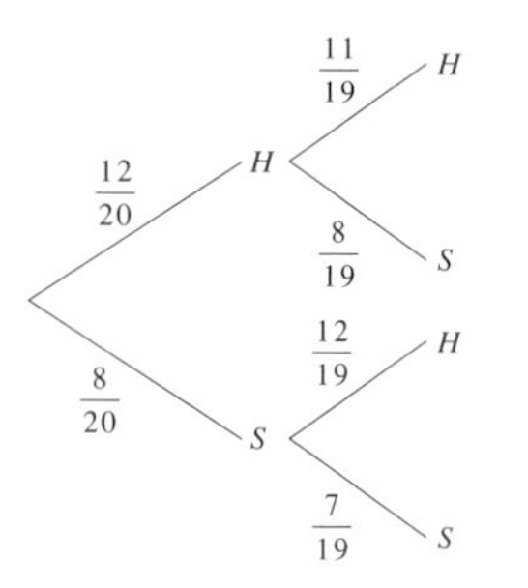

$P(\text{one of each type})$
$= P(HS) + P(SH)$ ✓
$= \frac{12}{20} \times \frac{8}{19} + \frac{8}{20} \times \frac{12}{19}$
$= \frac{48}{95}$ ✓
(3 marks)

18 There are 28 males who do not own a mobile. There are 150 people in the group.
$\text{P(male, no mobile)} = \frac{28}{150}$
Answer A

19 Based on averages, the probability of rain on any given day in October is $\frac{8}{31}$.
Assuming the events independent, the probability of rain on two consecutive days could be given by $\frac{8}{31} \times \frac{8}{31}$.
Answer D

20 If Jaz chooses a bag at random then $P(\text{bag } A) = P(\text{bag } B) = \frac{1}{2}$.
In bag A, 4 of the 7 apples are red and 3 are green.
So, for bag A, $P(\text{red}) = \frac{4}{7}$ and $P(\text{green}) = \frac{3}{7}$
In bag B, 3 of the 4 apples are red and 1 is green.
So, for bag B, $P(\text{red}) = \frac{3}{4}$ and $P(\text{green}) = \frac{1}{4}$

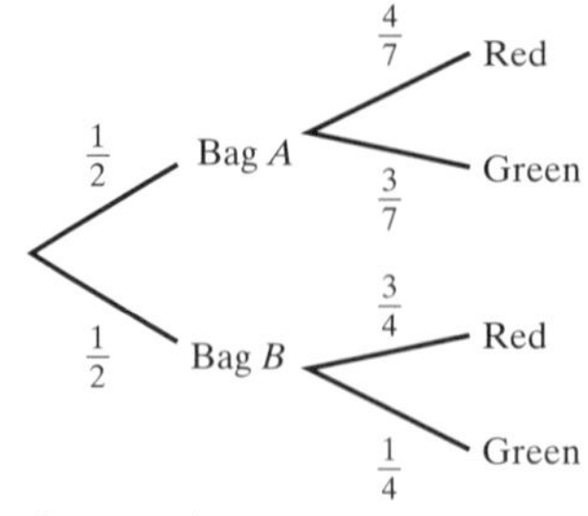

Answer A

21

1st toss	2nd toss	3rd toss	Sample space
H	H	H	HHH
		T	HHT
	T	H	HTH
		T	HTT
T	H	H	THH
		T	THT
	T	H	TTH
		T	TTT

✓

$P(\text{two heads, one tail}) = \frac{3}{8}$ ✓
(2 marks)

22 $P(\text{head on coin}) = \frac{1}{2}$
$P(3 \text{ on a die}) = \frac{1}{6}$
$P(\text{G from alphabet}) = \frac{1}{26}$
$P(\text{first prize}) = \frac{4}{100} = \frac{1}{25}$
The smallest probability is $\frac{1}{26}$.
The event least likely to occur is randomly selecting G from the alphabet.
Answer C

23 When rolling a die, each number is equally likely to occur.
$P(\text{a certain number}) = \frac{1}{6}$
In 72 rolls, expected frequency of a certain number $= \frac{1}{6} \times 72$
$= 12$
From the table the number that occurs 12 times is 2.
Answer B

24 $P(\text{male living with parents})$
$= \frac{155}{505} \times 100\%$
$= 30.693\,069\ldots\ \%$
$= 31\%$ [nearest per cent]
Answer A

25 Number of outcomes $= 6 \times 6$
$= 36$
Outcomes of a sum of 6 are $1 + 5$, $2 + 4$, $3 + 3$, $4 + 2$, and $5 + 1$
So there are 5 favourable outcomes.
$P(\text{sum of } 6) = \frac{5}{36}$
Answer D

26 **i** No, the commentator is not correct.
He or she did not take into account the probability that Michael will score exactly 100. The statement would be correct if the commentator had said the probability that Michael scored 100 points or less is $\frac{9}{40}$.
(1 mark)

ii $P(\text{more than 100 points in first game and second game})$
$= \frac{31}{40} \times \frac{31}{40}$
$= \frac{961}{1600}$
(1 mark)

27 **i**

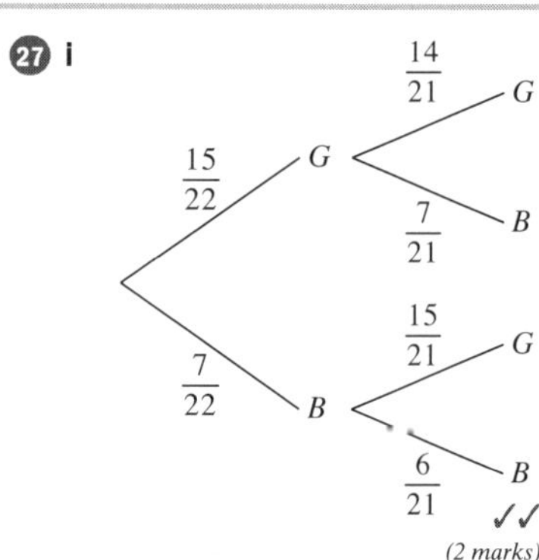

(2 marks)

ii $P(\text{same gender})$
$= P(GG) + P(BB)$ ✓
$= \frac{15}{22} \times \frac{14}{21} + \frac{7}{22} \times \frac{6}{21}$
$= \frac{6}{11}$ ✓
(2 marks)

28 On one die: ✓
$P(\text{rolling a } 6) = \frac{1}{6}$
$P(\text{not rolling a } 6) = \frac{5}{6}$
On two dice:
$P(\text{neither showing } 6) = \frac{5}{6} \times \frac{5}{6}$
$= \frac{25}{36}$
$P(\text{at least one } 6) = 1 - \frac{25}{36}$
$= \frac{11}{36}$
Answer B

29 Number of spins = 40

Total frequencies shown in table
= 2 + 4 + 6 + 10 + 12
= 34

Frequency for orange = 40 − 34
= 6

Relative frequency of orange
$= \frac{6}{40}$
$= \frac{3}{20}$

Answer A

30

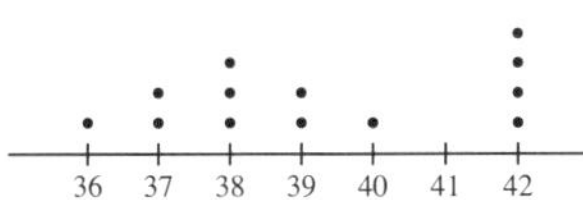

i 7 members can do more than 38 push-ups in one minute.

P(more than 38 push-ups)
$= \frac{7}{13}$

(1 mark)

ii Yes, the new member does change the probability. There are now 14 members.

P(more than 38 push-ups)
$= \frac{7}{14}$
$= \frac{1}{2}$

(1 mark)

31 **i**

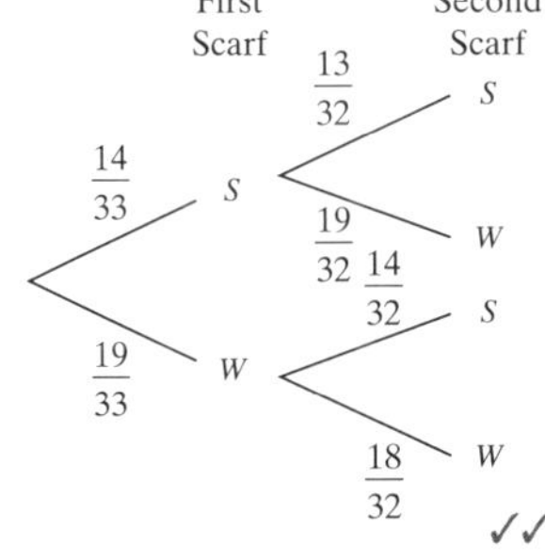

✓✓

(2 marks)

ii $P(SS) = \frac{14}{33} \times \frac{13}{32}$
$= \frac{91}{528}$

(1 mark)

iii P(different fabrics)
$= P(SW) + P(WS)$ ✓
$= \frac{14}{33} \times \frac{19}{32} + \frac{19}{33} \times \frac{14}{32}$
$= \frac{133}{264}$ ✓

(2 marks)

32 $0 \leq P(E) \leq 1$

Options B, C and D are all greater than 1.

Only option A could be the probability of an event occurring.

Answer A

33 If the coin is unbiased $P(\text{tail}) = \frac{1}{2}$.

So, for the 10th toss, $P(\text{tail}) = \frac{1}{2}$.

Answer B

34 **i** Total of the given frequencies
= 16 + 11 + 8 + 12 + 15
= 62

Total frequency = 72

So A = 72 − 62
= 10

(1 mark)

ii Relative frequency of 4 $= \frac{8}{72}$
$= \frac{1}{9}$

(1 mark)

iii If the die is unbiased, the probability of each number is $\frac{1}{6}$.

Expected number of times each number occurs
$= \frac{1}{6} \times 72$
= 12

So 5 was the number that occurred the expected number of times.

(1 mark)

35 **i** Total students = 319 + 261
= 580

[Or 330 + 250 = 580]

(1 mark)

ii P(female uses pre-paid) $= \frac{172}{319}$

(1 mark)

iii New number of males on a plan
= 103 + 10
= 113

New total number of males
= 261 + 10
= 271

Percentage of males on plan
$= \frac{113}{271} \times 100\%$
= 41.697 416 97…%
= 42% [nearest per cent]

(1 mark)

36

		Spinner B			
		1	1	2	3
Spinner A	1	2	2	3	4
	1	2	2	3	4
	3	4	4	X	6

i $X = 3 + 2$
$= 5$

(1 mark)

ii P(score less than 4) $= \frac{6}{12}$
$= \frac{1}{2}$

(1 mark)

iii To obtain a score of 3 when a 2 is obtained on spinner B, a 1 needs to be obtained on spinner A.

P(1 on Spinner A) $= \frac{2}{3}$

(1 mark)

37 Probability (red, green or blue)
$= \frac{1}{3} + \frac{1}{4} + \frac{1}{6}$
$= \frac{3}{4}$

Probability (yellow) $= 1 - \frac{3}{4}$
$= \frac{1}{4}$

So $\frac{1}{4}$ of the balls are yellow.

There are 12 yellow balls.

Total number of balls = 4 × 12
= 48

Answer C

38 Number of accurate tests
= 72 + 256
= 328

Probability (accurate test)
$= \frac{328}{347} \times 100\%$
= 94.524 4956…%
= 95% [nearest per cent]

Answer C

39 P(only one successful)
= P(Lou is successful, Ali is not)
+ P(Lou is not successful, Ali is)
= 0.7 × 0.4 + 0.3 × 0.6
= 0.28 + 0.18
= 0.46

Answer D

40 i Number of people
$= 5 + 15 + 10 + 3 + 1$
$= 34$

There were 34 people at the gym on Saturday. *(1 mark)*

ii Number wearing blue or green
$= 5 + 10$
$= 15$

Probability (blue or green)
$= \frac{15}{34}$ *(1 mark)*

41 "Most probably" implies very likely, therefore 80% best represents the probability.
Answer C

42 There are 5 numbers less than 6. The probability that the marker stops on a number less than 6 is $\frac{5}{20} = \frac{1}{4}$.

∴ If wheel is spun 120 times, you would expect it to land on a number less than 6,
$\frac{1}{4} \times 120 = 30$ times.

Answer C

43

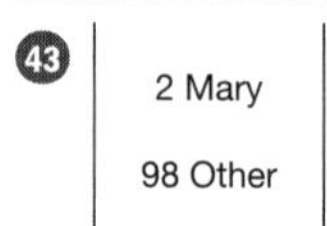

P(Mary wins a prize) $= \frac{2}{100}$
$= \frac{1}{50}$
$= 0.02$ ✓

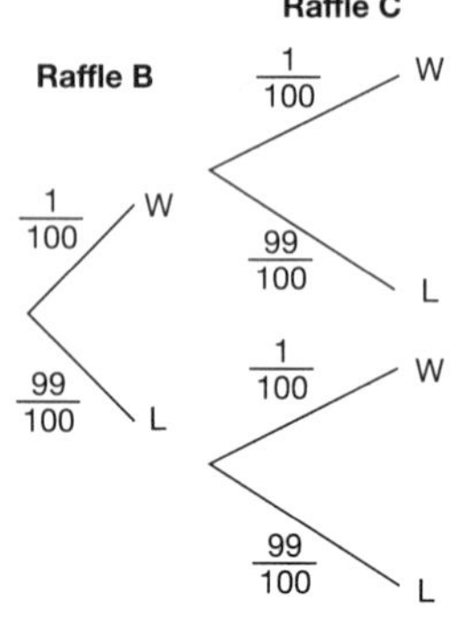

✓

P(Jane wins at least 1 prize)
= P(wins B, not C) or P(not B, wins C) or P(wins both)

$= \frac{1}{100} \times \frac{99}{100} + \frac{99}{100} \times \frac{1}{100} + \frac{1}{100} \times \frac{1}{100}$
$= 0.0199$ ✓

∴ Mary has a better chance of winning a prize by buying 2 tickets in the one raffle. ✓

(4 marks)

44 The sample space for the dice differences is:

		Die 1					
		1	2	3	4	5	6
Die 2	1	0	1	2	3	4	5
	2	1	0	1	2	3	4
	3	2	1	0	1	2	3
	4	3	2	1	0	1	2
	5	4	3	2	1	0	1
	6	5	4	3	2	1	0

✓

Theoretically, the results should be

Difference	*Frequency*
0	6
1	10
2	8
3	6
4	4
5	2
	36

✓

If the dice are rolled 18 times, you would expect half these frequencies:

Difference	*Frequency*
0	3
1	5
2	4
3	3
4	2
5	1

✓

∴ Experiment 2 has given results closer to what is expected and Juan is correct. ✓ *(4 marks)*

45 $P(B) = \frac{3}{8}$

(A): $P(B) = \frac{3}{11}$

(B): $P(B) = \frac{6}{17}$

(C): $P(B) = \frac{3}{11}$

(D): $P(B) = \frac{6}{16} = \frac{3}{8}$

Answer D

46 Experimental probability of rolling a 6 is $\frac{450}{1200}$.

∴ Other numbers appear 750 times.

Numbers 1–5 are equally likely, so they should appear 150 times each.

∴ The number 6 is expected to appear 3 times ($3 \times 150 = 450$) as often.

Answer C

47 i

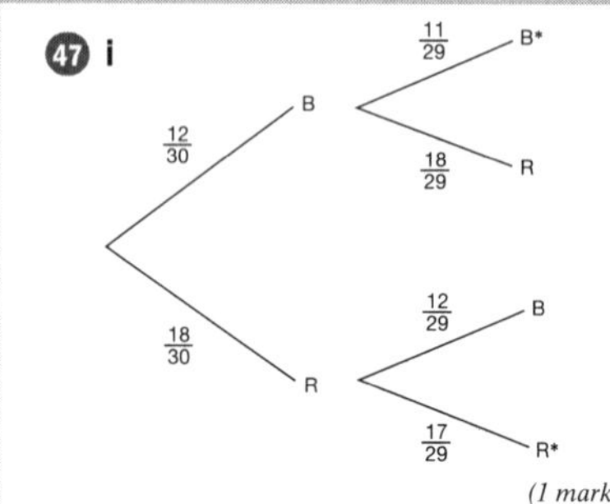

(1 mark)

ii P(Same colour)
= P(RR) or P(BB) ✓
$= \frac{18}{30} \times \frac{17}{29} + \frac{12}{30} \times \frac{11}{29}$
$= \frac{73}{145}$ ✓

(2 marks)

48 i

	Aged <40	*Aged ≥40*	*Totals*
Like	65	102 − 65 = 37	102
Dislike	73 − 31 = 42	31	175 − 102 = 73
Total	107	A = 37 + 31 = 68	175

$A = 68$

(1 mark)

ii $\frac{42}{175} = \frac{6}{25}$ ✓✓

(2 marks)

iii Percentage of critics liking movie
$= \frac{102}{175} \times 100$
$\doteqdot 58\%$

As this is less than 65%, the movie will not be a box office success.

(1 mark)

49 i Relative frequency
$= \frac{\text{frequency}}{\text{total frequency}}$
$= \frac{35\,000}{2\,000\,000}$
$= 0.0175$

(1 mark)

ii The distribution is negatively skewed as the frequency grows considerably for the higher scores.

(1 mark)

50 Relative frequency

$= \dfrac{\text{frequency of more than 3 red}}{\text{total frequency}}$

$= \dfrac{4}{19}$

Answer A

51 Using a tree diagram:

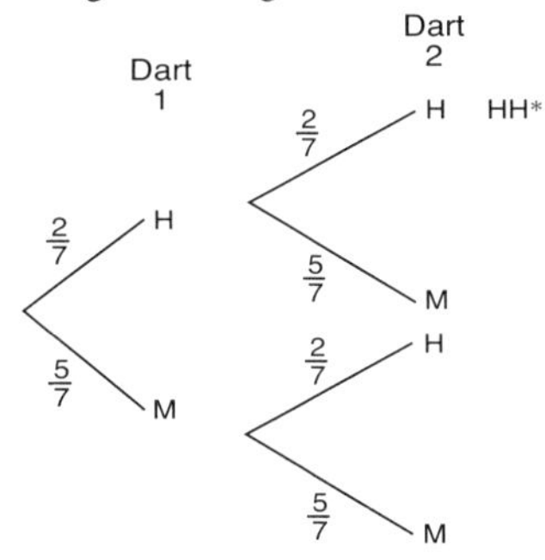

$P(HH) = \dfrac{2}{7} \times \dfrac{2}{7}$

$= \dfrac{4}{49}$

Answer B

52 In a bag of marbles, there are 3 red marbles and 1 white marble. The probability of choosing a red marble is exactly $\frac{3}{4}$.

(1 mark)

53 **i** $P(M) = \dfrac{2}{10} = \dfrac{1}{5}$

(1 mark)

ii

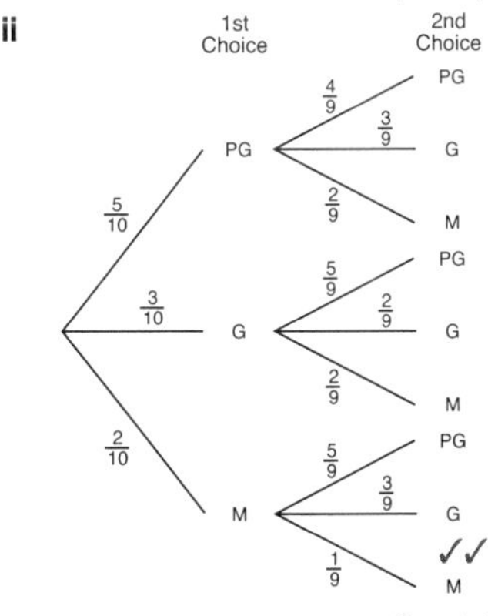

(2 marks)

iii P(2 DVDs with same rating)

= P(PG, PG) + P(G, G) + P(M, M) ✓

$= \dfrac{5}{10} \times \dfrac{4}{9} + \dfrac{3}{10} \times \dfrac{2}{9} + \dfrac{2}{10} \times \dfrac{1}{9}$

$= \dfrac{28}{90}$

$= \dfrac{14}{45}$ ✓

(2 marks)

54 P(Difference of 1)

$= \dfrac{10}{36}$ (from table)

$= \dfrac{5}{18}$

(1 mark)

55 Now $\dfrac{9}{10} > \dfrac{1}{2}$,

but $\dfrac{9}{10} \neq 1$

$\therefore$ not certain but very likely.

Answer A

56 Expected number

= no. of rolls × probability of 4

$= 60 \times \dfrac{1}{6}$

$= 10$

Answer B

57 $P(\text{Choosing green}) = \dfrac{4}{24}$

$= \dfrac{1}{6}$

Answer C

58 **i** $P(\text{first prize}) = \dfrac{3}{180}$

$= \dfrac{1}{60}$

(1 mark)

ii $P(\text{both prizes}) = \dfrac{3}{180} \times \dfrac{2}{179}$

$= \dfrac{1}{5370}$ ✓✓

(2 marks)

59 **i** $A = 200 - 74 - 12 - 16$

$= 98$

(1 mark)

ii From the table:

No. of carriers = 74 + 12 = 86
The test only recorded positive for 74 of these. ✓

$\therefore P(\text{Test showing this}) = \dfrac{74}{86}$

$= \dfrac{37}{43}$ ✓

$(\doteqdot 0.8605)$

(2 marks)

iii The test was inaccurate for 16 + 12 = 28 people.

(1 mark)

60 **i** Possible combinations of coins:

\$2	\$1	\$1	50c
50c	\$1	50c	50c
	50c	50c	50c
		50c	50c
			50c

(1 mark)

ii

1st coin		2nd coin		Total
3/11	\$2	2/10	\$2	\$4.00
		6/10	\$1	\$3.00
		2/10	50c	\$2.50*
6/11	\$1	3/10	\$2	\$3.00
		5/10	\$1	\$2.00
		2/10	50c	\$1.50
2/11	50c	3/10	\$2	\$2.50*
		6/10	\$1	\$1.50
		1/10	50c	\$1.00 ✓

P(Choosing exactly \$2.50)

$= \dfrac{3}{11} \times \dfrac{2}{10} + \dfrac{2}{11} \times \dfrac{3}{10}$ ✓

$= \dfrac{6}{55}$ ✓

(3 marks)

61

Radio Station	*Probability*
2AT	0.53
2BW	17% = 0.17
2CZ	$\frac{13}{25} = 0.52$
2DL	0.6

$\therefore$ 2DL reported the highest probability.

Answer D

62 There are 6 numbers and 3 of these are greater than 4.

$\therefore P(\text{number greater than 4}) = \dfrac{3}{6}$

$= \dfrac{1}{2}$

Answer D

63 Total number of male viewers

= 1372 + 2054

= 3426

Number of males voting for contestant 1 = 1372

$\therefore P(\text{male voting for contestant 1})$

$= \dfrac{1372}{3426}$

Answer C

64 **i** Each friend did not buy the same number of tickets. As Danielle has the most tickets, she has the best chance of winning.

(1 mark)

ii P(Khalid or Herman do not win)
$= 1 - P(\text{They do win})$ ✓
$= 1 - \left(\frac{5}{100} + \frac{26}{100}\right)$
$= 1 - \frac{31}{100}$
$= \frac{69}{100}$ ✓

(2 marks)

65 **i** $P(\text{Red tie}) = \frac{5}{12}$

(1 mark)

ii

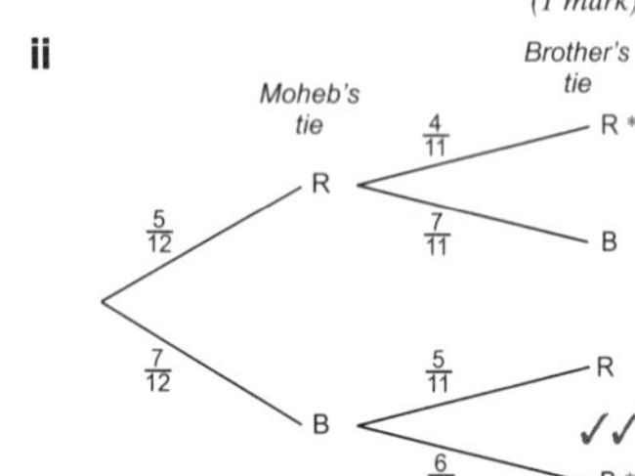

✓✓ *(2 marks)*

iii $P(RR)$ or $P(BB)$ ✓
$= \frac{5}{12} \times \frac{4}{11} + \frac{7}{12} \times \frac{6}{11}$
$= \frac{31}{66}$ ✓

(2 marks)

66 $25\% = \frac{25}{100}$
$= \frac{1}{4}$

Answer B

67 P(Only 1 die shows 6)
$= P(6, \tilde{6})$ or $P(\tilde{6}, 6)$
$= \frac{1}{6} \times \frac{5}{6} + \frac{5}{6} \times \frac{1}{6}$
$= \frac{10}{36}$
$= \frac{5}{18}$

Answer C

68 **i**

	Test indicated a lie	*Test did not indicate a lie*	*Total*
People who lied	40	10	50
People who did not lie	20	130	150

✓✓ *(2 marks)*

ii $130 + 40 = 170$

(1 mark)

iii $\frac{170}{200} \times 100 = 85\%$

(1 mark)

iv $\frac{20}{150} = \frac{2}{15}$

(1 mark)

69 After choosing 1 child, there are only 11 children left, so the correct answer could only be C or D.
If a boy was chosen first, there are only 4 boys left out of 11, hence only D can be correct.
Answer D

70

P(At least 2 brothers)
$= \frac{\text{No. of students with 2, 3 or 4 brothers}}{20}$
$= \frac{5}{20}$
$= 0.25$

Answer C

71 **i**

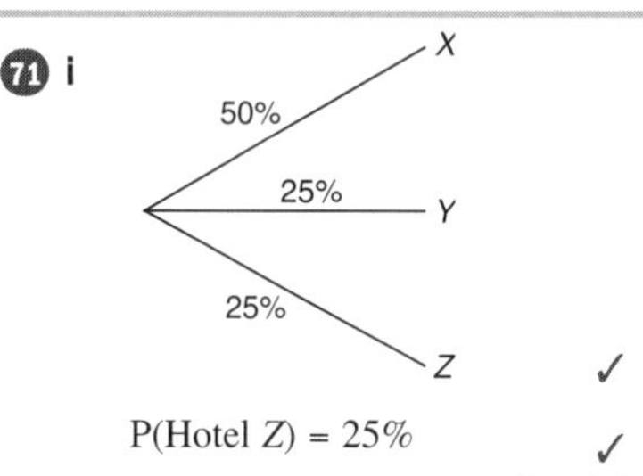

✓

P(Hotel Z) = 25% ✓

(2 marks)

ii **1**

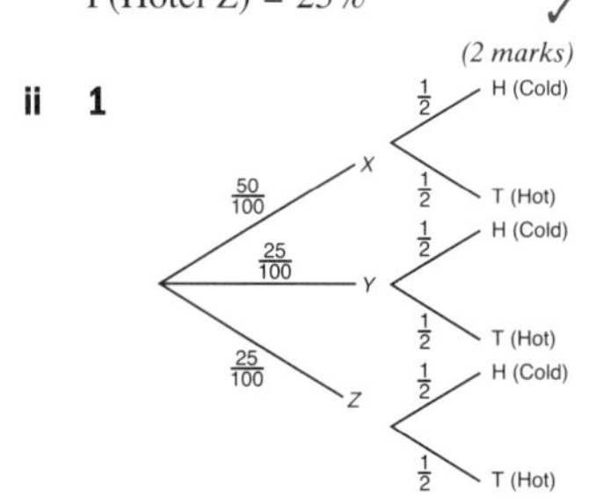

X Cold
X Hot
Y Cold
Y Hot
Z Cold ✓✓
Z Hot

(2 marks)

2 These combinations are not all equally likely as there is more chance of him staying in Hotel X than Hotels Y and Z.

(1 mark)

3
$P(X, \text{Cold}) = \frac{25}{100} \times \frac{1}{2} = \frac{1}{8}$

(1 mark)

72 $P(BB) = \frac{1}{6} \times \frac{1}{6}$
$= \frac{1}{36}$

Answer A

73 Let $P(\text{Tan wins}) = \frac{1}{x}$

$\therefore P(\text{Alex or Rob wins}) = \frac{2}{x}$

Someone must win
$\therefore \frac{1}{x} + \frac{2}{x} + \frac{2}{x} = 1$
$\frac{5}{x} = 1$
$\therefore x = 5$
$\therefore P(\text{Tan wins}) = \frac{1}{5}$

Answer B

74 **i** $P(\text{Jane on day 1}) = \frac{1}{12}$

(1 mark)

ii $P(\text{Jane on day 2}) = \frac{1}{12}$

(1 mark)

iii P(Jane not drawn)
$= \frac{11}{12} \times \frac{10}{11} \times \frac{9}{10}$
$= \frac{3}{4}$

(2 marks)

OR by using a tree diagram

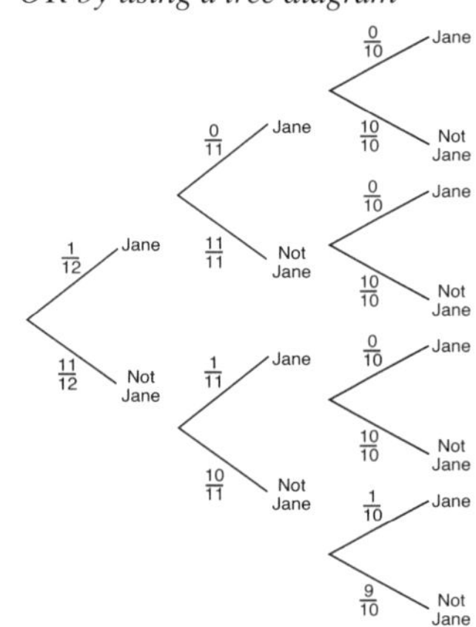

i $P(\text{Jane on day 1}) = \frac{1}{12}$

(1 mark)

ii P(Jane on day 2)
$= P$(Not Jane, Jane)
$= \frac{11}{12} \times \frac{1}{11}$
$= \frac{1}{12}$

(1 mark)

iii P(Jane not drawn)
$= P$(Not Jane, Not Jane, Not Jane) ✓
$= \frac{11}{12} \times \frac{10}{11} \times \frac{9}{10}$
$= \frac{9}{12}$
$= \frac{3}{4}$ ✓

(2 marks)

75

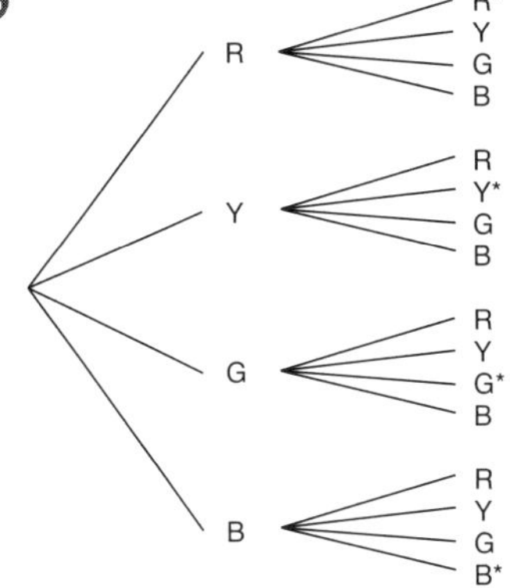

$P(\text{same colour}) = \frac{4}{16}$

$= \frac{1}{4}$

Answer C

76 No. of ways of placing the letters
$= 4 \times 3 \times 2 \times 1$
$= 24$

$\therefore P(\text{Correct letter}) = \frac{1}{24}$

Answer B

77 **i** $P(\text{Female}) = \frac{3}{8}$

(1 mark)

ii **1**

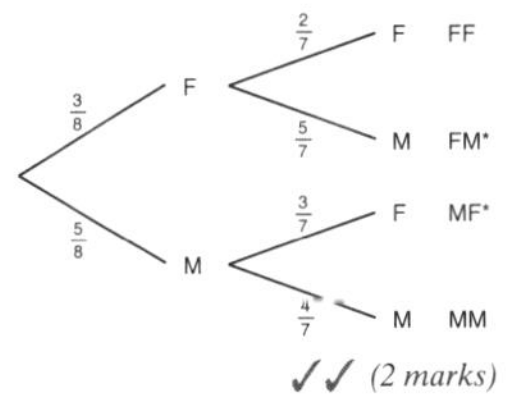

✓✓ *(2 marks)*

2 $P(\text{Exactly 1 female})$
$= P(\text{FM}) \text{ or } P(\text{MF})$ ✓
$= \frac{3}{8} \times \frac{5}{7} + \frac{5}{8} \times \frac{3}{7}$
$= \frac{30}{56}$
$= \frac{15}{28}$ ✓

(2 marks)

iii Yes, she has a good chance of remaining. ✓
Let A represent Antoinette remaining:

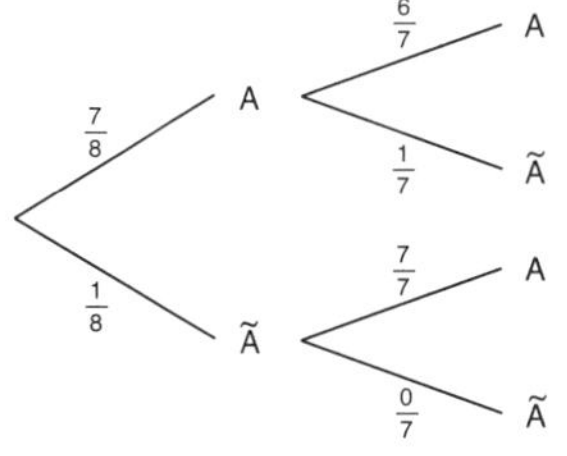

$\therefore P(\text{Antoinette remaining})$
$= P(\text{AA})$
$= \frac{7}{8} \times \frac{6}{7}$
$= \frac{3}{4}$ ✓

(2 marks)

1 In a park the only animals are goannas and emus. Let x be the number of goannas and let y be the number of emus.

The number of goannas plus the number of emus in the park is 31. Hence $x + y = 31$.

Each goanna has four legs and each emu has two legs. In total the emus and goannas have 76 legs.

By writing another relevant equation and graphing both equations on the grid on the following page, find the number of goannas and the number of emus in the park. *(4 marks)*

(Q34, **2021 HSC**) **Hard**

2 There are two tanks on a property, Tank A and Tank B. Initially, Tank A holds 1000 litres of water and Tank B is empty.

a Tank A begins to lose water at a constant rate of 20 litres per minute. The volume of water in Tank A is modelled by $V = 1000 - 20t$ where V is the volume in litres and t is the time in minutes from when the tank begins to lose water. On the grid below, draw the graph of this model and label it as Tank A. *(1 mark)* **Easy**

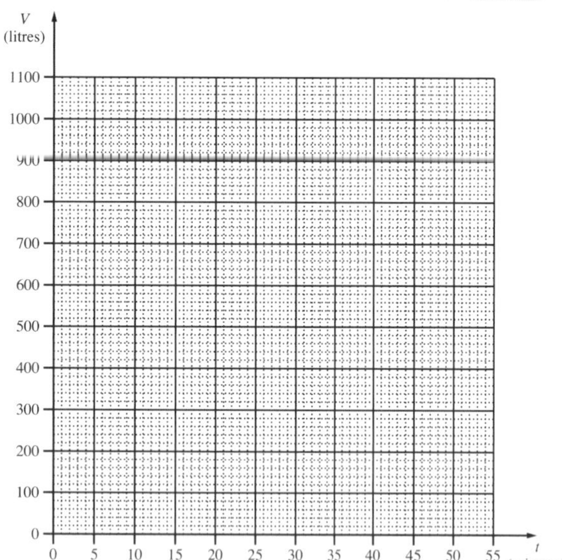

b Tank B remains empty until $t = 15$ when water is added to it at a constant rate of 30 litres per minute.
By drawing a line on the grid on the previous page, or otherwise, find the value of t when the two tanks contain the same volume of water. *(2 marks)* **Medium**

c Using the graphs drawn, or otherwise, find the value of t (where $t > 0$) when the total volume of water in the two tanks is 1000 litres. *(1 mark)* **Hard**

CQ (Q24, **2020 HSC**)

3 Last Saturday, Luke had 165 followers on social media. Rhys had 537 followers. On average, Luke gains another 3 followers per day and Rhys loses 2 followers per day. If x represents the number of days since last Saturday and y represents the number of followers, which pair of equations model this situation?

A	Luke: $y = 165x + 3$ Rhys: $y = 537x - 2$	**B**	Luke: $y = 165 + 3x$ Rhys: $y = 537 - 2x$
C	Luke: $y = 3x + 165$ Rhys: $y = 2x - 537$	**D**	Luke: $y = 3 + 165x$ Rhys: $y = 2 - 537x$

(1 mark)

(Q14, **2019 HSC**) **Easy**

4 A small business makes and sells bird houses.

Technology was used to draw straight-line graphs to represent the cost of making bird houses (C) and the revenue from selling bird houses (R). The x-axis displays the number of bird houses and the y-axis displays the cost/ revenue in dollars.

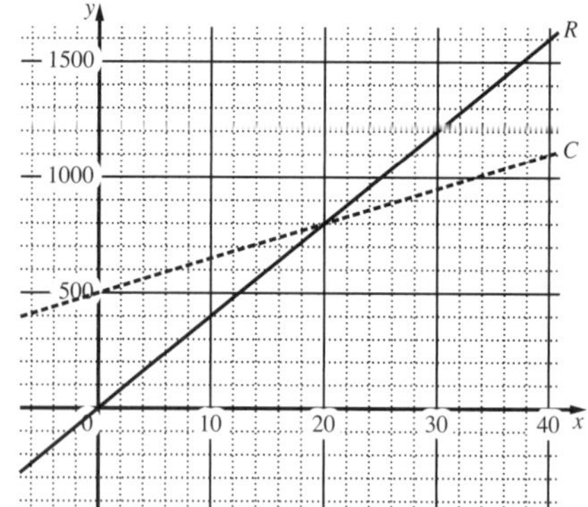

a How many bird houses need to be sold to break even? *(1 mark)* **Easy**

b By first forming equations for cost (C) and revenue (R), determine how many bird houses need to be sold to earn a profit of \$1900. *(3 marks)* **Hard**

(Q36, **2019 HSC**)

5 The cost, C, in dollars, of producing n kilograms of potatoes is given by $C = 1.2n + 24000$. The revenue R in dollars from selling n kilograms of potatoes is given by $R = 2.4n$.

The cost, C, for the production of n kilograms of potatoes is graphed below.

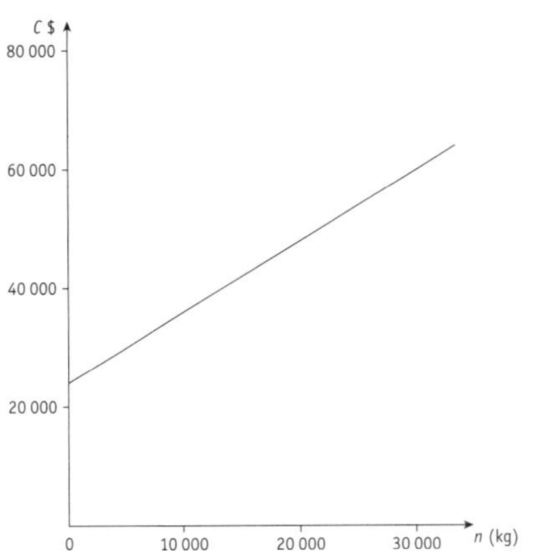

a On the graph above draw the graph for the equation $R = 2.4n$ to identify the break-even point. *(1 mark)* **Medium**

b What profit will be made if 30 000 kg of potatoes are grown and sold? *(2 marks)* **Medium**

c If a loss of $\$t$ occurs, write an expression to represent the mass of potatoes grown and sold. *(1 mark)* **Hard**

Bonus question (see page iv)

6 A meal delivery business produces and sells dinner packages.

The cost $\$C$ to make and deliver n dinner packages is modelled by the equation $C = 600 + 10n$.

The revenue $\$R$ is the amount of money received when selling n dinner packages.

The diagram below shows the graph of $C = 600 + 10n$, where the horizontal axis represents the number of packages and the vertical axis displays the cost, or revenue, in dollars.

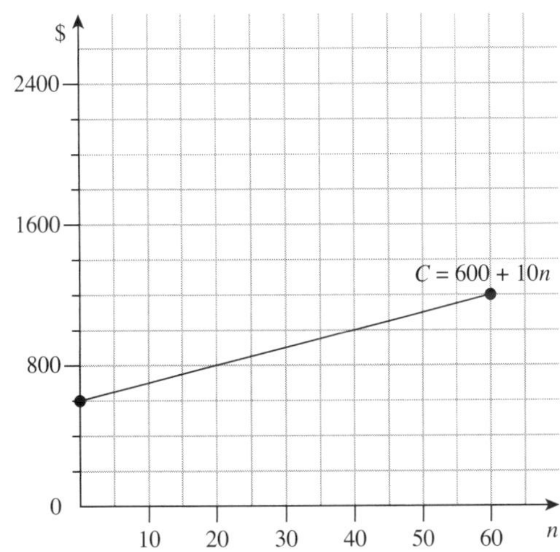

It is known that the company breaks even at 20 dinner packages.

a On the above diagram draw a straight line representing the graph of R. *(1 mark)* **Medium**

b If the equation for R is written in the form $R = kn$, what is the value of k? *(1 mark)* **Medium**

c How many dinner packages need to be sold for the company to make a profit of $900? *(2 marks)* **Medium**

Bonus question

7 The graph of the line with equation $y = 6 - 2x$ is shown.

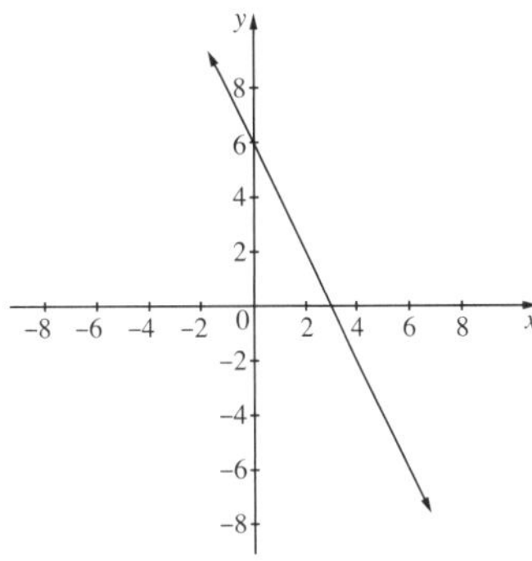

When the graph of the line with equation $y = x + 3$ is also drawn on this number plane, what will be the point of intersection of the two lines?

A $(0, 6)$ **B** $(1, 4)$
C $(2, 2)$ **D** $(3, 0)$ *(1 mark)*

(Q17, **2017 HSC**) **Medium**

8 A company manufactures phones. The company's income equation and cost equation are drawn on the same graph.

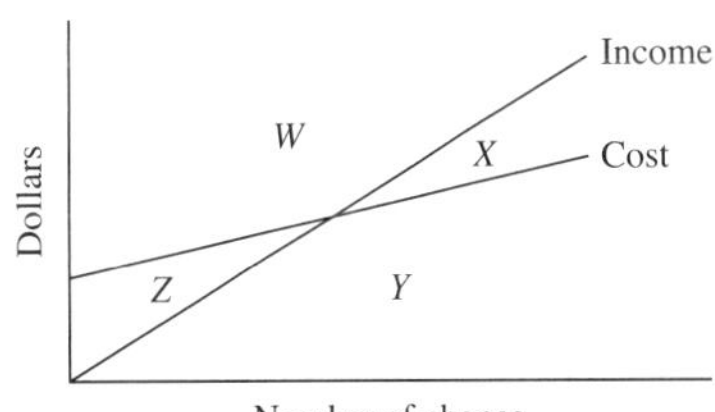

Which region of the graph is the profit zone?

A W **B** X
C Y **D** Z *(1 mark)*

(Q4, **2016 HSC**) **Easy**

9 A function centre hosts events for up to 500 people. The cost C, in dollars, for the centre to host an event, where x people attend, is given by:

$$C = 10000 + 50x$$

The centre charges $100 per person. Its income I, in dollars, is given by:

$$I = 100x$$

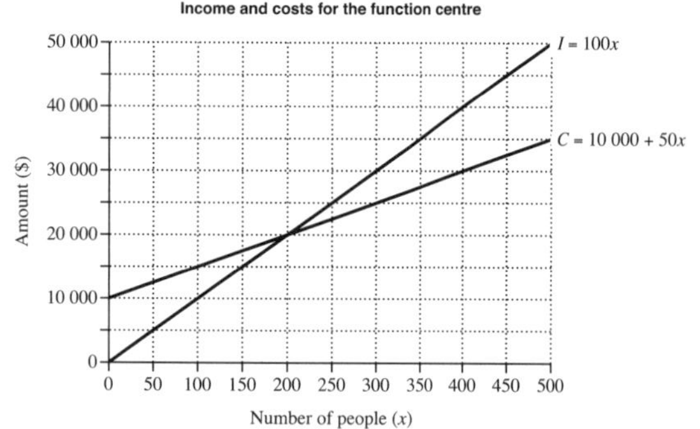

How much greater is the income of the function centre when 500 people attend an event, than its income at the breakeven point?

A $15000 **B** $20000
C $30000 **D** $40000 *(1 mark)*

(Q20, **2011 HSC**) **Medium**

---

10 Ashley makes picture frames as part of her business. To calculate the cost, C, in dollars, of making x frames, she uses the equation

$$C = 40 + 10x.$$

She sells the frames for $20 each and determines her income, I, in dollars, using the equation

$$I = 20x.$$

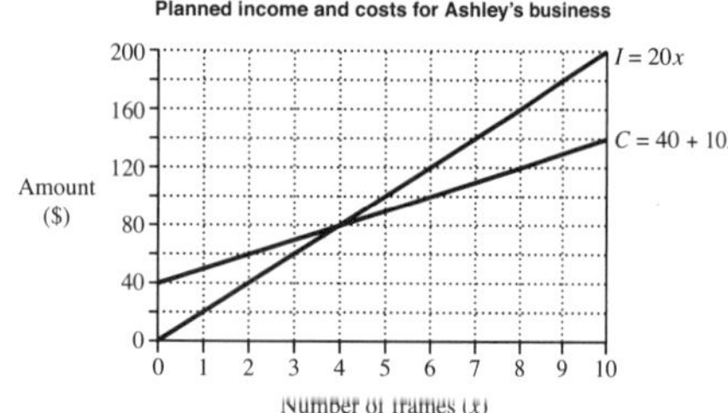

Use the graph to solve the two equations simultaneously for x and explain the significance of this solution for Ashley's business. *(2 marks)*

(Q24b, **2010 HSC**) **Medium**

---

11 Sue and Mikey are planning a fund-raising dance. They can hire a hall for $400 and a band for $300. Refreshments will cost them $12 per person.

i Write a formula for the cost ($C) of running the dance for x people. *(1 mark)* **Easy**

The graph shows planned income and costs when the ticket price is $20.

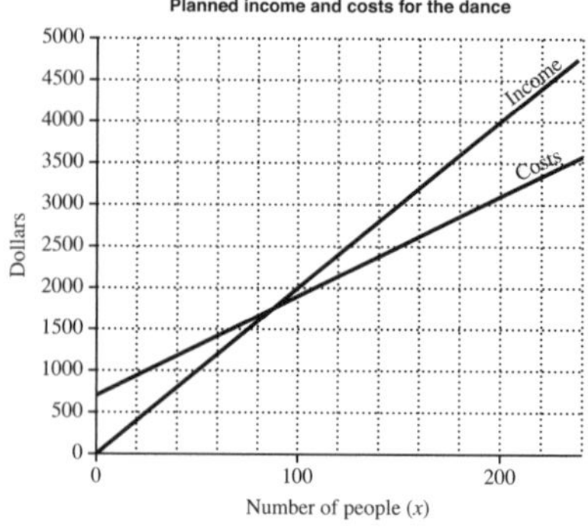

ii Estimate the minimum number of people needed at the dance to cover the costs. *(1 mark)* **Easy**

iii How much profit will be made if 150 people attend the dance? *(1 mark)* **Medium**

iv Sue and Mikey plan to sell 200 tickets. They want to make a profit of $1500. What should be the price of a ticket, assuming all 200 tickets will be sold? *(3 marks)* **Hard**

(Q28b, **2005 HSC**)

---

12 George drew a correct diagram that gave the solution to the simultaneous equations $y = 2x - 5$ and $y = x + 6$.

Which diagram did he draw?

A

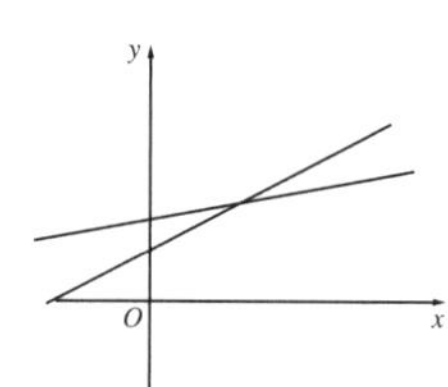

B

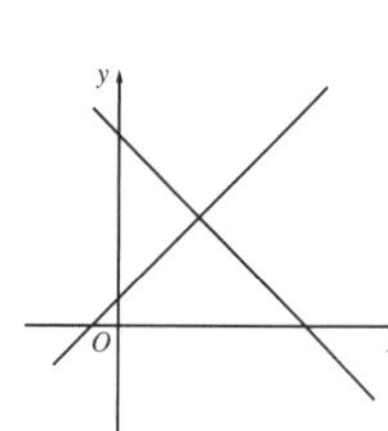

C

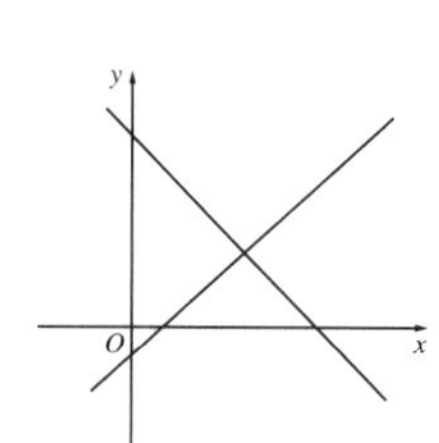

D

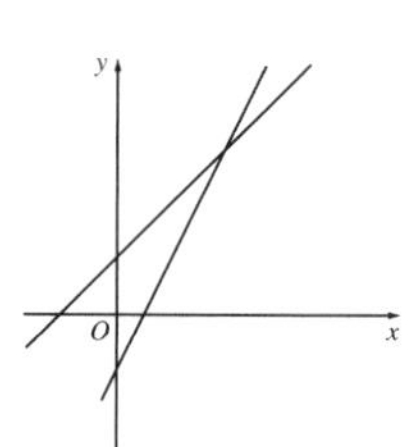

(1 mark)

(Q16, **2004 HSC**) **Medium**

---

13 In 2001, when Toby was in Year 9, he started earning money by juggling at children's parties. He charged $50 per party.

i Write a formula for the amount, Q (in dollars), that Toby had earned in 2001 after he had juggled at n parties. *(1 mark)* **Medium**

ii By the end of 2001 Toby had juggled at 30 parties.
Draw the straight line graph of Q against n, with n on the horizontal axis and Q on the vertical axis. Use your ruler to draw the axes, and mark a scale on each axis. *(3 marks)* **Hard**

iii Before Toby started juggling at parties he spent $300 on juggling equipment. On your graph in part **ii** draw a horizontal line through the point on the vertical axis where $Q = 300$. Give an interpretation of the point at which this horizontal line crosses the straight line graph of Q against n. *(1 mark)* **Hard**

(Q28b, **2002 HSC**)

Year 12 Simultaneous linear equations—Worked answers

1 Considering $x + y = 31$:
When $x = 0, y = 31$.
When $y = 0, x = 31$.
The total number of legs is 4 for each goanna plus 2 for each emu.
So $4x + 2y = 76$. ✓
When $x = 0$, $2y = 76$
$y = 38$
When $y = 0$, $4x = 76$
$x = 19$

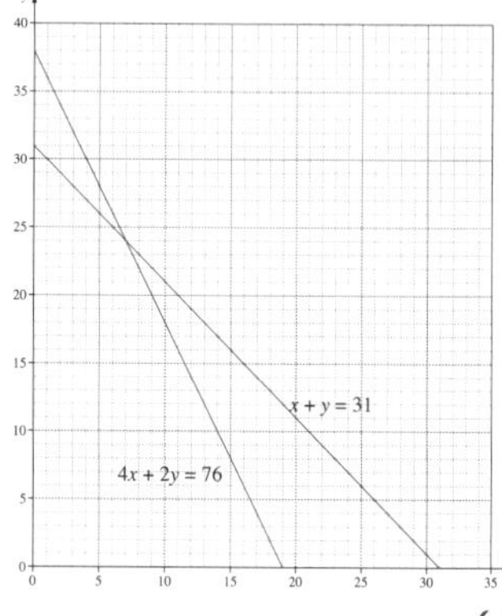

✓✓

The point of intersection is when $x = 7$ and $y = 24$.
So, from the graph, there are 7 goannas and 24 emus. ✓
(4 marks)

2 **a** $V = 1000 - 20t$
Vertical intercept = 1000
Gradient = –20

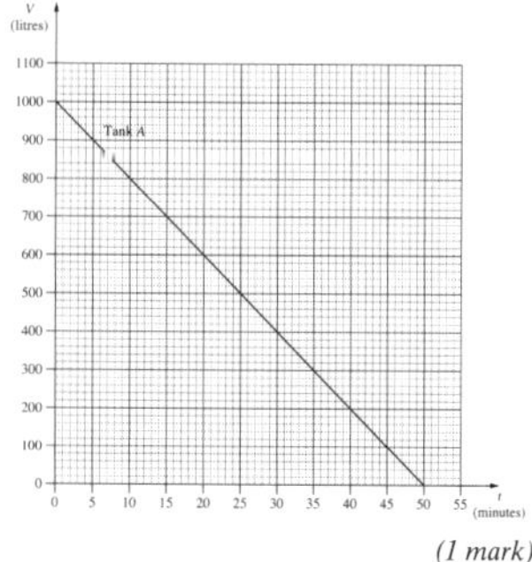

(1 mark)

b Tank B: $V = 0$ when = 15.
Gradient = 30

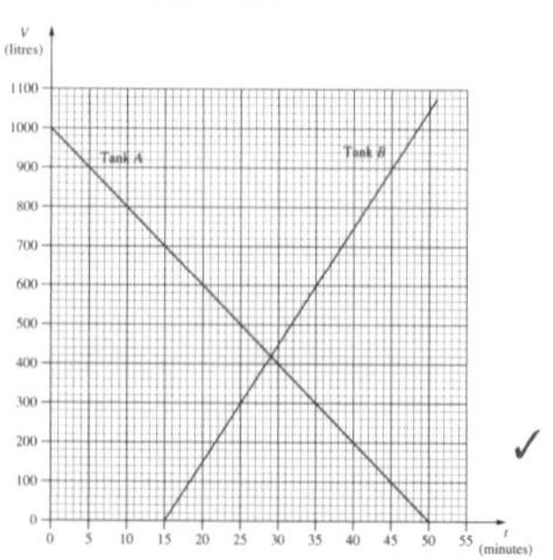

✓

The two tanks have the same volume when $t = 29$. ✓
(2 marks)

c From the graph the total volume of the two tanks will be 1000 L when $t = 45$.
(100 L in tank A and 900 L in tank B.)
(1 mark)

3 Luke has 165 followers plus 3 per day.
So for Luke $y = 165 + 3x$
Rhys has 537 followers less 2 per day.
So for Rhys $y = 537 - 2x$
Answer B

4 **a** 20 bird houses need to be sold to break even.
(1 mark)

b Revenue:
y-intercept = 0
Gradient $= \frac{800}{20}$
$= 40$ ✓
R: $y = 40x$
Cost: y-intercept = 500
Gradient $= \frac{300}{20}$
$= 15$
C: $y = 15x + 500$ ✓
Now profit = revenue – cost
$= 40x - (15x + 500)$
$= 40x - 15x - 500$
$= 25x - 500$
When the profit is \$1900,
$25x - 500 = 1900$
$25x = 2400$
$x = 96$
So 96 bird houses need to be sold to earn a profit of \$1900.
✓ *(3 marks)*

5 **a**

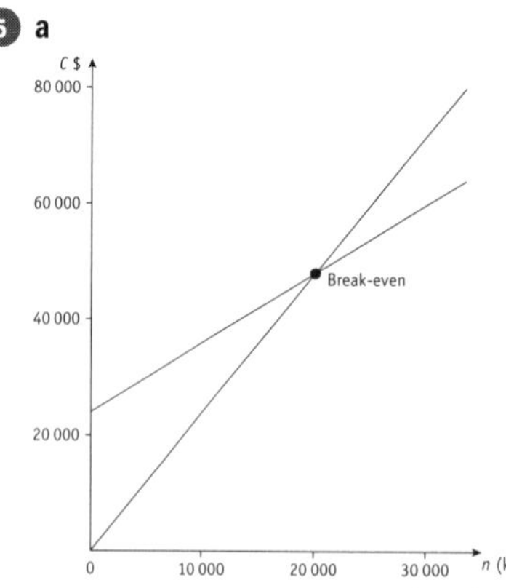

(1 mark)

b Substitute 30000:
$R = 2.4(30000)$
$= 72000$
$C = 1.2(30000) + 24000$
$= 60000$ ✓

Profit $= 72000 - 60000$
$= 12000$
∴ a profit of \$12000 ✓
(2 marks)

c Loss = $C - R$
$t = 1.2n + 24000 - 2.4n$
$= 24000 - 1.2n$
$1.2n = 24000 - t$
$n = \frac{24000 - t}{1.2}$
(1 mark)

6 **a**

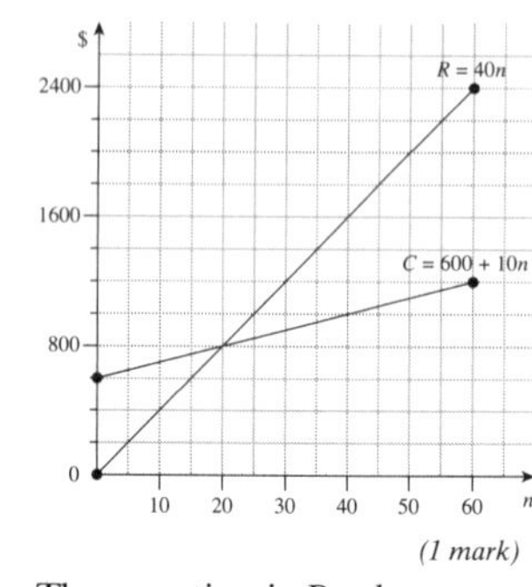

(1 mark)

b The equation is $R = kn$.
As the straight line passes through (20, 800), substitute $n = 20$ and $R = 800$ into equation:
$800 = k(20)$
$20k = 800$
$k = 40$
(1 mark)

c The cost is $C = 600 + 10n$ and the revenue is $R = 40n$.
As Profit
= Revenue – Cost,
$= 40n - (600 + 10n)$
$= 40n - 600 - 10n$
$= 30n - 600$ ✓
Substituting Profit = 900:
$900 = 30n - 600$
$30n = 900 + 600$
$30n = 1500$
$n = 50$
The company needs to sell 50 dinner packages. ✓
(2 marks)

7 At the point of intersection:
$x + 3 = 6 - 2x$
$3x + 3 = 6$
$3x = 3$
$x = 1$
When $x = 1, y = 1 + 3 = 4$.
The point of intersection is (1, 4).

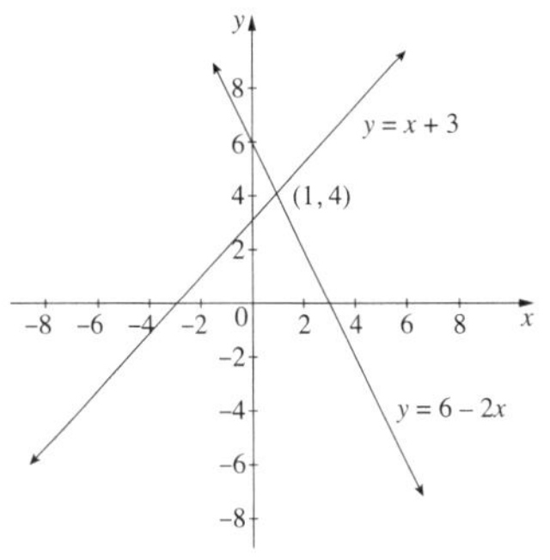

Answer B

8 The profit zone is X. [Where the income is greater than the cost]

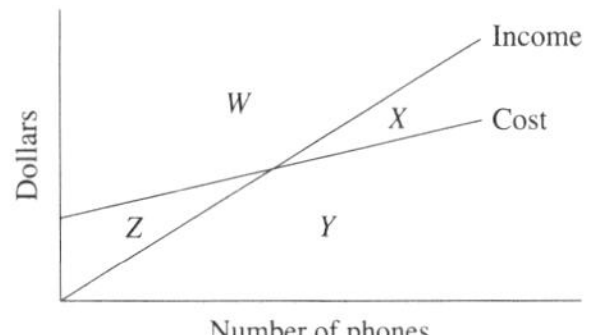

Answer B

9

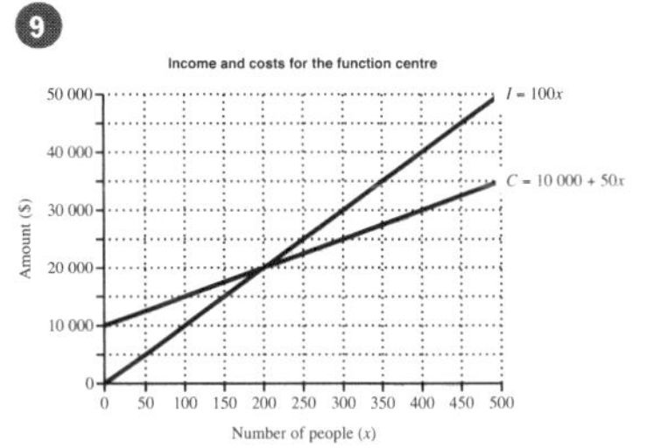

When 500 people attend, the income is \$50 000.

At the breakeven point, the income is \$20 000.

Difference = \$50 000 – \$20 000
= \$30 000

Answer C

10 The lines intersect when $x = 4$.
So the simultaneous solution of $I = 20x$ and $C = 40 + 10x$ is when $x = 4$. ✓
This is the break-even point for Ashley's business. If she makes 4 frames her costs will be \$80 and if she sells 4 frames her income will also be \$80.
If she makes and sells more than 4 frames she will make a profit, but if she sells fewer than 4 frames she will make a loss. ✓

(2 marks)

11 **i** $C = \$400 + \$300 + \$12 \times x$
$= \$700 + \$12x$

(1 mark)

ii The two lines drawn, meet at approximately $x = 90$.
∴ About 90 people will need to attend.

(1 mark)

iii If $x = 150$,
$C = \$700 + \$12 \times 150$
$= \$2500$
Income = \$3000 (from graph)
∴ Profit = \$3000 – \$2500
= \$500

(1 mark)

iv If $x = 200$,
$C = \$700 + \$12 \times 200$
$= \$3100$ ✓
Profit = \$1500
Now Income = Cost + Profit
= \$3100 + \$1500
= \$4600 ✓
∴ Price of ticket
= \$4600 ÷ 200
= \$23 ✓

(3 marks)

12 $y = 2x - 5$ has a positive gradient of 2 and a y-intercept of –5:

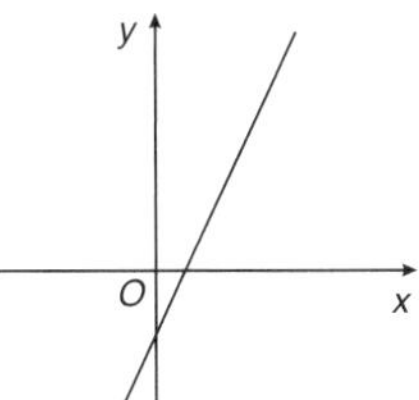

$y = x + 6$ has a positive gradient of 1 and a y-intercept of 6:

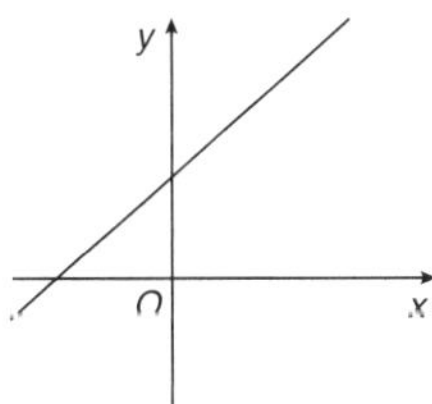

Answer D

13 **i** $Q = 50n$

(1 mark)

ii

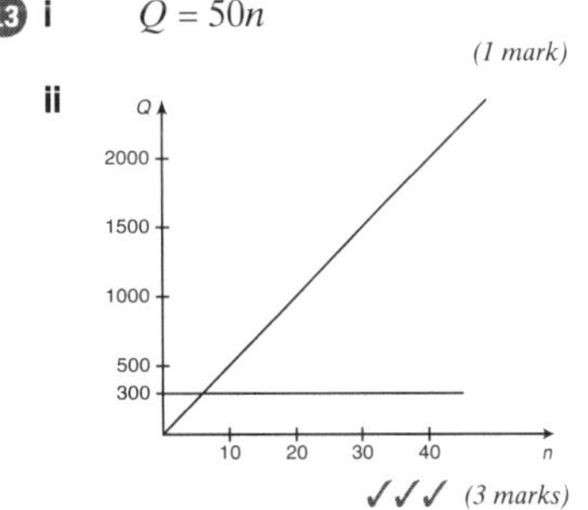

✓✓✓ *(3 marks)*

iii Where the two lines intersect is the breakeven point. Until he has done 6 parties, Toby has not made a profit.

(1 mark)

1 Which of the following best represents the graph of $y = 10(0.8)^x$?

A

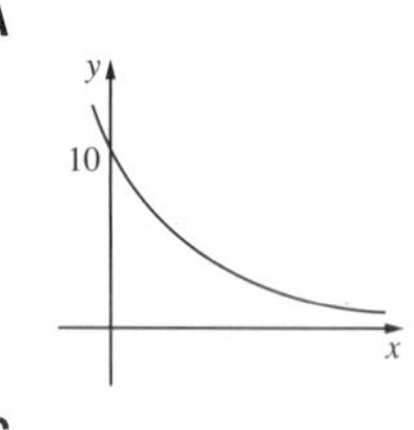

B

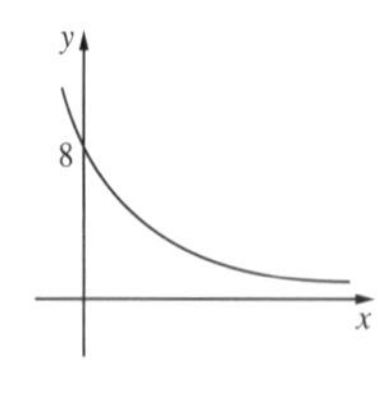

C

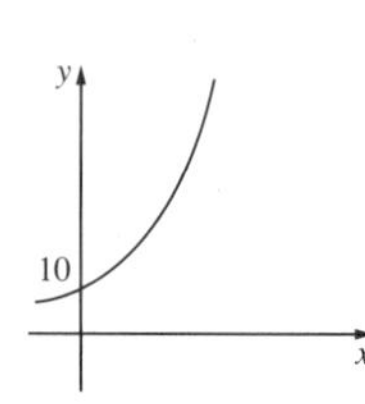

D

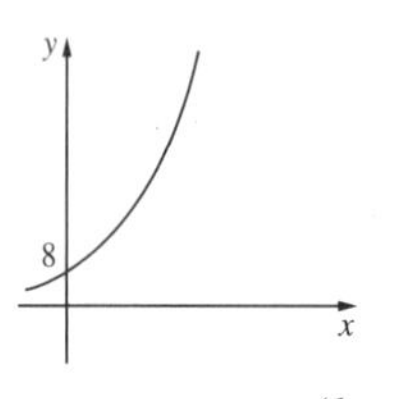

(1 mark)

(Q10, **2021 HSC**) **Medium**

---

2 The time taken to clean a warehouse varies inversely with the number of cleaners employed.

It takes 8 cleaners 60 hours to clean a warehouse.

Working at the same rate, how many hours would it take 10 cleaners to clean the same warehouse?

A 45 **B** 48

C 62 **D** 75 *(1 mark)*

(Q13, **2021 HSC**) **Medium**

---

3 A population, P, is to be modelled using the function $P = 2000(1.2)^t$, where t is the time in years.

a What is the initial population? *(1 mark)* **Medium**

b Find the population after 5 years. *(1 mark)* **Easy**

c On the axes below, draw the graph of the population against time, showing the points at $t = 0$ and at $t = 5$. *(2 marks)* **Medium**

(Q24, **2021 HSC**)

---

4 A publisher sells a book for \$10. At this price, 5000 copies of the book will be sold and the revenue raised will be $5000 \times 10 = \$50\,000$.

The publisher is considering increasing the price of the book. For every dollar the price of the book is increased, the publisher will sell 50 fewer copies of the book.

If the publisher charges $(10 + x)$ dollars for each book, a quadratic model for the revenue raised, R, from selling the books is

$$R = -50x^2 + 4500x + 50\,000.$$

A graph of this quadratic model for revenue is shown. A dashed line is used for values of x which are not relevant to the practical context of this problem.

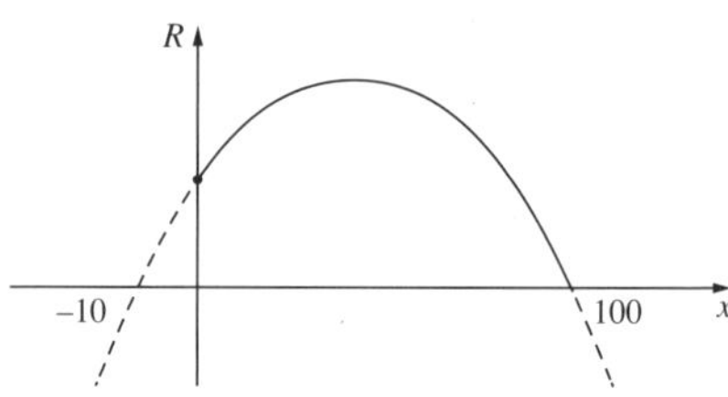

a By first finding a suitable value of x, find the price the publisher should charge for each book to maximise the revenue raised from sales of the book. *(2 marks)* **Hard**

b Find the value of the intercept of the parabola with the vertical axis. *(1 mark)* **Hard**

(Q35, **2021 HSC**)

---

5 Which of the following could represent the graph of $y = -x^2 + 1$?

A

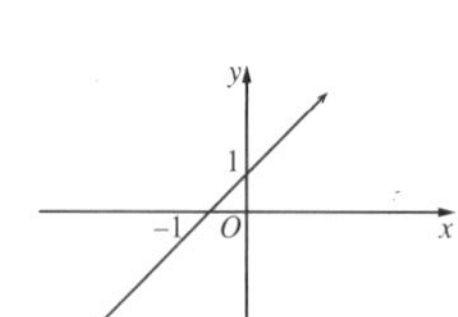

B

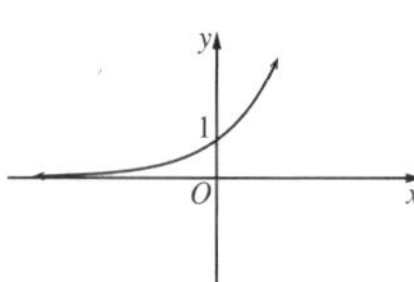

C

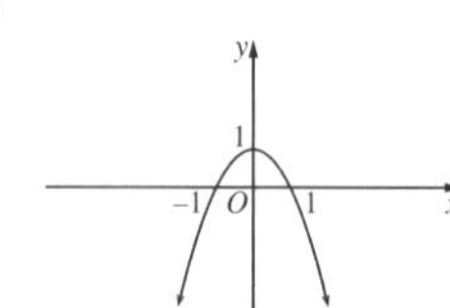

D

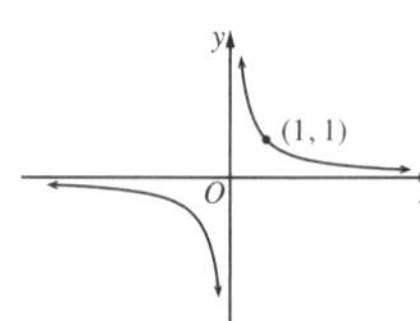

(1 mark)

(Q1, **2020 HSC**) **Medium**

---

6 A fence is to be built around the outside of a rectangular paddock. An internal fence is also to be built.

The side lengths of the paddock are x metres and y metres, as shown in the diagram.

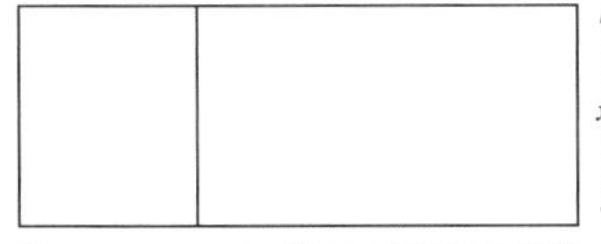

A total of 900 metres of fencing is to be used. Therefore $3x + 2y = 900$.

The area, A, in square metres, of the rectangular paddock is given by $A = 450x - 1.5x^2$. The graph of this equation is shown.

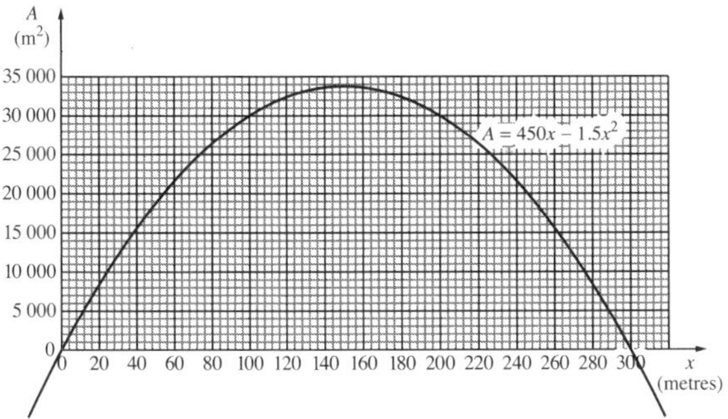

a If the area of the paddock is 30 000 m^2, what is the largest possible value of x? *(1 mark)* **Hard**

b Find the values of x and y so that the area of the paddock is as large as possible. *(2 marks)* **Hard**

c Using your values from part **b**, find the largest possible area of the paddock. *(1 mark)* **Medium**

(Q19, **2020 HSC**)

7 The graph shows the number of bacteria, y, at time n minutes. Initially (when $n = 0$) the number of bacteria is 1000.

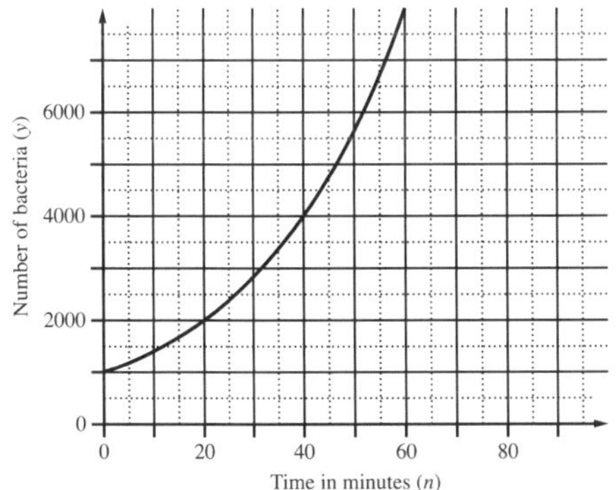

a Find the number of bacteria at 40 minutes. *(1 mark)* **Easy**

b The number of bacteria can be modelled by the equation $y = A \times b^n$, where A and b are constants. Use the guess and check method to find, to two decimal places, an upper and lower estimate for the value of b. The upper and lower estimates must differ by 0.01. *(2 marks)* **Hard**

(Q33, **2020 HSC**)

8 A rectangle has width w centimetres. The area of the rectangle, A, in square centimetres, is $A = 2w^2 + 5w$.

The graph of $A = 2w^2 + 5w$ is shown.

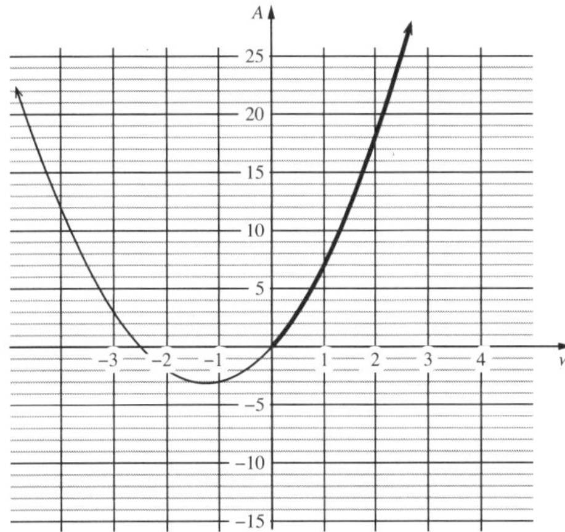

a Explain why, in this context, the model $A = 2w^2 + 5w$ only makes sense for the bold section of the graph. *(1 mark)* **Medium**

b The area of the rectangle is 18 cm^2. Calculate the perimeter of the rectangle. *(2 marks)* **Medium**

(Q31, **2019 HSC**)

9 The time taken for a car to travel between two towns at a constant speed varies inversely with its speed.

It takes 1.5 hours for the car to travel between the two towns at a constant speed of 80 km/h.

a Calculate the distance between the two towns. *(1 mark)* **Easy**

b By first plotting four points, draw the curve that shows the time taken to travel between the two towns at different constant speeds.

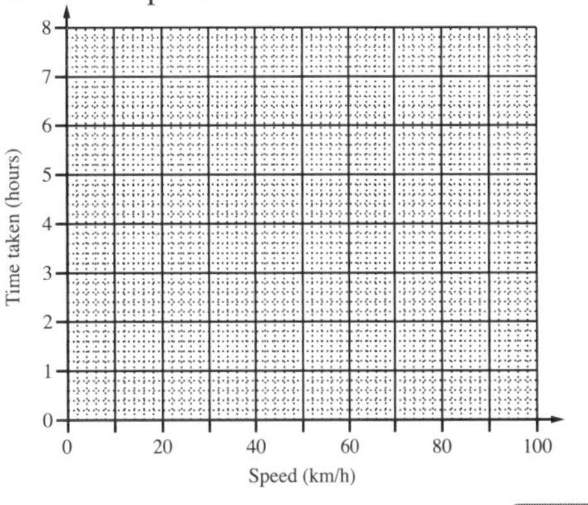

(3 marks) **Medium**

(Q33, **2019 HSC**)

10 Which graph best represents the equation $y = x^2 - 2$?

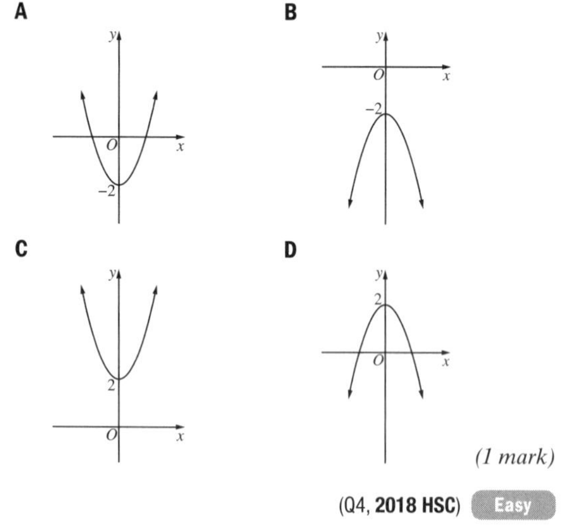

(1 mark)

(Q4, **2018 HSC**) Easy

11 When people walk in snow, the depth (D cm) of each footprint depends on both the area (A cm^2) of the shoe sole and the weight of the person.

The graph shows the relationship between the area of the shoe sole and the depth of the footprint in snow, for a group of people of the same weight.

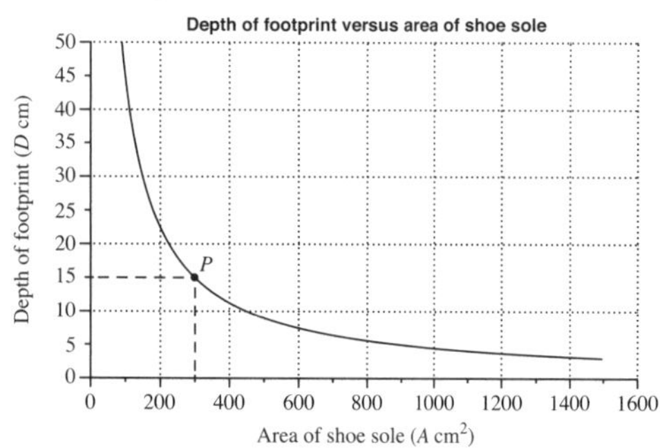

i The graph is a hyperbola because D is inversely proportional to A. The point P lies on the hyperbola. Find the equation relating D and A. *(2 marks)* Hard

ii A man from this group walks in snow and the depth of his footprint is 4 cm. Use your equation from part **i** to calculate the area of his shoe sole. *(1 mark)* Hard

(Q29c, **2018 HSC**)

12 A movie theatre has 200 seats. Each ticket currently costs \$8.

The theatre owners are currently selling all 200 tickets for each session. They decide to increase the price of tickets to see if they can increase the income earned from each movie session.

It is assumed that for each one dollar increase in ticket price, there will be 10 fewer tickets sold.

A graph showing the relationship between an increase in ticket price and the income is shown below.

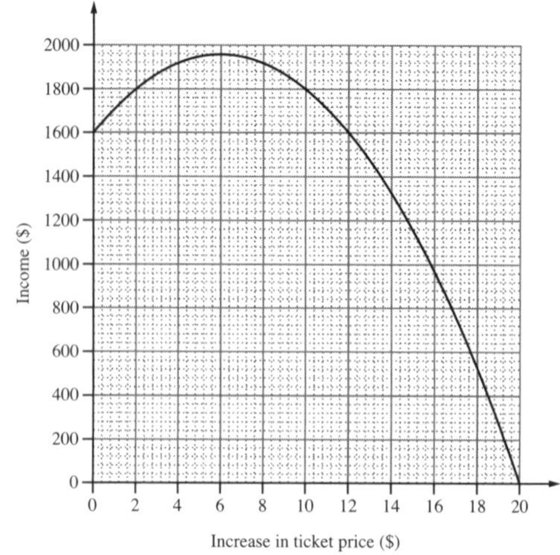

i What ticket price should be charged to maximise the income from a movie session? *(1 mark)* Medium

ii What is the number of tickets sold when the income is maximised? *(1 mark)* Medium

iii The cost to the theatre owners of running each session is \$500 plus \$2 per ticket sold. Calculate the profit earned by the theatre owners when the income earned from a session is maximised. *(2 marks)* Medium

(Q28e, **2017 HSC**)

13 The mass M kg of a baby pig at age x days is given by $M = A(1.1)^x$ where A is a constant. The graph of this equation is shown.

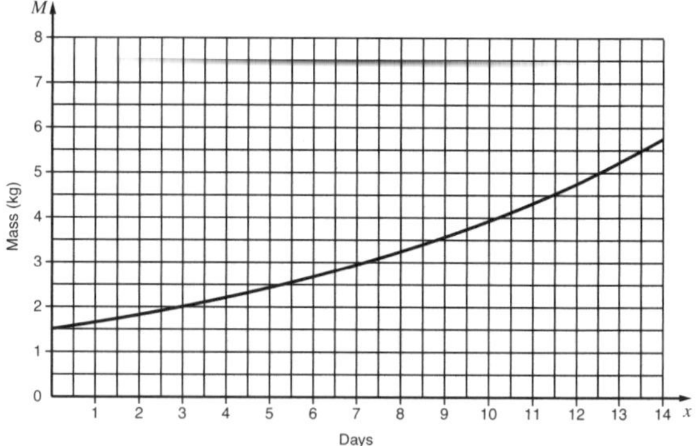

i What is the value of A? *(1 mark)* Hard

ii What is the daily growth rate of the pig's mass? Write your answer as a percentage. *(1 mark)* Hard

(Q29b, **2016 HSC**)

14 A charity seeks to raise money by telephoning people at random from a call centre and asking them to donate.

Over the years, this charity has found that the amount of money raised (\$$A$) is related to the number of telephone calls made (n). A graph of this relationship is shown.

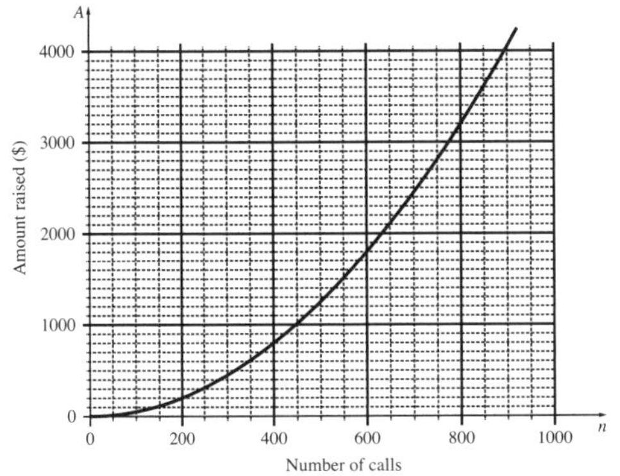

It costs the charity \$2100 per week to run the call centre. It also costs an average of 50 cents per telephone call.

i Write an equation to represent the total cost, C, of running the call centre for a week in which n phone calls are made. *(1 mark)* **Medium**

ii By graphing this equation on the axes above, determine the number of phone calls the charity needs to make in order to break even. *(2 marks)* **Hard**

(Q28f, **2015 HSC**)

15 A diver springs upwards from a diving board, then plunges into the water. The diver's height above the water as it varies with time is modelled by a quadratic function. Graphing software is used to produce the graph of this function.

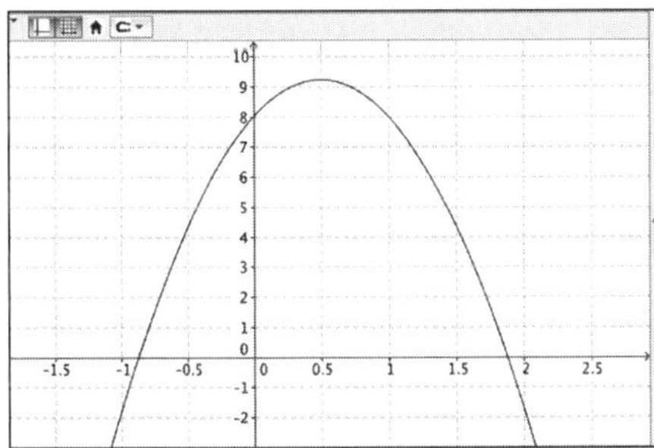

Explain how the graph could be used to determine how high above the height of the diving board the diver was when he reached the maximum height. *(2 marks*

(Q29e, **2015 HSC**) **Hard**

16 The diagram shows the graph of an equation.

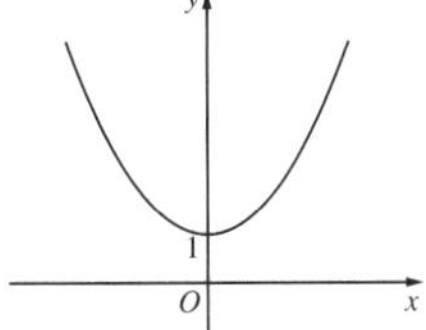

Which of the following equations does the graph best represent?

A $y = \dfrac{3}{x} + 1$ **B** $y = 3^x + 1$

C $y = 3x^2 + 1$ **D** $y = 3x^3 + 1$ *(1 mark)*

(Q3, **2014 HSC**) **Medium**

17 The cost of hiring an open space for a music festival is \$120 000. The cost will be shared equally by the people attending the festival, so that C (in dollars) is the cost per person when n people attend the festival.

i Complete the table below by filling in the THREE missing values. *(1 mark)* **Easy**

Number of people (n)	500	1000	1500	2000	2500	3000
Cost per person (C)				60	48	40

ii Using the values from the table, draw the graph showing the relationship between n and C. *(2 marks)* **Easy**

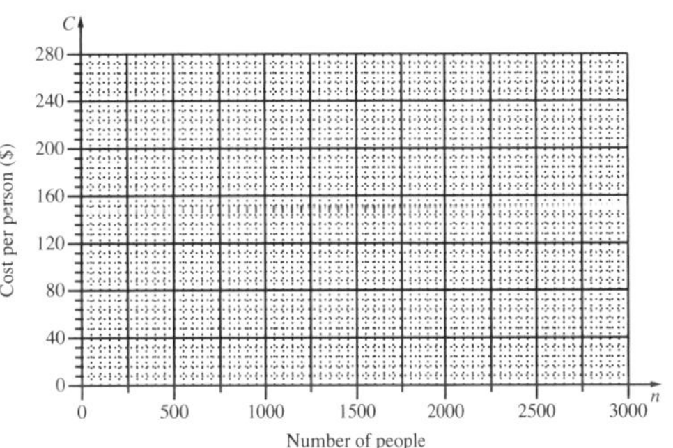

iii What equation represents the relationship between n and C? *(1 mark)* **Medium**

iv Give ONE limitation of this equation in relation to this context. *(1 mark)* **Hard**

v Is it possible for the cost per person to be \$94? Support your answer with appropriate calculations. *(1 mark)* **Hard**

(Q29a, **2014 HSC**)

18 Leanne wants to build a rectangular vegetable garden in her backyard. She has 20 metres of fencing and will use a wall as one side of the garden. The plan for her garden is shown, where x metres is the width of her garden.

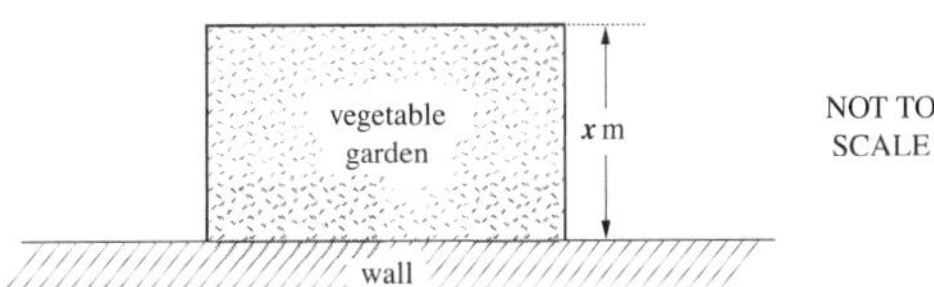

Which equation gives the area, A, of the vegetable garden?

A $A = 10x - x^2$ **B** $A = 10x - 2x^2$

C $A = 20x - x^2$ **D** $A = 20x - 2x^2$ *(1 mark)*

(Q22, **2013 HSC**) **Hard**

19 Wind turbines, such as those shown, are used to generate power.

In theory, the power that could be generated by a wind turbine is modelled using the equation

$$T = 20\,000w^3$$

where T is the theoretical power generated, in watts
w is the speed of the wind, in metres per second.

i Using this equation, what is the theoretical power generated by a wind turbine if the wind speed is 7.3 m/s? *(1 mark)* **Easy**

ii In practice, the actual power generated by a wind turbine is only 40% of the theoretical power.
If A is the actual power generated, in watts, write an equation for A in terms of w. *(1 mark)* **Hard**

The graph shows both the theoretical power generated and the actual power generated by a particular wind turbine.

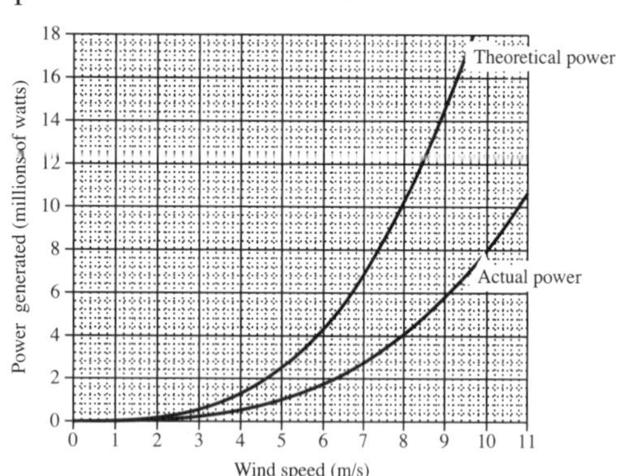

iii Using the graph, or otherwise, find the difference between the theoretical power and the actual power generated when the wind speed is 9 m/s. *(1 mark)* **Hard**

iv A particular farm requires at least 4.4 million watts of actual power in order to be self-sufficient. What is the minimum wind speed required for the farm to be self-sufficient? *(1 mark)* **Hard**

v A more accurate formula to calculate the power (P) generated by a wind turbine is

$$P = 0.61 \times \pi \times r^2 \times w^3$$

where r is the length of each blade, in metres
w is the speed of the wind, in metres per second.

Each blade of a particular wind turbine has a length of 43 metres. The turbine operates at a wind speed of 8 m/s.

Using the formula above, if the wind speed increased by 10%, what would be the percentage increase in the power generated by this wind turbine? *(3 marks)* **Hard**

(Q30a, **2013 HSC**)

20 The time taken to complete a journey varies inversely with the speed of a car. A car takes 6 hours to complete a journey when travelling at 60 km/h.

How long would the same journey take if the car were travelling at 100 km/h?

A 36 minutes
B 1 hour and 40 minutes
C 3 hours and 6 minutes
D 3 hours and 36 minutes *(1 mark)*

(Q15, **2012 HSC**) **Easy**

21 A golf ball is hit from point A to point B, which is on the ground as shown. Point A is 30 metres above the ground and the horizontal distance from point A to point B is 300 m.

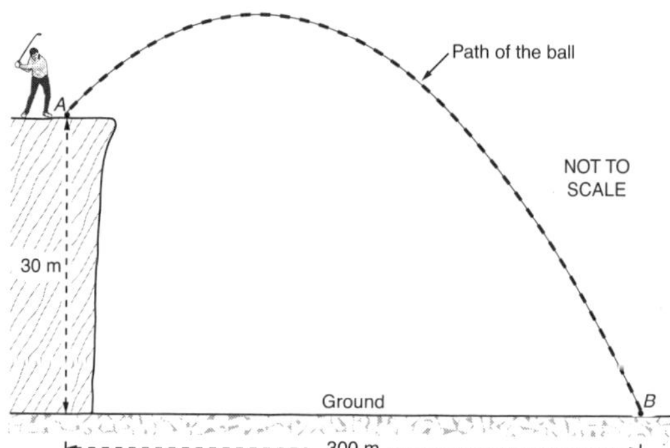

The path of the golf ball is modelled using the equation

$$h = 30 + 0.2d - 0.001d^2$$

where h is the height of the golf ball above the ground in metres, and

d is the horizontal distance of the golf ball from point A in metres.

The graph of this equation is drawn below.

i What is the maximum height the ball reaches above the ground? *(1 mark)* **Easy**

ii There are two occasions when the golf ball is at a height of 35 metres.
What horizontal distance does the ball travel in the period between these two occasions? *(1 mark)* **Medium**

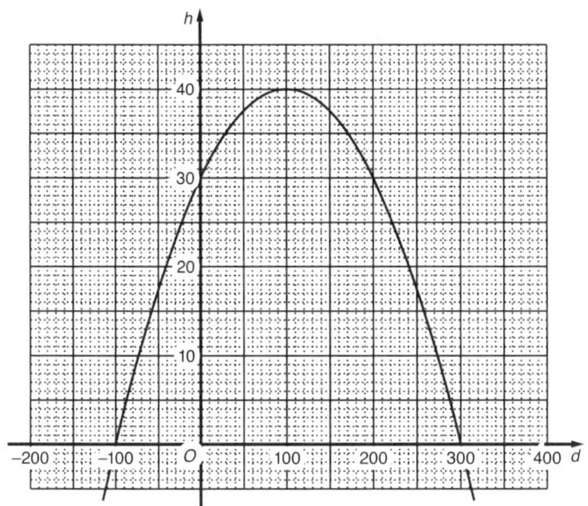

iii What is the height of the ball above the ground when it still has to travel a horizontal distance of 50 metres to hit the ground at point B? *(1 mark)* Easy

iv Only part of the graph applies to this model.
Find all values of d that are not suitable to use with this model, and explain why these values are not suitable. *(2 marks)* Hard

(Q30b, **2012 HSC**)

22 In 2010, the city of Thagoras modelled the predicted population of the city using the equation

$$P = A(1.04)^n.$$

That year, the city introduced a policy to slow its population growth. The new predicted population was modelled using the equation

$$P = A(b)^n.$$

In both equations, P is the predicted population and n is the number of years after 2010.

The graph shows the two predicted populations.

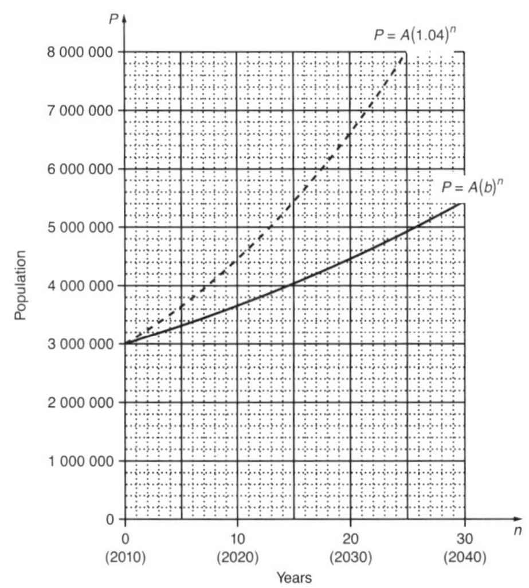

i Use the graph to find the predicted population of Thagoras in 2030 if the population policy had NOT been introduced. *(1 mark)* Medium

ii In each of the two equations given, the value of A is 3 000 000.
What does A represent? *(1 mark)* Medium

iii The guess-and-check method is to be used to find the value of b, in $P = A(b)^n$.

1. Explain, with or without calculations, why 1.05 is not a suitable first estimate for b. *(1 mark)* Hard
2. With $n = 20$ and $P = 4\,460\,000$, use the guess-and-check method and the equation $P = A(b)^n$ to estimate the value of b to two decimal places.
   Show at least TWO estimate values for b, including calculations and conclusions. *(2 marks)* Hard

iv The city of Thagoras was aiming to have a population under 7 000 000 in 2050.
Does the model indicate that the city will achieve this aim?
Justify your answer with suitable calculations. *(2 marks)* Hard

(Q30c, **2012 HSC**)

23 Which of the following graphs best represents the equation $y = a^x$, where a is a positive number greater than 1?

A

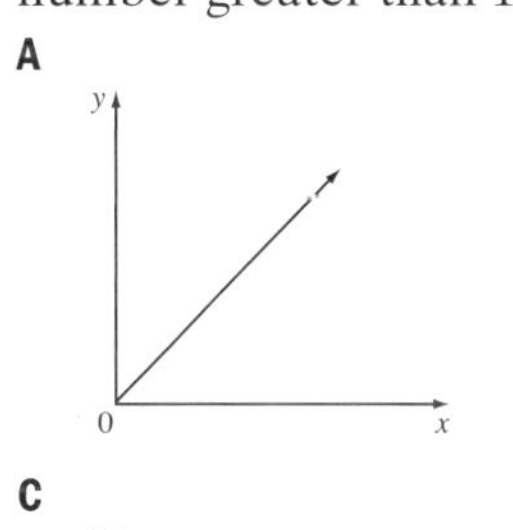

B

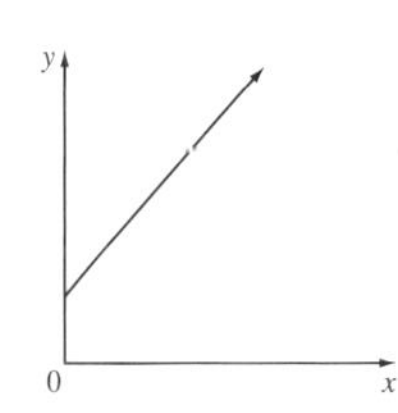

C

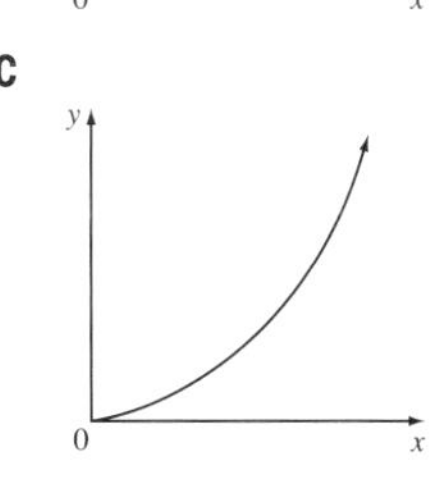

D

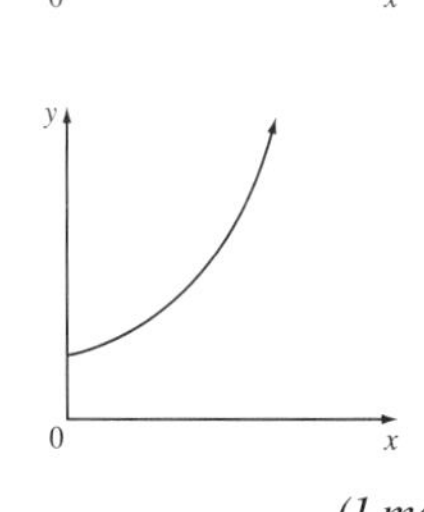

(1 mark)

(Q6, **2011 HSC**) Medium

24 Jack needs to find the number of years, t, it will take for a population of bats to first exceed 18 000.

He uses a 'guess-and-check' method to estimate t in the following equation

$$5 \times 3^t = 18000.$$

Here is his working:

Try $t = 9$

$5 \times 3^9 = 98\,415$

Conclusion: $t = 9$ is too big.

i Jack's next guess is $t = 6$. Show Jack's correct working for this guess, including the calculation and conclusion. *(1 mark)* **Easy**

ii Continue using the 'guess-and-check' method to find the number of years, t, it will take for the population to first exceed 18 000, if t is a whole number. Include the calculations and conclusions. *(2 marks)* **Easy**

(Q26b, **2011 HSC**)

25 The air pressure, P, in a bubble varies inversely with the volume, V, of the bubble.

i Write an equation relating P, V and a, where a is a constant. *(1 mark)* **Medium**

ii It is known that $P = 3$ when $V = 2$. By finding the value of the constant, a, find the value of P when $V = 4$. *(2 marks)* **Medium**

iii Sketch a graph to show how P varies for different values of V. Use the horizontal axis to represent volume and the vertical axis to represent air pressure. *(2 marks)* **Hard**

(Q28a, **2011 HSC**)

26 The number of hours that it takes for a block of ice to melt varies inversely with the temperature. At 30°C it takes 8 hours for a block of ice to melt.

How long will it take the same size block of ice to melt at 12°C?

A 3.2 hours **B** 20 hours

C 26 hours **D** 45 hours *(1 mark)*

(Q13, **2010 HSC**) **Medium**

27 Moivre's manufacturing company produces cans of Magic Beans. The can has a diameter of 10 cm and a height of 10 cm.

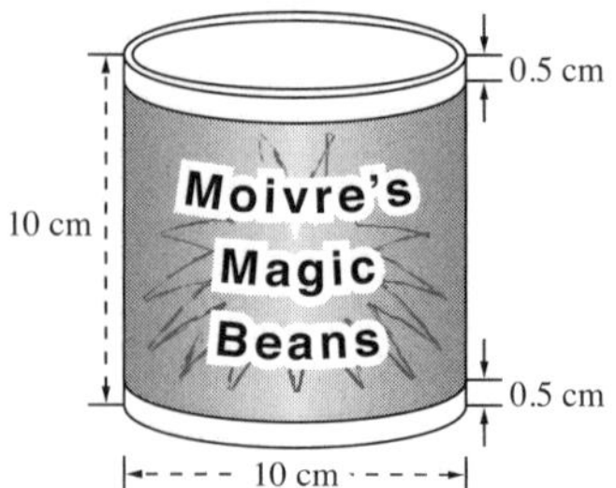

i Cans are packed in boxes that are rectangular prisms with dimensions 30 cm × 40 cm × 60 cm. What is the maximum number of cans that can be packed into one of these boxes? *(1 mark)* **Hard**

ii The shaded label on the can shown wraps all the way around the can with no overlap. What area of paper is needed to make the labels for all the cans in this box when the box is full? *(2 marks)* **Medium**

iii The company is considering producing larger cans. Monica says if you double the diameter of the can this will double the volume. Is Monica correct? Justify your answer with suitable calculations. *(2 marks)* **Medium**

iv The company wants to produce a can with a volume of 1570 cm^3, using the least amount of metal. Monica is given the job of determining the dimensions of the can to be produced. She considers the following graphs.

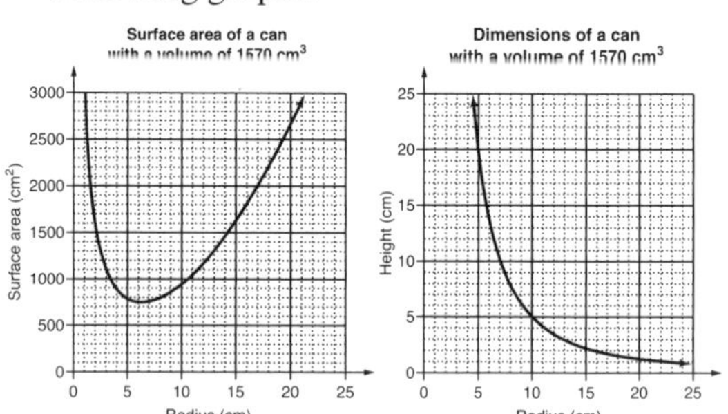

What radius and height should Monica recommend that the company use to minimise the amount of metal required to produce these cans? Justify your choice of dimensions with reference to the graphs and/or suitable calculations. *(2 marks)* **Hard**

(Q28b, **2010 HSC**)

28 The time for a car to travel a certain distance varies inversely with its speed.

Which of the following graphs shows this relationship?

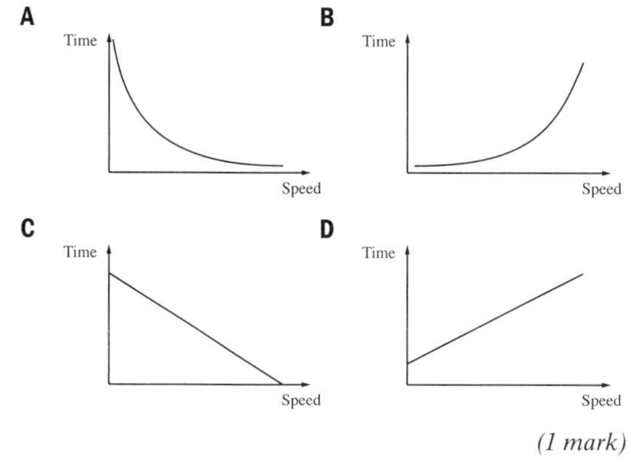

(1 mark)

(Q16, **2009 HSC**) **Medium**

29 Anjali is investigating stopping distances for a car travelling at different speeds. To model this she uses the equation

$$d = 0.01s^2 + 0.7s,$$

where d is the stopping distance in metres and s is the car's speed in km/h.

The graph of this equation is drawn below.

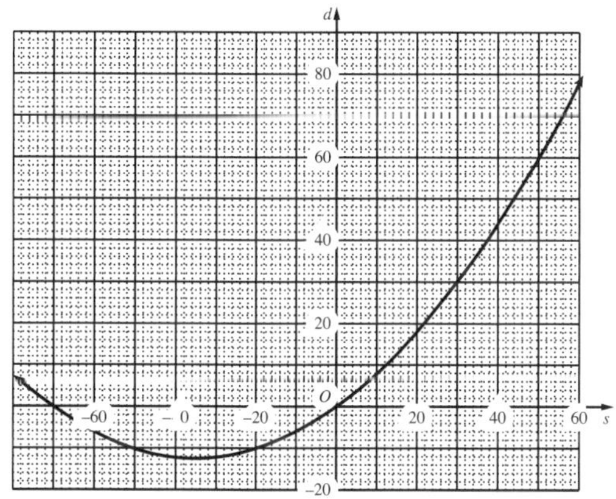

- **i** Anjali knows that only part of this curve applies to her model for stopping distances. In your writing booklet, using a set of axes, sketch the part of this curve that applies for stopping distances. *(1 mark)* **Medium**
- **ii** What is the difference between the stopping distances in a school zone when travelling at a speed of 40 km/h and when travelling at a speed of 70 km/h? *(2 marks)* **Medium**

(Q28a, **2009 HSC**)

30 Which graph best represents $y = 3^x$?

A

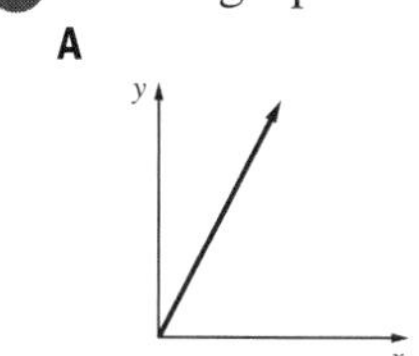

B

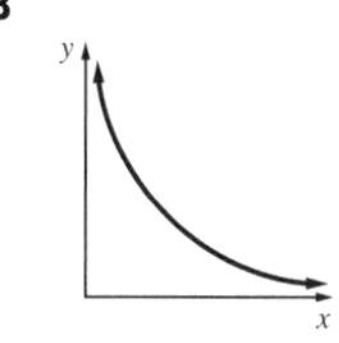

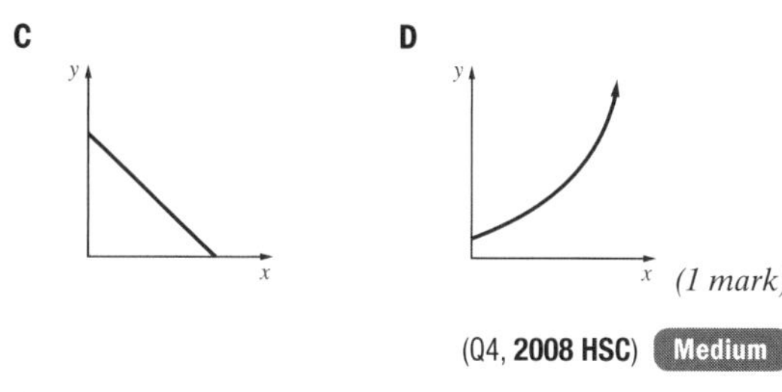

(1 mark)

(Q4, **2008 HSC**) **Medium**

31 The number of penguins, P, after t years in a new colony can be found using the following formula.

$$P = a \times 2^t$$

- **i** If there are 24 penguins after two years, find the value of a. *(2 marks)* **Easy**
- **ii** How many years will it take for the number of penguins to first exceed 1500? *(2 marks)* **Medium**

(Q25a, **2008 HSC**)

32 If pressure (p) varies inversely with volume (V), which formula correctly expresses p in terms of V and k, where k is a constant?

A $p = \frac{k}{V}$ **B** $p = \frac{V}{k}$

C $p = kV$ **D** $p = k + V$ *(1 mark)*

(Q15, **2007 HSC**) **Easy**

33 A rectangular playing surface is to be constructed so that the length is 6 metres more than the width.

- **i** Give an example of a length and width that would be possible for this playing surface. *(1 mark)* **Easy**
- **ii** Write an equation for the area (A) of the playing surface in terms of its length (l). *(1 mark)* **Easy**

A graph comparing the area of the playing surface to its length is shown.

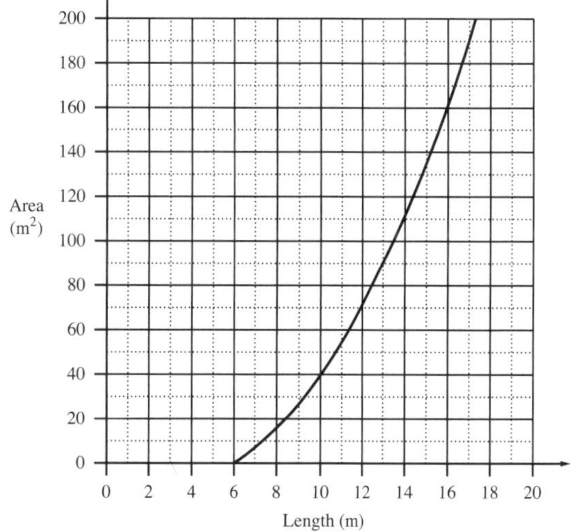

- **iii** Why are lengths of 0 metres to 6 metres impossible? *(1 mark)* **Medium**

iv What would be the dimensions of the playing surface if it had an area of 135 m^2? *(2 marks)* Easy

Company A constructs playing surfaces.

Company A charges

Size of playing surface	*Charges*
Up to and including 150 m^2	\$50 000
Greater than 150 m^2	\$50 000 plus a rate of \$300 per square metre for the area in excess of 150 m^2

v Draw a graph to represent the cost of using Company A to construct all playing surface sizes up to and including 200 m^2.
Use the horizontal axis to represent the area and the vertical axis to represent the cost. *(2 marks)* Hard

vi Company B charges a rate of \$360 per square metre regardless of size.
Which company would charge less to construct a playing surface with an area of 135 m^2? Justify your answer with suitable calculations. *(1 mark)* Medium

(Q27a, **2007 HSC**)

34 In 2004 there were 13.5 million registered motor vehicles in Australia. The number of registered motor vehicles is increasing at a rate of 2.3% per year.

Which expression represents the number (in millions) of registered motor vehicles, if y represents the number of years after 2004?

A $13.5 \times (1.023)^y$ **B** $13.5 \times (0.023)^y$
C $13.5 \times (1.023) \times y$ **D** $13.5 \times (0.023) \times y$

(1 mark)

(Q14, **2006 HSC**) Medium

35 The time (t) taken to clean a house varies inversely with the number (n) of people cleaning the house.

Which graph represents this relationship?

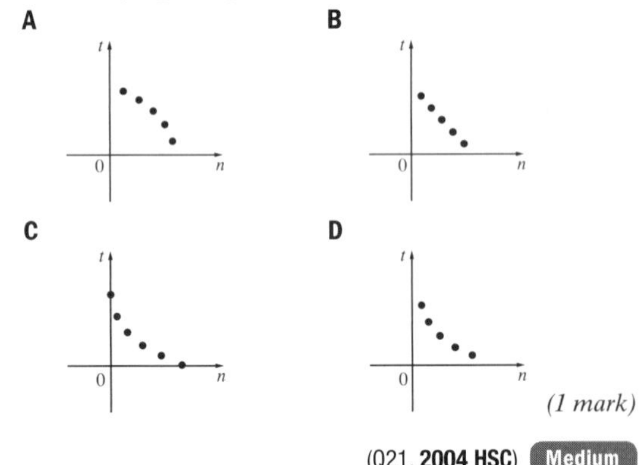

(1 mark)

(Q21, **2004 HSC**) Medium

36 **i** The number of bacteria in a culture grows from 100 to 114 in one hour.
What is the percentage increase in the number of bacteria? *(1 mark)* Easy

ii The bacteria continue to grow according to the formula $n = 100(1.14)^t$, where n is the number of bacteria after t hours.
What is the number of bacteria after 15 hours? *(1 mark)* Easy

Time in hours (t)	0	5	10	15
Number of bacteria (n)	100	193	371	?

iii Use the values of n from $t = 0$ to $t = 15$ to draw a graph of $n = 100(1.14)^t$. Use about half a page for your graph and mark a scale on each axis. *(4 marks)* Medium

iv Using your graph or otherwise, estimate the time in hours for the number of bacteria to reach 300. *(1 mark)* Medium

(Q26a, **2004 HSC**)

37 If an electrical current varies inversely with resistance, what is the effect on the current when the resistance is doubled?

A The current is doubled.
B The current is exactly the same.
C The current is halved.
D The current is squared. *(1 mark)*

(Q17, **2003 HSC**) Medium

38 In 2002, the population of Mexico was approximately 103 400 000.

i The growth rate of Mexico's population is estimated to be 1.57% per annum. If y represents the estimated number of people in Mexico at a time x years after 2002, write a formula relating x and y in the form $y = b(a^x)$. Use appropriate values for a and b in your formula. *(2 marks)* Medium

ii Using your formula, or otherwise, find an estimate for the size of Mexico's population two years after 2002.
Express your answer to the nearest thousand. *(2 marks)* Medium

(Q28b, **2003 HSC**)

39 A long rectangular sheet of metal 28 cm wide is to be made into a gutter by turning up sides of equal height x cm, perpendicular to the base.

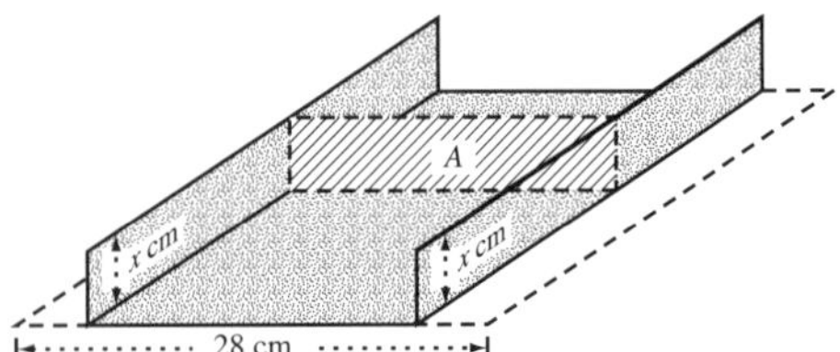

i Show that a formula for the cross-sectional area, A, of the gutter is
$A = 28x - 2x^2$. *(1 mark)* Hard

ii Explain why the formula in part **i** is only valid for values of x between 0 and 14. *(1 mark)* **Medium**

iii The graph of A against x, for values of x between 0 and 14, is a parabola, as shown.

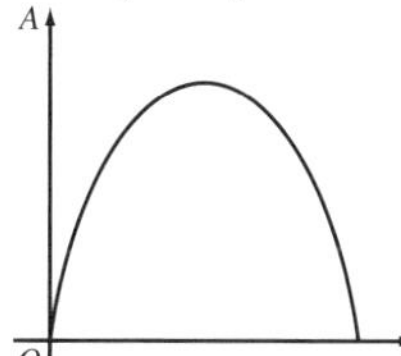

What is the maximum value of A? *(2 marks)* **Medium**

(Q28a, **2002 HSC**)

40 On the island of Wupetoi the unit of currency is the clam. The rate of inflation on Wupetoi has been constant for many years. Assuming the rate of inflation remains constant, the price of a surfboard will increase as shown in the graph.

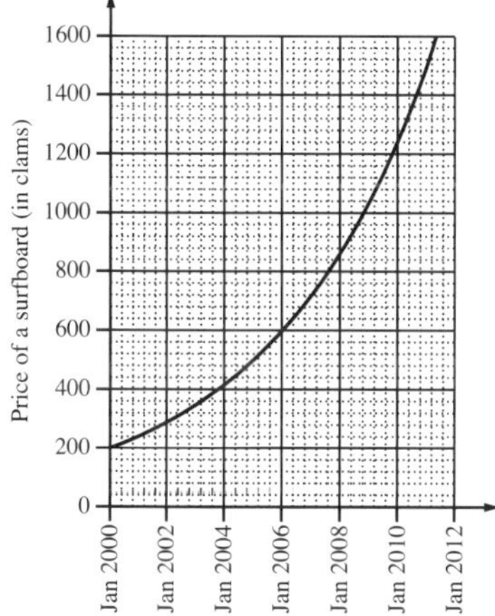

i The formula used to draw this graph was $P = A \times (1.2)^t$,
where P = price of a surfboard
and t = number of years after January 2000.

1. What is the value of A, and what does it represent? *(2 marks)* **Medium**
2. What annual rate of inflation has been assumed? *(1 mark)* **Hard**

ii In January 2000, Tana started saving a fixed number of clams each month, in order to buy a surfboard.
The straight line on the graph below represents Tana's savings.

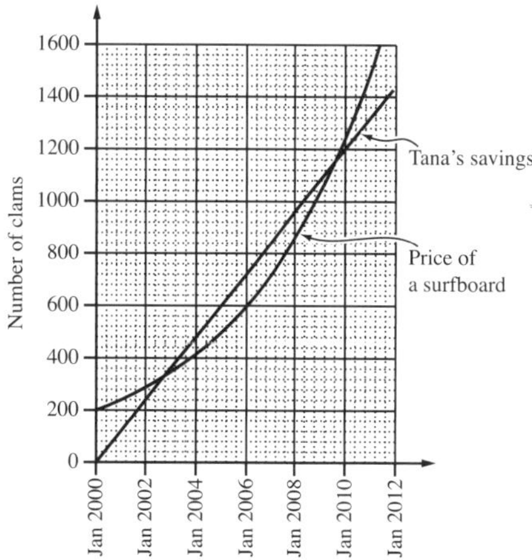

1. During which year will Tana first be able to afford a surfboard? Explain your answer. *(2 marks)* **Medium**
2. If Tana uses his savings to buy a surfboard in January 2006, how many clams will he have left? *(1 mark)* **Medium**
3. Write an equation that describes the relationship between Tana's savings in clams (c) and the number of months (n) after January 2000. *(2 marks)* **Hard**

(Q26b, **2001 HSC**)

41 Joe's pizzas are made in three different sizes.

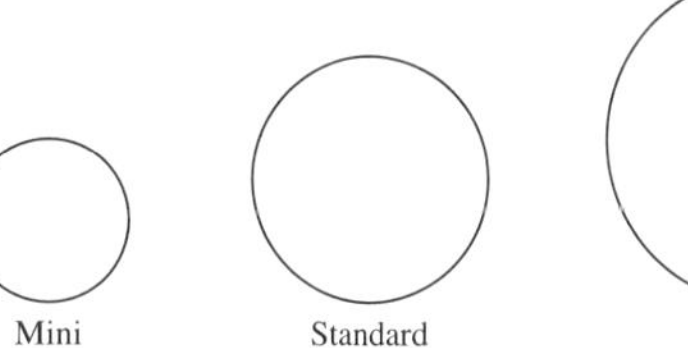

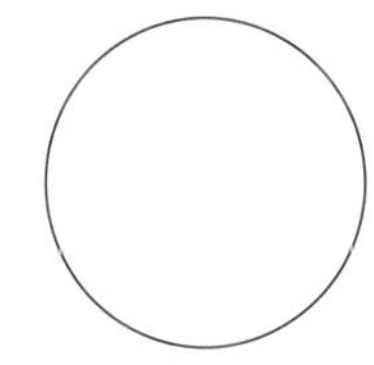

Joe puts olives on all his pizzas. The number of olives depends on the size of the pizza, as shown in the table.

Size	*Diameter, d* (cm)	*Number of olives, n*
Mini	20	8
Standard	30	18
Large	40	32

The relationship between the diameter of the pizza and the number of olives can be expressed by the formula:

$n = kd^2$, where k is a constant.

i Use a pair of values from the table to show that $k = 0.02$. *(1 mark)* **Medium**

ii Joe decides to make a mega-pizza, with diameter 52 cm. Use the formula to find the number of olives needed for a mega-pizza. *(1 mark)* **Medium**

iii Joe is asked to make a pizza in the shape of a square with sides of length 25 cm. He decides to use the same number of olives as would be needed on a round pizza with the same area. How many olives will be needed? *(3 marks)* **Hard**

(Q28a, **2001 HSC**)

42 Joe uses a microwave oven to heat lasagne. The time taken for heating is inversely proportional to the power setting (in watts). It takes ten minutes at a power setting of 240 watts to heat the lasagne.

How long would it take at a power setting of 500 watts? *(3 marks)*

(Q28b, **2001 HSC**) **Medium**

Year 12 Non-linear relationships—Worked answers

1 $y = 10(0.8)^x$
When $x = 0, y = 10$.
When $x = 1, y = 8$.
Answer A

2 It takes 8 cleaners 60 hours.
So it will take 1 cleaner
$60 \times 8 = 480$ hours.
It will take 10 cleaners
$480 \div 10 = 48$ hours.
Answer B

3 **a** $P = 2000(1.2)^t$
When $t = 0, P = 2000$.
The initial population is 2000.
(1 mark)

b $P = 2000(1.2)^t$
When $t = 5$,
$P = 2000(1.2)^5$
$= 4976.64$
After 5 years the population is 4977 to the nearest unit.
(1 mark)

c

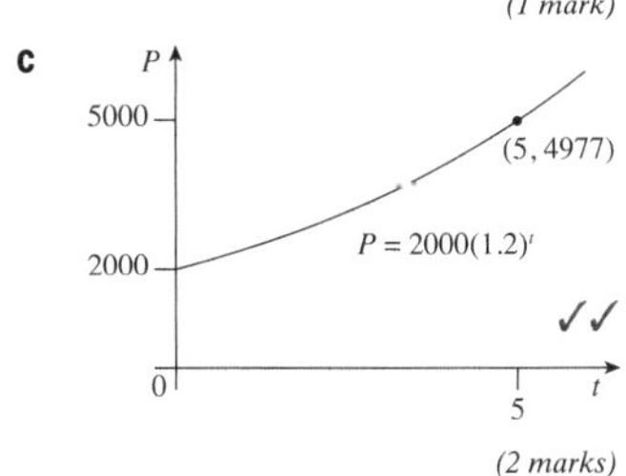

(2 marks)

4 **a** From the graph, $R = 0$ when $x = -10$ and 100.
The maximum value of R will occur midway between those values.

So $x = \dfrac{-10 + 100}{2}$
$= 45$ ✓
The maximum revenue will occur when the price is $(10 + 45)$ dollars or \$55. ✓
(2 marks)

b When $x = 0, R = 50000$.
The intercept of the parabola with the vertical axis is 50000.
(1 mark)

5 $y = -x^2 + 1$ is a parabola.
It passes through $(-1, 0)$, $(0, 1)$ and $(1, 0)$.

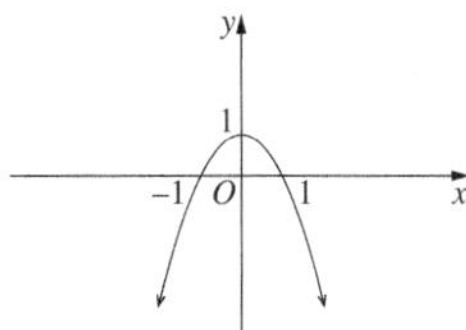

Answer C

6 **a** From the graph, when
$A = 30\,000, x = 100$ or
$x = 200$.
The largest possible value of x is 200.
(1 mark)

b From the graph the area is largest when $x = 150$. ✓
Now $3x + 2y = 900$
$2y = 900 - 3x$
$= 900 - 3 \times 150$
$= 450$
$y = 225$
The largest possible area occurs when $x = 150$ and $y = 225$. ✓
(2 marks)

c $A = xy$
$= 150 \times 225$
$= 33\,750$
The largest possible area is $33\,750$ m^2.
(1 mark)

7 **a** When $n = 40, y = 4000$
The number of bacteria at 40 minutes is 4000.
(1 mark)

b $y = A \times b^n$
When $n = 0, y = 1000$
$1000 = A \times b^0$
$= A \times 1$
So $A = 1000$.
$y = 1000 \times b^n$ ✓
When $n = 40, y = 4000$
$4000 = 1000 \times b^{40}$
So $b^{40} = 4$.
Now $1^{40} = 1$ and 2^{40}
$\approx 1.1 \times 10^{12}$
So b lies between 1 and 2 and is much closer to 1.
$1.1^{40} = 45.259\ldots$
So b lies between 1 and 1.1 (and is closer to 1).
$1.03^{40} = 3.262\ldots$
$1.04^{40} = 4.801\ldots$
So b lies between 1.03 and 1.04. ✓
(2 marks)

8 **a** w is the width of the rectangle. The width cannot be negative so the model only makes sense for non-negative values of w.
(1 mark)

b From the graph,
when $A = 18, w = 2$. ✓
Now Area = length × width
So $l \times 2 = 18$
$l = 9$
$P = 2(l + w)$
$= 2 \times (9 + 2)$
$= 22$
The perimeter of the rectangle is 22 cm. ✓
(2 marks)

9 **a** Distance $= 1.5 \times 80$ km
$= 120$ km
(1 mark)

b [At 20 km/h it will take 6 hours, at 40 km/h 3 hours, at 60 km/h 2 hours to travel 120 km.
So four possible points are (20, 6), (40, 3), (60, 2) and (80, 1.5).] ✓✓

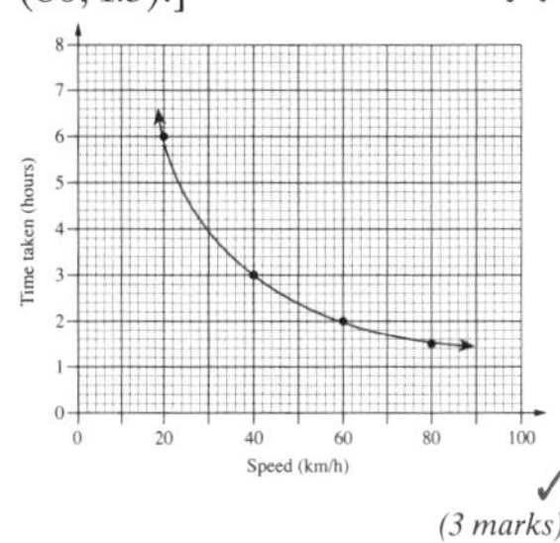

✓
(3 marks)

10 $y = x^2 - 2$ is concave up and has y-intercept -2.

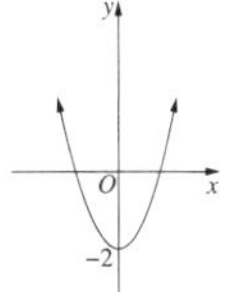

Answer A

11 **i** D is inversely proportional to A.
$\therefore D = \dfrac{k}{A}$ where k is a constant. ✓
At P, $A = 300$ and $D = 15$
$15 = \dfrac{k}{300}$
$k = 4500$
$\therefore D = \dfrac{4500}{A}$ ✓
(2 marks)

ii $D = \dfrac{4500}{A}$
When $D = 4$,
$4 = \dfrac{4500}{A}$
$4A = 4500$
$A = 1125$
The area of the shoe sole is 1125 cm^2.
(1 mark)

12 i From the graph, the maximum income occurs when the ticket price is increased by $6.

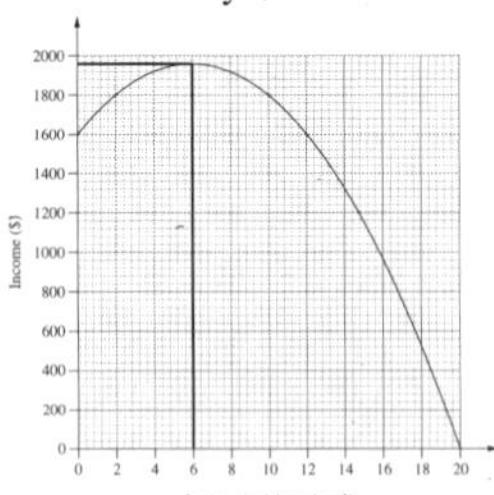

Price charged = $8 + $6
= $14

(1 mark)

ii Number of tickets
= 200 – 6 × 10
= 140

(1 mark)

iii Maximum income
= 140 × $14
= $1960 ✓
Cost to owners
= $500 + 140 × $2
= $780
Profit earned = $1960 – $780
= $1180 ✓

(2 marks)

13 i $A = 1.5$
[The initial value from the graph]

(1 mark)

ii Growth rate = 10%
[From the equation:
1.1 = 110% so the rate of increase is 10%.]

(1 mark)

14 i $C = \$(2100 + 0.5n)$

(1 mark)

ii [The y-intercept is 2100.
When $n = 800$, $C = 2500$.]

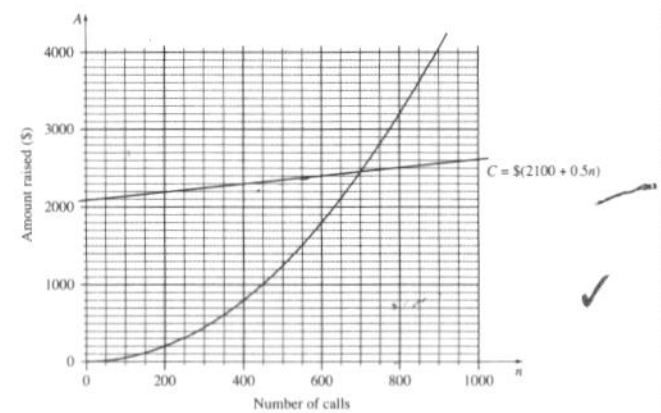

✓

From the graph, 700 calls are needed to break even. ✓

(2 marks)

15 The curve should cut the y-axis at the height of the diving board. The difference between the maximum height and the height at which the curve cuts the y-axis is how high the diver is above the height of the diving board when he reached the maximum height. ✓✓

(2 marks)

16 The graph is a parabola. The only equation of a parabola is $y = 3x^2 + 1$.

Answer C

17 i $120\,000 \div 500 = 240$

$120\,000 \div 1000 = 120$

$120\,000 \div 1500 = 80$

(n)	500	1000	1500	2000	2500	3000
(C)	240	120	80	60	48	40

(1 mark)

ii

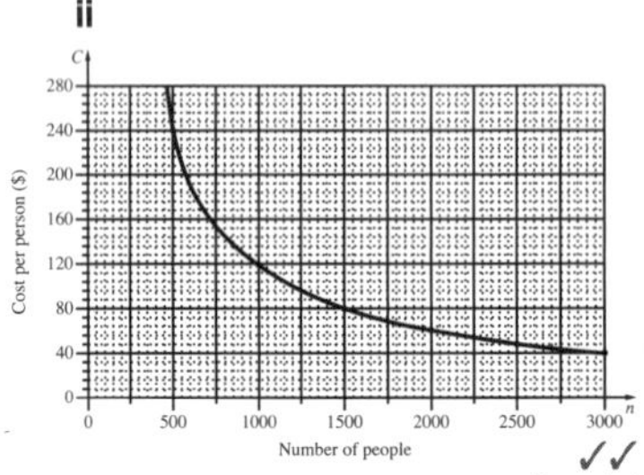

✓✓

(2 marks)

iii $C = \dfrac{120\,000}{n}$

(1 mark)

iv [Possible answers]

- n can only have positive integral values.
- There must be a limit to the number of people that can attend.
- There must be a limit to the price or no-one will choose to attend.
- The equation would be difficult to put into practice. People would not commit to attend without knowing the cost, but the cost cannot be calculated without knowing the number of people.

(1 mark)

v If $C = 94$,

$$94 = \frac{120\,000}{n}$$

$$94n = 120\,000$$

$$n = 1276.595\,74\ldots$$

So as n is not an exact number, the cost per person cannot be exactly $94.

[But, if 1277 people attend, the cost per person will be $93.97 which could be rounded to $94.]

(1 mark)

18 One side of the garden is a wall and does not need a fence.

Let the length of the garden be l m.

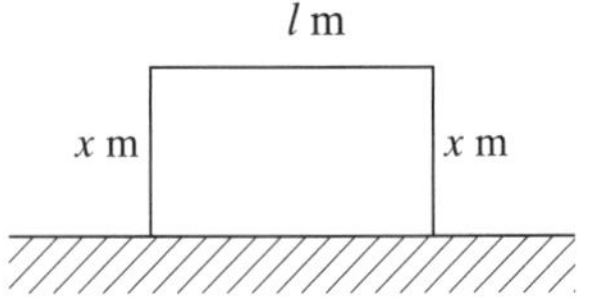

There is 20 m of fencing.

so $l + 2x = 20$
$l = 20 - 2x$
Area = length × width
$A = (20 - 2x) \times x$
$= 20x - 2x^2$

Answer D

19 i $T = 20000w^3$
When $w = 7.3$,
$T = 20000 \times 7.3^3$
$= 7\,780\,340$
The theoretical power generated is 7 780 340 watts.

(1 mark)

ii $A = 40\%$ of T
$= 0.4 \times 20000w^3$
$= 8000w^3$

(1 mark)

iii From the graph, when the wind speed is 9 m/s the actual power is 5.8 million watts and the theoretical power is 14.6 million watts.
Now 14.6 – 5.8 = 8.8
So the difference is 8.8 million watts.

(1 mark)

iv From the graph, to produce 4.4 million watts of actual power the wind speed needs to be 8.2 m/s.
So the minimum wind speed for the farm to be self-sufficient is 8.2 m/s.

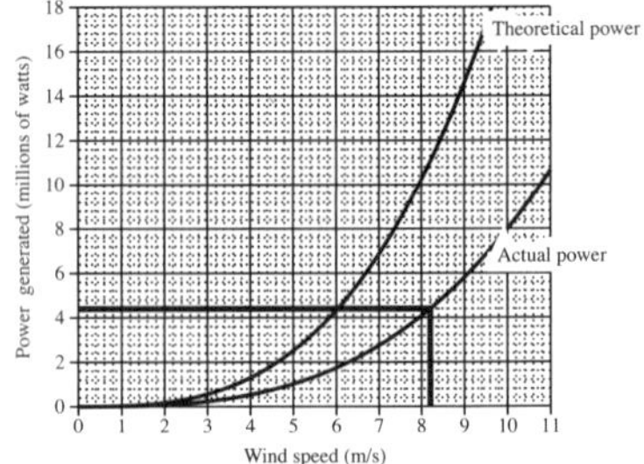

(1 mark)

v $P = 0.61 \times \pi \times r^2 \times w^3$
When $r = 43$ and $w = 8$,

$P = 0.61 \times \pi \times 43^2 \times 8^3$
$= 1\,814\,205.92\ldots$ ✓

If w increases by 10%,

new $w = 1.1 \times 8$
$= 8.8$

$P = 0.61 \times \pi \times 43^2 \times 8.8^3$
$= 2\,414\,708.08\ldots$ ✓

Increase
$= 2\,414\,708.08\ldots - 1\,814\,05.92\ldots$
$= 600\,502.1596\ldots$

% increase
$= \dfrac{600502.1596\ldots}{1814205.92\ldots} \times 100\%$
$= 33.1\%$

The percentage increase in power would be 33.1%. ✓

(3 marks)

$T = \dfrac{D}{S}$

$6 = \dfrac{D}{60}$

$D = 6 \times 60$
$= 360$

So $T = \dfrac{360}{S}$

When $S = 100$,

$T = \dfrac{360}{100}$
$= 3.6$

So the journey takes 3.6 hours or 3 hours and 36 minutes.

Answer D

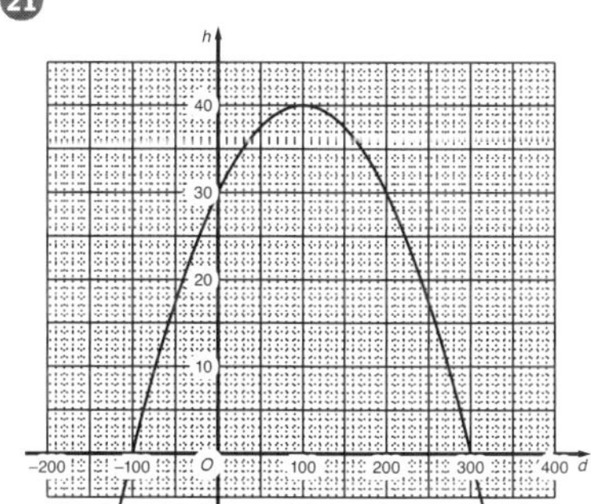

i From the graph, the maximum height reached by the ball is 40 m.

(1 mark)

ii The ball is at a height of 35 m when it is at a horizontal distance of 30 m and again when it is at a horizontal distance of 170 m.

Difference = 170 − 30
= 140 m

The horizontal distance travelled by the ball in that time is 140 m.

(1 mark)

iii The ball hits the ground at a distance of 300 m.

50 m short of that distance is 250 m.

The height of the ball is 17.5 m.

(1 mark)

iv If $d < 0$ the graph is not suitable. ✓

The ball was hit forwards and negative values of d apply to distances behind where the ball was hit.

If $d > 300$ the graph is not suitable. The ball hits the ground at a horizontal distance of 300 m and the graph suggests the height will then be negative but the ball will not go below the ground. ✓

(2 marks)

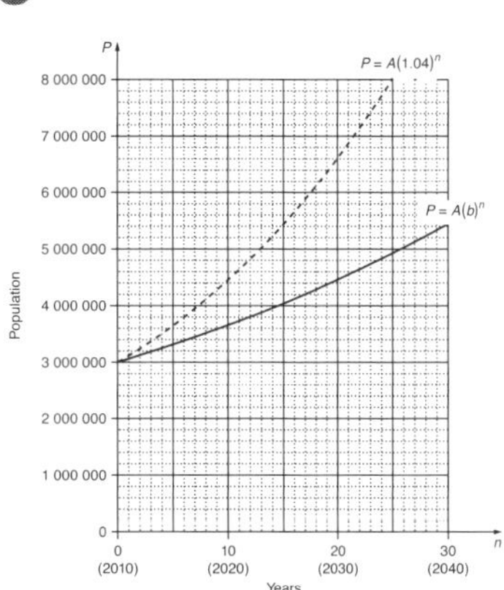

- - - Predicted population if the policy had not been introduced
—— Predicted population with the policy introduced

i In 2030, $n = 20$

From the graph, when $n = 20$,

$P = 6\,600\,000$

If the policy is not introduced the predicted population in 2030 is 6 600 000.

(1 mark)

ii $A = 3\,000\,000$ represents the population of Thagoras in 2010.

(1 mark)

iii 1 The population growth was slower using $P = A(b)^n$. So the value of b must be smaller than in the original equation in order for P to be smaller. That is $b < 1.04$. So a value of 1.05 is not suitable.

(1 mark)

2 When $n = 20$,
$P = 4\,460\,000$
$P = 3\,000\,000(b)^{20}$
Try $b = 1.03$
$P = 3\,000\,000(1.03)^{20}$
$= 5\,418\,333.704\ldots.$

This value of P is too high so b needs to be smaller. ✓

Try $b = 1.01$
$P = 3\,000\,000(1.01)^{20}$
$= 3\,660\,570.12\ldots$

This value of P is too small so b needs to be larger.

Try $b = 1.02$
$P = 3\,000\,000(1.02)^{20}$
$= 4\,457\,842.188\ldots$
$= 4\,460\,000$
[3 significant figures]

So the value of b should be 1.02 to 2 decimal places. ✓

(2 marks)

iv In 2050, $n = 40$

$P = 3\,000\,000(1.02)^{40}$
$= 6\,624\,118.991\ldots$ ✓

This is less than 7 000 000 so the model does suggest that the city will achieve its goal.

✓ *(2 marks)*

23 The graph of $y = a^x$ is an exponential curve (not a straight line). When $x = 0$, $y = 1$.

The graph that best represents $y = a^x$ is [D].

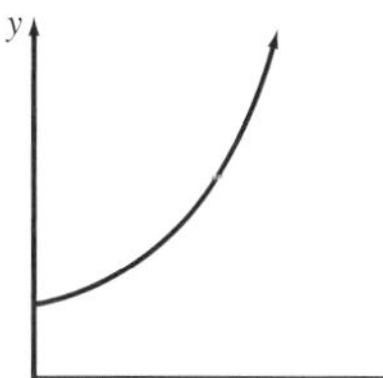

Answer D

24 i Try $t = 6$

$5 \times 3^6 = 3645$

Conclusion: $t = 6$ is too small.

(1 mark)

ii Try $t = 8$

$5 \times 3^8 = 32\,805$

Conclusion: $t = 8$ is too big. ✓

Try $t = 7$

$5 \times 3^7 = 10\,935$

Conclusion: $t = 7$ is too small.

Now t is a whole number.

So, the first value of t for which 5×3^t is bigger than 18 000 is $t = 8$. ✓

(2 marks)

25 i $P = \dfrac{a}{V}$

(1 mark)

ii $P = 3$ when $V = 2$

$P = \dfrac{a}{V}$

$3 = \dfrac{a}{2}$

$a = 3 \times 2$
$= 6$ ✓

So $P = \dfrac{6}{V}$

When $V = 4$,

$P = \dfrac{6}{4}$

$= 1.5$ ✓

(2 marks)

iii

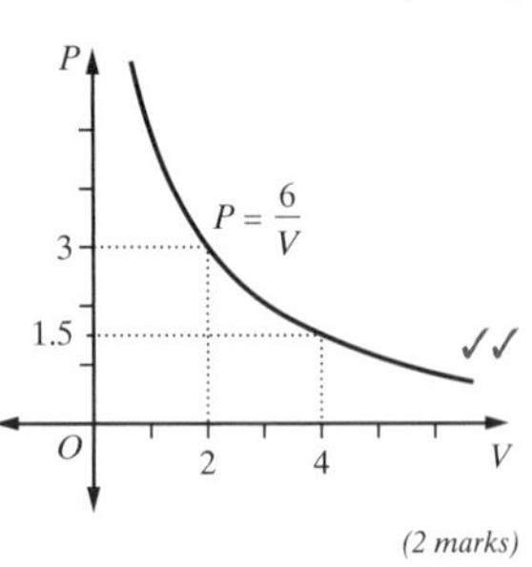

(2 marks)

26 Let h be the number of hours and $T°$ the temperature.

Now $h \propto \dfrac{1}{T}$

So $h = \dfrac{k}{T}$ for some constant k

When $T = 30, h = 8$

$8 = \dfrac{k}{30}$

$k = 8 \times 30$

$k = 240$

So $h = \dfrac{240}{T}$

When $T = 12$,

$h = \dfrac{240}{12}$

$= 20$

It would take 20 hours for the ice to melt at 12°C.

Answer B

27 i The cans have diameter 10 cm and height 10 cm. So 3 cans can be placed across the width of 30 cm, 4 cans can be placed along the length of 40 cm and 6 cans can be placed up the height of 60 cm.

Number of cans $= 3 \times 4 \times 6$

$= 72$

(1 mark)

ii Height of each label

$= (10 - 2 \times 0.5)$ cm

$= 9$ cm

Area of a closed cylinder

$A = 2\pi rh + 2\pi r^2$

Area of each label:

$A = 2\pi rh$

$= 2 \times \pi \times 5 \times 9$

$= 282.743\,33\ldots$ cm^2 ✓

Total area for 72 labels:

Area $= 72 \times 282.743\,33\ldots$ cm^2

$= 20\,357.52\ldots$ cm^2

$= 20\,358$ cm^2 [nearest cm^2]

(or $= 2.04$ m^2 ✓ [2 decimal places])

(2 marks)

iii Volume of existing can:

$V = \pi r^2 h$

$= \pi \times 5^2 \times 10$

$= 785.398\,1634\ldots$ cm^3

The volume of the existing can is about 785 cm³. ✓

If the diameter doubles to 20 cm, the new radius is 10 cm.

Volume of larger can:

$V = \pi r^2 h$

$= \pi \times 10^2 \times 10$

$= 3141.592\,654\ldots$ cm^3

Monica is not correct. The volume of the larger can is much more than twice the volume of the original can. (It is in fact 4 times the volume of the original can.)

✓ *(2 marks)*

iv To minimise the surface area Monica should recommend a can with radius of about 6.3 cm (from the first graph). ✓

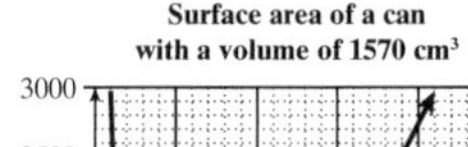

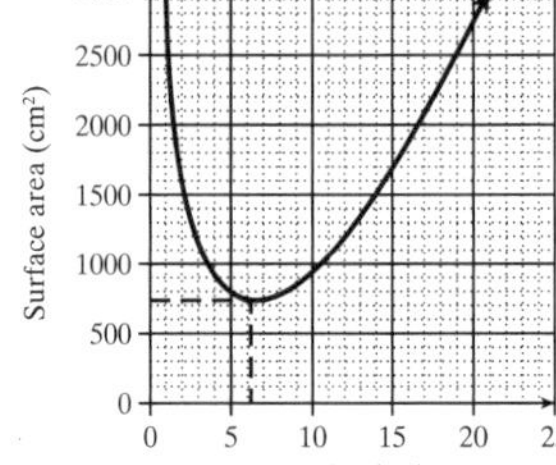

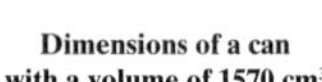

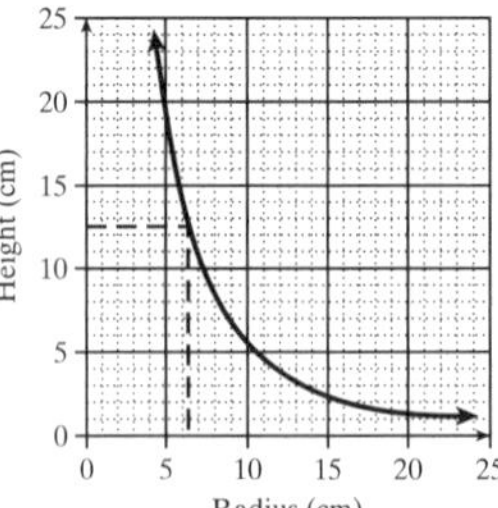

If the radius is 6.3 cm, the height is about 12.6 cm (from the second graph).

Monica should recommend a can with radius 6.3 cm and height 12.6 cm. ✓

(2 marks)

28 Time varying inversely with speed means that $t = \dfrac{d}{s}$ where d is a certain distance.

This equation can be represented by a hyperbola, noting that as speed becomes larger, the time becomes smaller.

Answer A

29 i

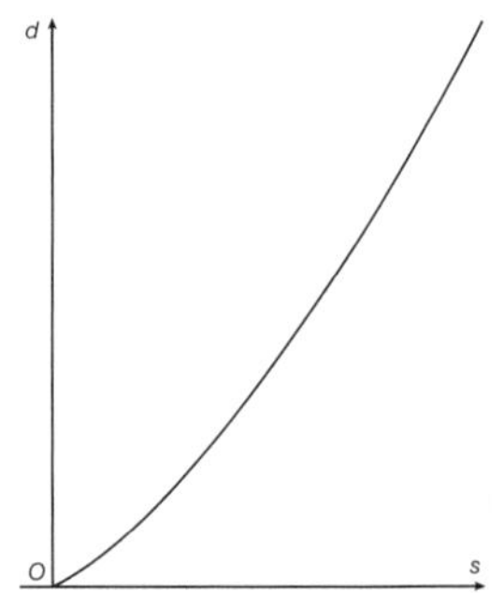

The car's speed (s) can only be positive.

(1 mark)

ii When $s = 40$ km/h, $d = 44$ m ✓

When $s = 70$ km/h,

$d = 0.01(70)^2 + 0.7(70)$

$= 98$ m

∴ The difference between stopping distances is

$98 - 44 = 54$ m. ✓

(2 marks)

30 Using GDC, graph $y = 3^x$.

or

$y = 3^x$ is not linear so correct answer is B or D.

When $x = 1$, $y = 3^1 = 3$

When $x = 2$, $y = 3^2 = 9$

Answer D

31 i $P = a \times 2^t$

If $t = 2$ and $P = 24$,

$24 = a \times 2^2$ ✓

$24 = 4a$

$\therefore a = 6$ ✓

(2 marks)

ii $P = 1500, a = 6$

$\therefore 1500 = 6 \times 2^t$

$\dfrac{1500}{6} = 2^t$

$250 = 2^t$ ✓

By guess and check:

$2^7 = 128$

$2^8 = 256$

∴ It will take nearly 8 years. ✓

(2 marks)

32 $p = \frac{k}{V}$

(N.B. If p varies inversely with V, then as V increases, p decreases.)

Answer A

33 i Width = 10 m
Length = 16 m
(1 mark)

ii Let the length be l, then the width is $l - 6$.
$\therefore A = l(l - 6)$
(1 mark)

iii Since the width is 6 metres less than the length, the length cannot be smaller than 6 metres. It cannot even be 6 metres as the width would then be 0 metres.
(1 mark)

iv If $A = 135$ m^2, then $l = 15$ m ✓ (from graph) and width = 15 – 6 = 9 m.
$\therefore$ Dimensions of playing surface are 15 m × 9 m. ✓
(2 marks)

v If playing surface = 200 m^2,
Charge = \$50000 + 50 × \$300
= \$65000 ✓

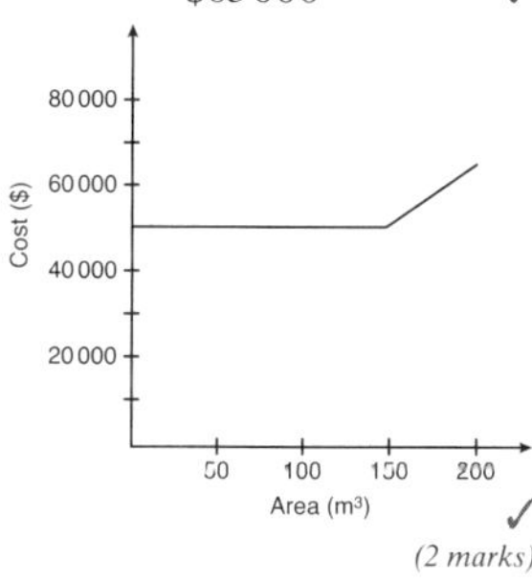

✓
(2 marks)

vi Company A would charge \$50000.
Company B:
Cost = 135 × \$360
= \$48600
$\therefore$ Company B charges less by \$1400.
(1 mark)

34 Each year, the number of vehicles is 2.3% more than the year before. This is similar to compound interest.

Using the formula,
$A = P(1 + r)^n$
$r = 2.3\% = 0.023$
$\therefore A = 13.5 \times (1 + 0.023)^y$
$= 13.5 \times (1.023)^y$

Answer A

35 $t = \frac{k}{n}$

This represents a hyperbola.

As the number of people (n) increases, the time (t) it takes to clean the house will decrease, but not at a constant rate.

Answer D

36 i Percentage increase
$= \frac{114 - 100}{100} \times 100$
$= 14\%$
(1 mark)

ii $n = 100(1.14)^t$

When $t = 15$,
$n = 100(1.14)^{15}$
$= 713.793\ldots$
$\doteqdot 714$ bacteria
(1 mark)

iii

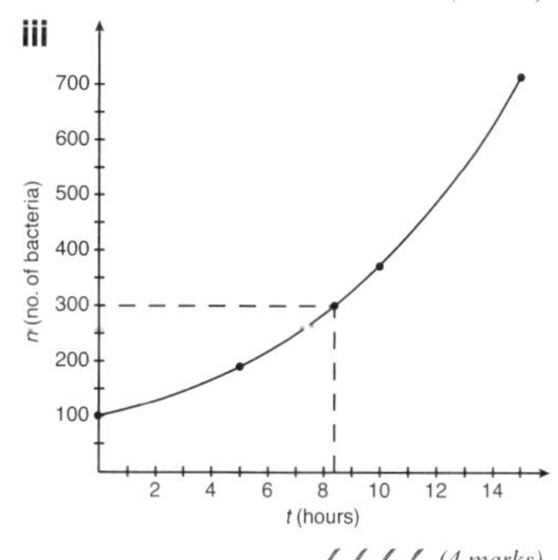

✓✓✓✓ *(4 marks)*

iv Using the equation,
$300 = 100(1.14)^t$
$3 = 1.14^t$
By guess and check, and using the graph, when $t = 8.4$
$1.14^{8.4} = 3.006\ldots$

$\therefore$ It takes about 8.4 hours for the number of bacteria to reach 300.
(1 mark)

37 $E = \frac{k}{R}$

Where E is electric current and R is resistance.

Let $R = 10$ and $k = 20$

$\therefore E = \frac{20}{10} = 2$

If R is doubled, then

$E = \frac{20}{20} = 1$

$\therefore E$ is halved

Answer C

38 i The formula of the equation required is similar to the formula for compound interest, i.e. $A = P(1 + r)^n$.

In this case, P is the present population, r is the growth rate and n is the number of years.

$y = 103\,400\,000\left(1 + \frac{1.57}{100}\right)^x$ ✓

$y = 1.034 \times 10^8(1.0157)^x$ ✓
(2 marks)

ii If $x = 2$,

$y = 1.034 \times 10^8 \times 1.0157^2$ ✓
$= 106\,672\,247.1$
$= 106\,670\,000$ ✓
to nearest thousand
(2 marks)

39 i A = width of gutter × height of gutter
Now width = $28 - 2x$
and height = x
$\therefore A = (28 - 2x) \times x$
$= 28x - 2x^2$
(1 mark)

ii The height of the gutter must be greater than 0 cm, therefore x must be greater than 0 cm and it must be less than 14 cm ($\frac{1}{2} \times 28$ cm), otherwise the width $(28 - 2x)$ will be less than 0 cm. Even 14 cm is not possible as this would not form a rectangular gutter.

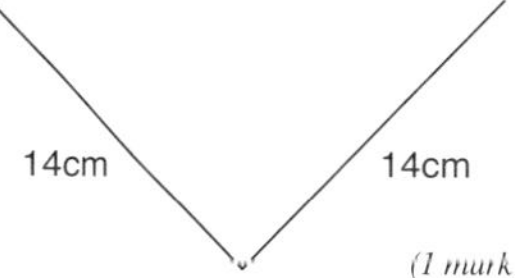

(1 mark)

iii Maximum value of A occurs when $x = 7$ cm (halfway between 0 and 14 cm).

$\therefore A = 28 \times 7 - 2 \times 7^2$ ✓
$= 196 - 98$
$= 98$ cm^2 ✓
(2 marks)

40 i 1 $A = 200$ ✓
It is the price of a surfboard in Jan 2000, i.e. at the beginning (when $t = 0$). ✓
(2 marks)

2 20%
($P = A \times (1.2)^t$ corresponds to the compound interest formula,
$A = P(1 + r)^n$
where $1 + r = 1.2$
$\therefore r = 0.2$
$= 0.2 \times 100\%$
$= 20\%$)
(1 mark)

ii **1** During 2002, Tana will have just saved enough for a surfboard. This can be seen by the intersection of the two graphs. ✓ ✓

(2 marks)

2 $720 - 600 = 120$ clams

(1 mark)

3 $y = mx + b$ (straight line)

$\therefore c = mn + b$

When $n = 0, c = 0 \therefore b = 0$

$\therefore c = mn$

When $n = 24$ months (2 years), ✓

$c = 240$

$\therefore 240 = m \times 24$

$\therefore m = 240 \div 24$

$= 10$

$\therefore c = 10n$ ✓

(2 marks)

41 **i**

$n = kd^2$

$n = 8, d = 20$

$\therefore 8 = k \times 20^2$

$8 = k \times 400$

$k = \frac{8}{400}$

$\therefore k = 0.02$

(1 mark)

ii If $d = 52$,

$n = 0.02 \times 52^2$

$= 54.08$

$\therefore$ Approx. 54 olives are required.

(1 mark)

iii Area of square pizza

$= 25 \times 25$

$= 625 \text{ cm}^2$ ✓

Area of round pizza $= \pi r^2$

$\therefore 625 = \pi r^2$

$r^2 = 625 \div \pi$

$= 198.94\ldots$

$\therefore r = 14.10\ldots$

$\therefore$ diameter $= 28.20\ldots$ cm ✓

No. of olives needed

$= kd^2$

$= 0.02 \times (28.20\ldots)^2$

$= 15.9\ldots$

$\therefore$ Approx. 16 olives are needed. ✓

(3 marks)

42

$T = \frac{k}{P}$ ✓

If $T = 10$, $P = 240$

$\therefore 10 = \frac{k}{240}$

$\therefore k = 2400$

$\therefore T = \frac{2400}{P}$ ✓

Now if $P = 500$,

$T = \frac{2400}{500}$

$= 4.8$ minutes ✓

(3 marks)

1 Consider the diagram below.

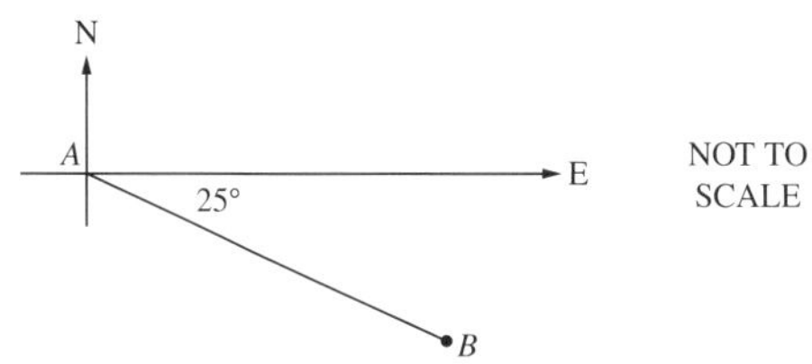

What is the true bearing of A from B?

A 025° **B** 065°
C 115° **D** 295° *(1 mark*

(Q14, **2021 HSC**) **Hard**

2 A right-angled triangle XYZ is cut out from a semicircle with centre O. The length of the diameter XZ is 16 cm and $\angle YXZ = 30°$, as shown on the diagram.

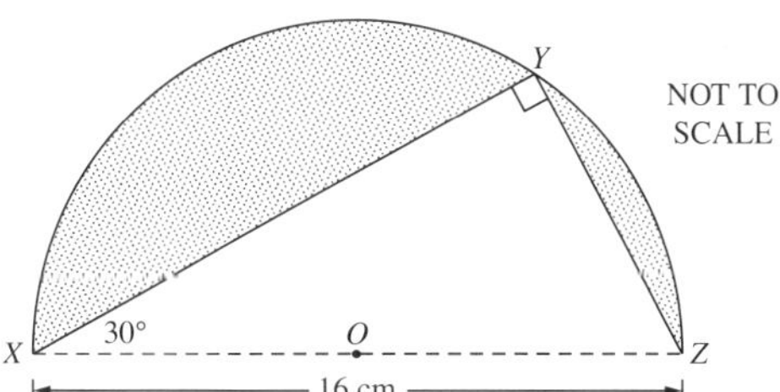

a Find the length of XY in centimetres, correct to two decimal places. *(2 marks)* **Medium**

b Hence, find the area of the shaded region in square centimetres, correct to one decimal place. *(3 marks)* **Hard**

CQ (Q32, **2021 HSC**)

3 The diagram shows a triangle ABC where $AC = 25$ cm, $BC = 16$ cm, $\angle BAC = 28°$ and angle ABC is obtuse.

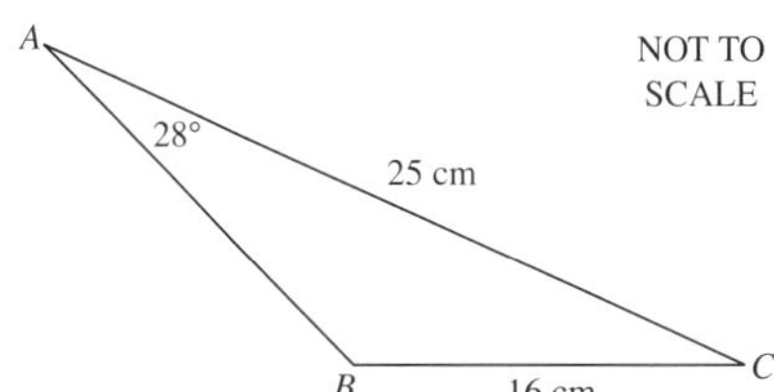

Find the size of the obtuse angle ABC correct to the nearest degree. *(3 marks)*

CQ (Q37, **2021 HSC**) **Hard**

4 The diagram shows a compass radial survey of the field $ABCD$.

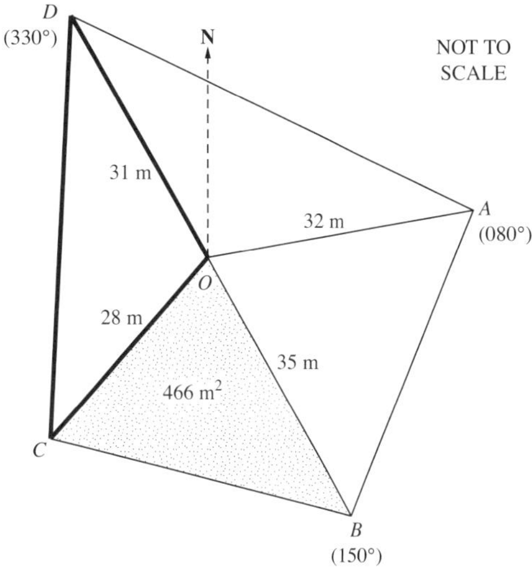

a Triangle COB has an area of 466 m^2. Find the size of acute angle COB, correct to the nearest degree. *(2 marks)* **Hard**

b A farmer wants to put a fence around the triangle DOC. Find the length of fencing required. Give your answer in metres correct to one decimal place. *(3 marks)* **Hard**

(Q39, **2021 HSC**)

5 Consider the triangle shown.

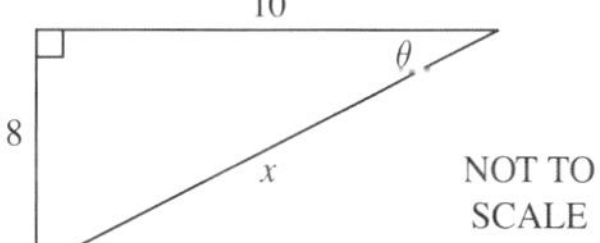

a Find the value of θ, correct to the nearest degree. *(2 marks)* **Easy**

b Find the value of x, correct to one decimal place. *(2 marks)* **Medium**

(Q16, **2020 HSC**)

6 Mr Ali, Ms Brown and a group of students were camping at the site located at P. Mr Ali walked with some of the students on a bearing of 035° for 7 km to location A. Ms Brown, with the rest of the students, walked on a bearing of 100° for 9 km to location B.

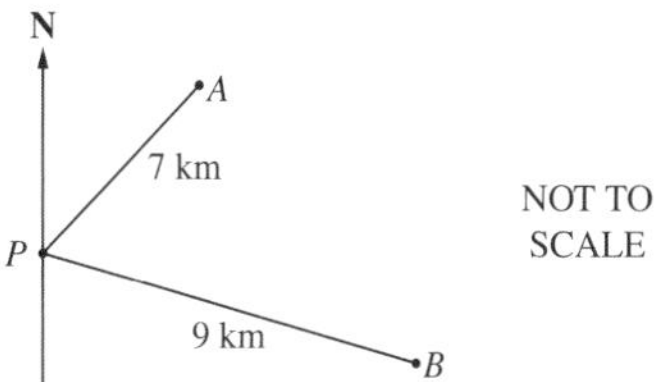

a Show that the angle APB is 65°. *(1 mark)* **Easy**

b Find the distance AB. *(2 marks)* **Medium**

c Find the bearing of Ms Brown's group from Mr Ali's group. Give your answer correct to the nearest degree. *(2 marks)* **Hard**

CQ (Q31, **2020 HSC**)

7 The diagram shows a regular decagon (ten-sided shape with all sides equal and all interior angles equal). The decagon has centre O.

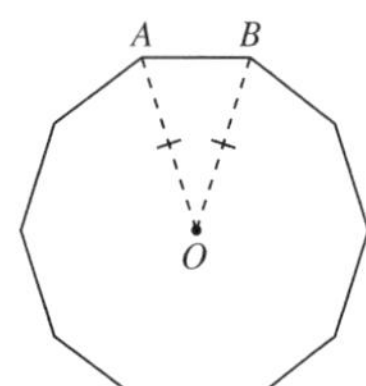

The perimeter of the shape is 80 cm.

By considering triangle OAB, calculate the area of the ten-sided shape. Give your answer in square centimetres correct to one decimal place. *(4 marks)*

CQ (Q32, **2020 HSC**) **Hard**

8 Which compass bearing is the same as a true bearing of 110°?

A S20°E **B** S20°W

C S70°E **D** S70°W *(1 mark)*

(Q4, **2019 HSC**) **Medium**

9 An owl is 7 metres above ground level, in a tree. The owl sees a mouse on the ground at an angle of depression of 32°.

How far must the owl fly in a straight line to catch the mouse, assuming the mouse does not move?

A 3.7 m **B** 5.9 m

C 8.3 m **D** 13.2 m *(1 mark)*

(Q12, **2019 HSC**) **Hard**

10 The diagram shows a triangle with sides of length x cm, 11 cm and 13 cm and an angle of 80°.

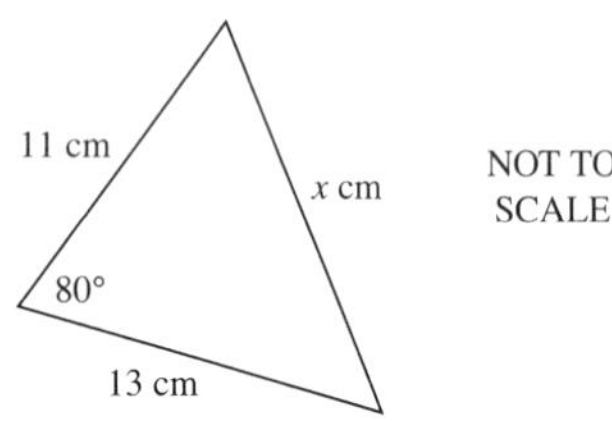

Use the cosine rule to calculate the value of x, correct to two significant figures. *(3 marks)*

(Q17, **2019 HSC**) **Medium**

11 Two right-angled triangles, ABC and ADC, are shown.

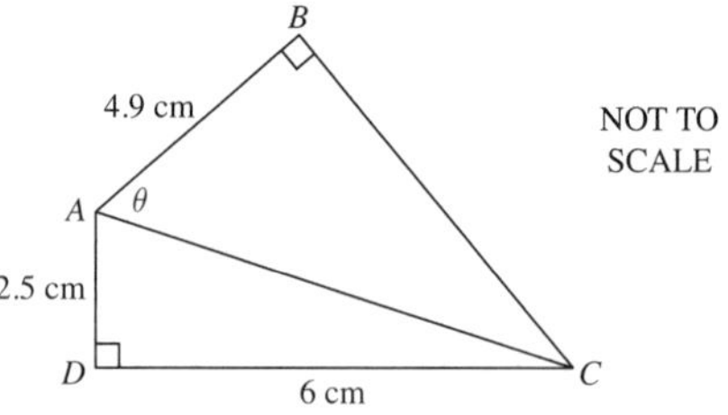

Calculate the size of angle θ, correct to the nearest minute. *(3 marks)*

(Q22, **2019 HSC**) **Medium**

12 A compass radial survey shows the positions of four towns A, B, C and D relative to point O.

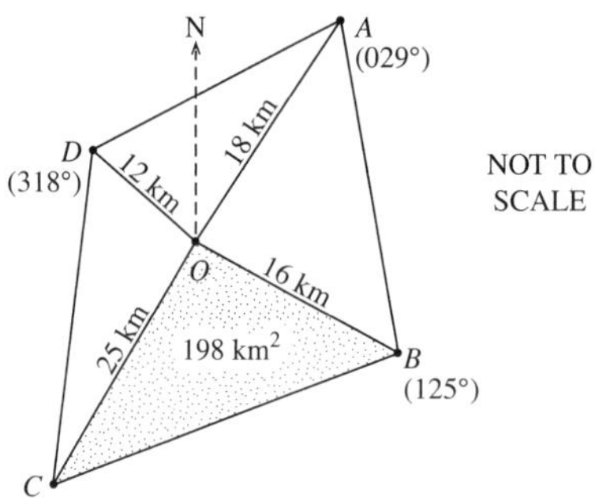

The area of triangle BOC is 198 km^2.

Calculate the bearing of town C from point O, correct to the nearest degree. *(3 marks)*

(Q35, **2019 HSC**) **Hard**

13 The diagram shows three towns P, Q and R. Town Q is due east of Town P. The bearing of Town R from Town Q is 234°. The distance between P and Q is 170 km and between Q and R is 265 km.

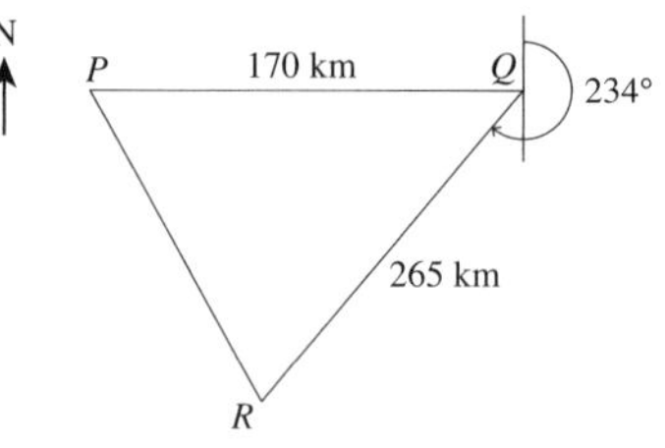

a A plane flies at an average speed of 280 km/h between the three towns.
How long does the flight from P to R take? Give your answer to the nearest minute. *(3 marks)* **Medium**

b A second plane takes off from R and flies east to town S which is directly south of Q. If this plane flew for 48 minutes, what was the average speed of the flight, to the nearest km/h? *(2 marks)* **Hard**

Bonus question (see page iv)

14 The diagram shows the positions of towns A, B and C.

Town A is due north of town B and $\angle CAB = 34°$.

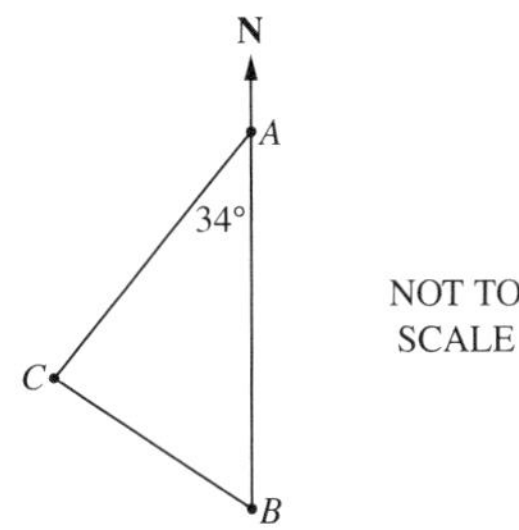

NOT TO SCALE

What is the bearing of town C from town A?

A 034° **B** 146°
C 214° **D** 326° *(1 mark)*

(Q7, **2018 HSC**) Easy

15 The diagram shows a triangle with side lengths 8 m, 9 m and 10 m.

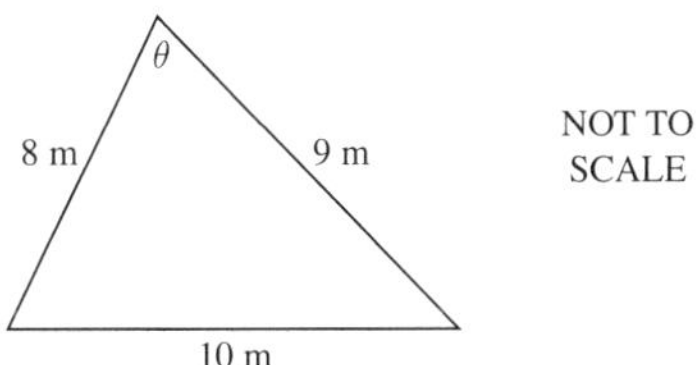

NOT TO SCALE

What is the value of θ, marked on the diagram, to the nearest degree?

A 49° **B** 51°
C 59° **D** 72° *(1 mark)*

(Q12, **2018 HSC**) Easy

16 The diagram shows two triangles.

Triangle ABC is right-angled, with $AB = 13$ cm and $\angle ABC = 62°$.

In triangle ACD, $AD = x$ cm and $\angle DAC = 40°$. The area of triangle ACD is 30 cm^2.

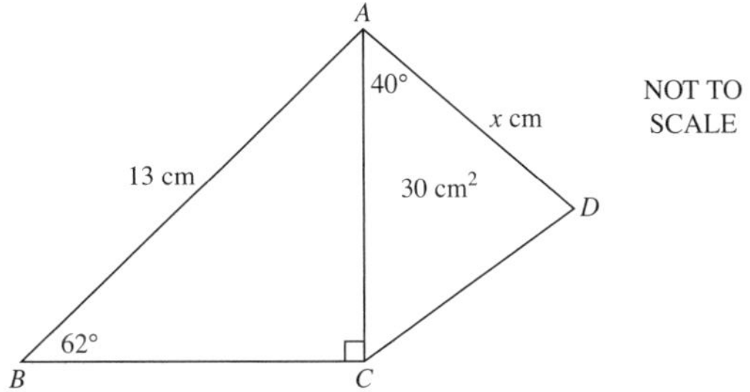

NOT TO SCALE

What is the value of x, correct to one decimal place? *(3 marks)*

(Q30c, **2018 HSC**) Hard

17 The diagram shows a right-angled triangle.

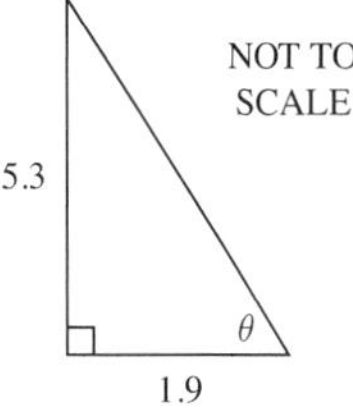

What is the value of θ, to the nearest minute?

A 70°16' **B** 70°17'
C 70°27' **D** 70°28' *(1 mark)*

(Q8, **2017 HSC**) Easy

18 A sewer pipe needs to be placed into the ground so that it has a 2° angle of depression. The length of the pipe is 15 000 mm.

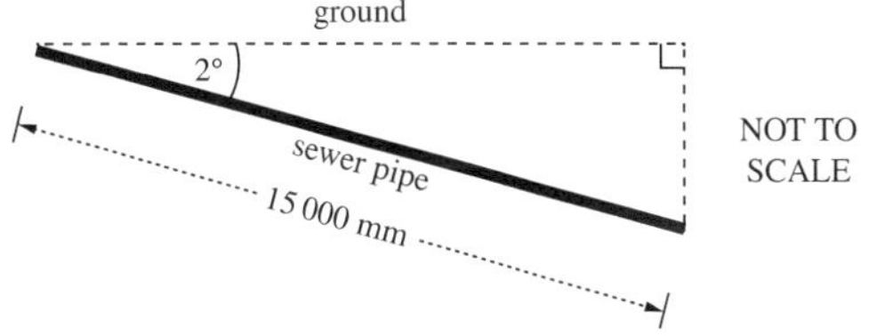

NOT TO SCALE

How much deeper should one end of the pipe be compared to the other end? Answer to the nearest mm. *(2 marks)*

(Q26d, **2017 HSC**) Medium

19 The diagram shows the location of three schools. School A is 5 km due north of school B, school C is 13 km from school B and $\angle ABC$ is 135°.

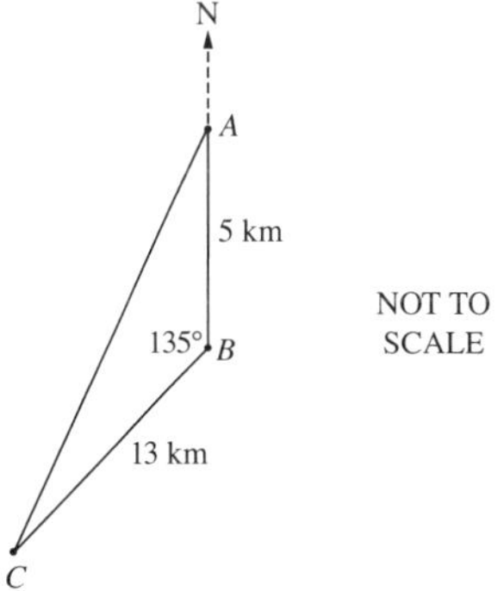

NOT TO SCALE

i Calculate the shortest distance from school A to school C, to the nearest kilometre. *(2 marks)* Medium

ii Determine the bearing of school C from school A, to the nearest degree. *(3 marks)* Hard

(Q30c, **2017 HSC**)

20 The diagram shows towns A, B and C. Town B is 40 km due north of town A. The distance from B to C is 18 km and the bearing of C from A is 025°. It is known that $\angle BCA$ is obtuse.

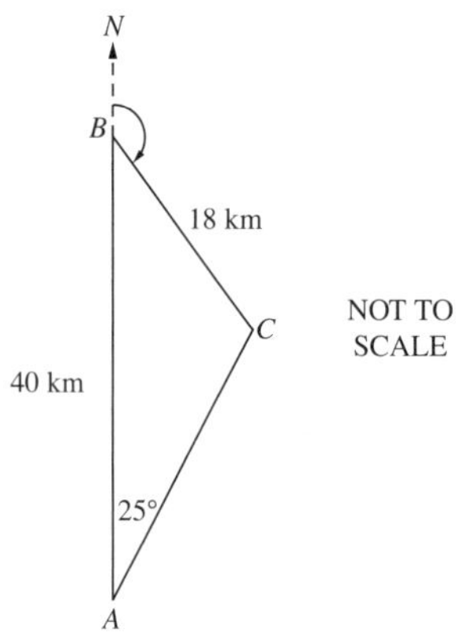

What is the bearing of C from B?

A 070° **B** 095°
C 110° **D** 135° *(1 mark)*

(Q25, **2016 HSC**) **Hard**

21 A school playground consists of part of a circle, with centre O, and a rectangle as shown in the diagram. The radius OB of the circle is 45 m, the width BC of the rectangle is 20 m and $\angle AOB$ is 100°.

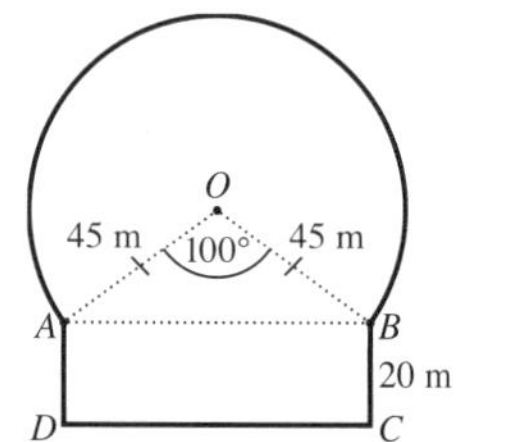

What is the area of the whole playground, correct to the nearest square metre? *(5 marks)*

(Q30c, **2016 HSC**) **Hard**

22 The diagram shows a radial survey of a field $ABCD$.

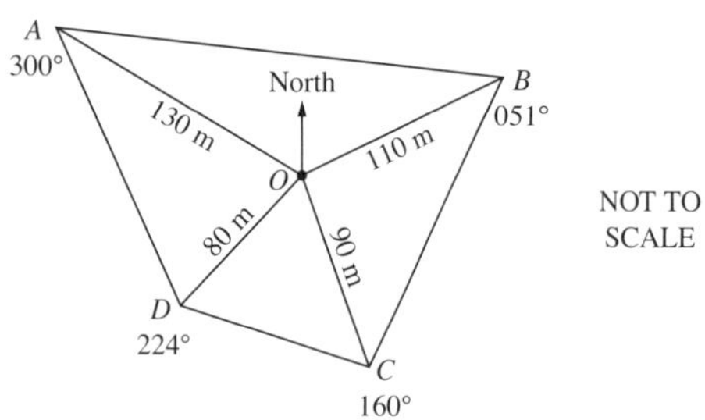

In triangle AOB, what is the size of $\angle AOB$?

A 51° **B** 111°
C 125° **D** 249° *(1 mark)*

(Q7, **2015 HSC**) **Easy**

23 From the top of a cliff 67 metres above sea level, the angle of depression of a buoy is 42°.

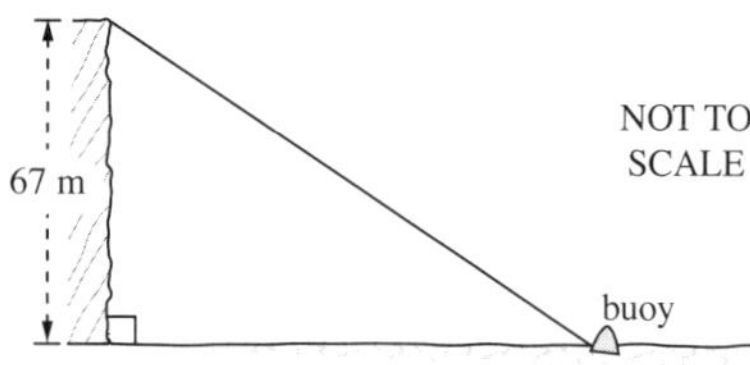

How far is the buoy from the base of the cliff, to the nearest metre?

A 60 m **B** 74 m
C 90 m **D** 100 m *(1 mark)*

(Q9, **2015 HSC**) **Medium**

24 The area of the triangle shown is 250 cm^2.

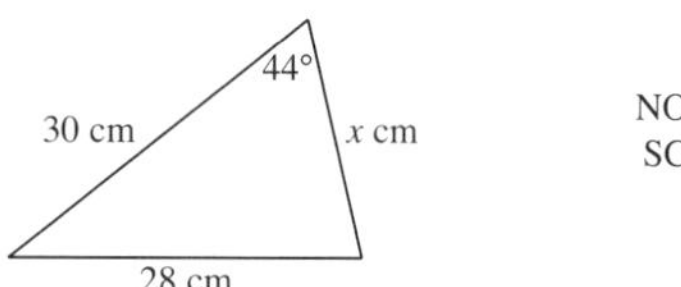

What is the value of x, correct to the nearest whole number?

A 11 **B** 18
C 22 **D** 24 *(1 mark)*

(Q22, **2015 HSC**) **Medium**

25 From point S, which is 1.8 m above the ground, a pulley at P is used to lift a flat object F. The lengths SP and PF are 5.4 m and 2.1 m respectively. The angle PSC is 108°.

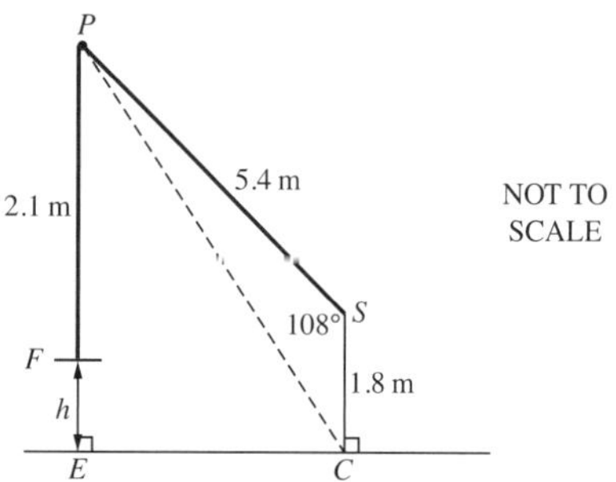

i Show that the length PC is 6.197 m, correct to 3 decimal places. *(1 mark)* **Medium**

ii Calculate h, the height of the object above the ground. *(4 marks)* **Hard**

(Q30e, **2015 HSC**)

26 The following information is given about the locations of three towns X, Y and Z:

- X is due east of Z
- X is on a bearing of 145° from Y
- Y is on a bearing of 060° from Z.

Which diagram best represents this information?

A **B**

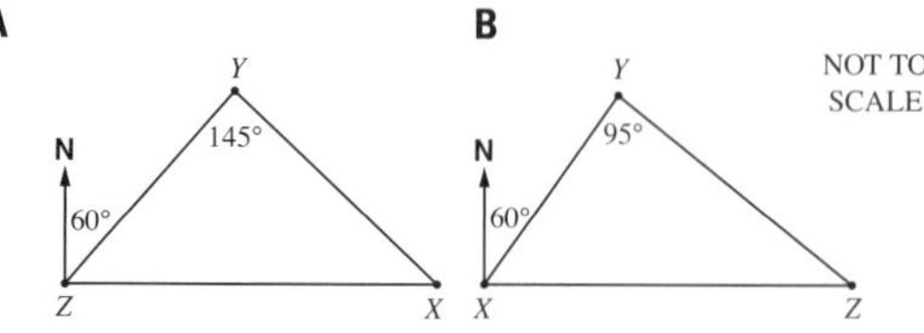

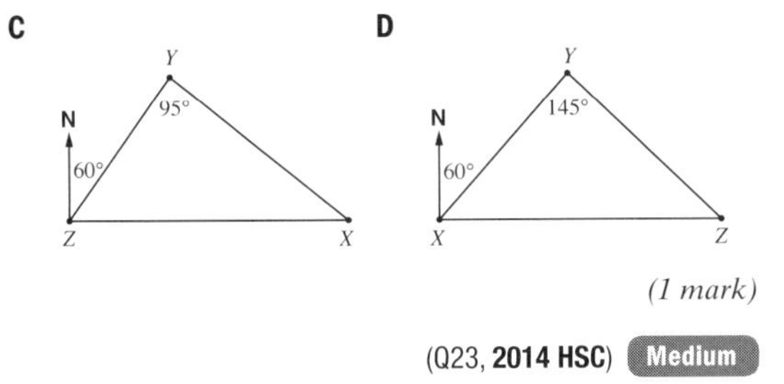

(1 mark)

(Q23, **2014 HSC**) Medium

27 Calculate the value of h correct to two decimal places.

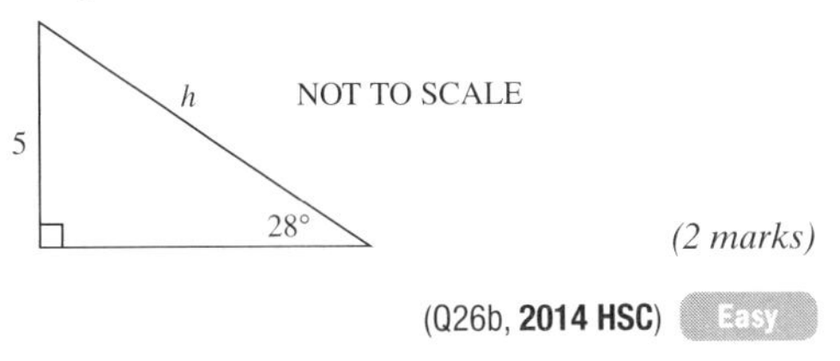

(2 marks)

(Q26b, **2014 HSC**) Easy

28 A radial compass survey of a sports centre is shown in the diagram.

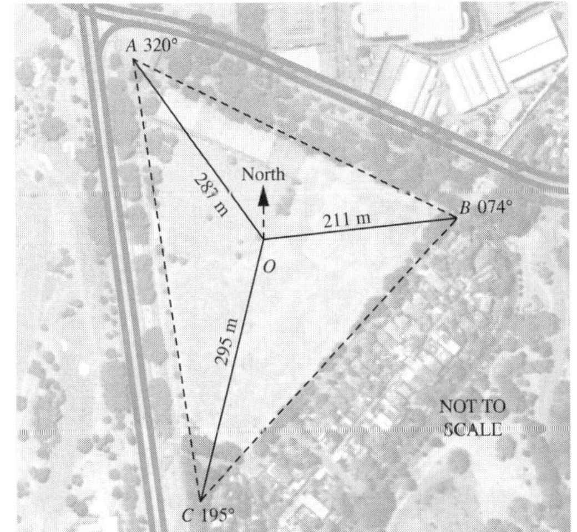

i Show that the size of angle AOB is 114°. *(1 mark)* Medium

ii Calculate the length of the boundary AB, to the nearest metre. *(2 marks)* Medium

iii Find the area of triangle AOB in hectares, correct to two significant figures. *(3 marks)* Medium

(Q28b, **2014 HSC**)

29 What is the value of θ, to the nearest degree?

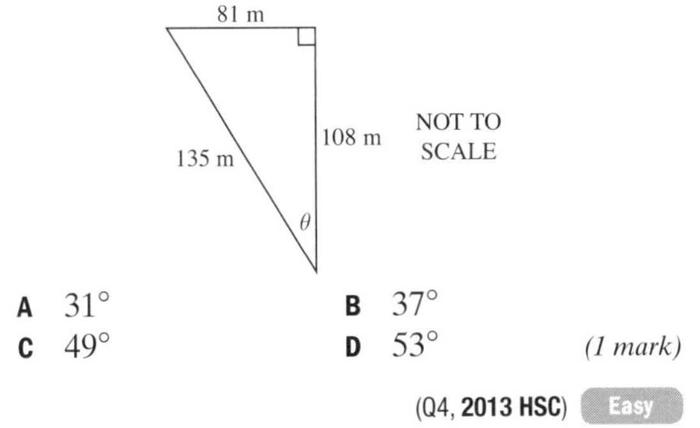

A 31° **B** 37°

C 49° **D** 53° *(1 mark)*

(Q4, **2013 HSC**) Easy

30 What is the value of θ, to the nearest degree?

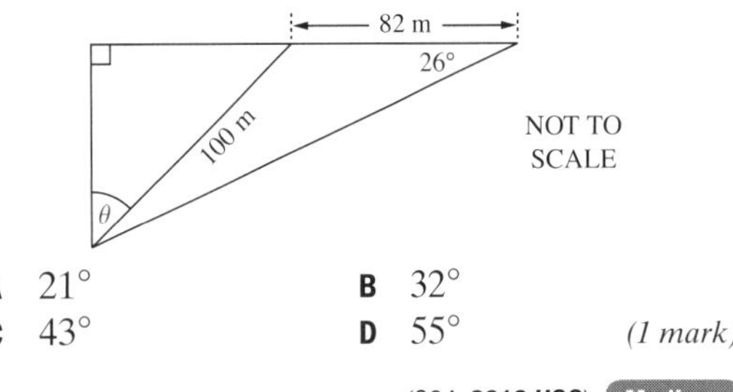

A 21° **B** 32°

C 43° **D** 55° *(1 mark)*

(Q24, **2013 HSC**) Medium

31 Triangle PQR is shown.

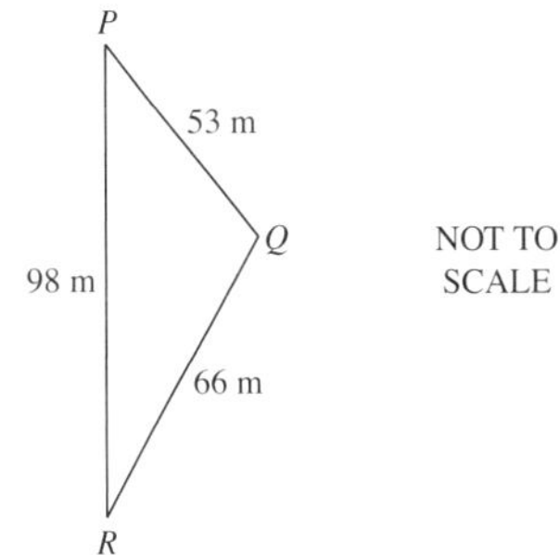

Find the size of angle Q, to the nearest degree. *(2 marks)*

(Q26a, **2013 HSC**) Easy

32 A compass radial survey of the field $ABCD$ has been conducted from O.

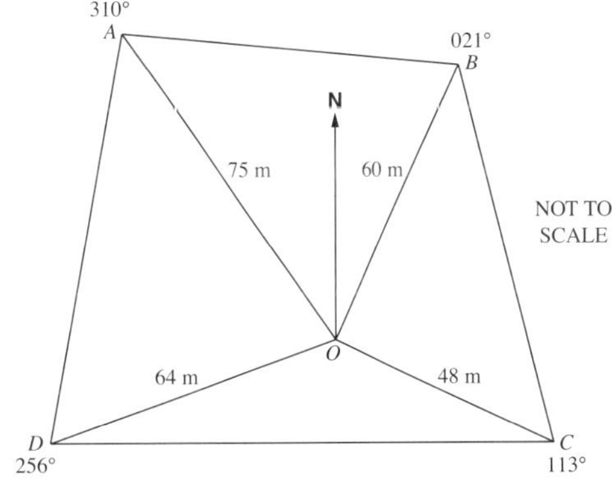

Find the area of the section ABO, to the nearest square metre. *(2 marks)*

(Q28a, **2013 HSC**) Medium

33 Which expression could be used to calculate the value of x in this triangle?

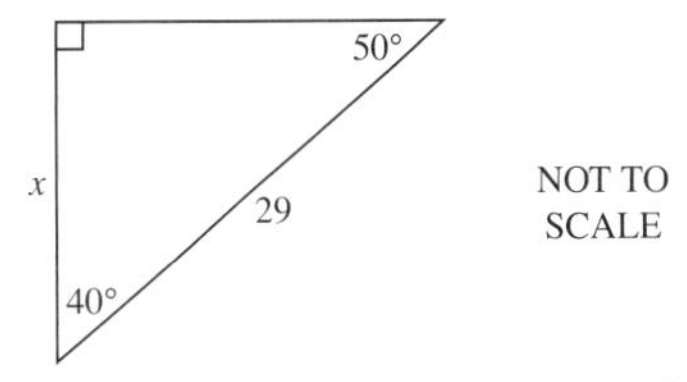

A $29 \times \cos 40°$ **B** $29 \times \cos 50°$

C $\dfrac{\cos 40°}{29}$ **D** $\dfrac{\cos 50°}{29}$ *(1 mark)*

(Q4, **2012 HSC**) Easy

34 What is the area of this triangle, to the nearest square metre?

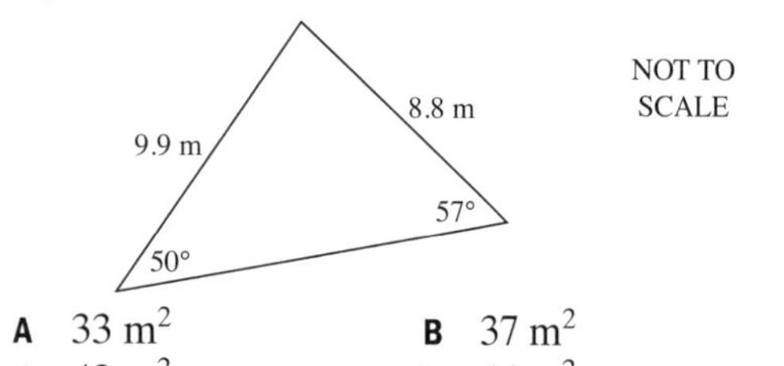

A 33 m^2 **B** 37 m^2
C 42 m^2 **D** 44 m^2 *(1 mark)*

(Q10, **2012 HSC**) Easy

35 Town B is 80 km due north of Town A and 59 km from Town C. Town A is 31 km from Town C.

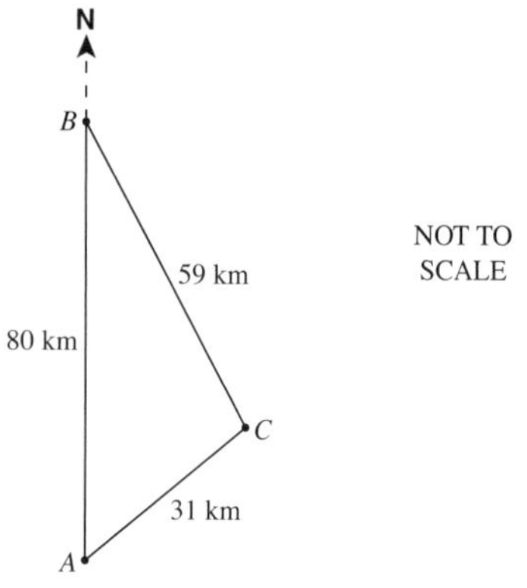

What is the bearing of Town C from Town B?

A 019° **B** 122°
C 161° **D** 341° *(1 mark)*

(Q20, **2012 HSC**) Medium

36 A disability ramp is to be constructed to replace steps, as shown in the diagram. The angle of inclination for the ramp is to be 5°.

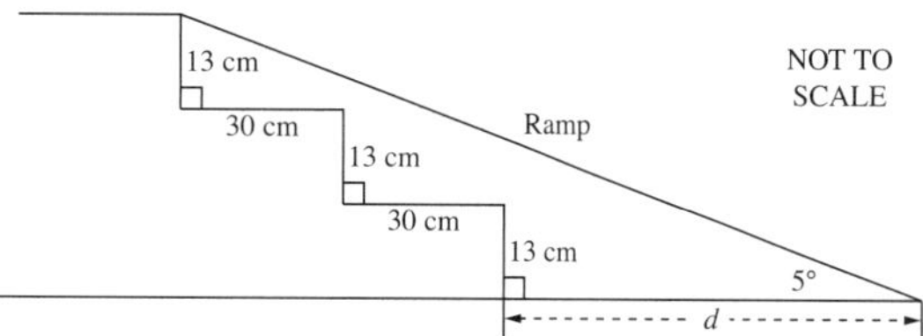

Calculate the extra distance, d, that the ramp will extend beyond the bottom step. Give your answer to the nearest centimetre. *(3 marks)*

(Q27d, **2012 HSC**) Hard

37 Raj cycles around a course. The course starts at E, passes through F, G and H and finishes at E. The distances EH and GH are equal.

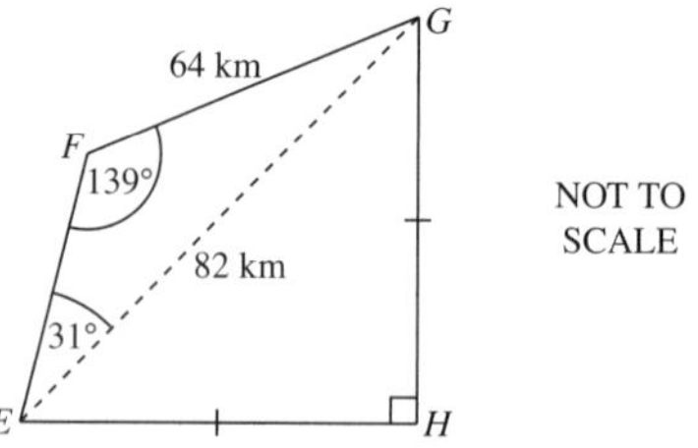

i What is the length of EF, to the nearest kilometre? *(2 marks)* Medium

ii What is the total distance that Raj cycles, to the nearest kilometre? *(3 marks)* Hard

(Q29c, **2012 HSC**)

38 The angle of depression from a kookaburra's feet to a worm on the ground is 40°. The worm is 15 metres from a point on the ground directly below the kookaburra's feet.

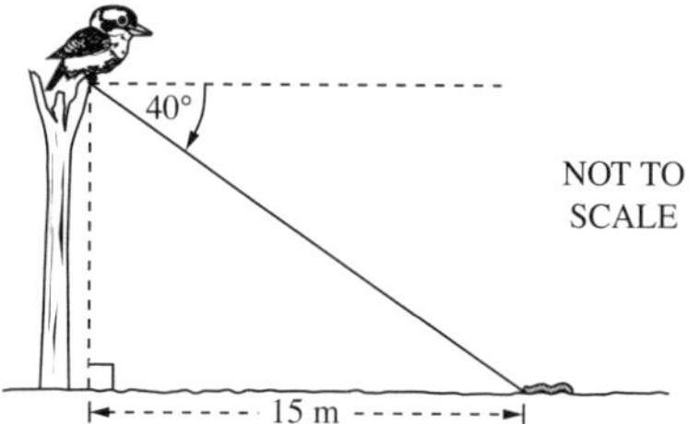

How high above the ground are the kookaburra's feet, correct to the nearest metre?

A 10 m **B** 11 m
C 13 m **D** 18 m *(1 mark)*

(Q4, **2011 HSC**) Easy

39 Two trees on level ground, 12 metres apart, are joined by a cable. It is attached 2 metres above the ground to one tree and 11 metres above the ground to the other.

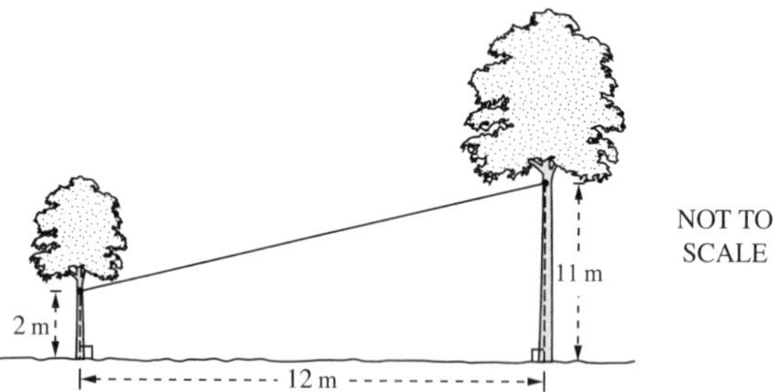

What is the length of the cable between the two trees, correct to the nearest metre?

A 9 m **B** 12 m
C 15 m **D** 16 m *(1 mark)*

(Q9, **2011 HSC**) 

40 A ship sails 6 km from A to B on a bearing of 121°. It then sails 9 km to C. The size of angle ABC is 114°.

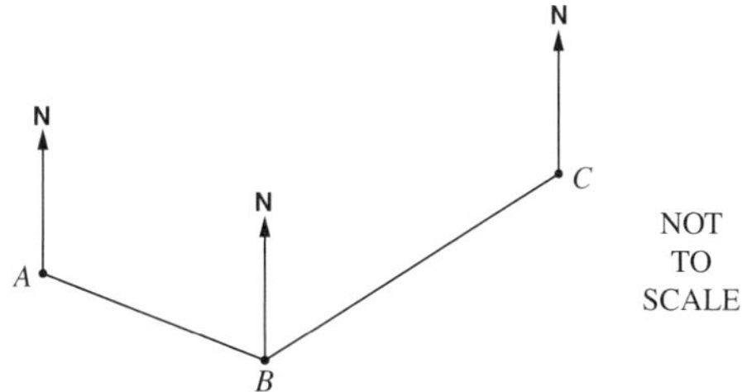

Copy the diagram into your writing booklet and show all the information on it.

i What is the bearing of C from B? *(1 mark)* **Hard**

ii Find the distance AC. Give your answer correct to the nearest kilometre. *(2 marks)* **Medium**

iii What is the bearing of A from C? Give your answer correct to the nearest degree. *(3 marks)* **Hard**

(Q24c, **2011 HSC**)

41 Three towns P, Q and R are marked on the diagram.

The distance from R to P is 76 km. $\angle RQP = 26°$ and $\angle RPQ = 46°$.

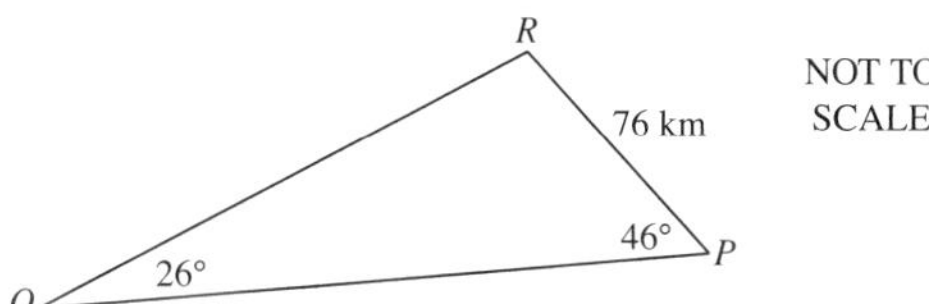

What is the distance from P to Q to the nearest kilometre?

A 100 km **B** 125 km

C 165 km **D** 182 km *(1 mark)*

(Q9, **2010 HSC**) **Easy**

42 A plane flies on a bearing of 150° from A to B.

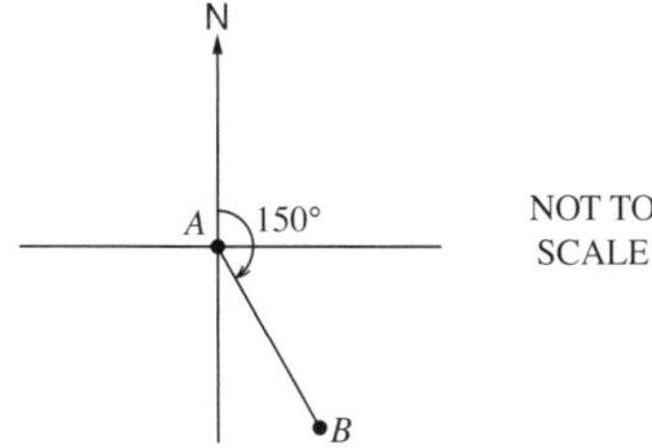

What is the bearing of A from B?

A 30° **B** 150°

C 210° **D** 330° *(1 mark)*

(Q10, **2010 HSC**) **Medium**

43 The base of a lighthouse, D, is at the top of a cliff 168 metres above sea level. The angle of depression from D to a boat at C is 28°. The boat heads towards the base of the cliff, A, and stops at B. The distance AB is 126 metres.

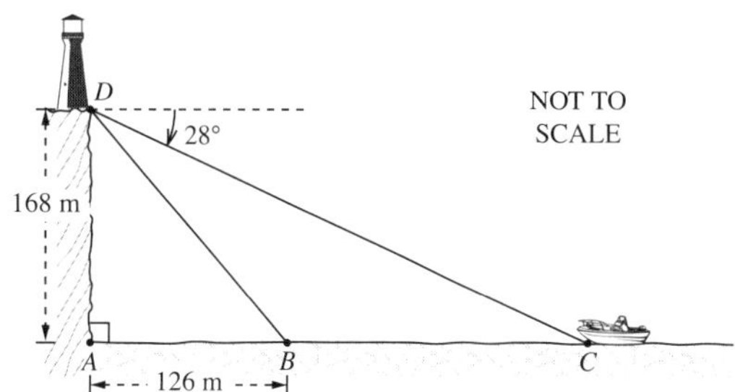

i What is the angle of depression from D to B, correct to the nearest minute? *(3 marks)* **Hard**

ii How far did the boat travel from C to B, correct to the nearest metre? *(2 marks)* **Hard**

(Q24d, **2010 HSC**)

44 Find the area of triangle ABC, correct to the nearest square metre.

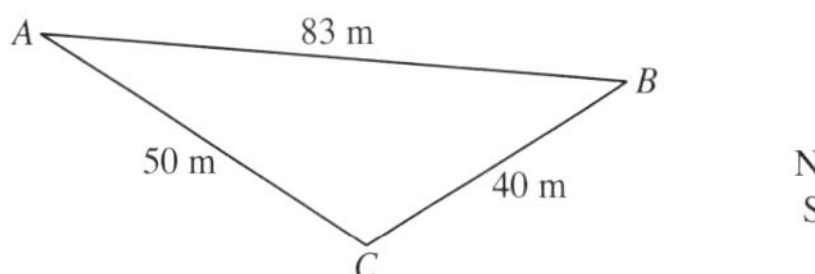

(3 marks)

(Q26d, **2010 HSC**) **Hard**

45 Which is the correct expression for the value of x in this triangle?

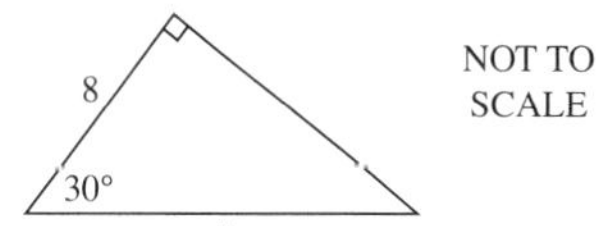

A $\dfrac{8}{\cos 30°}$ **B** $\dfrac{8}{\sin 30°}$

C $8 \times \cos 30°$ **D** $8 \times \sin 30°$ *(1 mark)*

(Q4, **2009 HSC**) **Easy**

46 In the diagram, AD and DC are equal to 30 cm.

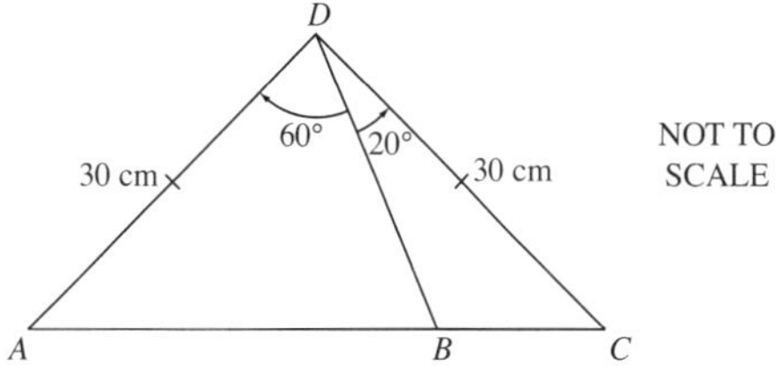

What is the length of AB to the nearest centimetre?

A 28 cm **B** 31 cm

C 34 cm **D** 39 cm *(1 mark)*

(Q22, **2009 HSC**) **Hard**

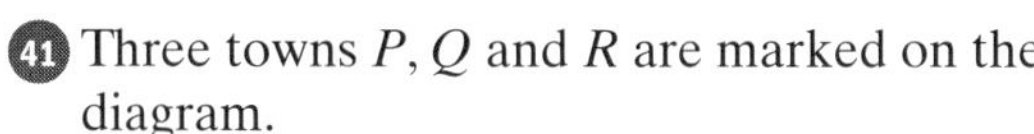

47 The point A is 25 m from the base of a building. The angle of elevation from A to the top of the building is 38°.

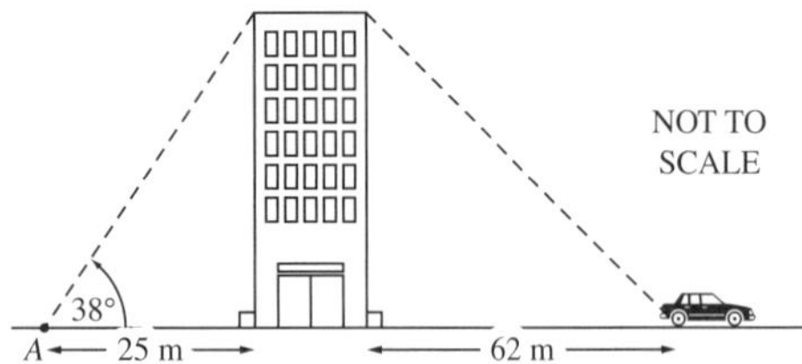

i Show that the height of the building is approximately 19.5 m. *(1 mark)* Easy

ii A car is parked 62 m from the base of the building.
What is the angle of depression from the top of the building to the car?
Give your answer to the nearest degree. *(2 marks)* Medium

(Q23a, **2009 HSC**)

48 A yacht race follows the triangular course shown in the diagram. The course from P to Q is 1.8 km on a true bearing of 058°. At Q the course changes direction. The course from Q to R is 2.7 km and $\angle PQR = 74°$.

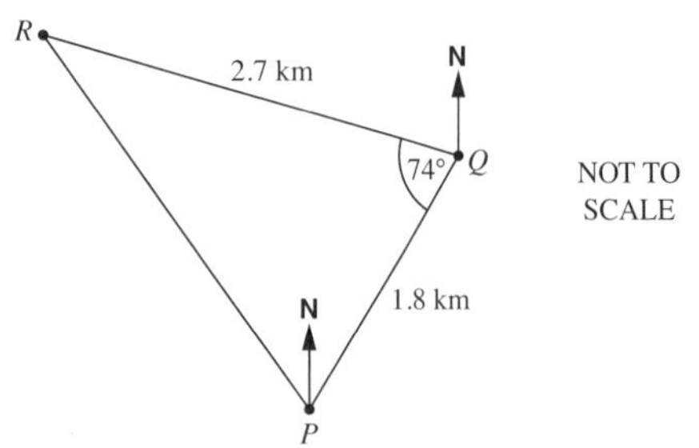

i What is the bearing of R from Q? *(1 mark)* Easy

ii What is the distance from R to P? *(2 marks)* Medium

iii The area inside this triangular course is set as a 'no-go' zone for other boats while the race is on. What is the area of this 'no-go' zone? *(1 mark)* Medium

(Q27b, **2009 HSC**)

49 What is the size of the smallest angle in this triangle?

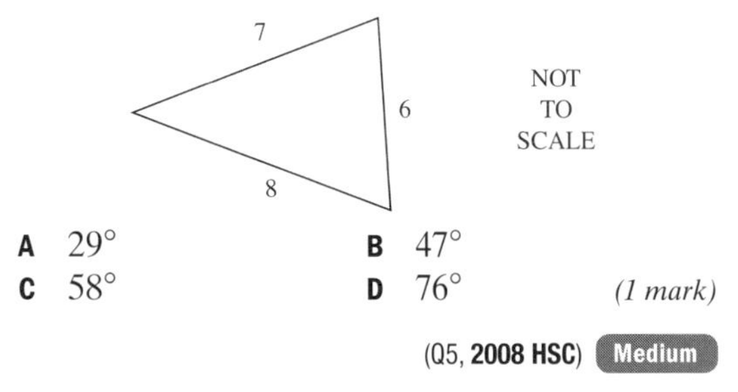

A 29° **B** 47°
C 58° **D** 76° *(1 mark)*

(Q5, **2008 HSC**) Medium

50 Danni is flying a kite that is attached to a string of length 80 metres. The string makes an angle of 55° with the horizontal.

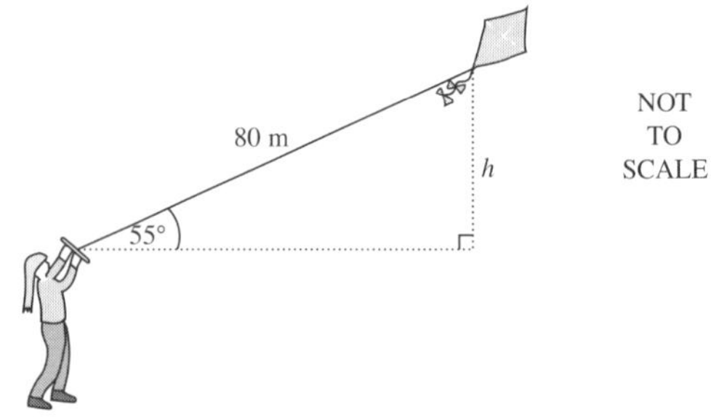

How high, to the nearest metre, is the kite above Danni's hand?

A 46 m **B** 66 m
C 98 m **D** 114 m *(1 mark)*

(Q14, **2008 HSC**) Easy

51 The diagram shows the position of Q, R and T relative to P.

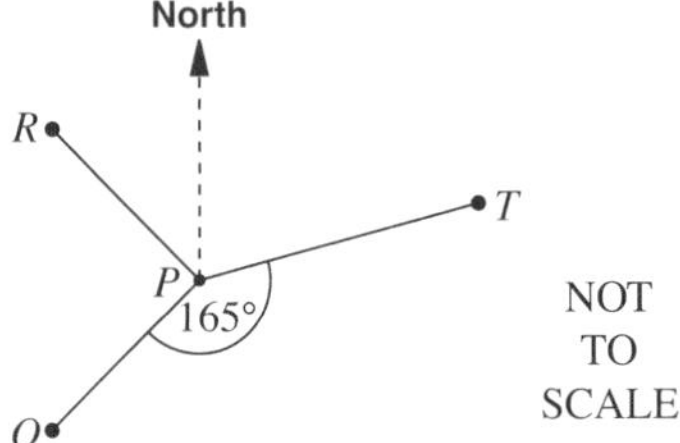

In the diagram,
Q is SW of P
R is NW of P
$\angle QPT$ is 165°

What is the bearing of T from P?

A 060° **B** 075°
C 105° **D** 120° *(1 mark)*

(Q17, **2008 HSC**) Medium

52 Pieces of cheese are cut from cylindrical blocks with dimensions as shown.

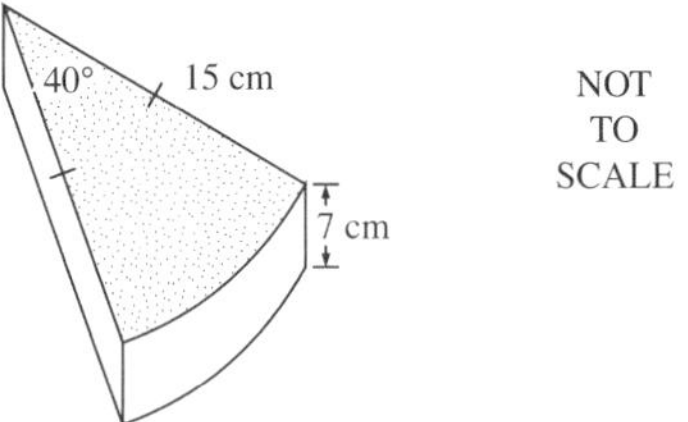

Twelve pieces are packed in a rectangular box. There are three rows with four pieces of cheese in each row. The curved surface is face down with the pieces touching as shown.

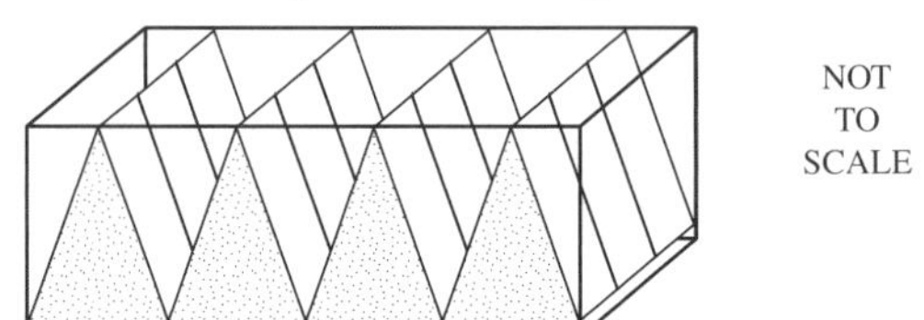

i What are the dimensions of the rectangular box? *(4 marks)* Hard

To save packing space, the curved section is removed.

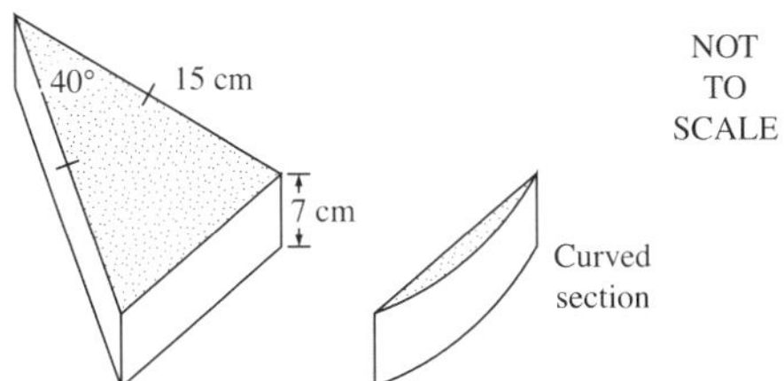

ii What is the volume of the remaining triangular prism of cheese? Answer to the nearest cubic centimetre. *(2 marks)* **Hard**

(Q25c, **2008 HSC**)

53 What is the length of the side MN in the following triangle, correct to two decimal places?

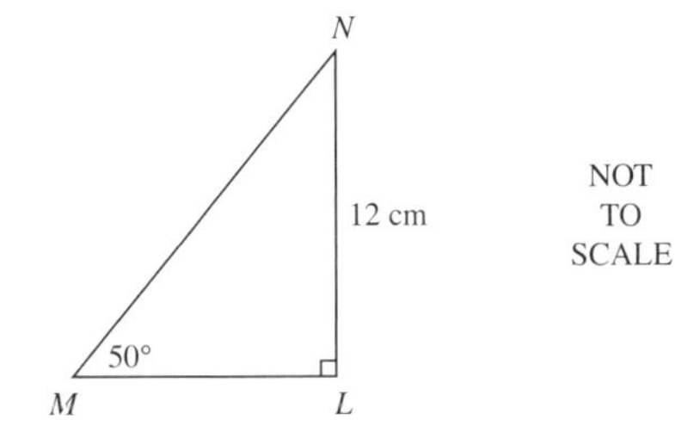

A 9.19 cm **B** 10.07 cm
C 15.66 cm **D** 18.67 cm *(1 mark)*

(Q8, **2007 HSC**) **Easy**

54 The angle of depression from J to M is 75°. The length of JK is 20 m and the length of MK is 18 m.

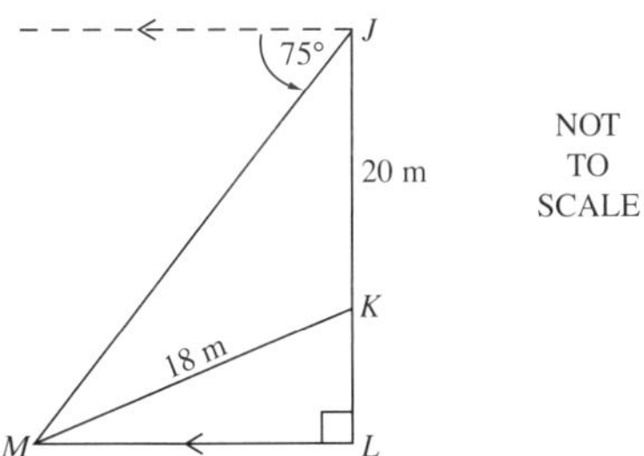

Copy or trace this diagram into your writing booklet and calculate the angle of elevation from M to K. Give your answer to the nearest degree. *(3 marks)*

(Q25b, **2007 HSC**) **Hard**

55 The diagram shows information about the locations of towns A, B and Q.

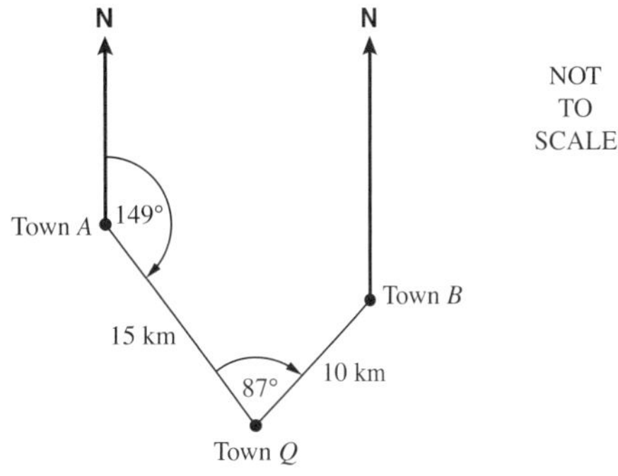

i It takes Elina 2 hours and 48 minutes to walk directly from Town A to Town Q. Calculate her walking speed correct to the nearest km/h. *(1 mark)* **Hard**

ii Elina decides, instead, to walk to Town B from Town A and then to Town Q. Find the distance from Town A to Town B. Give your answer to the nearest km. *(2 marks)* **Hard**

iii Calculate the bearing of Town Q from Town B. *(1 mark)* **Hard**

(Q26a, **2007 HSC**)

56 The angle of depression of the base of the tree from the top of the building is 65°. The height of the building is 30 m.

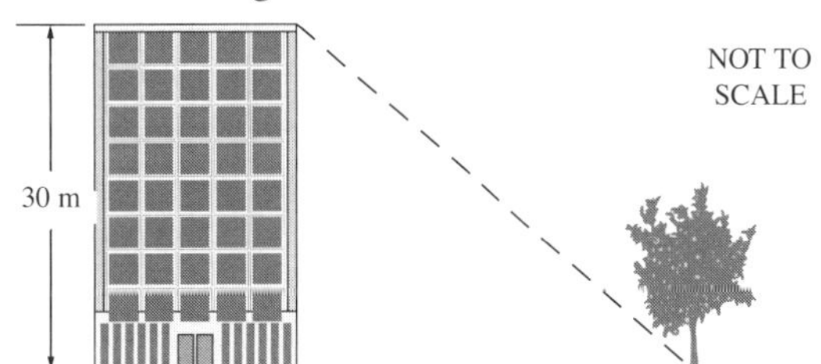

How far away is the base of the tree from the building, correct to one decimal place?

A 12.7 m **B** 14.0 m
C 33.1 m **D** 64.3 m *(1 mark)*

(Q3, **2006 HSC**) **Easy**

57 What is the area of this triangle, to the nearest square metre?

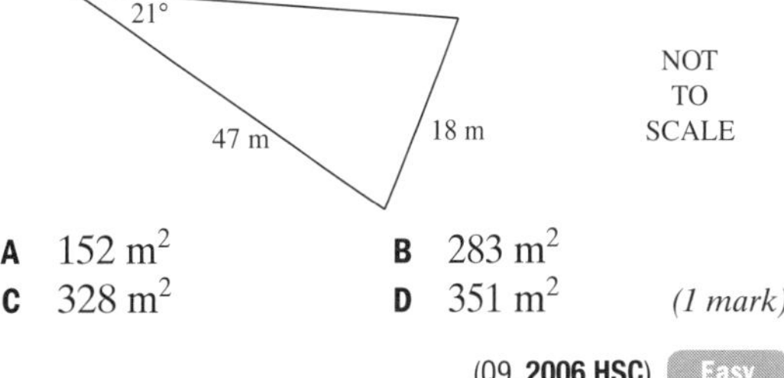

A 152 m^2 **B** 283 m^2
C 328 m^2 **D** 351 m^2 *(1 mark)*

(Q9, **2006 HSC**) **Easy**

58 What is the bearing of A from B?

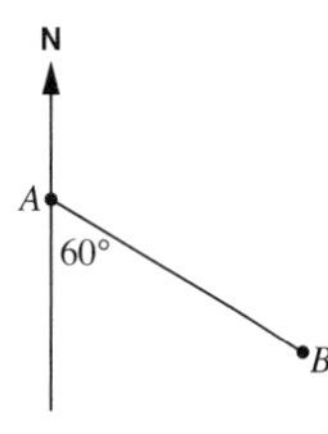

A 060° **B** 120°
C 150° **D** 300° *(1 mark)*

(Q13, **2006 HSC**) Medium

59 A 130 cm long garden rake leans against a fence. The end of the rake is 44 cm from the base of the fence.

i If the fence is vertical, find the value of θ to the nearest degree.

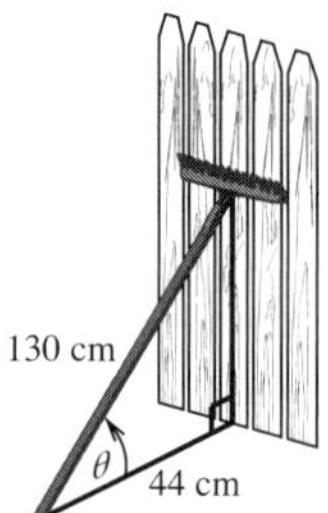

NOT TO SCALE

(2 marks) Easy

ii The fence develops a lean and the rake is now at an angle of 53° to the ground. Calculate the new distance (x cm) from the base of the fence to the head of the rake. Give your answer to the nearest centimetre.

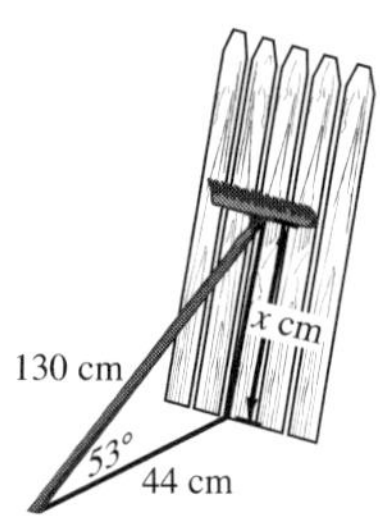

NOT TO SCALE

(2 marks) Medium

(Q24b, **2006 HSC**)

60 Which formula should be used to calculate the distance between Toby and Frankie?

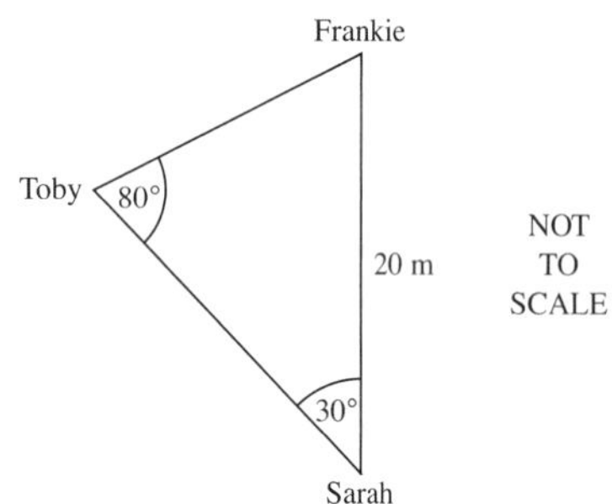

A $\dfrac{a}{\sin A} = \dfrac{b}{\sin B}$

B $c^2 = a^2 + b^2$

C $A = \dfrac{1}{2}ab\sin C$

D $c^2 = a^2 + b^2 - 2ab\cos C$ *(1 mark)*

(Q5, **2005 HSC**) Easy

61 If $\tan\theta = 85$, what is the value of θ, correct to the nearest minute?

A 11° 25' **B** 11° 26'
C 89° 19' **D** 89° 20' *(1 mark)*

(Q8, **2005 HSC**) Easy

62

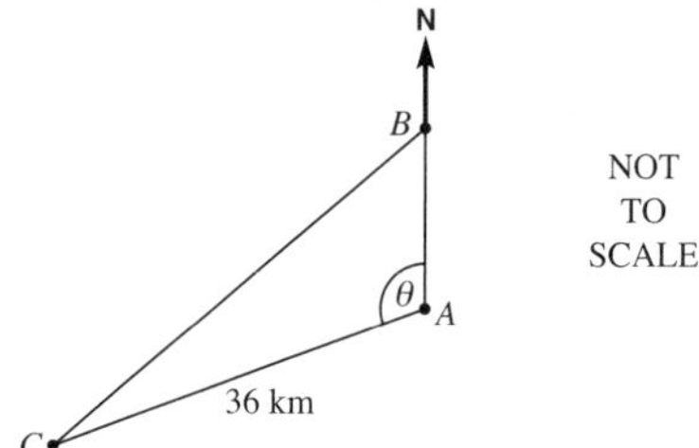

i Use Pythagoras' theorem to show that ΔABC is a right-angled triangle. *(1 mark)* Easy

ii Calculate the size of $\angle ABC$ to the nearest degree. *(2 marks)* Easy

(Q25b, **2005 HSC**)

63 The bearing of C from A is 250° and the distance of C from A is 36 km.

N
B
θ
A
36 km
C

NOT TO SCALE

i Explain why θ is 110°. *(1 mark)* Medium

ii If B is 15 km due north of A, calculate the distance of C from B, correct to the nearest kilometre. *(3 marks)* Hard

(Q27c, **2005 HSC**)

64 What is the correct expression for tan 20° in this triangle?

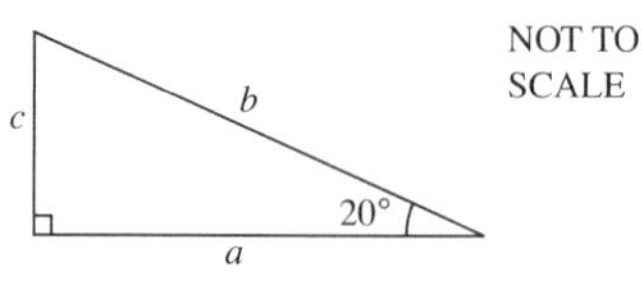

A $\dfrac{a}{b}$ **B** $\dfrac{a}{c}$

C $\dfrac{c}{b}$ **D** $\dfrac{c}{a}$ *(1 mark)*

(Q5, **2004 HSC**) Easy

65 What is the area of the triangle to the nearest square metre?

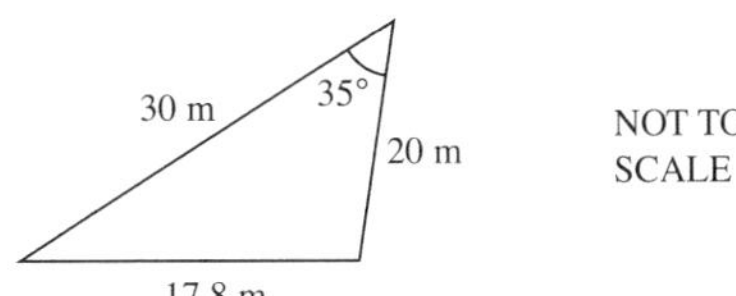

A 102 m^2 **B** 153 m^2
C 172 m^2 **D** 178 m^2 *(1 mark)*

(Q9, **2004 HSC**) Easy

66 The diagram shows a radial survey of a piece of land.

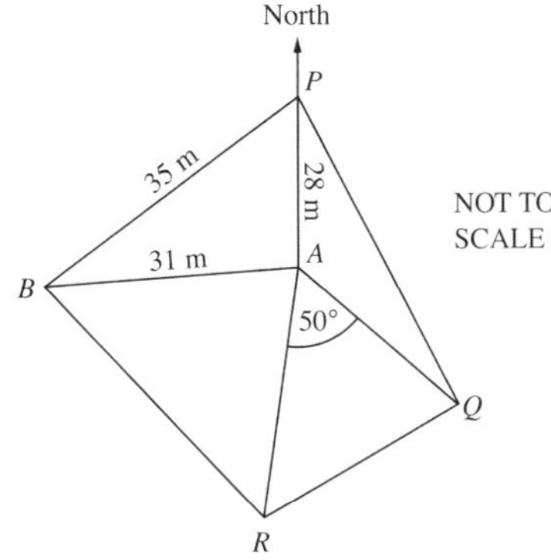

i Q is south-east of A. What is the size of angle PAQ? *(1 mark)* Easy
ii What is the bearing of R from A? *(1 mark)* Easy
iii Find the size of angle PAB to the nearest degree. *(3 marks)* Medium

(Q24b, **2004 HSC**)

67 Which equation should be used to obtain the value of x in this triangle?

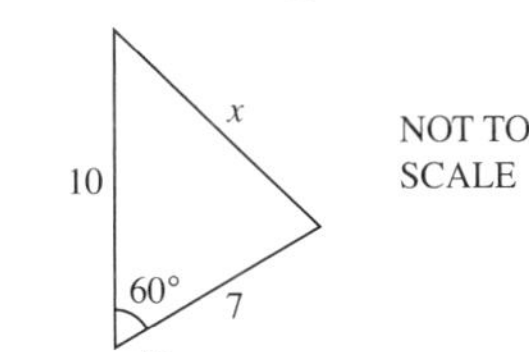

A $\dfrac{x}{\sin 60^\circ} = \dfrac{7}{\sin 10^\circ}$

B $x^2 = 10^2 + 7^2 - 2 \times 10 \times 7 \cos 60^\circ$

C $\cos 60^\circ = \dfrac{x^2 + 10^2 - 7^2}{2 \times 10 \times 7}$

D $x^2 = 10^2 - 7^2$ *(1 mark)*

(Q14, **2003 HSC**) Easy

68

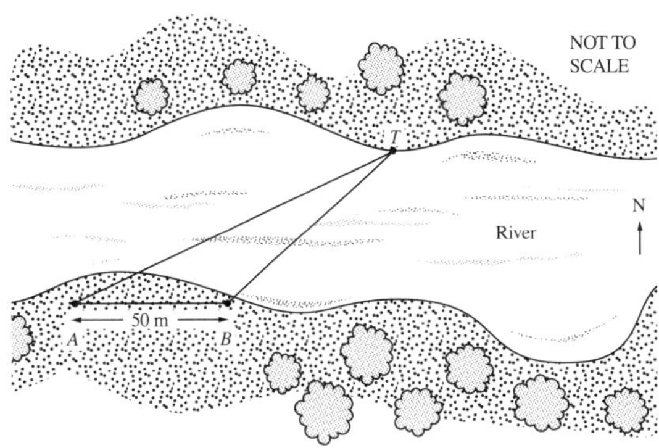

In the diagram above, the following measurements are given:

$\angle TAB = 30^\circ$.
B is 50 m due east of A.
The bearing of T from B is 020°.

Copy or trace ΔABT into your answer booklet.

i Explain why $\angle ABT$ is 110°. *(1 mark)* Easy
ii Calculate the distance BT (to the nearest metre). *(3 marks)* Hard

(Q26b, **2003 HSC**)

69 The diagram shows a radio mast AD with two of its supporting wires, BE and CE. The point B is half-way between A and C.

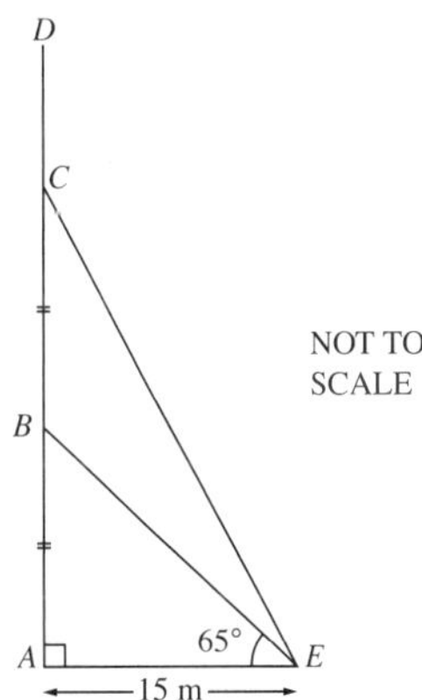

i Calculate the height AB in metres, correct to one decimal place. *(2 marks)* Easy
ii Calculate the distance CE in metres, correct to one decimal place. *(2 marks)* Medium

(Q27c, **2003 HSC**)

70 In the diagram X, Y and Z represent the locations of three towns. The town Y is due east of X, and the bearing of Z from Y is 046°.

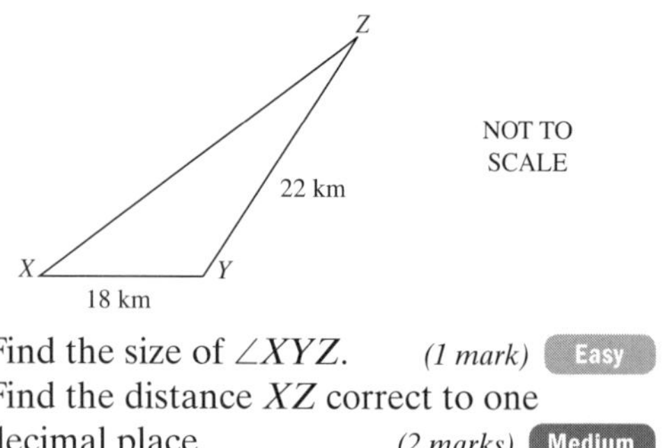

i Find the size of $\angle XYZ$. *(1 mark)* Easy

ii Find the distance XZ correct to one decimal place. *(2 marks)* Medium

iii What is the bearing of Y from Z? *(1 mark)* Medium

(Q27a, **2002 HSC**)

71

NOT TO SCALE

i Find the perimeter of $\triangle PQR$. (Give your answer to one decimal place.) *(3 marks)* Medium

ii Find the size of $\angle QPS$ to the nearest degree. *(1 mark)* Medium

(Q27b, **2002 HSC**)

---

72 Calculate the length of AD (to the nearest metre).

NOT TO SCALE

A 25 m **B** 134 m
C 190 m **D** 214 m *(1 mark)*

(Q20, **2001 HSC**) Medium

---

73 The following notebook entry was made during a radial survey of a field.

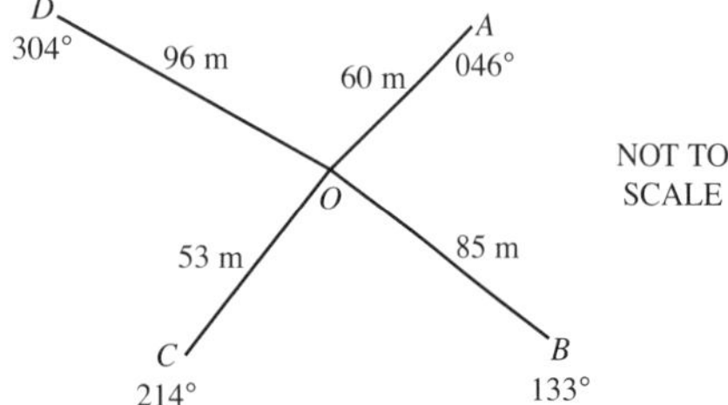

i What is the size of $\angle AOB$? *(1 mark)* Easy

ii Calculate the area of triangle AOB. Round your answer to the nearest square metre. *(2 marks)* Medium

iii Find the distance from A to B. *(2 marks)* Medium

(Q24a, **2001 HSC**)

1 True bearing of A from B is $270° + 25° = 295°$.

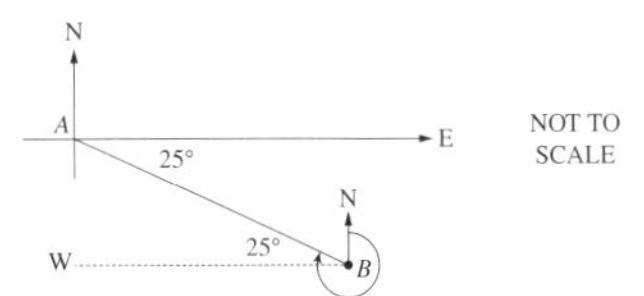

Answer D

2 **a** $\cos 30° = \dfrac{XY}{16}$ ✓

$XY = 16 \cos 30°$

$= 13.8564064\ldots$

$= 13.86$ cm (2 d.p.) ✓

(2 marks)

b Area of triangle:

$A = \dfrac{1}{2}ab \sin C$

$= \dfrac{1}{2} \times 16 \times 13.8564064\ldots \times \sin 30°$

$= 55.4256258\ldots \text{ cm}^2$ ✓

Area of semicircle:

$A = \dfrac{1}{2}\pi r^2$

$= \dfrac{1}{2} \times \pi \times 8^2$

$= 100.530964\ldots \text{ cm}^2$ ✓

Shaded area

$= 100.530964\ldots - 55.425628\ldots$

$= 45.1053390\ldots$

$= 45.1 \text{ cm}^2$ (1 d.p.) ✓

(3 marks)

3 By the sine rule:

$\dfrac{\sin B}{b} = \dfrac{\sin A}{a}$

$\dfrac{\sin B}{25} = \dfrac{\sin 28°}{16}$ ✓

$\sin B = \dfrac{25 \sin 28°}{16}$

Now $\sin^{-1}\left(\dfrac{25 \sin 28°}{16}\right)$

$= 47.18477\ldots°$

$= 47°$ (nearest degree) ✓

But $\angle ABC$ is obtuse.

So $\angle B = 180° - 47°$

$= 133°$

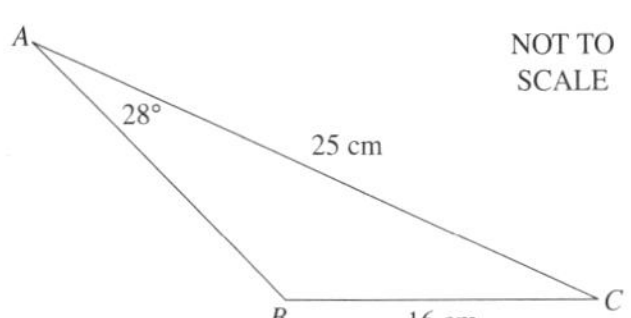

The obtuse angle is 133° to the nearest degree. ✓

(3 marks)

4 **a** Area of triangle is

$A = \dfrac{1}{2}ab \sin C$

Let $\angle COB = \theta$

So $\dfrac{1}{2} \times 28 \times 35 \times \sin \theta$

$= 466$ ✓

$490 \sin \theta = 466$

$\sin \theta = \dfrac{466}{490}$

$\theta = 71.9933062\ldots°$

$= 72°$ (nearest degree)

The size of the acute angle COB is 72° to the nearest degree. ✓ *(2 marks)*

b Bearing of C from O

$= 150° + 72°$

$= 222°$

So $\angle DOC = 330° - 222°$

$= 108°$ ✓

Using the cosine rule:

$c^2 = a^2 + b^2 - 2ab \cos C$

$CD^2 = 28^2 + 31^2 - 2 \times 28 \times 31 \times \cos 108°$

$= 2281.45350\ldots$

$CD = \sqrt{2281.4530\ldots}$

$= 47.7645632\ldots$

$= 47.8$ m (1 d.p.) ✓

Perimeter of ΔDOC

$= (31 + 28 + 47.8)$ m

$= 106.8$ m

The length of fencing required is 106.8 m correct to one decimal place. ✓ *(3 marks)*

5

a $\tan \theta = \dfrac{8}{10}$ ✓

$\theta = \tan^{-1}\left(\dfrac{8}{10}\right)$

$= 38.6598082\ldots°$

$= 39°$ (nearest degree)

✓ *(2 marks)*

b $x^2 = 8^2 + 10^2$ ✓

$= 164$

$x = \sqrt{164}$ $(x > 0)$

$= 12.8062484\ldots$

$= 12.8$ (1 decimal place) ✓

(2 marks)

6 **a**

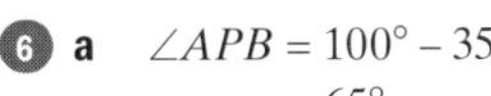

$\angle APB = 100° - 35°$

$= 65°$

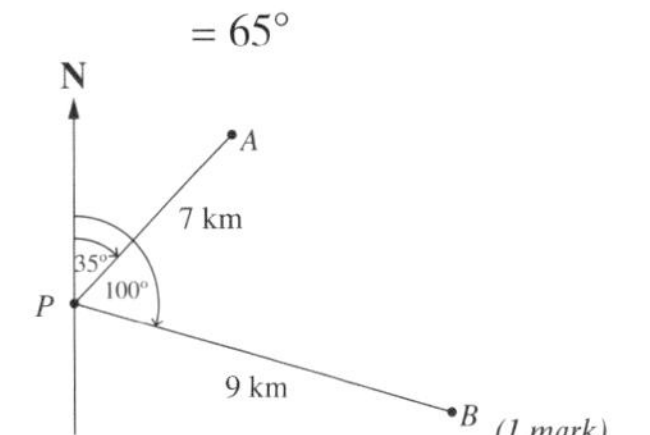

(1 mark)

b By the cosine rule:

$AB^2 = 7^2 + 9^2 - 2 \times 7 \times 9 \times \cos 65°$ ✓

$= 76.7500990\ldots$

AB

$= \sqrt{76.750\,0990\ldots}$ $(AB > 0)$

$= 8.76071338\ldots$

$= 8.8$ (1 d.p.)

The distance AB is 8.8 km correct to one decimal place. ✓

(2 marks)

c In ΔAPB,

$\dfrac{\sin A}{a} = \dfrac{\sin P}{p}$

$\dfrac{\sin A}{9} = \dfrac{\sin 65°}{8.760\,713\,38\ldots}$

✓ $\sin A = \dfrac{9 \sin 65°}{8.760\,713\,38\ldots}$

$A = \sin^{-1}\left(\dfrac{9 \sin 65°}{8.760\,713\,38\ldots}\right)$

$= 68.6010202\ldots°$

$= 69°$ (nearest degree) ✓

Let X be a point due North of A.

$\angle XAP = 180° - 35°$

(co-int. angles, parallel lines)

$= 145°$

$\angle XAB = 360° - (145 + 69)°$

$= 146°$

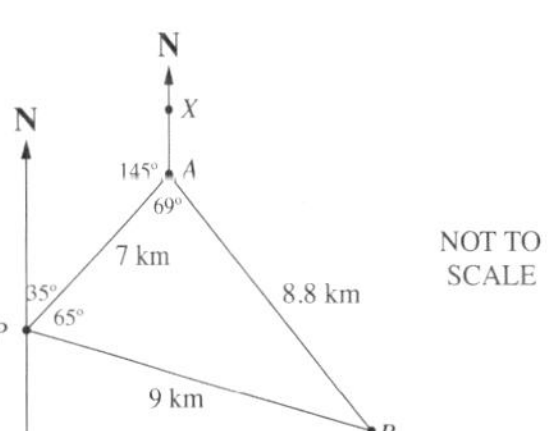

The bearing of Ms Brown's group from Mr Ali's group was 146°, to the nearest degree. ✓

(2 marks)

7 Perimeter = 80 cm

So each side is (80 ÷ 10) cm.

$\therefore AB = 8$ cm ✓

$\angle AOB = 360° \div 10$

$= 36°$

Let P be the midpoint of AB.

As ΔAOB is isosceles,

$\angle AOP = 18°$.

In ΔAOP,

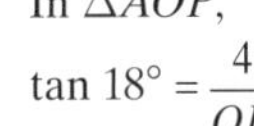

$\tan 18° = \dfrac{4}{OP}$

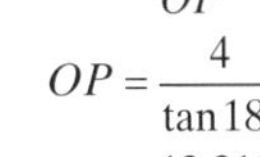

$OP = \dfrac{4}{\tan 18°}$

$= 12.3107341\ldots$ ✓

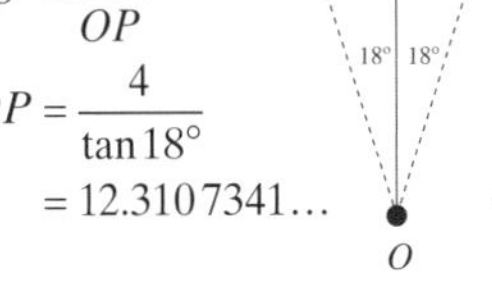

In ΔAOB,

$\text{Area} = \frac{1}{2} \times 8 \times 12.3107341\ldots$

$= 49.2429365\ldots \text{ cm}^2$ ✓

Area of decagon

$= 10 \times 49.2429365\ldots$

$= 492.429365\ldots$

$= 492.4 \text{ cm}^2$ (1 d.p.) ✓

(4 marks)

8 $110° = 180° - 70°$

Compass bearing is S70°E

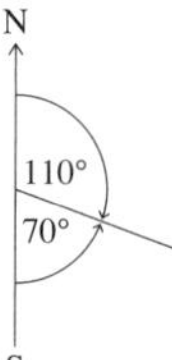

Answer C

9 Let x m be the distance the owl must fly.

$\sin 32° = \frac{7}{x}$

$x = \frac{7}{\sin 32°}$

$= 13.209559\ldots$

$= 13.2$ m (1 d.p.)

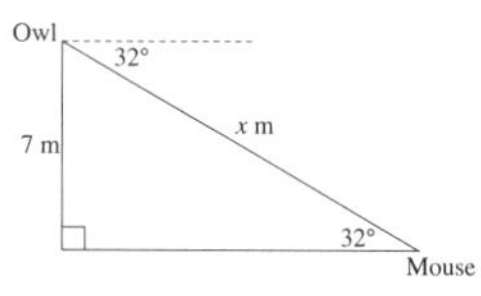

Answer D

10 $c^2 = a^2 + b^2 - 2ab\cos C$

$x^2 = 11^2 + 13^2 - 2 \times 11 \times 13 \times \cos 80°$ ✓

$= 240.336621\ldots$

$x = \sqrt{240.336621\ldots}$ $(x > 0)$

$= 15.5027939\ldots$ ✓

$= 16$ (2 sig. figs) ✓

(3 marks)

11 In ΔADC,

$AC^2 = 2.5^2 + 6^2$

$= 42.25$

So $AC = 6.5$ cm ✓✓

In ΔABC,

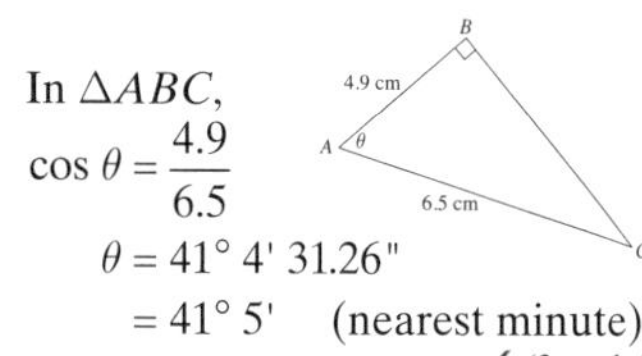

$\cos\theta = \frac{4.9}{6.5}$

$\theta = 41°\ 4'\ 31.26''$

$= 41°\ 5'$ (nearest minute)

✓ *(3 marks)*

12 Let θ be the size of $\angle BOC$.

$A = \frac{1}{2}ab\sin C$

So $\frac{1}{2} \times 16 \times 25 \times \sin\theta = 198$ ✓

$\sin\theta = \frac{198}{200}$

$\theta = 81.8903855\ldots°$

$= 82°$ (nearest degree) ✓

The bearing of C from O is $(125 + 82)°$ or $207°$ to the nearest degree. ✓ *(3 marks)*

13 **a** Let x = distance from P to Q.

$x^2 = 170^2 + 265^2 - 2 \times 170 \times 265 \times \cos 36°$ ✓

$= 26232.56881\ldots$

$x = 161.9647147\ldots$

$= 161.9647$ (4 dec. pl.) ✓

$\text{Time} = 161.9647 \div 280$

$= 0.578445409\ldots$

$= 35$ minutes ✓

(nearest whole)

(3 marks)

b

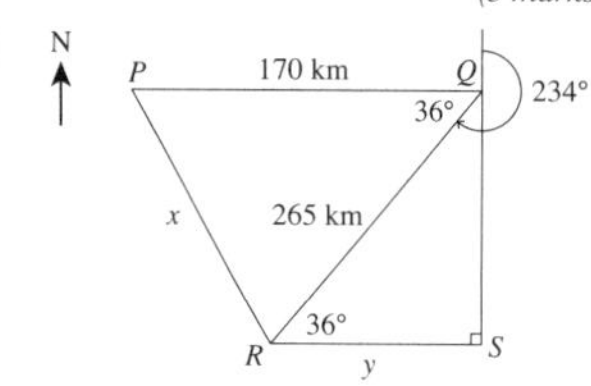

Let y = distance from R to S.

$\frac{y}{265} = \cos 36°$

$y = 265 \times \cos 36°$

$= 214.3895035\ldots$

$= 214.3895$ (4 dec. pl.) ✓

$\text{Speed} = 214.3895 \div \frac{48}{60}$

$= 267.9868794\ldots$

$= 268$ (nearest whole) ✓

$\therefore$ speed is 268 km/h. *(2 marks)*

14 The bearing of C from A is $(180 + 34)°$ or $214°$.

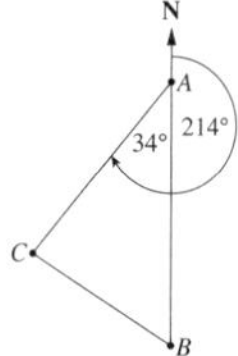

Answer C

15 $\cos C = \frac{a^2 + b^2 - c^2}{2ab}$

$\cos\theta = \frac{8^2 + 9^2 - 10^2}{2 \times 8 \times 9}$

$\theta = \cos^{-1}\left(\frac{8^2 + 9^2 - 10^2}{2 \times 8 \times 9}\right)$

$= 71.7900431\ldots°$

$= 72°$ (nearest degree)

Answer D

16 In ΔABC,

$\sin 62° = \frac{AC}{13}$

$AC = 13\sin 62°$

$= 11.4783187\ldots$ cm ✓

In ΔACD,

$\text{Area} = \frac{1}{2}cd\sin A$

$30 = \frac{1}{2} \times x \times 11.4783187\ldots \times \sin 40°$ ✓

$= x \times 3.68906052\ldots$

$x = 30 \div 3.68906052\ldots$

$= 8.13215175\ldots$

$= 8.1$ (1 d.p.) ✓

(3 marks)

17 $\tan\theta = \frac{5.3}{1.9}$

$\theta = \tan^{-1}\left(\frac{5.3}{1.9}\right)$

$= 70°16'39.8''$

$= 70°17'$ (nearest minute)

Answer B

18 Let x mm be the depth of the pipe.

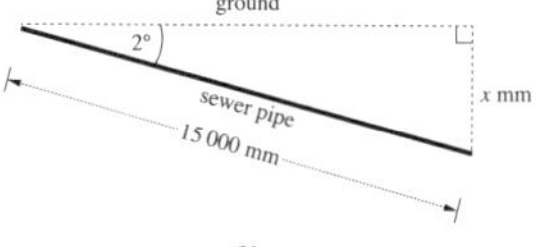

$\sin 2° = \frac{x}{15000}$ ✓

$x = 15000 \times \sin 2°$

$= 523.49245\ldots$

$= 523$ (nearest unit)

The pipe should have one end 523 mm, to the nearest millimetre, deeper than the other. ✓

(2 marks)

19 **i** By the cosine rule in ΔABC:

$b^2 = a^2 + c^2 - 2ac\cos B$

$= 13^2 + 5^2 - 2 \times 13 \times 5 \times \cos 135°$ ✓

$= 285.92388\ldots$

$b = \sqrt{285.92388\ldots}$ $(b > 0)$

$= 16.9092838\ldots$

$= 17$ (nearest unit)

The shortest distance from A to C is 17 km, to the nearest kilometre. ✓

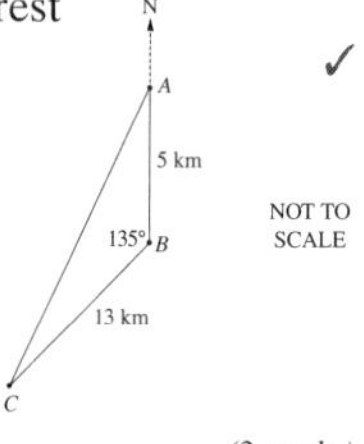

(2 marks)

ii By the sine rule in ΔABC:

$\frac{\sin A}{a} = \frac{\sin B}{b}$

$$\frac{\sin A}{13} = \frac{\sin 135°}{16.909\,2838...}$$ ✓

$$\sin A = \frac{13 \sin 135°}{16.909\,2838...}$$

$$A = \sin^{-1}\left(\frac{13 \sin 135°}{16.909\,2838...}\right)$$

$= 32.931\,0687...°$

$= 33°$ (nearest degree) ✓

Now A is due north of B.

So the bearing of B from A is 180°.

The bearing of C from $A = 180° + 33°$

$= 213°$.

The bearing of school C from school A is 213° to the nearest degree. ✓

(3 marks)

20 $\frac{\sin C}{c} = \frac{\sin A}{a}$

$\frac{\sin C}{40} = \frac{\sin 25°}{18}$

$\sin C = \frac{40 \sin 25°}{18}$

$= 0.939\,1516...$

$C = 70°$ or $110°$ (nearest degree)

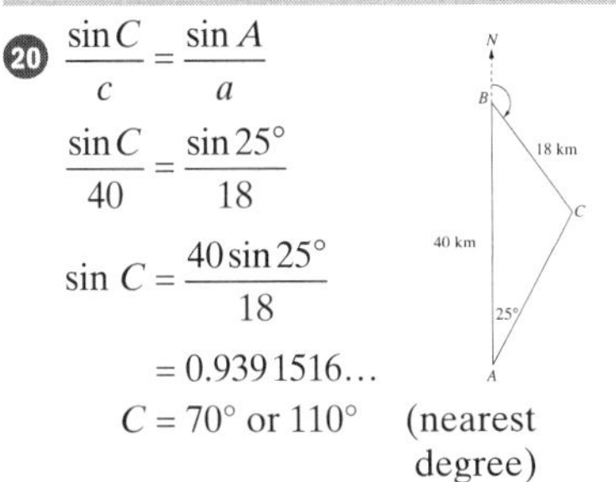

But the angle is obtuse so $\angle BCA = 110°$.

So, using the exterior angle of the triangle, the bearing of C from B is $110° + 25° = 135°$

Answer D

21 By the cosine rule in ΔAOB:

$c^2 = a^2 + b^2 - 2ab \cos C$

$AB^2 = 45^2 + 45^2 - 2 \times 45 \times 45 \times \cos 100°$ ✓

$= 4753.2751...$

$AB = 68.943\,9998...$ m ✓

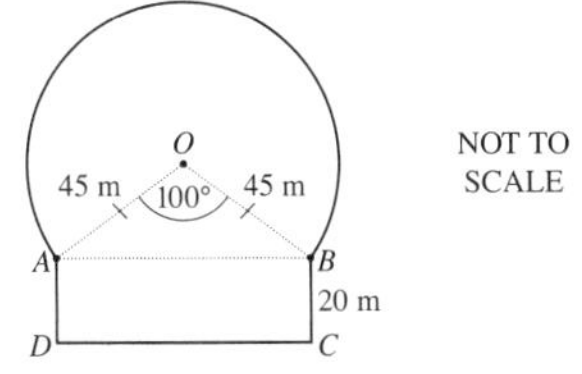

Area of rectangle $ABCD$

$= 68.943\,9998... \times 20$

$= 1378.879\,99...$ m^2

Area of triangle AOB:

$A = \frac{1}{2}ab \sin C$

$= \frac{1}{2} \times 45 \times 45 \times \sin 100°$

$= 997.117\,849...$ m^2 ✓

Angle in major sector

$= 360° - 100° = 260°$

Area of major sector:

$A = \frac{\theta}{360}\pi r^2$

$= \frac{260}{360} \times \pi \times 45^2$

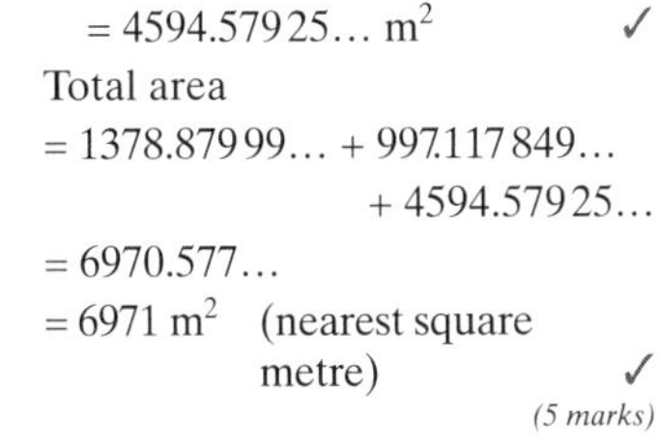

$= 4594.579\,25...$ m^2 ✓

Total area

$= 1378.879\,99... + 997.117\,849... + 4594.579\,25...$

$= 6970.577...$

$= 6971$ m^2 (nearest square metre) ✓

(5 marks)

22 Let X be a point due North of O.

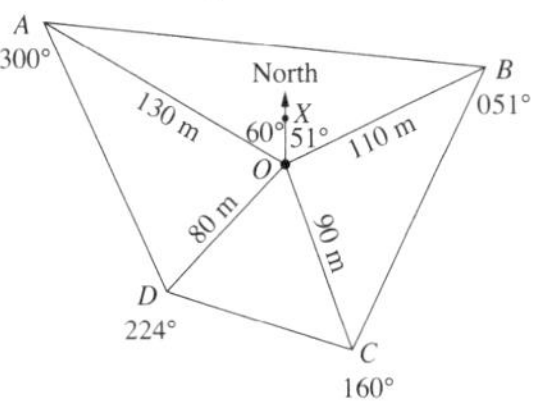

$\angle AOX = 360° - 300° = 60°$

$\angle XOB = 51°$

$\angle AOB = 60° + 51° = 111°$

Answer B

23 Angle at top of triangle

$= 90° - 42° = 48°$

Let x m be the distance from the buoy to the cliff.

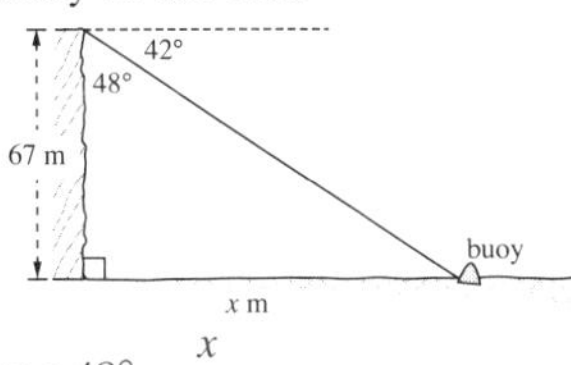

$\tan 48° = \frac{x}{67}$

$x = 67 \times \tan 48°$

$= 74.411\,038...$

$= 74$ (nearest metre)

∴ the distance of the buoy from the base of the cliff is 74 m to the nearest metre.

Answer B

24 Area $= \frac{1}{2}ab \sin C$

$250 = \frac{1}{2} \times 30 \times x \times \sin 44°$

$x = \frac{250}{15 \sin 44°}$

$= 23.992\,6089...$

$= 24$ (nearest whole number)

Answer D

25 **i** $c^2 = a^2 + b^2 - 2ab \cos C$

In ΔPSC,

$PC^2 = 1.8^2 + 5.4^2 - 2 \times 1.8 \times 5.4 \times \cos 108°$

$= 38.407\,29....$

$PC = \sqrt{38.407\,29...}$ ($PC > 0$)

$= 6.197\,361\,56...$

$= 6.197$ m (3 d.p.)

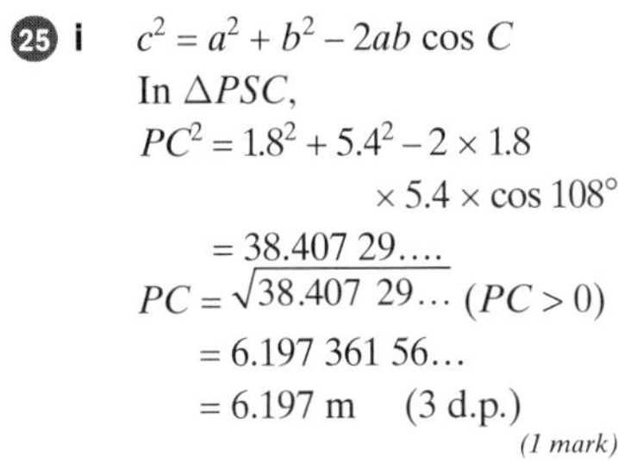

(1 mark)

ii Let $\angle PCS = \theta$

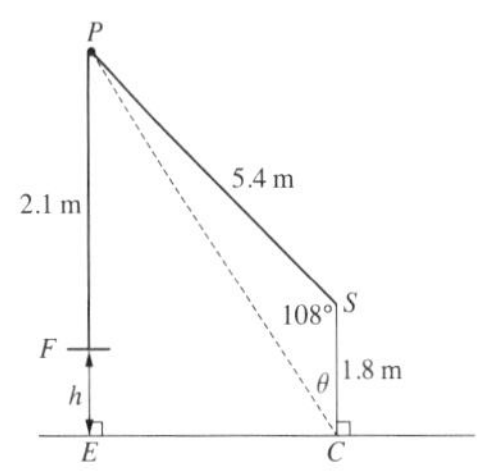

By the sine rule in ΔPSC,

$\frac{\sin \theta}{5.4} = \frac{\sin 108°}{6.197...}$ ✓

$\sin \theta = 0.828\,69...$

$\theta = 55.9646...°$ ✓

$\angle PCE = 90° - 55.9646...°$

$= 34.035...°$

In ΔPEC,

$\sin 34.035...° = \frac{PE}{6.197...}$ ✓

$PE = 3.468\,6917...$

$PE = 3.47$ m (2 d.p.)

$h = 3.47 - 2.1$

$= 1.37$ m

The object is 1.37 m above the ground, to the nearest centimetre. ✓

(4 marks)

26 X is due east of Z.

[So options B and D are eliminated.]

Let P be a point due north of Y.

$\angle PYX = 145°$

$\angle PYZ = 180° - 60°$

$= 120°$

(co-interior angles, parallel lines)

$\angle XYZ = 360° - (120° + 145°)$

$= 95°$

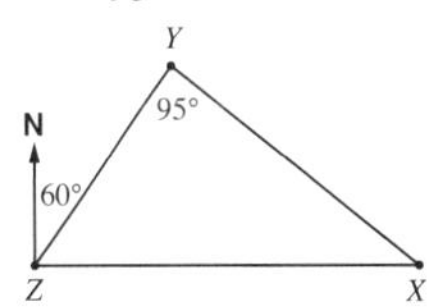

Answer C

27

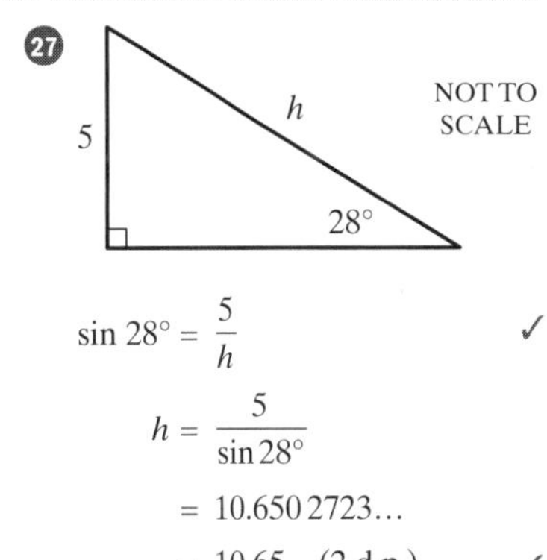

$\sin 28° = \frac{5}{h}$ ✓

$h = \frac{5}{\sin 28°}$

$= 10.650\,2723...$

$= 10.65$ (2 d.p.) ✓

(2 marks)

28 **i** The bearing of A from O is 320° and the bearing of B from O is 074°.

$\angle AOB = (360 - 320)° + 74°$
$= 114°$

(1 mark)

ii $c^2 = a^2 + b^2 - 2ab\cos C$

$AB^2 = 211^2 + 287^2 - 2 \times 211 \times 287 \times \cos 114°$
$= 176\,151.501\ldots$ ✓

$AB = \sqrt{176\,151.501\ldots}$ $(AB > 0)$
$= 419.704\,06\ldots$ ✓
$= 420$ m (nearest metre)

(2 marks)

iii Area $= \frac{1}{2}ab\sin C$
$= \frac{1}{2} \times 211 \times 287 \times \sin 114°$ ✓
$= 27\,660.7861\ldots$ m^2 ✓
$= 2.766\,078\,61\ldots$ ha
$= 2.8$ ha (2 sig. figs) ✓

(3 marks)

29

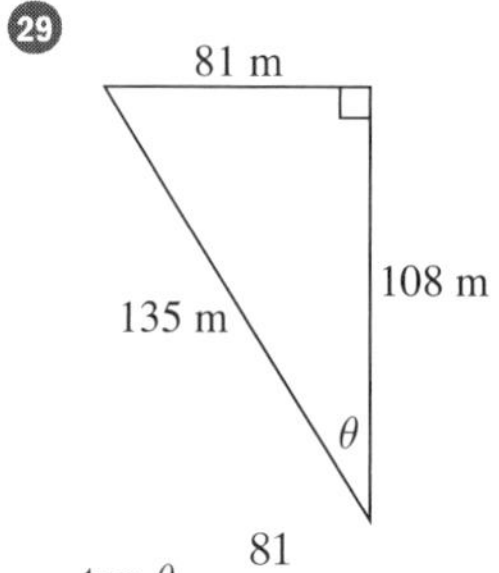

$\tan\theta = \frac{81}{108}$

$\theta = 36.869\,89\ldots°$
$= 37°$ [nearest degree]

[Or use $\sin\theta = \frac{81}{135}$
or $\cos\theta = \frac{108}{135}$]

Answer B

30

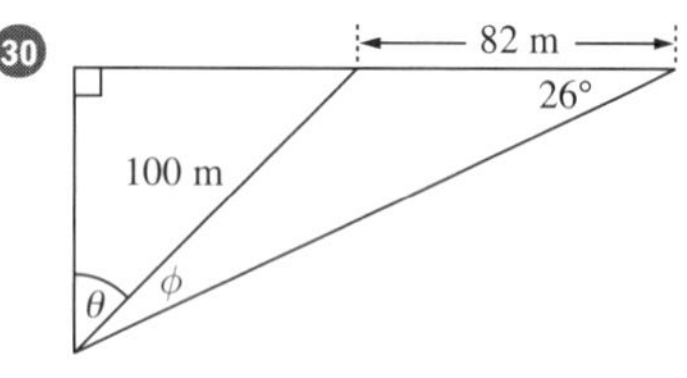

By the sine rule:

$\frac{\sin\phi}{82} = \frac{\sin 26°}{100}$

$\sin\phi = \frac{82\sin 26°}{100}$

$\phi = 21.067\,30\ldots°$
$= 21°$ [nearest degree]

Now $\theta + 21° + 26° + 90°$
$= 180°$ (angle sum of triangle)

$\theta + 137° = 180°$
$\theta = 43°$

Answer C

31

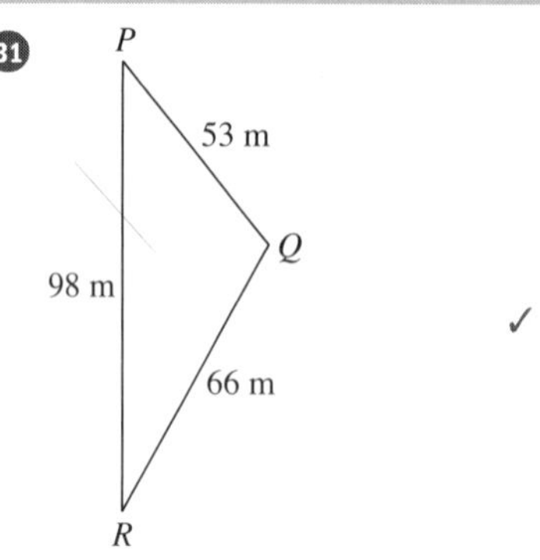

✓

$\cos Q = \frac{p^2 + r^2 - q^2}{2pr}$
$= \frac{66^2 + 53^2 - 98^2}{2 \times 66 \times 53}$ ✓

$Q = \cos^{-1}\left\{\frac{66^2 + 53^2 - 98^2}{2 \times 66 \times 53}\right\}$
$= 110.403\,4074\ldots°$
$= 110°$ [nearest degree] ✓

(2 marks)

32 $\angle AOB = (360 - 310)° + 21°$
$= 71°$ ✓

$A = \frac{1}{2}ab\sin C$
$= \frac{1}{2} \times 60 \times 75 \times \sin 71°$
$= 2127.416\,795\ldots$

The area will be 2127 m^2, to the nearest square metre. ✓

(2 marks)

33

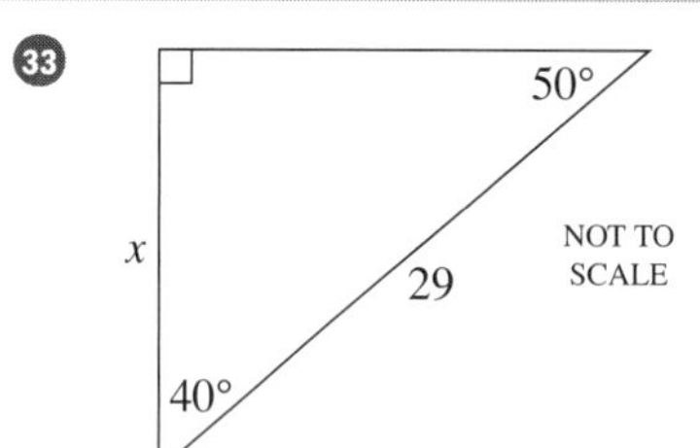

$\cos 40° = \frac{x}{29}$

$x = 29 \times \cos 40°$

Answer A

34 Third angle $= 180° - (50 + 57)°$
$= 73°$

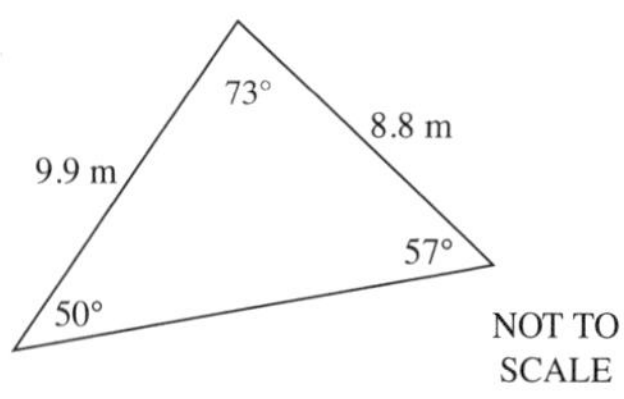

$A = \frac{1}{2}ab\sin C$
$= \frac{1}{2} \times 9.9 \times 8.8 \times \sin 73°$
$= 41.656\,635\,17\ldots$

The area is 42 m^2, to the nearest square metre.

Answer C

35 Using the cosine rule in ΔABC,

$\cos B = \frac{59^2 + 80^2 - 31^2}{2 \times 59 \times 80}$
$= 0.944\,9152\ldots.$

$B = 19.105\,8926\ldots°$
$= 19°$ [nearest degree]

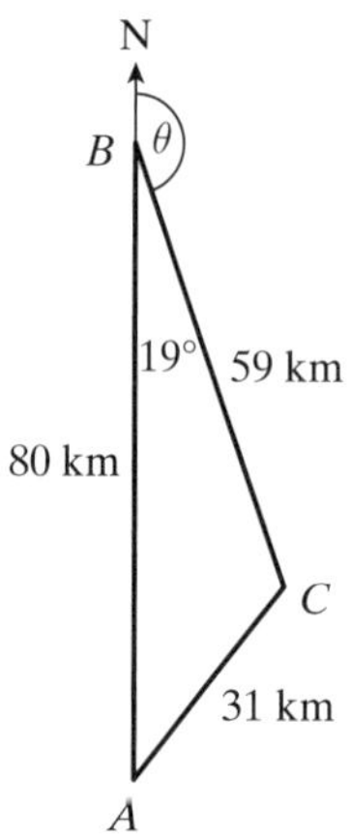

$\theta = 180° - 19°$
$= 161°$

The bearing of C from B is 161°.

Answer C

36 Vertical height of the ramp
$= 3 \times 13$ cm
$= 39$ cm

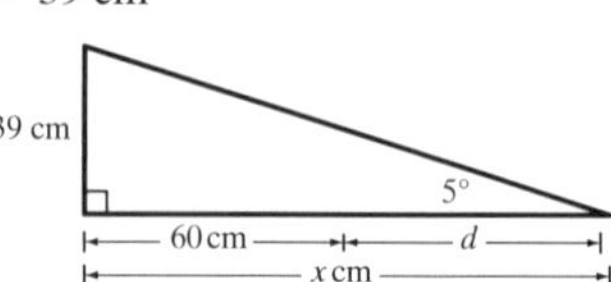

Let x cm be the horizontal length of the ramp.

$\tan 5° = \frac{39}{x}$

$x = \frac{39}{\tan 5°}$
$= 445.772\,0398\ldots$
$= 446$ [nearest unit] ✓

So the horizontal length of the ramp is 446 cm, to the nearest centimetre.

$d \times 60 \text{ cm} = 446 \text{ cm}$

$d = 386 \text{ cm}$ ✓

(3 marks)

37 **i**

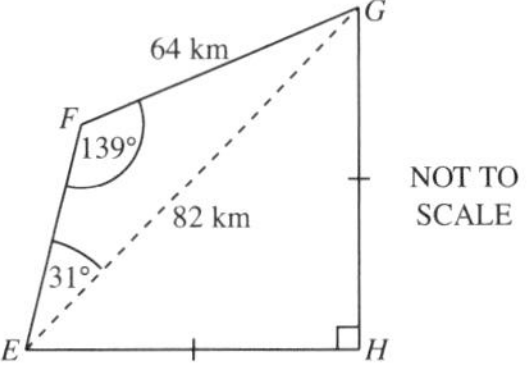

$\angle FGE = 180° - (139 + 31)°$
$= 10°$ ✓

By the sine rule in ΔEFG,

$$\frac{EF}{\sin 10°} = \frac{82}{\sin 139°}$$

$$EF = \frac{82 \sin 10°}{\sin 139°}$$

$= 21.704\,069\ldots$ ✓
$= 22 \text{ km}$ [nearest km]

{Or: $\frac{EF}{\sin 10°} = \frac{64}{\sin 31°}$

$$EF = \frac{64 \sin 10°}{\sin 31°}$$

$= 21.577\,984\ldots$
$= 22 \text{ km}$ [nearest km]}

Or: By the cosine rule in ΔEFG,

$EF^2 = 64^2 + 82^2 - 2 \times 64 \times 82 \times \cos 10°$
$= 483.457\,82\ldots$

$EF = 21.987\,674\ldots$ $(EF > 0)$
$= 22 \text{ km}$ [nearest km]}

(2 marks)

ii Let the length of GH be x km.

$EH = GH = x \text{ km}$

By Pythagoras' theorem in ΔGEH,

$x^2 + x^2 = 82^2$ ✓
$2x^2 = 6724$
$x^2 = 3362$
$x = 57.982\,756\ldots$ $[x > 0]$

The length of GH is 58 km to the nearest kilometre. ✓

Total distance
$= (2 \times 57.982\ldots + 64 + 21.704\ldots) \text{ km}$
$= 201.669\,58\ldots \text{ km}$
$= 202 \text{ km}$ ✓
[nearest kilometre]

(3 marks)

38

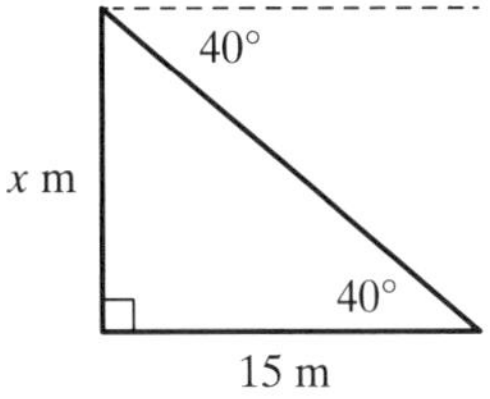

$$\tan 40° = \frac{x}{15}$$

$x = 15 \times \tan 40°$
$= 12.586\,494\,47\ldots$
$= 13$
[nearest whole number]

The kookaburra's feet are 13 m above the ground, to the nearest metre.

Answer C

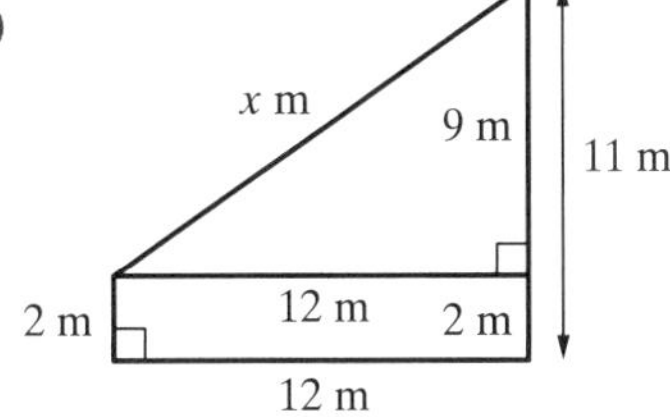

By Pythagoras' theorem:

$x^2 = 9^2 + 12^2$
$= 81 + 144$
$= 225$
$x = \sqrt{225}$ $(x > 0)$
$= 15$

The length of the cable is 15 m.

Answer C

40

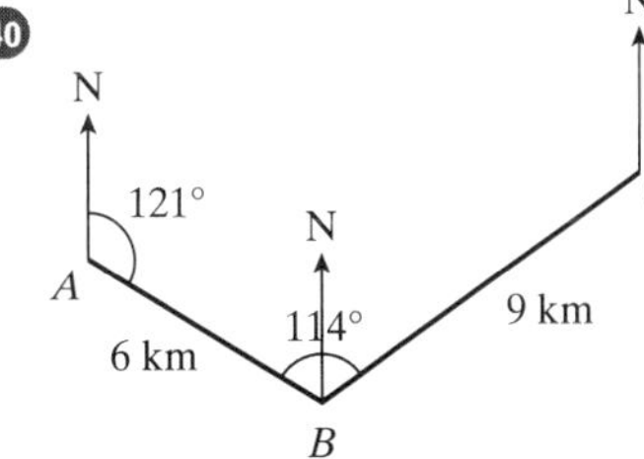

i Let X be a point due North of B.

$\angle ABX = 180° - 121°$
(co-interior angles, parallel lines)
$= 59°$

$\angle XBC = 114° - 59°$
$= 55°$

$\therefore$ the bearing of C from B is 055°.

(1 mark)

ii Using the cosine rule in ΔABC,

$b^2 = a^2 + c^2 - 2ac \cos B$
$= 9^2 + 6^2 - 2 \times 9 \times 6 \times \cos 114°$ ✓
$= 160.927\,55\ldots$

$b = \sqrt{160.927\,55\ldots}$ $(b > 0)$
$= 12.685\,722\ldots$
$= 13$
[nearest whole number]

The distance AC is 13 km to the nearest kilometre. ✓

(2 marks)

iii Using the sine rule in ΔABC,

$$\frac{\sin C}{c} = \frac{\sin B}{b}$$

$$\frac{\sin C}{6} = \frac{\sin 114°}{12.685\,722\ldots}$$ ✓

$$\sin C = \frac{6 \times \sin 114°}{12.685\,722\ldots}$$

$= 0.432\,082\,028\ldots$
$C = 25.599\,763\,64\ldots°$
$= 26°$ [nearest degree] ✓

N
N
C
A
26°
55°
59°
55°
9 km
6 km
Y
B

Let Y be a point due south of C.

$\angle BCY = 55°$ (alternate angles, parallel lines)

Required bearing
$= (180 + 55 + 26)°$
$= 261°$

The bearing of A from C is 261°. ✓

(3 marks)

41 The angles of a triangle add to 180°.

So $\angle QRP = 180° - (26 + 46)°$
$= 108°$

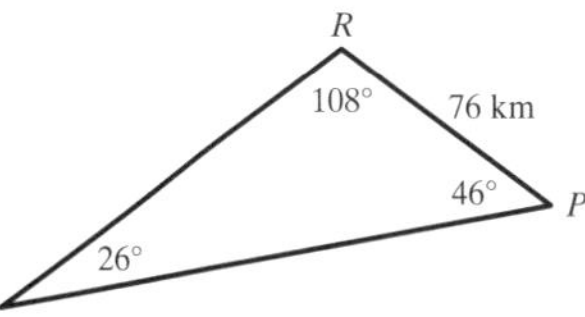

By the sine rule:

$$\frac{a}{\sin A} = \frac{b}{\sin B}$$

$$\frac{PQ}{\sin 108°} = \frac{76}{\sin 26°}$$

$$PQ = \frac{76 \times \sin 108°}{\sin 26°}$$

$= 164.883\,788\ldots$
$= 165 \text{ km}$
[nearest kilometre]

The distance from P to Q is 165 km to the nearest kilometre.

Answer C

42 Let P be a point due north of B.

$\angle PBA = 30°$ (alternate angles, parallel lines)

Reflex $\angle PBA = 360° - 30°$
$= 330°$

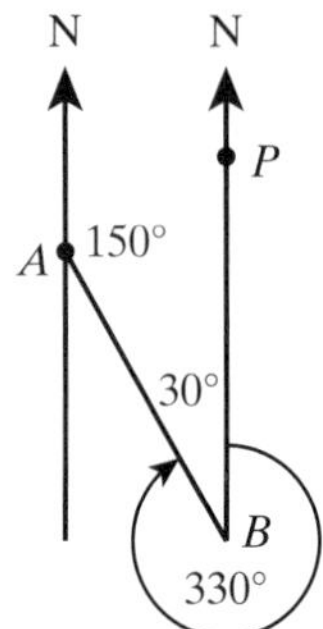

So the bearing of A from B is 330°.

Answer D

43 i

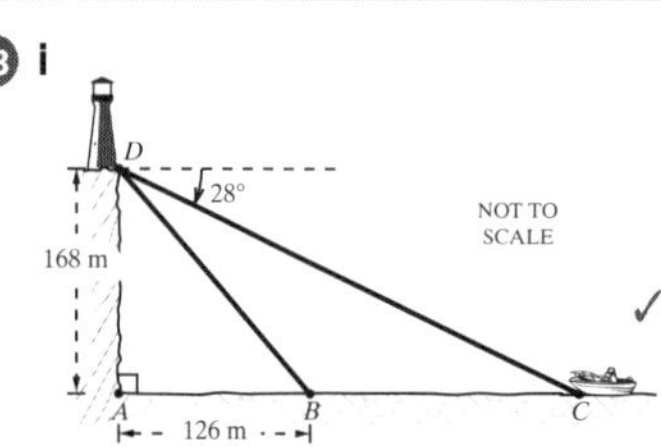

✓

The angle of depression from D to B
$= \angle DBA$ (alternate angles, parallel lines)

In ΔDBA,

$\tan B = \dfrac{168}{126}$ ✓

$B = 53.130\,102\,35...°$
$= 53° 8'$ [nearest minute]
$= 53°$ [nearest degree]

So the angle of depression is 53°8′ correct to the nearest minute or 53° to the nearest degree. ✓

(3 marks)

ii $\angle ADC = 90° - 28°$
$= 62°$

In ΔADC,

$\tan 62° = \dfrac{AC}{168}$ ✓

$AC = 168 \times \tan 62°$
$= 315.962\,04...$ m

$BC = 315.962\,04... - 126$
$= 189.962\,04...$
$= 190$ m ✓ [nearest metre]

(2 marks)

44

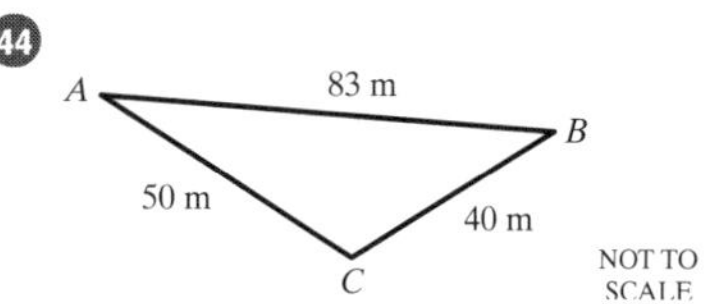

$\cos C$

$= \dfrac{a^2 + b^2 - c^2}{2ab}$

$= \dfrac{40^2 + 50^2 - 83^2}{2 \times 40 \times 50}$ ✓

$= -0.697\,25$

$C = 134.206\,78...°$ ✓

$A = \dfrac{1}{2}ab \sin C$

$= \dfrac{1}{2} \times 40 \times 50 \times \sin 134.206\,78...°$

$= 716.828\,039...$

$= 717 \text{ m}^2$ [nearest square metre]

The area is 717 m² to the nearest square metre. ✓

(3 marks)

45 $\cos 30° = \dfrac{8}{x}$

$\therefore x = \dfrac{8}{\cos 30°}$

Answer A

46 $\angle DAC = \angle DCA$, since ΔADC is isosceles

$\therefore \angle DAC = \angle DCA = \dfrac{180° - 80°}{2}$

(Angle sum of $\Delta ADC = 180°$)

$= 50°$

$\therefore \angle DBA = 180° - 60° - 50°$ (angle sum ΔDBA)

$= 70°$

Using the sine rule,

$\dfrac{AB}{\sin 60°} = \dfrac{30}{\sin 70°}$

$\therefore AB = \dfrac{30 \sin 60°}{\sin 70°}$

$= 27.6481...$ cm

$= 28$ cm to nearest cm

Answer A

47

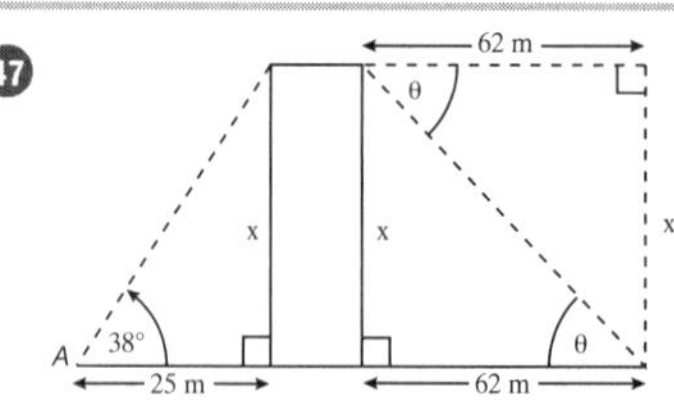

i Let x be the height of the building.

$\therefore \tan 38° = \dfrac{x}{25}$

$x = 25 \tan 38°$
$= 19.5321...$ m
$= 19.5$ m

(1 mark)

ii Let θ be the angle of depression which is equal to the angle of elevation.

$\therefore \tan \theta = \dfrac{x}{62}$ ✓

$= \dfrac{19.5321...}{62}$

$\therefore \theta = 17.48...°$
$= 17°$ to nearest degree ✓

(2 marks)

48 i

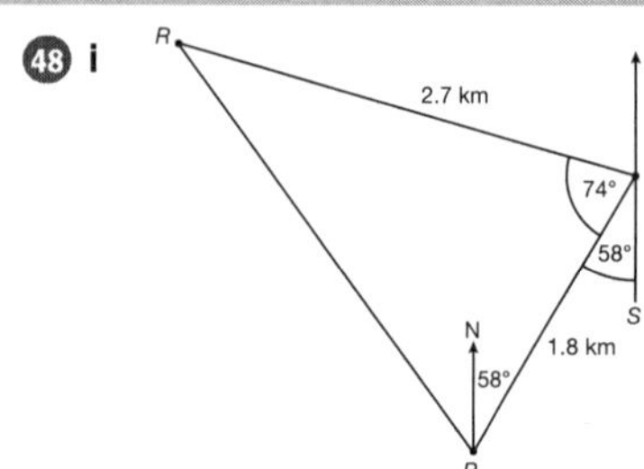

$\angle NPQ = \angle PQS = 58°$

$\therefore$ Bearing of R from Q is $(180 + 58 + 74)° = 312°$.

(1 mark)

ii Using the cosine rule,

$RP^2 = 2.7^2 + 1.8^2 - 2(2.7)(1.8) \cos 74°$ ✓

$= 7.8508...$

$RP = 2.8019...$

$\doteqdot 2.8$ km. ✓

(2 marks)

iii $A = \dfrac{1}{2} ab \sin C$

$= \dfrac{1}{2} \times 2.7 \times 1.8 \times \sin 74°$

$= 2.3358...$

$\doteqdot 2.3 \text{ km}^2$

(1 mark)

49 The smallest angle is opposite the smallest side.

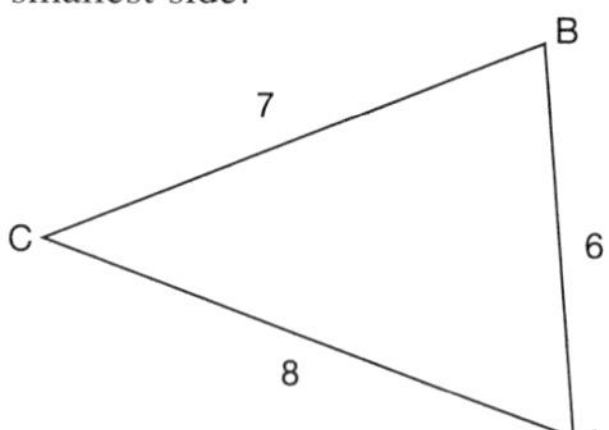

Using the Cosine rule:

$\cos C = \dfrac{a^2 + b^2 - c^2}{2ab}$

$= \dfrac{7^2 + 8^2 - 6^2}{2 \times 7 \times 8}$

$= 0.6875$

$\therefore C = 47°$

Answer B

50 $\sin 55° = \dfrac{h}{80}$

$\therefore \quad h = 80 \sin 55°$
$= 65.5321 \ldots$
$\doteqdot 66$ m

Answer B

51

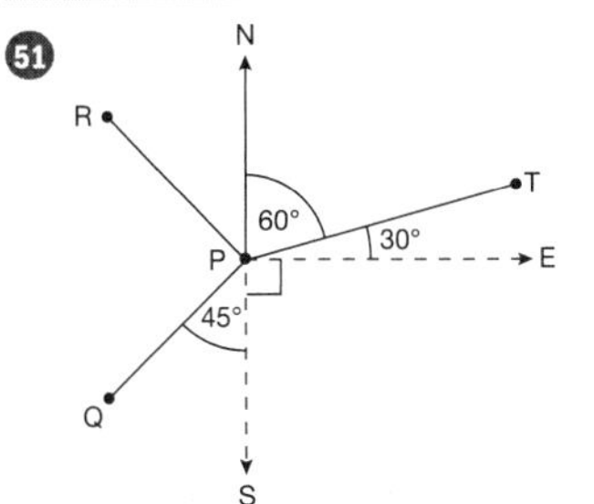

Now $\angle SPQ = 45°$
(as Q is SW of P)
and $\angle EPS = 90°$
$\therefore \angle TPE = 165° - 90° - 45°$
$= 30°$
$\therefore \angle NPT = 90° - 30°$
$= 60°$

Answer A

52 **i**

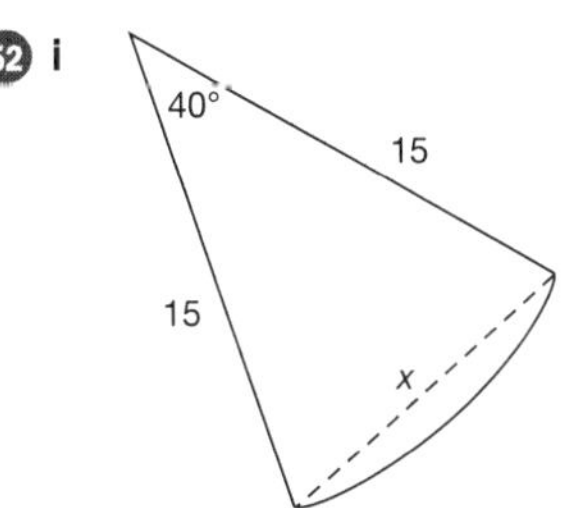

Let x be the length of the straight chord.
$\therefore$ Using the Cosine rule,
$x^2 = 15^2 + 15^2 - 2 \times 15 \times 15 \times \cos 40°$
$= 105.2800$
$x = 10.2606 \ldots$ cm ✓

Length of rectangular box
$= 4 \times 10.2606$
$\doteqdot 41$ cm ✓

Depth of rectangular box
$= 3 \times 7$
$= 21$ cm ✓

Height of rectangular box
$= 15$ cm

$\therefore$ Dimensions of rectangular box are 41 cm × 15 cm × 21 cm. ✓

(4 marks)

ii Volume of remaining prism
= area of triangle × height
$= \frac{1}{2} ab \sin C \times$ height ✓
$= \frac{1}{2} \times 15 \times 15 \times \sin 40° \times 7$
$= 506.1952 \ldots$
$= 506$ cm^3 to nearest cm^3. ✓

(2 marks)

53 $\sin 50° = \dfrac{12}{MN}$

$MN = \dfrac{12}{\sin 50°}$
$= 15.6648\ldots$
$\doteqdot 15.66$ cm

Answer C

54

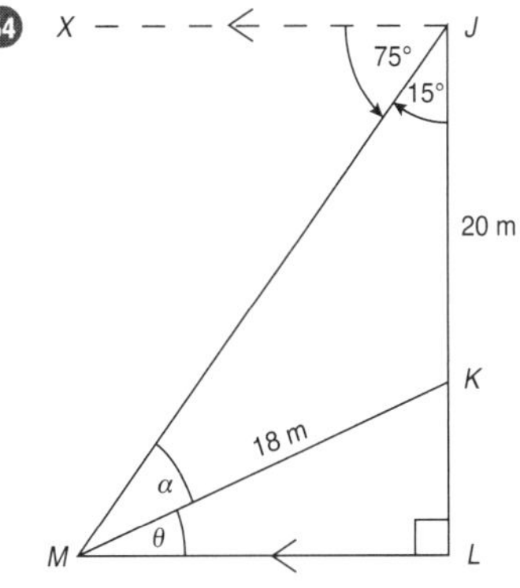

Let θ = angle of elevation from M to K.
and $\alpha = \angle JMK$
Now $\angle JML = 75° = \angle XJM$
$\therefore \quad \angle MJK = 90° - 75°$
$= 15°$ ✓

$\therefore$ By the sine rule in ΔMJK,

$\dfrac{\sin\alpha}{20} = \dfrac{\sin 15°}{18}$ ✓

$\sin\alpha = \dfrac{20 \sin 15°}{18}$
$= 0.2875 \ldots$
$\therefore \alpha = 17°$ to nearest degree
$\therefore \theta = 75° - 17°$
$= 58°$ ✓

(3 marks)

55

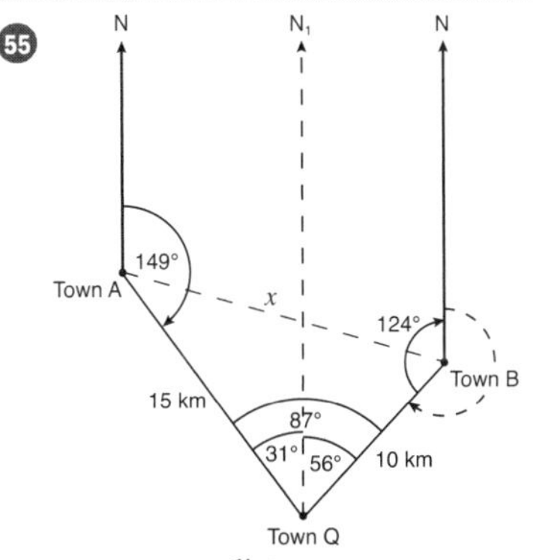

i speed $= \dfrac{\text{distance}}{\text{time}}$

$= \dfrac{15 \text{ km}}{2 \text{ h } 48 \text{ min}}$
$= 5.3571 \ldots$ km/h
$= 5$ km/h
to the nearest km/h.

N.B. To enter 2 h 48 min into calculator, use the degrees and minutes button.

(1 mark)

ii Let x = distance from A to B.
$\therefore$ Using the cosine rule in ΔABQ,
$x^2 = 15^2 + 10^2 - 2 \times 15 \times 10 \times \cos 87°$ ✓
$= 309.299 \ldots$
$\therefore x = 17.5869 \ldots$
$= 18$ km to nearest km. ✓

(2 marks)

iii $\angle AQN_1 = 180° - 149°$
$= 31°$

$\therefore \angle N_1QB = 87° - 31°$
$= 56°$

$\therefore \angle QBN = 180° - 56°$
$= 124°$

$\therefore$ The bearing of Q from B is $360° - 124° = 236°$.

(1 mark)

56

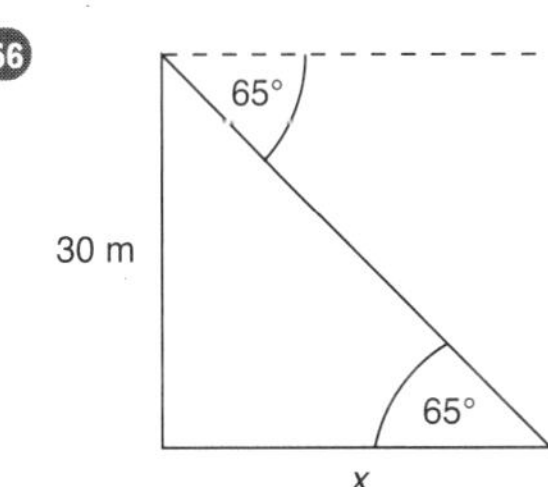

$\tan 65° = \dfrac{30}{x}$

$\therefore x = \dfrac{30}{\tan 65°}$
$= 13.9892$
$= 14.0$ m to 1 decimal place

Answer B

57 $A = \frac{1}{2} ab \sin C$

$= \frac{1}{2} \times 39 \times 47 \times \sin 21°$
$= 328.4442\ldots$
$= 328$m^2 to nearest m^2

Answer C

58

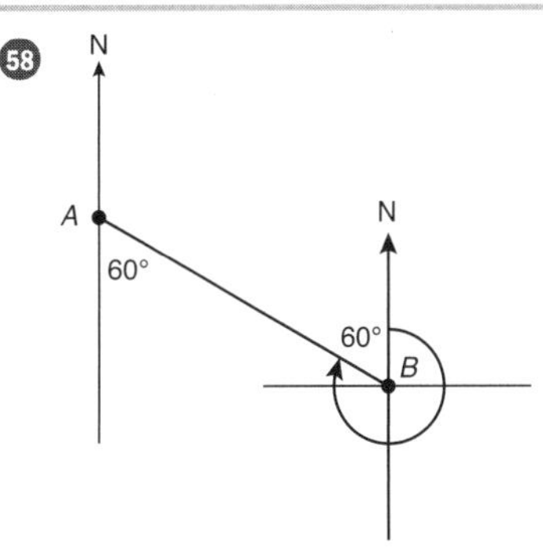

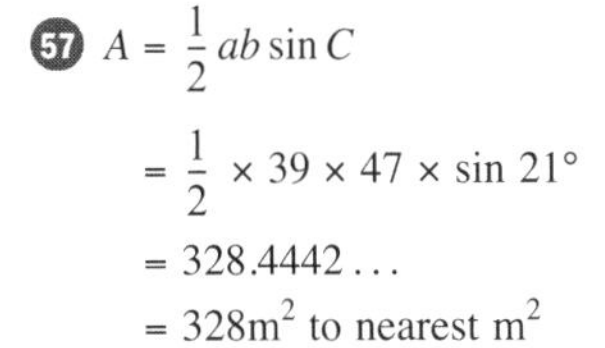

Bearing $= 360° - 60°$
$= 300°$

Answer D

59 i

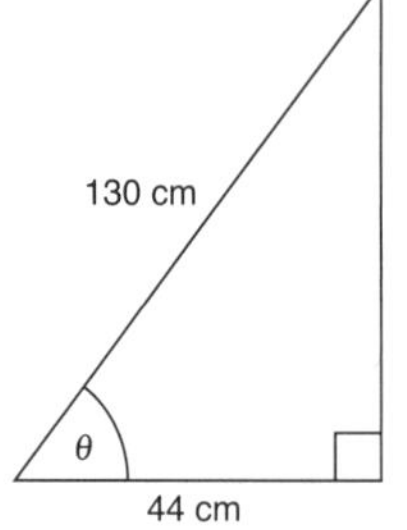

$\cos\theta = \dfrac{44}{130}$ ✓

$\therefore \theta = 70.21\ldots°$
$= 70°$ to nearest degree ✓

(2 marks)

ii

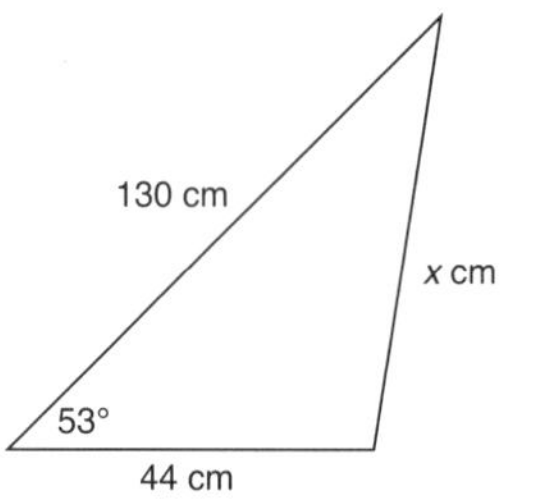

Using the cosine rule,
$x^2 = 130^2 + 44^2 - 2(130)(44)\cos 53°$ ✓
$= 11\,951.2361\ldots$

$x = \sqrt{11\,951.2361\ldots}$
$= 109.3217\ldots$
$= 109$ cm to nearest cm ✓

(2 marks)

60

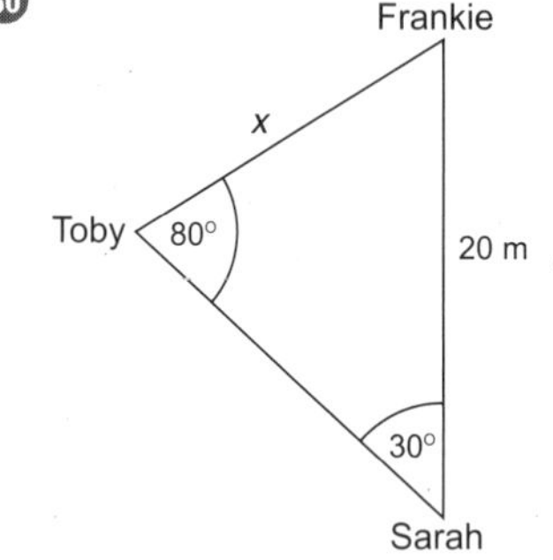

Using the sine rule
$\dfrac{x}{\sin 30°} = \dfrac{20}{\sin 80°}$

Answer A

61 $\tan\theta = 85$
$\therefore \theta = 89°19'33.47\ldots''$
$= 89°20'$ to the nearest minute.

Answer D

62 i Using Pythagoras' Theorem,
$AB^2 = BC^2 + AC^2$

Now $AB^2 = 13^2$
$= 169$

and $BC^2 + AC^2 = 5^2 + 12^2$
$= 25 + 144$
$= 169$

$\therefore \angle ABC$ is a right-angled triangle.

(1 mark)

ii $\tan\angle ABC = \dfrac{12}{5}$ ✓

$\therefore \angle ABC = 67°$ to nearest degree ✓

(2 marks)

63 i

$\theta = 360° - 250°$
$= 110°$

(1 mark)

ii $a^2 = b^2 + c^2 - 2bc\cos A$
$= 36^2 + 15^2 - 2 \times 36 \times 15 \times \cos 110°$ ✓ ✓
$= 1890.3817\ldots$ ✓

$\therefore a = \sqrt{1890.3817}$
$= 43.4785\ldots$
$= 43$ km to the nearest km

$\therefore$ The distance of C from B is 43 km to the nearest km. ✓

(3 marks)

64 $\tan 20° = \dfrac{\text{opposite side}}{\text{adjacent side}}$

$= \dfrac{c}{a}$

Answer D

65 Area $= \dfrac{1}{2}ab\sin C$

$= \dfrac{1}{2} \times 30 \times 20 \times \sin 35°$
$= 172.0729\ldots$
$= 172$ m^2 to nearest m^2

Answer C

66 i South-east $= 90° + 45°$
$\therefore \angle PAQ = 135°$

(1 mark)

ii $\angle PAR = 135° + 50°$
$= 185°$

$\therefore$ The bearing is 185° T or S 5° W.

(1 mark)

iii Using cosine rule in ΔPAB,

$\cos\angle PAB$
$= \dfrac{28^2 + 31^2 - 35^2}{2 \times 28 \times 31}$ ✓
$= \dfrac{520}{1736}$
$= 0.2995\ldots$ ✓

$\therefore \angle PAB$
$= 72.57\ldots°$
$= 73°$ to nearest degree ✓

(3 marks)

67 Using the cosine rule:
$x^2 = 10^2 + 7^2 - 2 \times 10 \times 7 \times \cos 60°$

Answer B

68

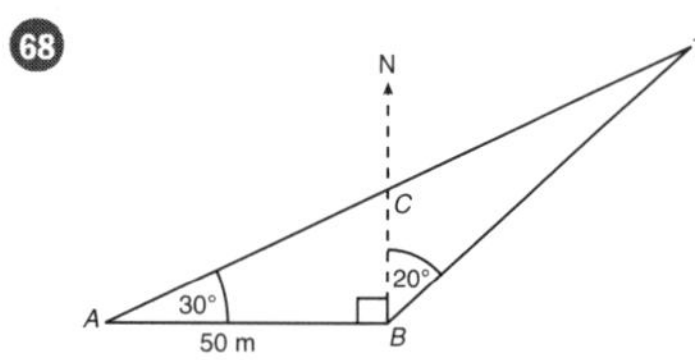

i $\angle ABT = \angle ABC + \angle CBT$
$= 90° + 20°$
$= 110°$

(1 mark)

ii By the sine rule:
$\dfrac{BT}{\sin A} = \dfrac{AB}{\sin T}$

Now $\angle BTA$
$= 180° - 30° - 110°$
$= 40°$ ✓

$\therefore \dfrac{BT}{\sin 30°} = \dfrac{50}{\sin 40°}$ ✓

$\therefore BT = \dfrac{50\sin 30°}{\sin 40°}$
$= 38.893\ldots$ m
$= 39$ m to the nearest m.

✓ *(3 marks)*

69 i

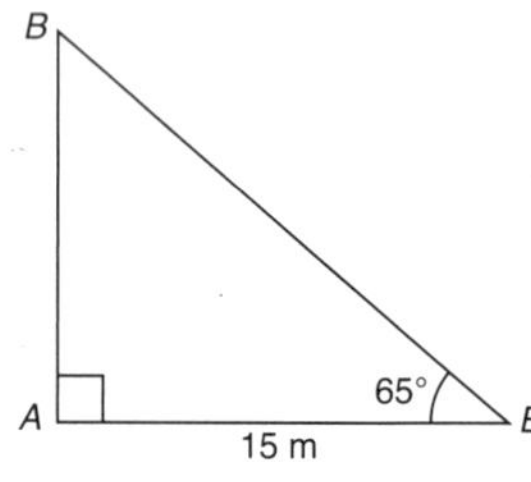

$\tan 65° = \dfrac{AB}{15}$ ✓

$\therefore AB = 15\tan 65°$
$= 32.1676\ldots$ m
$= 32.2$ m to 1 decimal place ✓

(2 marks)

ii

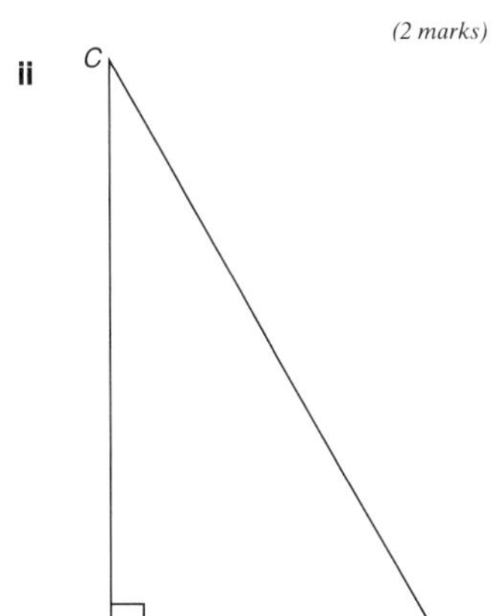

Now $AC = 2 \times 32.1676...$

$\therefore CE^2 = AE^2 + AC^2$ (Pythagoras' theorem)

$= 15^2 + (2 \times 32.1676...)^2$

$= 4364.01...$ ✓

$\therefore CE = 66.0607...$ m

$\therefore CE = 66.1$ m ✓ to nearest decimal place

(2 marks)

70

i $\angle XYZ = 90° + 46°$

$- 136°$

(1 mark)

ii $XZ^2 = XY^2 + YZ^2 - 2(XY)(YZ)\cos\angle XYZ$

$= 18^2 + 22^2 - 2(18)(22)\cos 136°$ ✓

$= 1377.7171...$

$\therefore XZ = 37.1176...$

$= 37.1$ km ✓ to 1 decimal place

(2 marks)

iii Bearing of Y from Z is $180° + 46° = 226°$.
(or S46°W)

(1 mark)

71 **i** $QS^2 = 5^2 - 4^2$ (Pythagoras' Theorem)

$= 25 - 16$

$= 9$

$\therefore QS = 3$ cm ✓

$SP^2 = 7^2 - 3^2$ (Pythagoras' Theorem)

$= 49 - 9$

$= 40$

$\therefore SP = \sqrt{40}$ ✓

Perimeter $= 7 + 5 + 4 + \sqrt{40}$

$= 22.3245...$

$= 22.3$ cm ✓ to 1 decimal place

(3 marks)

ii $\sin\angle QPS = \dfrac{QS}{QP}$

$= \dfrac{3}{7}$

$\therefore \angle QPS = 25.37...°$

$= 25°$ to nearest degree

(1 mark)

72

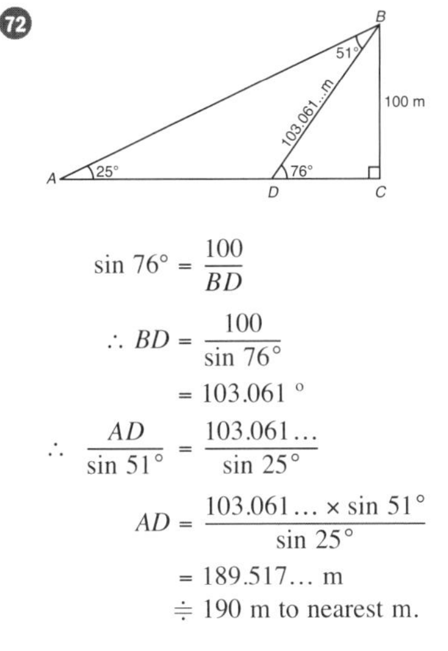

$\sin 76° = \dfrac{100}{BD}$

$\therefore BD = \dfrac{100}{\sin 76°}$

$= 103.061$ °

$\therefore \dfrac{AD}{\sin 51°} = \dfrac{103.061...}{\sin 25°}$

$AD = \dfrac{103.061... \times \sin 51°}{\sin 25°}$

$= 189.517...$ m

$\doteqdot 190$ m to nearest m.

Answer C

73 **i**

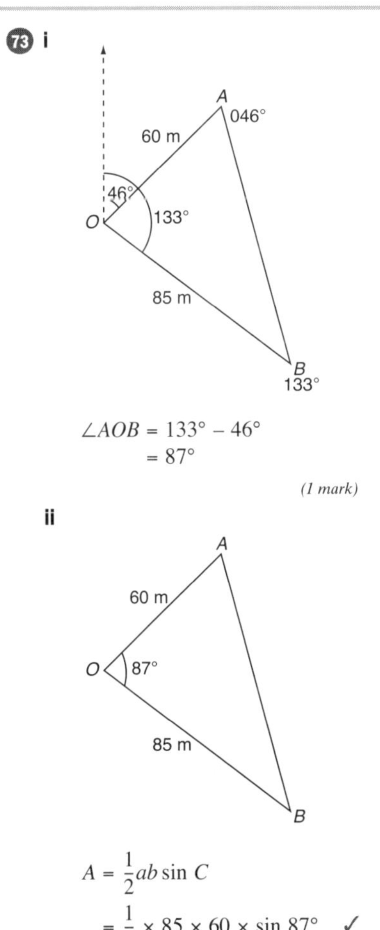

$\angle AOB = 133° - 46°$

$= 87°$

(1 mark)

ii

$A = \dfrac{1}{2}ab\sin C$

$= \dfrac{1}{2} \times 85 \times 60 \times \sin 87°$ ✓

$= 2546.50...$ m^2

$= 2547$ m^2 to nearest m^2 ✓

(2 marks)

iii $AB^2 = OA^2 + OB^2 - 2(OA)(OB)\cos\angle AOB$

$= 60^2 + 85^2 - 2(60)(85)\cos 87°$ ✓

$= 10\,291.173...$

$\therefore AB = \sqrt{10\,291.173...}$

$= 101.445$ m

$\doteqdot 101$ m ✓

(2 marks)

1 A total of 11 400 people entered a running race. The ratio of professional runners to amateurs was 3:16. All the professional runners completed the race while 600 of the amateurs did not complete the race.

For those who completed the race, what is the ratio, in simplest form, of professional runners to amateurs?

A 1:2 **B** 1:5
C 1:8 **D** 1:19 *(1 mark)*

(Q15, **2021 HSC**) Medium

2 The fuel consumption for a car is 6.7 litres /100 km. On a road trip, the car travels a distance of 1560 km and the fuel cost is $1.45 per litre.

What is the total fuel cost for this trip? *(2 marks)*

(Q18, **2021 HSC**) Easy

3 A rectangular sportsground has been drawn to scale on a 1-cm grid as shown. The scale used is 1:3000.

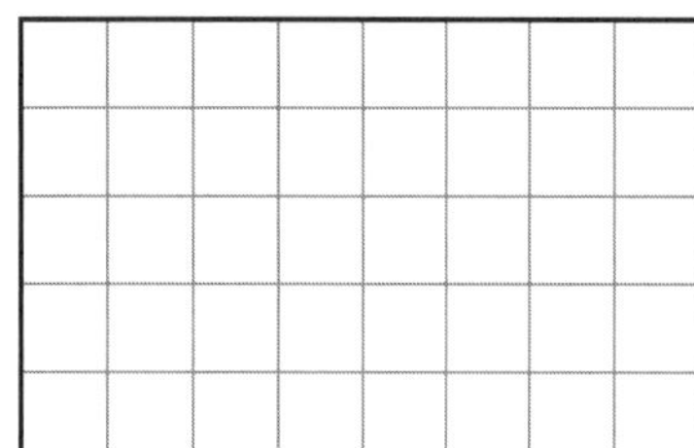

Kerry took 12 minutes to walk around the perimeter of this sportsground.

What was Kerry's average speed in kilometres per hour? *(4 marks)*

(Q25, **2021 HSC**) Hard

4 The distance between Bricktown and Koala Creek is 75 km. A person travels from Bricktown to Koala Creek at an average speed of 50 km/h.

How long does it take the person to complete the journey?

A 40 minutes **B** 1 hour 25 minutes
C 1 hour 30 minutes **D** 1 hour 50 minutes
(1 mark)

(Q3, **2020 HSC**) Easy

5 Ayla wishes to estimate the number of trees on a square block of land measuring 1000 m by 1000 m. She counts the number of trees on a 5 m by 5 m section of the block and finds there are 8 trees.

Based on this, estimate the number of trees on the entire square block of land. *(2 marks)*

(Q17, **2020 HSC**) Medium

6 In a tropical drink, the ratio of pineapple juice to mango juice to orange juice is 15:9:4.

a How much orange juice is needed if the tropical drink is to contain 3 litres of pineapple juice? *(2 marks)* Medium

b The internal dimensions of a drink container, in the shape of a rectangular prism, are shown.

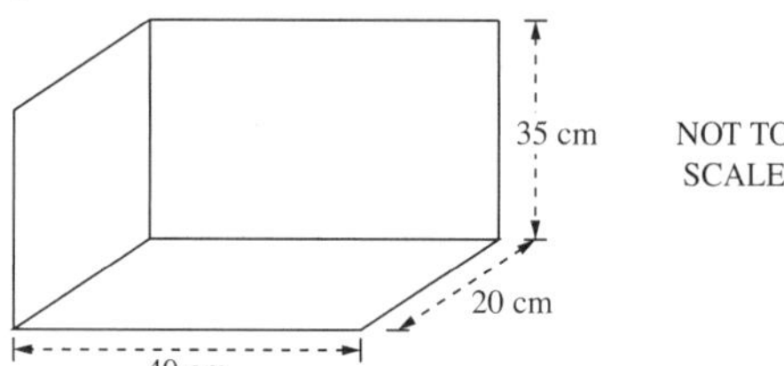

To completely fill the container with the tropical drink, how many litres of mango juice are required? *(3 marks)* Medium

(Q23, **2020 HSC**)

7 The shaded region on the diagram represents a garden. The scale is 1 cm = 5 m.

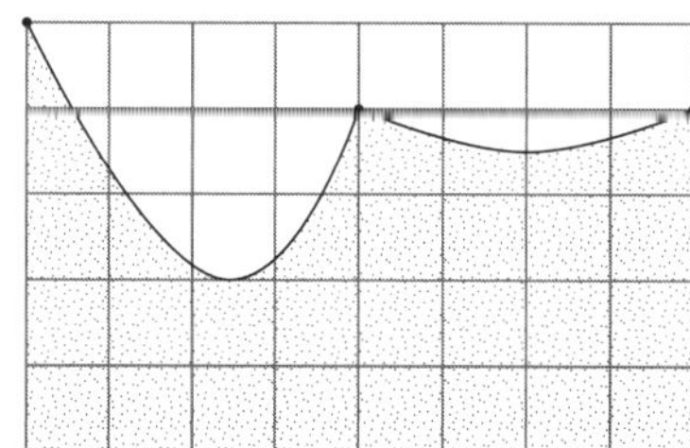

a Use two applications of the trapezoidal rule to calculate the approximate area of the garden. *(3 marks)* Medium

b Should the answer to part **a** be more than, equal to or less than the actual area of the garden? Referring to the diagram above, briefly explain your answer. *(2 marks)* Hard

(Q27, **2020 HSC**)

8 Sugar is sold in four different sized packets. Which is the best buy?

A 100 g for $0.40 **B** 500 g for $1.65
C 1 kg for $3.50 **D** 2 kg for $6.90
(1 mark)

(Q2, **2019 HSC**) 

9 Andrew, Brandon and Cosmo are the first three batters in the school cricket team. In a recent match, Andrew scored 30 runs, Brandon scored 25 runs and Cosmo scored 40 runs.

a What is the ratio of Andrew's to Brandon's to Cosmo's runs scored, in simplest form? *(2 marks)* **Easy**

b In this match, the ratio of the total number of runs scored by Andrew, Brandon and Cosmo to the total number of runs scored by the whole team is 19:36.
How many runs were scored by the whole team? *(2 marks)* **Medium**

(Q18, **2019 HSC**)

10 A map is drawn to scale, on 1-cm grid paper, showing the positions of a supermarket and a cinema. A reservoir is also shown.

a It takes 10 minutes to walk in a straight line from the cinema to the supermarket at a constant speed of 3 km/h. Show that the scale of the map is 1 cm = 100 m. *(3 marks)* **Medium**

b The reservoir is initially empty. During a storm 20 mm of rain falls on the reservoir.

With the aid of one application of the trapezoidal rule, estimate the amount of water in the reservoir immediately after the storm. Assume that all rain which falls over the reservoir is stored.

Give your answer in cubic metres. *(3 marks)* **Hard**

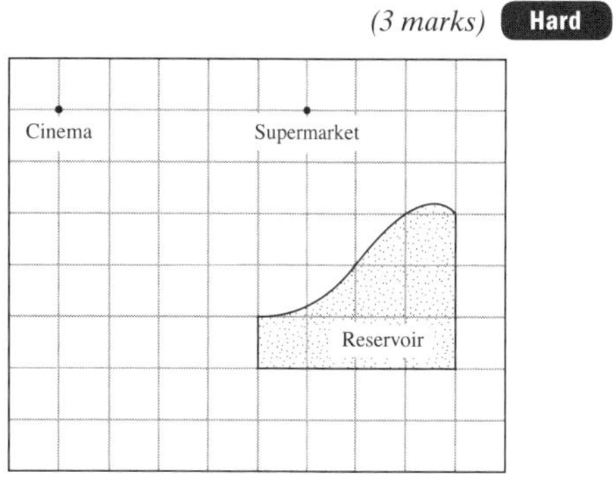

(Q41, **2019 HSC**)

11 Steve has purchased a flat block of farmland that adjoins a river. A scale drawing of the block is shown right.

a By measuring, determine the scale used, expressed in the form
1 cm =  m. *(1 mark)* **Easy**

b Using two applications of the trapezoidal rule, find an approximation for the area of the land. *(2 marks)* **Medium**

c In a thunderstorm, 35 mm of rain fell on Steve's land. Find the volume of water that fell, in kilolitres. *(2 marks)* **Hard**

Bonus question (see page iv)

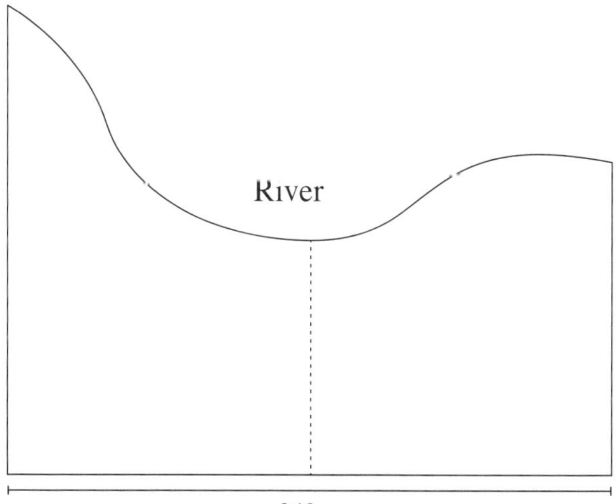

12 The table shows the number of Calories per 100 grams for three macronutrients.

Source of energy	Number of Calories per 100 grams
Carbohydrate	400
Protein	400
Fat	900

Use the conversion 1 Calorie = 4.184 kJ to find the number of kilojoules in 12.4 g of carbohydrates, to the nearest kilojoule? *(2 marks)*

Bonus question **Medium**

13 Michael's hot water unit has a rating of 3.6 kW and he has a 15-minute shower every morning. At that time of the day, electricity costs Michael an average of 30 cents per kWh.

a What is the cost of electricity each week? *(1 mark)* **Medium**

b Michael has calculated that 6% of his annual electricity is used when he showers. To save money, he plans to reduce the length of his shower by 5 minutes per day. What will be the new annual cost of his electricity bill? *(2 marks)* **Hard**

Bonus question

14 A biologist caught a random sample of 56 parrots in a national park. She tagged them and then released them. She later returned to the park and caught a random sample of 47 parrots. In this sample 29 had been tagged.

Using the capture/recapture technique, what is the estimated number of parrots in the park?

A 35 **B** 74

C 91 **D** 132 *(1 mark)*

(Q10, **2018 HSC**) Easy

15 A field diagram of a block of land has been drawn to scale. The shaded region $ABFG$ is covered in grass.

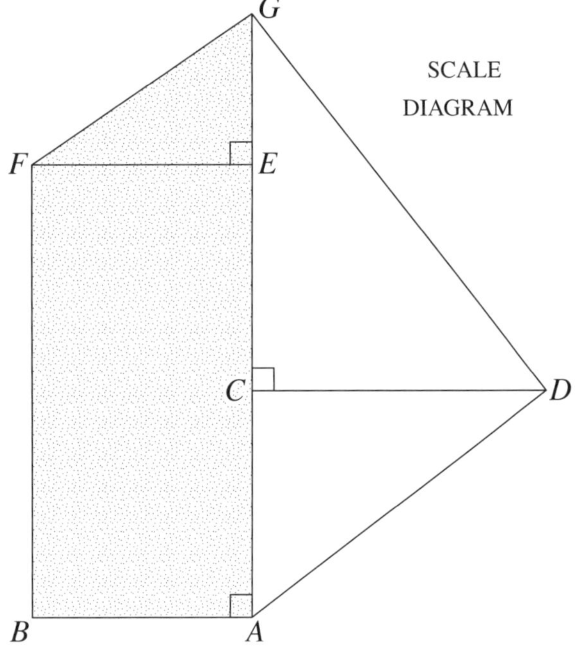

The actual length of AG is 24 m.

i Show that the scale of the diagram is 1 cm = 3 m. *(1 mark)* Medium

ii How much fertiliser would be needed to fertilise the grassed area $ABFG$ at the rate of 26.5 g /m^2? *(3 marks)* Medium

(Q26g, **2018 HSC**)

16 A car is travelling at 95 km/h. How far will it travel in 2 hours and 30 minutes?

A 38 km

B 41.3 km

C 218.5 km

D 237.5 km *(1 mark)*

(Q2, **2017 HSC**) Easy

17 A farmer needed to estimate the number of goats on his property. He tagged 80 of his goats. Later, he collected a random sample of 45 goats and found that 16 of these had tags.

Estimate the number of goats the farmer has on his property. *(2 marks)*

(Q26c, **2017 HSC**) Medium

18 Peta's car uses fuel at the rate of 5.9 L /100 km for country driving and 7.3 L /100 km for city driving. On a trip, she drives 170 km in the country and 25 km in the city. Calculate the amount of fuel she used on this trip. *(2 marks)*

(Q26c, **2016 HSC**) Easy

19 Jacob has a large jar of silver coins. He adds 20 gold coins into the jar. He then seals the jar and shakes it to ensure that the gold coins are mixed in thoroughly with the silver coins. Jacob then opens the jar and takes a handful of coins. In his hand he has 33 silver coins and 4 gold coins.

i Based on Jacob's handful, if a coin is selected at random from the jar, what is the probability that it is a gold coin? *(1 mark)* Easy

ii Jacob returns the handful of coins to the jar. Estimate the total number of coins in the jar. *(2 marks)* Medium

(Q28a, **2016 HSC**)

20 A farmer used the 'capture-recapture' technique to estimate the number of chickens he had on his farm. He captured, tagged and released 18 of the chickens. Later, he caught 26 chickens at random and found that 4 had been tagged.

What is the estimate for the total number of chickens on this farm? *(2 marks)*

(Q26a, **2015 HSC**) 

21 The image shows a rectangular farm shed with a flat roof.

The width of the shed indicated by the dotted line was measured using an online ruler tool, and found to be approximately 12 metres.

i By measurement and calculation, show that the area of the roof of the shed is approximately 216 m^2. *(2 marks)* **Medium**

ii All the rain that falls onto this roof is diverted into a cylindrical water tank which has a diameter of 3.6 m. During a storm, 5 mm of rain falls onto the roof. Calculate the increase in the depth of water in the tank due to the rain that falls onto the roof during the storm. *(3 marks)* **Hard**

(Q29c, **2015 HSC**)

22 Heather's car uses fuel at the rate of 6.6 L per 100 km for long-distance driving and 8.9 L per 100 km for short-distance driving.

She used the car to make a journey of 560 km, which included 65 km of short-distance driving.

Approximately how much fuel did Heather's car use on the journey?

A 37L
B 38L
C 48L
D 50L *(1 mark)*

(Q22, **2014 HSC**) 

23 Joel mixes petrol and oil in the ratio 40:1 to make fuel for his leaf blower.

i Joel pours 5 litres of petrol into an empty container to make fuel for his leaf blower.

How much oil should he add to the petrol to ensure that the fuel is in the correct ratio? *(1 mark)* **Medium**

ii Joel has 4.1 litres of fuel left in his container after filling his leaf blower. He wishes to use this fuel in his lawnmower. However, his lawnmower requires the petrol and oil to be mixed in the ratio 25:1.

How much oil should he add to the container so that the fuel is in the correct ratio for his lawnmower? *(3 marks)* **Hard**

(Q30c, **2013 HSC**)

24 The capture-recapture technique was used to estimate a population of seals in 2012.

- 60 seals were caught, tagged and released.
- Later, 120 seals were caught at random.
- 30 of these 120 seals had been tagged.

The estimated population of seals in 2012 was 11% less than the estimated population for 2008.

What was the estimated population for 2008? *(2 marks)*

(Q26f, **2012 HSC**) **Hard**

25 A map has a scale of 1:500000.

i Two mountain peaks are 2 cm apart on the map.
What is the actual distance between the two mountain peaks, in kilometres? *(1 mark)* **Medium**

ii Two cities are 75 km apart.
How far apart are the two cities on the map, in centimetres? *(1 mark)* **Medium**

(Q27c, **2012 HSC**)

26 A train departs from Town A at 3.00 pm to travel to Town B. Its average speed for the journey is 90 km/h, and it arrives at 5.00 pm. A second train departs from Town A at 3.10 pm and arrives at Town B at 4.30 pm.

What is the average speed of the second train?

A 135 km/h
B 150 km/h
C 216 km/h
D 240 km/h *(1 mark)*

(Q21, **2011 HSC**) **Medium**

27 Part of the floor plan of a house is shown. The plan is drawn to scale.

i What is the width of the stairwell, in millimetres? *(1 mark)* **Easy**

ii What are the internal dimensions of the bathroom, in millimetres? *(1 mark)* **Easy**

iii What is the length AB, the internal length of the rumpus room, in millimetres? *(1 mark)* **Medium**

iv There are three identical windows to be purchased for this rumpus room.
Use the floor plan to determine the width of the windows to be purchased. Give your answer in millimetres. *(1 mark)* **Hard**

(Q24a, **2011 HSC**)

All measurements are in millimetres.

RUMPUS ROOM

28 The elevation and floor plan of a building are shown.

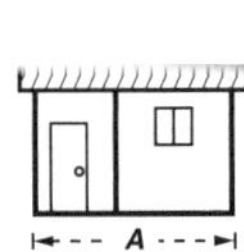

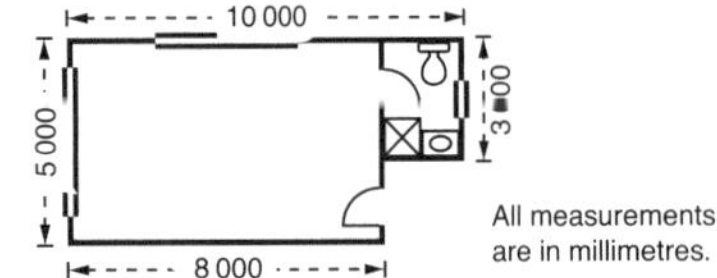

i What does this symbol ⊠ represent on the plan? *(1 mark)* **Easy**

ii What is the value of ***A*** on the elevation? *(1 mark)* **Easy**

iii Calculate the area of the floor of this building in square metres. *(2 marks)* **Medium**

(Q23b, **2010 HSC**)

29 Huong used the 'capture–recapture' technique to estimate the number of trout living in a dam.

- She caught, tagged and released 20 trout.
- Later she caught 36 trout at random from the same dam.
- She found that 8 of these 36 trout had been tagged.

What estimate should Huong give for the total number of trout living in this dam, based on her use of the 'capture–recapture' technique?

A 56 **B** 90

C 160 **D** 162 *(1 mark)*

(Q18, **2009 HSC**) **Medium**

30 An alcoholic drink has 5.5% alcohol by volume. The label on a 375 mL bottle says it contains 1.6 standard drinks.

i How many millilitres of alcohol are in a 375 mL bottle? *(1 mark)* **Easy**

ii It is recommended that a fully-licensed male driver should have a maximum of one standard drink every hour.
Express this as a rate in millilitres per minute, correct to one decimal place. *(2 marks)* **Medium**

(Q23c, **2008 HSC**)

31 A scientific study uses the 'capture-recapture' technique.

In the first stage of the study, 24 crocodiles were caught, tagged and released.

Later, in the second stage of the study, some crocodiles were captured from the same area. Eighteen of these were found to be tagged, which was 40% of the total captured during the second stage.

i How many crocodiles were captured in total during the second stage of the study? *(1 mark)* **Medium**

ii Calculate the estimate for the total population of crocodiles in this area. *(2 marks)* **Medium**

(Q23c, **2007 HSC**)

32 Peter rides his bike at a speed of 27 km/h. What is this speed in m/s?

A 7.5 **B** 18.75
C 97.2 **D** 450 *(1 mark)*

(Q11, **2006 HSC**) Easy

33 The diagram is a scale drawing of a butterfly.

What is the actual wingspan of the butterfly?

A 2.5 cm **B** 3 cm
C 15 cm **D** 18.75 cm *(1 mark)*

(Q4, **2005 HSC**) 

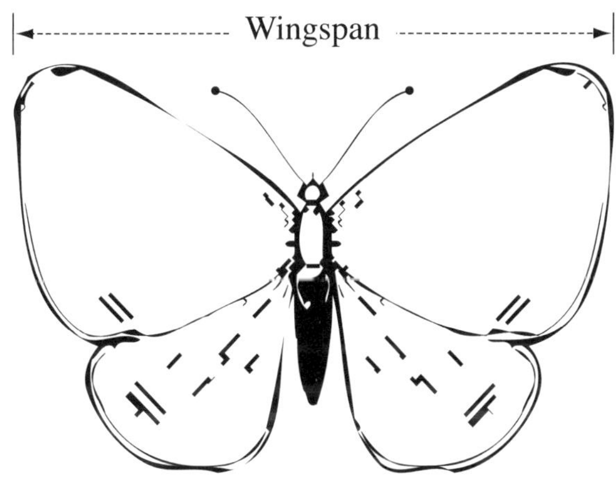

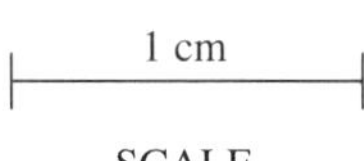

34 Yousef used the 'capture-recapture' technique to estimate the number of kangaroos living in a particular area.

- He caught, tagged and released 50 kangaroos.
- Later, he caught 200 kangaroos at random from the same area.
- He found that 5 of these 200 kangaroos had been tagged.

What is the correct estimate for the total number of kangaroos living in this area, using the 'capture-recapture' technique?

A 245 **B** 250
C 2000 **D** 10 000 *(1 mark)*

(Q21, **2005 HSC**) **Medium**

35 During a ten-minute period, Kath is exercising and Jim is resting.

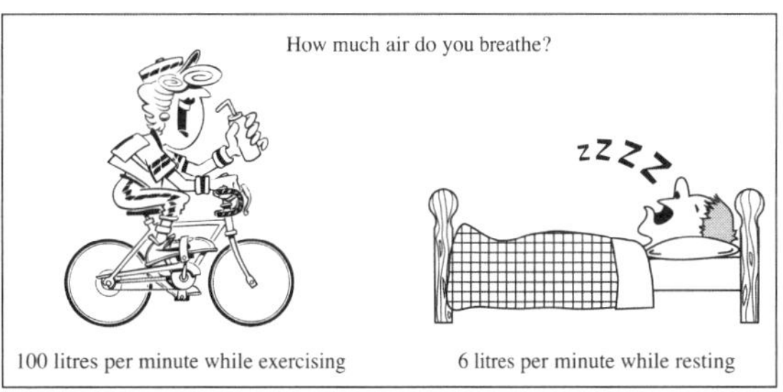

How much more air would Kath breathe than Jim during this time?

A 40 litres **B** 94 litres
C 940 litres **D** 1060 litres *(1 mark)*

(Q13, **2004 HSC**) 

36 Keryn is designing a new watering system for the shrubs in her garden. She knows that each shrub needs 1.2 litres of water per day. To minimise evaporation, Keryn designs a system to drip water into a tube that takes the water to the roots.

i What is the number of litres of water required daily for 13 shrubs? *(1 mark)* Easy

ii Keryn pays 94.22 cents per kilolitre for water. Calculate the total cost of watering 13 shrubs for one week. *(1 mark)* Easy

iii Keryn knows that 1 mL = 15 drops. Find the number of drops that one shrub needs daily. *(1 mark)* Medium

iv How many drops per minute are required for one shrub if the system is in use for 10 hours per day? *(1 mark)* Medium

(Q23a, **2003 HSC**)

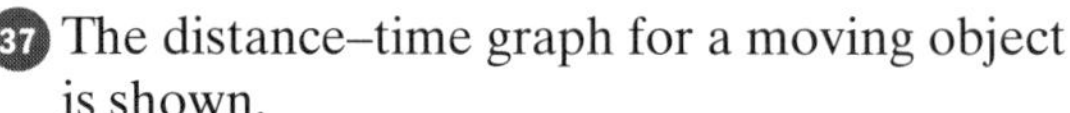

37 The distance–time graph for a moving object is shown.

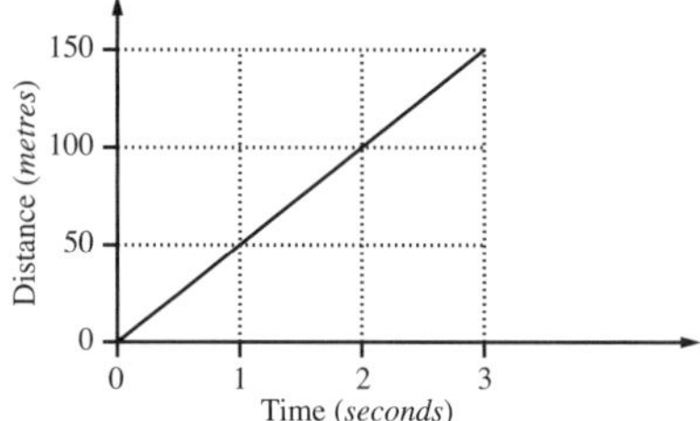

What is the speed of the object in *kilometres per hour*?

A 3 km/h **B** 14 km/h

C 50 km/h **D** 180 km/h *(1 mark)*

(Q17, **2001 HSC**) Medium

38 This is a site plan, drawn to scale, of Lot 3, General Drive.

i A fence is to be erected along all boundaries of Lot 3 except for the boundary on General Drive.
How many metres of fencing will be required? *(2 marks)* Medium

ii Lot 3 is in the shape of a trapezium. By measurement and calculation, determine the actual area of Lot 3 in square metres. *(2 marks)* Medium

(Q24c, **2001 HSC**)

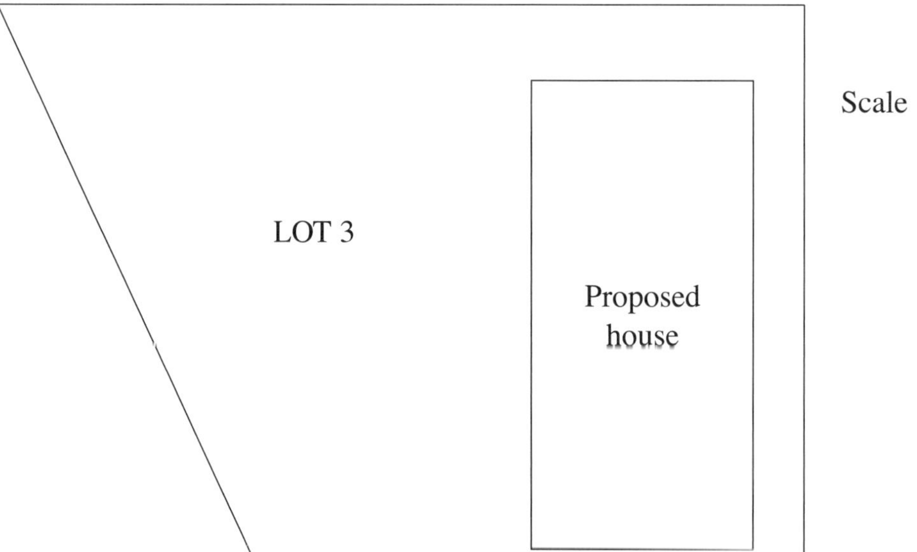

Year 12 Rates and ratio—Worked answers

1 11 400 runners.
Total parts in ratio = 3 + 16 = 19
Each part is 11 400 ÷ 19 = 600.
Now, if 600 amateurs did not finish, there is one fewer part for the amateurs.
New ratio = 3:15
= 1:5
Answer B

2 Fuel consumption = 6.7 L/100 km
Fuel used = 6.7 × 15.6
= 104.52 L ✓
Cost = 104.52 × \$1.45
= \$151.554
= \$151.55 ✓
(nearest cent)
(2 marks)

3 Scale is 1:3000
So 1 cm represents 3000 cm or 30 m. ✓
Perimeter = 26 squares ✓
Actual distance = 26 × 30 m
= 780 m ✓
Average speed = 780 m in 12 min
= 65 m/min
= 3900 m/h
= 3.9 km/h ✓
(4 marks)

4 Time = (75 ÷ 50) h
= 1.5 h
= 1 hour 30 minutes
Answer C

5 Area of block = 1000 × 1000
= 1 000 000 m^2 ✓
Area of section = 5 × 5
= 25 m^2
Number of sections in block
= 1 000 000 ÷ 25
= 40 000
Estimated number of trees ✓
= 40 000 × 8
= 320 000 ✓
(2 marks)

6 **a** The drink has 4 parts of orange juice for every 15 parts of pineapple juice. ✓
Orange juice required
= (3 ÷ 15 × 4) L
= 0.8 L
So 800 mL of orange juice is required. ✓
(2 marks)

b Volume = 40 × 20 × 35
= 28 000 cm^3 ✓
Capacity = 28 litres
Tropical drink:
Total parts = 15 + 9 + 4
= 28 ✓
So each part is a litre.
Mango is 9 parts so 9 L of mango juice is required. ✓
(3 marks)

7 **a** Scale is 1 cm = 5 m

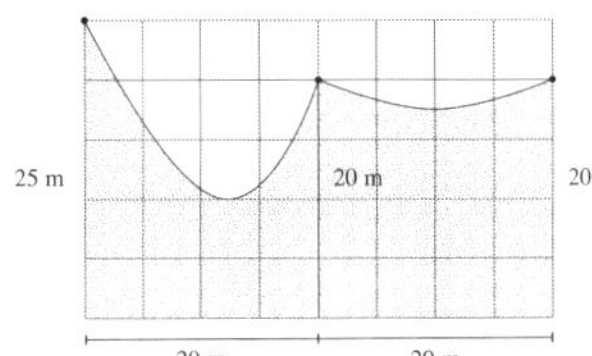

Trapezoidal rule:

$A \approx \frac{h}{2}(d_f + d_l)$ ✓

$A \approx \frac{20}{2}(25 + 20) + \frac{20}{2}(20 + 20)$ ✓

$= 850$ ✓

Approximate area is 850 m^2.
(3 marks)

b The answer to part **a** will be more than the actual area. ✓
The trapezoidal rule finds the area of a trapezium and in both applications the actual area is less than the area of the corresponding trapezium (or square in the case of the second application).
The actual area found by the trapezoidal rule is:

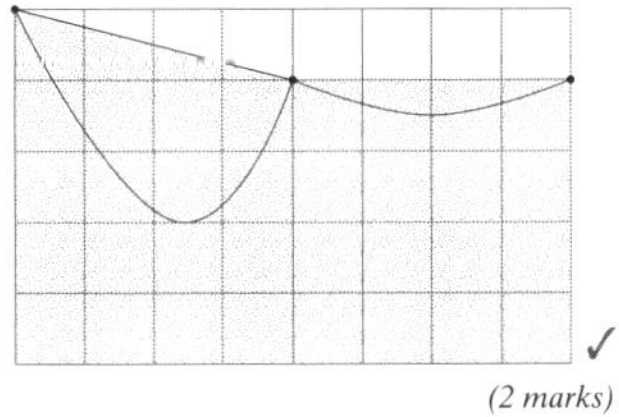

✓
(2 marks)

8 Try each option.
A: \$0.40 per 100 g
B: \$0.33 per 100 g
C: \$0.35 per 100 g
D: \$0.345 per 100 g
\$0.33 per 100 g is the best buy.
Answer B

9 **a** Ratio = 30:25:40 ✓
= 6:5:8 ✓
(2 marks)

b Runs scored by Andrew, Brandon and Cosmo
= 30 + 25 + 40
= 95 ✓
Let x be the number of runs scored by the whole team.
So 95:x = 19:36

$\frac{95}{x} = \frac{19}{36}$

$19x = 3420$

$x = 180$

So the number of runs scored by the whole team was 180.
Or: 6 + 5 + 8 = 19
Andrew, Brandon and Cosmo scored 5 × 19 runs.
So the whole team scored 5 × 36 runs or 180 runs. ✓
(2 marks)

10 **a** On the map the distance from the supermarket to the cinema is 5 units or 5 cm. ✓
Now 3 km/h
= 3000 m/h
= 3000 m in 60 min
= 500 m in 10 min ✓
So 5 cm represents 500 m.
So the scale is 1 cm = 100 m. ✓
(3 marks)

b Using the scale,
$h = 400$, $d_f = 100$ and $d_l = 300$

$A \approx \frac{h}{2}(d_f + d_l)$

$= \frac{400}{2}(100 + 300)$

$= 80\,000$ ✓

So the area is approximately 80 000 m^2.
Now 20 mm = 0.02 m
$V = 80\,000 \times 0.02$
$= 1600$ m^3 ✓
The amount of water in the reservoir will be approximately 1600 m^3. ✓

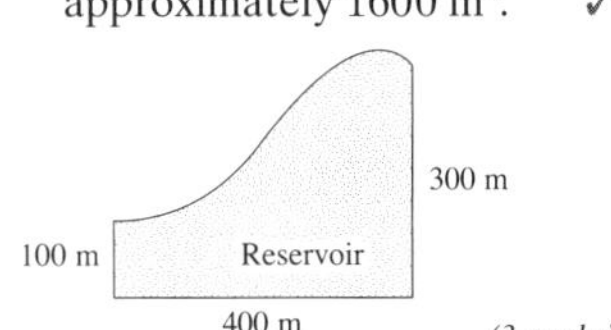

(3 marks)

11 **a** 8 cm = 240 m
1 cm = 30 m
(1 mark)

b

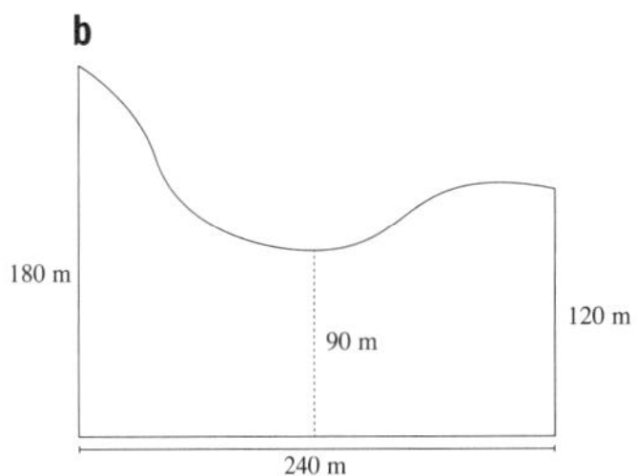

Area $\approx \frac{120}{2}[180 + 90] + \frac{120}{2}[90 + 120]$ ✓

$= 28\,800$

∴ the area is 28 800 m^2. ✓
(2 marks)

c Volume = $28\,800 \times 0.035$ ✓
$= 1008 \text{ m}^3$
∴ the volume was 1008 kL. ✓
(2 marks)

12 Number
$= 12.4 \div 100 \times 400 \times 4.184$ ✓
$= 207.5264\ldots$
$= 208$ ✓
∴ 208 kJ *(2 marks)*

13 **a** Cost = $3.6 \times 0.25 \times 7 \times 0.3$
$= 1.89$
∴ the cost is \$1.89 per week.
(1 mark)

b Existing annual bill
$= 3.6 \times 0.25 \times 0.3 \times 365 \div 6 \times 100$
$= 1642.50$
As 15 minutes – 10 minutes
= 5 minutes ✓
(or $\frac{1}{12}$ of an hour),
Annual savings from shorter shower
$= 3.6 \times \frac{1}{12} \times 0.3 \times 365$
$= 32.85$
New annual bill
$= 1642.50 - 32.85$
$= 1609.65$
∴ new annual cost will be \$1609.65. ✓
(2 marks)

14 Let x be the estimated number of parrots.
Now $\frac{29}{47} = \frac{56}{x}$
$29x = 2632$
$x = 90.758\,6206\ldots$
So the estimated number of parrots in the park is 91.
Answer C

15 **i** By measurement AG is 8 cm long.
So 8 cm = 24 m
1 cm = (24 ÷ 8) m
1 cm = 3 m *(1 mark)*

ii By measurement BA = 3 cm.
So actual length
$= 3 \times 3$ m
= 9 m ✓
By measurement BF = 6 cm
So actual length = 6×3 m
= 18 m
Area of $ABFG$:
$A = \frac{h}{2}(a + b)$
$= \frac{9}{2}(18 + 24)$
$= 189 \text{ m}^2$ ✓
Amount of fertiliser
$= 189 \times 26.5$ g
= 5008.5 g
= 5.0085 kg
So 5 kg of fertiliser would be needed. ✓
(3 marks)

16 2 h 30 min = 2.5 h
Distance = 2.5×95 km
= 237.5 km
Answer D

17 Let x be the number of goats.
Now $\frac{16}{45} = \frac{80}{x}$ ✓
$16x = 3600$
$x = 225$
There are about 225 goats on the property. ✓
(2 marks)

18 Country fuel used = 1.7×5.9 L
= 10.03 L ✓
City fuel used = 0.25×7.3 L
= 1.825 L
Total fuel used = (10.03 + 1.825) L
= 11.855 L ✓
(2 marks)

19 **i** $P(\text{gold}) = \frac{4}{37}$ *(1 mark)*

ii Let x be the total number of coins.
Now $\frac{4}{37} = \frac{20}{x}$ ✓
$4x = 740$
$x = 185$
There are about 185 coins in the jar. ✓
(2 marks)

20 Let x be the total number of chickens.
$\frac{x}{18} = \frac{26}{4}$ ✓
$x = \frac{26}{4} \times 18$
$= 117$
The estimate for the number of chickens is 117. ✓
(2 marks)

21 **i** By measuring, the length of the dotted line is 4 cm.
So 4 cm represents (approximately) 12 m.
1 cm represents 3 m.
By measuring, the length of the shed is 6 cm so the actual length would be about 18 m.
Area = 18 m × 12 m ✓
$= 216 \text{ m}^2$
So, the area of the roof would be approximately 216 m^2. ✓
(2 marks)

ii Now 5 mm = 0.005 m
Volume of water on roof
$= 216 \text{ m}^2 \times 0.005$ m
$= 1.08 \text{ m}^3$ ✓
The tank has radius 1.8 m.
$V = \pi r^2 h$
$1.08 = \pi \times 1.8^2 \times h$ ✓
$h = \frac{1.08}{\pi \times 1.8^2}$
$= 0.106\,103\,29\ldots$
= 0.11 m (2 d.p.)
So the depth of water in the tank would increase by about 11 cm. ✓
(3 marks)

22 65 km at 8.9 L/100 km
= 65 ÷ 100 × 8.9 L
= 5.785 L
Remainder at 6.6 L/100 km
= (560 − 65) ÷ 100 × 6.6 L
= 32.67 L
Total fuel = 5.785 + 32.67
= 38.455
≈ 38 L
Answer B

23 **i** petrol : oil = 40 : 1
So the amount of oil is $\frac{1}{40}$ of the amount of petrol.
5 L is 5000 mL
Required oil
= (5000 ÷ 40) mL
= 125 mL
(1 mark)

ii Remaining fuel = 4.1 L
= 4100 mL ✓
This fuel has petrol and oil in the ratio 40 : 1. So there is 4000 mL of petrol and 100 mL of oil in the mixture.
For the lawnmower the amount of oil should be $\frac{1}{25}$ of the amount of petrol. ✓
So the required oil
= (4000 ÷ 25) mL
= 160 mL
Joel should add an extra 60 mL of oil to his container. ✓
(3 marks)

24 The fraction of caught seals that had been tagged $= \frac{30}{120}$

$= \frac{1}{4}$

So $\frac{1}{4}$ of the population in 2012 was 60 seals. ✓

Population in 2012 $= 60 \times 4$
$= 240$

Now $100\% - 11\% = 89\%$.

So 240 is 89% of 2008 population.

$89\% = 240$

$1\% = 240 \div 89$
$= 2.696\,629\,213\ldots$

$100\% = 2.696\,629\,213\ldots \times 100$
$= 269.662\,9213\ldots$
$= 270$ [nearest unit]

The estimated population in 2008 was 270. ✓

(2 marks)

25 **i** Scale 1:500 000

1 cm represents 500 000 cm

But 500 000 cm = 5000 m
= 5 km

So 1 cm represents 5 km.

2 cm represents 10 km.

The actual distance between the two mountain peaks is 10 km.

(1 mark)

ii 1 cm represents 5 km.

Now $75 \div 5 = 15$

So 15 cm would represent 75 km.

The two towns will be 15 cm apart on the map.

(1 mark)

26 First train: Speed = 90 km/h

Time taken = 2 h

Distance travelled $= 2 \times 90$ km
$= 180$ km

Second train: distance = 180 km

Time taken = 1 h 20 min
$= 1\frac{1}{3}$ h

Speed $= \left(180 \div 1\frac{1}{3}\right)$ km/h
$= 135$ km/h

Answer A

27

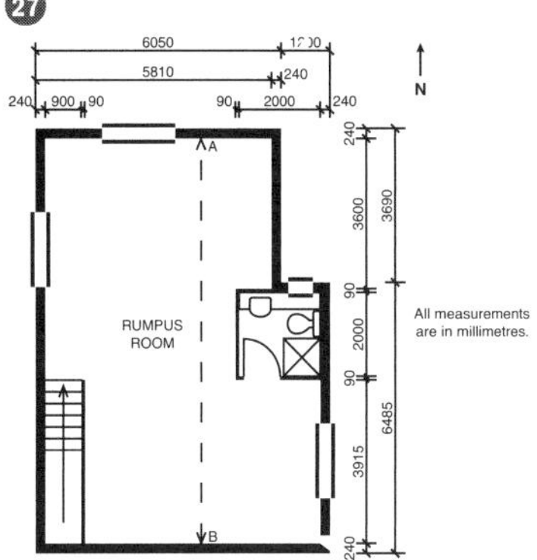

i The width of the stairwell is 900 mm.

(1 mark)

ii The bathroom is 2000 mm by 2000 mm.

(1 mark)

iii Length AB
$= (3600 + 90 + 2000 + 90 + 3915)$ mm
$= 9695$ mm

(1 mark)

iv The plan is drawn to scale.
For the stairwell, 9 mm represents 900 mm.
So 1 mm represents 100 mm.
By measurement, each of the required windows is 17.5 mm on the plan.
Actual width $= 17.5 \times 100$ mm
$= 1750$ mm

(1 mark)

28 **i** The symbol ☒ represents the shower recess.

(1 mark)

ii A mm is the width of the building.

$A = 5000$

(1 mark)

iii 1000 mm = 1 m

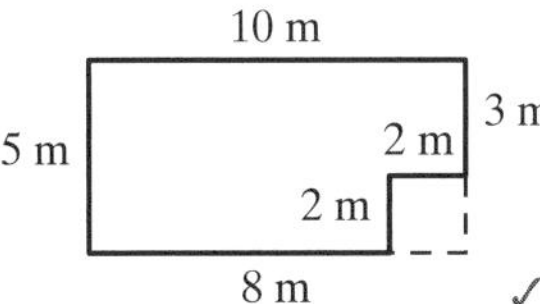

✓

Area $= 10 \times 5 - 2 \times 2$
$= 50 - 4$
$= 46\text{ m}^2$ ✓

(2 marks)

29 $\frac{\text{total in dam}}{\text{total tagged}} = \frac{\text{number in sample}}{\text{number tagged}}$

$\therefore \frac{\text{total in dam}}{20} = \frac{36}{8}$

total in dam $= \frac{36}{8} \times 20$
$= 90$

Answer B

30 **i** Millilitres of alcohol
$= \frac{5.5}{100} \times 375$ mL
$= 20.625$ mL (by calc.)

(1 mark)

ii 1.6 standard drinks contain 375 mL
$\therefore$ 1 standard drink contains

$\frac{375}{1.6} = 234.375$ mL ✓

$\therefore$ A male driver can have 234.375 mL per hour
= 234.375 mL per 60 minutes
= 3.9062 . . . mL per minute
= 3.9 mL per minute to 1 decimal place. ✓

(2 marks)

31 **i** 18 crocodiles = 40% of total

$\therefore \frac{18}{40} \times 100 = 100\%$ of total

$\therefore 45 = 100\%$ of total

45 crocodiles were captured.

(1 mark)

ii First stage:
24 tagged
x total

Second stage:
18 tagged
45 total

$\therefore \frac{x}{24} = \frac{45}{18}$ ✓

$\therefore x = \frac{45}{18} \times 24$
$= 60$

$\therefore$ Estimate of total population is 60 crocodiles. ✓

(2 marks)

32 27 km/h
= 27 000 m/60 minutes
= 27 000 m/3600 seconds
= (27 000 ÷ 3600) m/s
= 7.5 m/s

Answer A

33 By measurement of the scale, 2.5 cm represents 1 cm.
By measurement, the wingspan = 7.5 cm
$\therefore$ Actual measurement
$= 7.5 \div 2.5$
= 3 cm

Answer B

34 Total tagged = 50
Sample size = 200
Sample tagged = 5

$$\therefore \frac{\text{total no.}}{\text{sample size}} = \frac{\text{total tagged}}{\text{sample tagged}}$$

$$\therefore \frac{\text{total no.}}{200} = \frac{50}{5}$$

$$\therefore \text{total no.} = \frac{50}{5} \times 200 = 2000$$

Answer C

35 Kath: 100 L/minute
∴ 1000 L/10 minutes
Jim: 6 L/minute
∴ 60 L/10 minutes
∴ Kath breathes
1000 – 60 = 940 litres
more air than Jim.
Answer C

36 **i** 13 × 1.2 = 15.6 L

(1 mark)

ii Amount of water required weekly
= 15.6 L × 7
= 109.2 L

Total cost per week

$= \frac{109.2}{1000}$ kL × 94.22c/kL
= 10.288. . . cents
≑ 10.29 cents

(1 mark)

iii 1.2 L = 1200 mL
∴ number of drops required
= 1200 × 15
= 18 000 drops

(1 mark)

iv 18 000 drops/10 hours
= 1800 drops/1 hour
$= \frac{1800}{60}$ drops/1 min
= 30 drops/min

(1 mark)

37 From graph,
speed = 50 m/s
= 50 × 60 m/1 min
= 3000 m/1 min
= 3 km/min
= 3 × 60 km/h
= 180 km/h

Answer D

38 **i** By measurement,

Fencing required
= 9.4 + 11 + 8.5
= 28.9 cm ✓

Scale 1: 250

∴ Fencing required
= 28.9 × 250
= 7225 cm
= 72.25 m ✓

(2 marks)

ii Area of trapezium
$= \frac{1}{2}h(a + b)$

h = 8.5 cm
(by measurement)
actual h = 8.5 × 250
= 2125 cm
= 21.25 m

a = 11 cm
(by measurement)
actual a = 11 × 250
= 2750 cm
= 27.5 m ✓

b = 7 cm
(by measurement)
actual b = 7 × 250
= 1750 cm
= 17.5 m

∴ Area of trapezium
$= \frac{1}{2} \times 21.25 \times (27.5 + 17.5)$
= 478.125 m^2
≑ 480 m^2 ✓

(2 marks)

1 Peter currently earns $21.50 per hour. His hourly wage will increase by 2.1% compounded each year for the next four years.

What will his hourly wage be after four years?

A $21.50\,(1.21)^4$
B $21.50\,(1.021)^4$
C $21.50 + 21.50 \times 0.21 \times 4$
D $21.50 + 21.50 \times 0.021 \times 4$ *(1 mark)*

(Q5, **2021 HSC**) Easy

2 Nina plans to invest $35 000 for 1 year. She is offered two different investment options.

Option A: Interest is paid at 6% per annum compounded monthly.

Option B: Interest is paid at r% per annum simple interest.

a Calculate the future value of Nina's investment after 1 year if she chooses Option A. *(2 marks)* Medium

b Find the value of r in Option B that would give Nina the same future value after 1 year as for Option A. Give your answer correct to two decimal places. *(2 marks)* Hard

(Q26, **2021 HSC**)

3 Ariana owns 1500 shares in a company. The market price for each share is $27. Ariana's total dividend from these shares is $810.

Calculate the dividend yield for her shares. *(2 marks)*

(Q30, **2021 HSC**) Medium

4 Joan invests $200. She earns interest at 3% per annum, compounded monthly. What is the future value of Joan's investment after 1.5 years?

A $209.07 **B** $209.19
C $279.51 **D** $311.93 *(1 mark)*

(Q4, **2020 HSC**) Medium

5 The inflation rate over the year from January 2019 to January 2020 was 2%.

The cost of a school jumper in January 2020 was $122.

Calculate the cost of the jumper in January 2019 assuming that the only change in the cost of the jumper was due to inflation. *(2 marks)*

(Q21, **2020 HSC**) Medium

6 Jana owns a share portfolio. Details of her share portfolio at 30 June 2020 are given in the table.

Company name	*Number of shares in Jana's portfolio*	*Dividend yield (per annum)*	*Market price per share*
ABC	200	6.0%	$5.50
XYZ	?	4.0%	$6.00

Jana received a total annual dividend of $149.52 from her share portfolio.

Calculate the number of shares Jana has in company XYZ on 30 June 2020. *(3 marks)*

(Q29, **2020 HSC**) Medium

7 Chris opens a bank account and deposits $1000 into it. Interest is paid at 3.5% per annum, compounding annually.

Assuming no further deposits or withdrawals are made, what will be the balance in the account at the end of two years?

A $1070.00 **B** $1071.23
C $1822.50 **D** $2070.00 *(1 mark)*

(Q3, **2019 HSC**) Easy

8 The graphs show the future values over time of $P, invested at three different rates of compound interest.

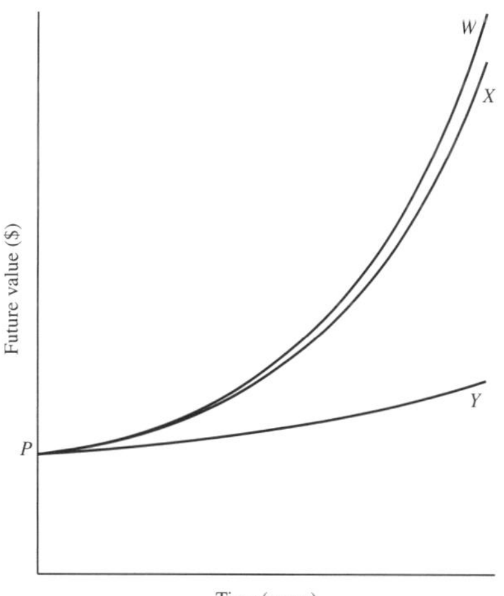

Which of the following correctly identifies each graph?

A

W	5% pa, compounding annually
X	10% pa, compounding annually
Y	10% pa, compounding quarterly

B

W	5% pa, compounding annually
X	10% pa, compounding quarterly
Y	10% pa, compounding annually

C

W	10% pa, compounding quarterly
X	10% pa, compounding annually
Y	5% pa, compounding annually

D

W	10% pa, compounding annually
X	10% pa, compounding quarterly
Y	5% pa, compounding annually

(1 mark)

(Q13, **2019 HSC**) **Medium**

9 A person owns 1526 shares with a market value of \$8.75 per share. The total dividend received for these shares is \$1068.20.

Calculate the percentage dividend yield. *(2 marks)*

(Q21, **2019 HSC**) **Medium**

10 Lucas invests \$80 000 for 5 years, at an interest rate of 6% per annum, compounded annually. Ethan invests \$2500 at the end of each month for 5 years, at an interest rate of 6% per annum, compounded monthly. Financial gain is defined as the difference between the final value of an investment and the total contribution. Who will have the better financial gain after 5 years, and by what amount? Use the table below and justify your answer with suitable calculations.

Period	**Table of future value interest factors**							
	Interest rate per period							
	0.50%	**1.00%**	**1.50%**	**2.00%**	**3.00%**	**4.00%**	**5.00%**	**6.00%**
5	5.0503	5.1010	5.1523	5.2040	5.3091	5.4163	5.5256	5.6371
6	6.0755	6.1520	6.2296	6.3081	6.4684	6.6330	6.8019	6.9753
7	7.1059	7.2135	7.3230	7.4343	7.6625	7.8983	8.1420	8.3938
8	8.1414	8.2857	8.4328	8.5830	8.8923	9.2142	9.5491	9.8975
12	12.3356	12.6825	13.0412	13.4121	14.1920	15.0258	15.9171	16.8699
24	25.4320	26.9735	28.6335	30.4219	34.4265	39.0826	44.5020	50.8156
36	39.3361	43.0769	47.2760	51.9944	63.2759	77.5983	95.8363	119.1209
48	54.0978	61.2226	69.5652	79.3535	104.4084	139.2632	188.0254	256.5645
60	69.7700	81.6697	96.2147	114.0515	163.0534	237.9907	353.5837	533.1282
72	86.4089	104.7099	128.0772	158.0570	246.6672	396.0566	650.9027	1089.6286

(3 marks)

Bonus question (see page iv) **Hard**

11 A company with a share price of \$3.42 declares a dividend of 14 cents. What is the dividend yield as a percentage, correct to two decimal places? *(1 mark)*

Bonus question **Medium**

12 The table shows the compounded values of \$1 at different interest rates over different periods.

Compounded values of \$1

Number of periods	*Interest rate per period*				
	1%	2%	3%	4%	5%
2	1.0201	1.0404	1.0609	1.0816	1.1025
4	1.0406	1.0824	1.1255	1.1699	1.2155
6	1.0615	1.1262	1.1941	1.2653	1.3401
8	1.0829	1.1717	1.2668	1.3686	1.4775
10	1.1046	1.2190	1.3439	1.4802	1.6289
12	1.1268	1.2682	1.4258	1.6010	1.7959

Amy hopes to have \$21 000 in 2 years to buy a car. She opens an account today which pays interest of 4% pa, compounded quarterly.

Using the table, which expression calculates the minimum single sum that Amy needs to invest today to ensure she reaches her savings goal?

A 21000×1.0816 **B** $21000 \div 1.0816$

C 21000×1.0829 **D** $21000 \div 1.0829$

(1 mark)

(Q19, **2018 HSC**) **Medium**

13 A single amount of \$10 000 is invested for 4 years, earning interest at the rate of 3% per annum, compounded monthly.

Which expression will give the future value of the investment?

A $10000 \times (1+0.03)^4$

B $10000 \times (1+0.03)^{48}$

C $10000 \times \left(1+\frac{0.03}{12}\right)^4$

D $10000 \times \left(1+\frac{0.03}{12}\right)^{48}$ *(1 mark)*

(Q10, **2017 HSC**) **Easy**

14 Sam purchased 500 company shares at \$3.20 per share. Brokerage fees were 1.5% of the purchase price.

Sam is paid a dividend of 26 cents per share, then immediately sells the shares for \$4.80 each.

If he pays no further brokerage fees, what is Sam's total profit? *(3 marks)*

(Q26e, **2017 HSC**) **Medium**

15 The table shows the future value of an investment of \$1000, compounding yearly, at varying interest rates for different periods of time.

Future values of an investment of \$1000

Number of years	Interest rate per annum				
	1%	2%	3%	4%	5%
1	1010.00	1020.00	1030.00	1040.00	1050.00
2	1020.10	1040.40	1060.90	1081.60	1102.50
3	1030.30	1061.21	1092.73	1124.86	1157.63
4	1040.60	1082.43	1125.51	1169.86	1215.51
5	1051.01	1104.08	1159.27	1216.65	1276.28

Based on the information provided, what is the future value of an investment of \$2500 over 3 years at 4% pa?

A \$1124.86 **B** \$2812.15
C \$3624.86 **D** \$5312.15 *(1 mark)*

(Q8, **2016 HSC**) **Easy**

16 What amount must be invested now at 4% per annum, compounded quarterly, so that in five years it will have grown to \$60 000?

A \$8919 **B** \$11 156
C \$49 173 **D** \$49 316 *(1 mark)*

(Q17, **2015 HSC**) **Easy**

17 A family currently pays \$320 for some groceries.

Assuming a constant annual inflation rate of 2.9%, calculate how much would be paid for the same groceries in 5 years' time. *(2 marks)*

(Q26d, **2015 HSC**) **Easy**

18 Chandra and Sascha plan to have \$20 000 in an investment account in 15 years time for their grandchild's university fees.

The interest rate for the investment account will be fixed at 3% per annum compounded monthly.

Calculate the amount that they will need to deposit into the account now in order to achieve their plan. *(3 marks)*

(Q30a, **2014 HSC**) **Medium**

19 Kimberley has invested \$3500.

Interest is compounded half-yearly at a rate of 2% per half-year.

Compounded values of \$1

Period	Interest rate per period					
	1%	2%	3%	4%	5%	6%
1	1.010	1.02	1.03	1.04	1.05	1.06
2	1.020	1.040	1.061	1.082	1.103	1.124
3	1.030	1.061	1.093	1.125	1.158	1.191
4	1.041	1.082	1.126	1.170	1.216	1.262
5	1.051	1.104	1.159	1.217	1.276	1.338
6	1.062	1.126	1.194	1.265	1.340	1.419
7	1.072	1.149	1.230	1.316	1.407	1.504
8	1.083	1.172	1.267	1.369	1.477	1.594

Use the table to calculate the value of her investment at the end of 4 years. *(2 marks)*

(Q26e, **2013 HSC**) **Medium**

20 Adhele has 2000 shares. The current share price is \$1.50 per share. Adhele is paid a dividend of \$0.30 per share.

i What is the current value of her shares? *(1 mark)* **Easy**

ii Calculate the dividend yield. *(1 mark)* **Medium**

(Q28d, **2013 HSC**)

21 Tracy invests some money for 2 years at 4% per annum, compounded quarterly.

Compounded values of \$1

Period	Interest rate per period				
	1%	2%	3%	4%	5%
1	1.010	1.020	1.030	1.040	1.050
2	1.020	1.040	1.061	1.082	1.103
3	1.030	1.061	1.093	1.125	1.158
4	1.041	1.082	1.126	1.170	1.216
5	1.051	1.104	1.159	1.217	1.276
6	1.062	1.126	1.194	1.265	1.340
7	1.072	1.149	1.230	1.316	1.407
8	1.083	1.172	1.267	1.369	1.477

Which figure from the table should Tracy use to calculate the value of her investment at the end of 2 years?

A 1.020 **B** 1.082
C 1.083 **D** 1.369 *(1 mark)*

(Q9, **2012 HSC**) **Medium**

22 A house was purchased in 1984 for \$35 000. Assume that the value of the house has increased by 3% per annum since then.

Which expression gives the value of the house in 2009?

A $35000(1 + 0.03)^{25}$ **B** $35000(1 + 3)^{25}$
C $35000 \times 25 \times 0.03$ **D** $35000 \times 25 \times 3$
(1 mark)

(Q6, **2009 HSC**) **Easy**

23 Lilly and Rose each have money to invest and choose different investment accounts.

The graph shows the values of their investments over time.

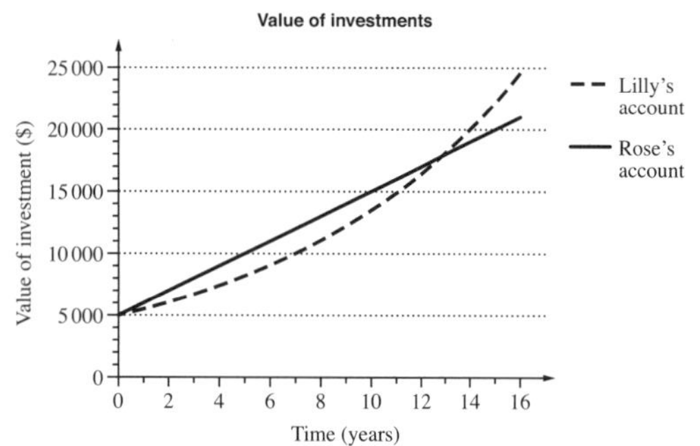

i How much was Rose's original investment? *(1 mark)* Easy

ii At the end of 6 years, which investment will be worth the most and by how much? *(2 marks)* Medium

iii Lilly's investment will reach a value of $20000 first.
How much longer will it take Rose's investment to reach a value of $20000? *(1 mark)* Medium

(Q23a, **2007 HSC**)

---

24 Last year, Helen bought 150 shares at $2.00 per share. They are now worth $2.50 per share. Helen receives a dividend of $0.10 per share.

What is the dividend yield?

A 4% **B** 20%
C $15 **D** $75 *(1 mark)*

(Q13, **2005 HSC**) 

---

25 Minh invests $24000 at an interest rate of 4.75% per annum, compounded monthly. What is the value of the investment after 3 years? *(2 marks)*

(Q24a, **2003 HSC**) Medium

---

26 Richard has 2000 shares with a current market value of $4.80 each. During the past twelve months, Richard received a total dividend of $240.

What is the current dividend yield on these shares?

A 0.025% **B** 2%
C 2.5% **D** 40% *(1 mark)*

(Q7, **2002 HSC**) Medium

---

27 **i** Katherine invests $50000 with Standard Credit Union for a term of 5 years. Her investment earns interest at 3.1% per annum compounded annually.
How much will Katherine's investment be worth at the end of 5 years?
Give your answer to the nearest dollar. *(2 marks)* Easy

ii Katherine's sister Liz also has $50000 to invest for a term of 5 years. She invests with General Bank. Her investment earns interest at 3% per annum, compounded monthly.
Which sister makes the better investment? Justify your answer with appropriate calculations. *(2 marks)* Medium

(Q23b, **2002 HSC**)

Year 12 Investments—Worked answers

1 $PV = \$21.50, r = 0.021, n = 4$
$FV = PV(1+r)^n$
$= \$21.50(1+0.021)^4$
$= \$21.50(1.021)^4$
Answer B

2 **a** Option A:
$PV = \$35\,000$,
$r = \frac{0.06}{12} = 0.005, n = 12$ ✓
$FV = PV(1+r)^n$
$= \$35\,000(1+0.005)^{12}$
$= \$37\,158.7234\ldots$
$= \$37\,158.72$ (nearest cent) ✓
(2 marks)

b Interest
$= \$37\,158.72 - \$35\,000$
$= \$2158.72$
Option B:
$I = Prn$
$\$2158.72 = \$35\,000 \times \frac{r}{100} \times 1$ ✓
$2158.72 = 350r$
$r = \frac{2158.72}{350}$
$= 6.167\,771\,42\ldots$
$= 6.17$ (2 d.p.) ✓
(2 marks)

3 Dividend per share
$= \$810 \div 1500$
$= \$0.54$ ✓
Dividend yield $= \frac{\$0.54}{\$27} \times 100\%$
$= 2\%$ ✓
(2 marks)

4 $FV = PV(1+r)^n$
$= \$200\left(1+\frac{0.03}{12}\right)^{18}$
$= \$209.193\,824\ldots$
$= \$209.19$ (nearest cent)
Answer B

5 $PV = \frac{FV}{(1+r)^n}$
$= \frac{\$122}{(1+0.02)^1}$ ✓
$= \$119.607\,843\ldots$
$= \$119.61$ (nearest cent)
The cost of the jumper in 2019 was \$119.61 to the nearest cent. ✓
(2 marks)

6

Company name	Number of shares in Jana's portfolio	Dividend yield (per annum)	Market price per share
ABC	200	6.0%	\$5.50
XYZ	?	4.0%	\$6.00

Total value of ABC shares
$= 200 \times \$5.50$
$= \$1100$
Dividend = 6% of \$1100
$= \$66.00$ ✓
Dividend from XYZ shares
$= \$149.52 - \$66.00$
$= \$83.52$ ✓
So \$83.52 is 4% of their value.
Value is $\$83.52 \div 4 \times 100$ or \$2088.
Number of shares $= \$2088 \div \$6$
$= 348$
So Jana has 348 shares in XYZ. ✓
(3 marks)

7 $PV = \$1000, \quad r = 0.035, \quad n = 2$
$FV = PV(1+r)^n$
$= \$1000(1+0.035)^2$
$= \$1071.225$
The balance will be \$1071.23.
Answer B

8 The future value will be least at 5% p.a. compounding annually. The future value will be greatest at 10% p.a. compounding quarterly. So Y will be 5% p.a. compounding annually, W will be 10% p.a. compounding quarterly and X 10% p.a. compounding annually.

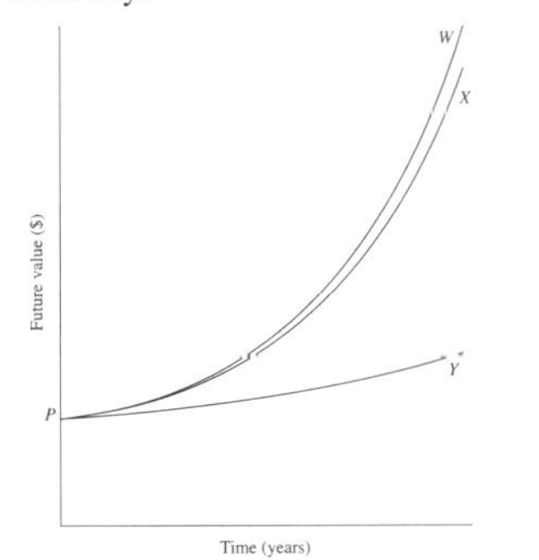

Answer C

9 Dividend per share
$= \$1068.20 \div 1526$
$= \$0.70$ ✓
Percentage dividend yield
$= \frac{\$0.70}{\$8.75} \times 100\%$
$= 8\%$ ✓
(2 marks)

10 Lucas: $FV = PV(1+r)^n$
$= 80\,000 \times 1.06^5$
$= 107\,058.05$ (2 dec. pl.) ✓
Financial gain
$= 107\,058.05 - 80\,000$
$= 27\,058.05$
∴ Lucas's gain is \$27 058.05.

Ethan: 6% p.a.
= 0.5% per month, for 5 years
= 60 months
$FV = 2500 \times 69.7700$ (from table)
$= 174\,425$ ✓
Financial gain
$= 174\,425 - 2500 \times 60$
$= 24\,425$
∴ Ethan's gain is \$24 425.
∴ Lucas's gain is \$2633.05 more than Ethan's gain. ✓
(3 marks)

11 Dividend yield
$= \frac{0.14}{3.42} \times 100\%$
$= 4.093\,567\,251\ldots$
$= 4.09$ (2 dec. pl.)
(1 mark)

12 2 years $= 2 \times 4$ quarters
So the number of periods is 8.
4% p.a. = 1% per quarter
From the table the required factor is 1.0829.
So required amount × 1.0829
= \$21 000
Amount $= \$(21\,000 \div 1.0829)$
Answer D

13 $PV = 10\,000, \quad r = \frac{0.03}{12}$,
$n = 4 \times 12 = 48$
$FV = PV(1+r)^n$
$= 10\,000 \times \left(1+\frac{0.03}{12}\right)^{48}$
Answer D

14 Price of shares $= 500 \times \$3.20$
$= \$1600$ ✓
Brokerage = 1.5% of \$1600
$= 0.015 \times \$1600$
$= \$24$
Total cost $= \$1600 + \$24$
$= \$1624$
Total dividend $= 500 \times \$0.26$
$= \$130$ ✓
Total sales $= 500 \times \$4.80$
$= \$2400$
Total received $= \$130 + \$2400$
$= \$2530$
Profit $= \$2530 - \$1624$
$= \$906$ ✓
(3 marks)

15 $FV = \$(2.5 \times 1124.86)$
$= \$2812.15$
Answer B

16 $PV = \frac{FV}{(1+r)^n}$

$= \frac{\$60000}{\left(1+\frac{0.04}{4}\right)^{5\times4}}$

$= \$49\,172.6682...$

$= \$49\,173$ (nearest dollar)

Answer C

17 $FV = PV(1+r)^n$

$= \$320(1+0.029)^5$ ✓

$= \$369.170\,38...$

$= \$369$ (nearest dollar)

Around \$369 would be paid for the groceries in 5 years' time. ✓

(2 marks)

18 $PV = \frac{FV}{(1+r)^n}$

$= \frac{\$20\,000}{\left(1+\frac{0.03}{12}\right)^{15\times12}}$ ✓✓

$= \$12\,759.7264...$

$= \$12\,760$ (nearest dollar)

✓ *(3 marks)*

19 4 years is 8 half-years so the period is 8.

The interest rate per period is 2%. ✓

Value of investment

$= 1.172 \times \$3500$

$= \$4102$ ✓

(2 marks)

20 **i** Current value $= 2000 \times \$1.50$

$= \$3000$

(1 mark)

ii Dividend yield $= \frac{0.30}{1.50} \times 100\%$

$= 20\%$

(1 mark)

21 4% pa – 1% per quarter

So the interest rate per period is 1%.

2 years = 8 quarters

So the number of periods is 8.

From the table, the required figure is 1.083.

Answer C

22 $A = P(1+r)^n$

$\therefore A = 35\,000(1+0.03)^{25}$

Answer A

23 **i** Rose's original investment was \$5000, (when $t = 0$).

(1 mark)

ii At the end of 6 years, Rose has \$11 000 and Lilly has \$9000. ✓

$\therefore$ Rose's investment is worth the most by \$2000. ✓

(2 marks)

iii Lilly's investment takes 14 years while Rose's takes 15 years.

$\therefore$ It will take Rose 1 year longer. *(1 mark)*

24 Dividend yield

$= \frac{\text{dividend}}{\text{market value}} \times 100$

$= \frac{0.10}{2.50} \times 100$

$= 4\%$

Answer A

25 $A = P(1+r)^n$

$P = \$24\,000$

$r = 4.75\%$ p.a.

$= \frac{4.75}{12}\%$ p. month

$= 0.3958\dot{3}\%$ p. month

$n = 3$ years

$= 36$ months ✓

$\therefore A = 24\,000\left(1+\frac{0.3958\dot{3}}{100}\right)^{36}$

$= \$27\,667.8901...$

$= \$27\,667.89$ ✓

(2 marks)

26 Total market value

$= 2000 \times \$4.80$

$= \$9600$

Total dividend

$= \$240$

Dividend yield

$= \frac{240}{9600} \times 100$

$= 2.5\%$.

Answer C

27 **i** $A = P(1+r)^n$

$= 50\,000\,(1+0.031)^5$ ✓

$= \$58\,245.63$

$= \$58\,246$ ✓

to nearest dollar

(2 marks)

ii $A = P(1+r)^n$

$P = \$50\,000$

$r = 3 \div 12 = 0.25\%$ per month

$n = 5 \times 12 = 60$ months ✓

$\therefore\ A = 50\,000\,(1+0.0025)60$

$= \$58\,080.84$

$= \$58\,081$

to nearest dollar

$\therefore$ Katherine's investment is marginally better by \$165. ✓

(2 marks)

1 Three years ago an appliance was valued at \$2467. Its value has depreciated by 15% each year, based on the declining-balance method.

What is its salvage value today, to the nearest dollar?

A \$952 **B** \$1110
C \$1357 **D** \$1515

(Q4, **2021 HSC**) Easy

2 An asset is depreciated using the declining-balance method with a rate of depreciation of 8% per half year. The asset was bought for \$10 000. What is the salvage value of the asset after 5 years?

A \$1749.01 **B** \$4182.12
C \$4343.88 **D** \$6590.82 *(1 mark)*

(Q11, **2020 HSC**) Medium

3 Nisa has a credit card on which interest at 17% per annum, compounded daily, is charged on the amount owing.

At the beginning of the month, Nisa owes \$500 on her credit card. She makes no other purchases using the credit card, but fifteen days later, she repays \$250.

Assuming that interest is charged for the fifteen days, calculate the amount owing on the credit card immediately after the \$250 payment is made. *(3 marks)*

(Q22, **2020 HSC**) 

4 Ashley has a credit card with the following conditions:

- There is no interest-free period.
- Interest is charged at the end of each month at 18.25% per annum, compounding daily, from the purchase date (included) to the last day of the month (included).

Ashley's credit card statement for April is shown, with some figures missing.

Statement period: 1 April to 30 April

Date	Details	Amount (\$)
1 April	Opening balance	0
20 April	Furniture	3700
30 April	Interest charged	***
30 April	Closing balance	***

Minimum payment: ______

The minimum payment is calculated as 2% of the closing balance on 30 April. Calculate the minimum payment. *(3 marks)*

(Q27, **2019 HSC**) Hard

5 A new car is bought for \$24 950. Each year the value of the car is depreciated by the same percentage.

The table shows the value of the car, based on the declining-balance method of depreciation, for the first three years.

End of year	*Value*
1	\$21 457.00
2	\$18 453.02
3	\$15 869.60

What is the value of the car at the end of 10 years? *(3 marks)*

(Q37, **2019 HSC**) Medium

6 A car is purchased for \$23 900.

The value of the car is depreciated by 11.5% each year using the declining-balance method.

What is the value of the car after three years? *(2 marks)*

(Q26h, **2018 HSC**) Easy

7 Andrew borrowed \$20 000 to be repaid in equal monthly repayments of \$243 over 10 years. Having made this monthly repayment for 4 years, he increased his monthly repayment to \$281. As a result, Andrew paid off the loan one year earlier.

How much less did he repay altogether by making this change? *(2 marks)*

(Q29e **2018 HSC**) Easy

8 A new car was bought for \$19 900 and one year later its value had depreciated to \$16 300.

What is the approximate depreciation, expressed as a percentage of the purchase price?

A 18% **B** 22%
C 78% **D** 82% *(1 mark)*

(Q11, **2017 HSC**) 

9 Michelle borrows \$100 000. The interest rate charged is 12% per annum compounded monthly. The monthly payment is \$1029 and the first repayment is made after one month.

What is the amount outstanding immediately after the SECOND monthly repayment is made? *(3 marks)*

(Q28c, **2017 HSC**) **Medium**

10 Ariana is charged compound interest at the rate of 0.036% per day on outstanding credit card balances. She has $780 outstanding for 24 days.

How much compound interest is she charged?

A $6.74 **B** $6.77
C $786.74 **D** $786.77 *(1 mark)*

(Q17, **2016 HSC**) **Medium**

11 A piece of machinery, initially worth $56 000, depreciates at 8% per annum.

Which graph best shows the salvage value of this piece of machinery over time?

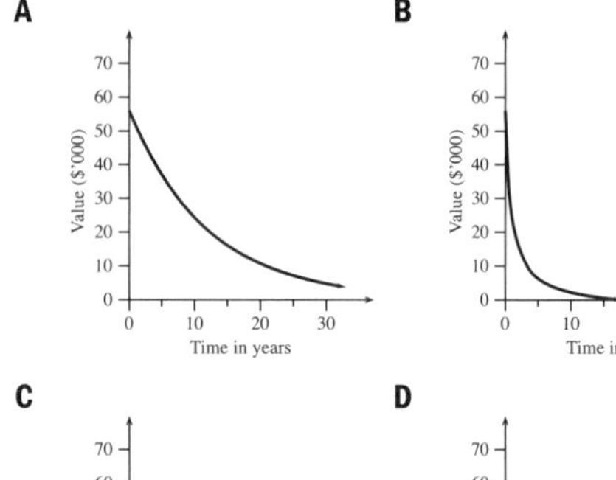

(1 mark)

(Q10, **2015 HSC**) **Easy**

12 Jamal borrowed $350 000 to be repaid over 30 years, with monthly repayments of $1880. However, after 10 years he made a lump sum payment of $80 000. The monthly repayment remained unchanged. The graph shows the balances owing over the period of the loan.

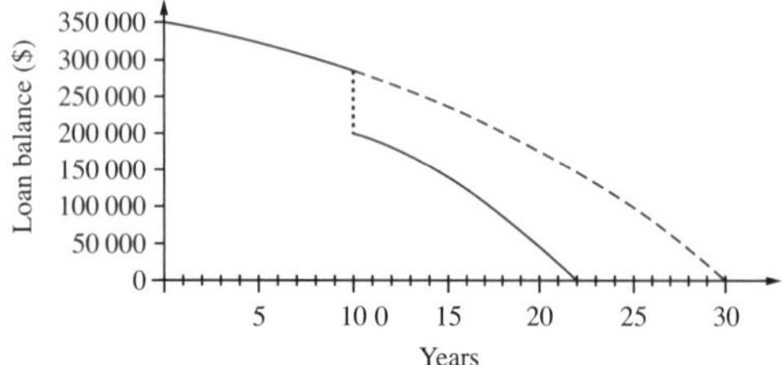

Over the period of the loan, how much less did Jamal pay by making the lump sum payment? *(2 marks)*

(Q29b, **2015 HSC**) **Medium**

13 A car is bought for $19 990. It will depreciate at 18% per annum.

Using the declining balance method, what will be the salvage value of the car after 3 years, to the nearest dollar?

A $8968 **B** $9195
C $11 022 **D** $16 392 *(1 mark)*

(Q9, **2014 HSC**) **Easy**

14 Zheng has purchased a computer for $5000 for his company. He wants to compare two different methods of depreciation over two years for the computer.

Method 1: Straight-line with $1250 depreciation per annum.

Method 2: Declining balance with 35% depreciation per annum.

Which method gives the greatest depreciation over the two years? Justify your answer with suitable calculations. *(3 marks)*

(Q28e, **2013 HSC**) **Easy**

15 A machine was bought for $25 000.

Which graph best represents the salvage value of the machine over 10 years using the declining balance method of depreciation?

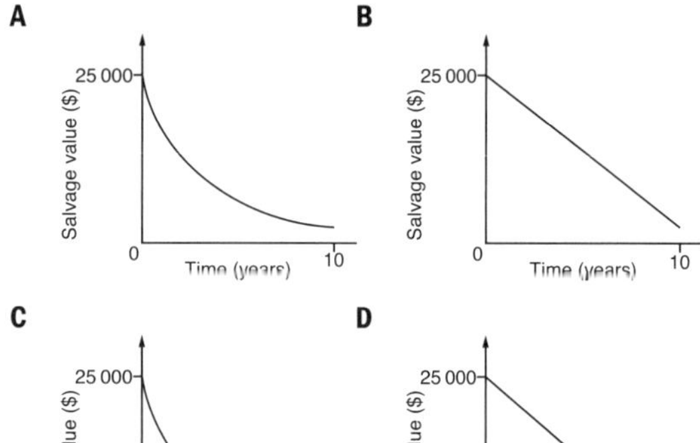

(1 mark)

(Q16, **2012 HSC**) **Medium**

16 A $400 000 loan can be repaid by making either monthly or fortnightly repayments. The graph shows the loan balances over time using these two different methods of repayment.

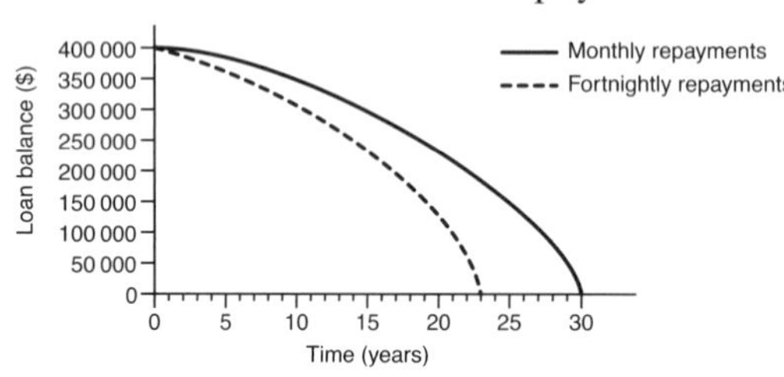

The monthly repayment is $2796.86 and the fortnightly repayment is $1404.76.

What is the difference in the total interest paid using the two different methods of repayment, to the nearest dollar?

A $51 596 **B** $166 823

C $210 000 **D** $234 936 *(1 mark)*

(Q24, **2012 HSC**) Medium

---

17 Jim buys a photocopier for $22 000. Its value is depreciated using the declining balance method at the rate of 15% per annum.

What is its value at the end of 3 years? *(2 marks)*

(Q26b, **2012 HSC**) Easy

---

18 Norman and Pat each bought the same type of tractor for $60 000 at the same time. The value of their tractors depreciated over time.

The salvage value *S*, in dollars, of each tractor, is its depreciated value after *n* years.

Norman drew a graph to represent the salvage value of his tractor.

Norman's graph of the salvage value of his tractor

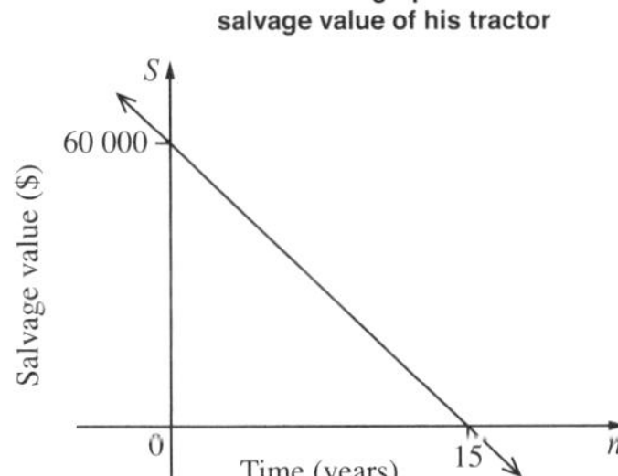

i Find the gradient of the line shown in the graph. *(1 mark)* Hard

ii What does the value of the gradient represent in this situation? *(1 mark)* Medium

iii Write down the equation of the line shown in the graph. *(1 mark)* Hard

iv Find all the values of *n* that are not suitable for Norman to use when calculating the salvage value of his tractor. Explain why these values are not suitable. *(2 marks)* Hard

Pat used the declining balance formula for calculating the salvage value of her tractor. The depreciation rate that she used was 20% per annum.

v What did Pat calculate the salvage value of her tractor to be after 14 years? *(2 marks)* Medium

vi Using Pat's method for depreciation, describe what happens to the salvage value of her tractor for all values of *n* greater than 15. *(1 mark)* Medium

(Q28b, **2011 HSC**)

19 The table shows monthly home loan repayments with interest rate changes from February to October 2009.

Monthly home loan repayments

	A	B	C	D	E
1					
2	Dates	Feb 2009	Apr 2009	Jun 2009	Oct 2009
3	*Increase/Decrease*	*−1.0%*	*−0.1%*	*0.05%*	*0.25%*
4	***Rate***	***5.85%***	***5.75%***	***5.80%***	***6.05%***
5	$1 000	$6.35	$6.29	$6.32	$6.47
6	$50 000	$318	$315	$316	$324
7	$100 000	$635	$629	$632	$647
8	$150 000	$953	$944	$948	$971
9	$200 000	$1 270	$1 258	$1 264	$1 295
10	$250 000	$1 588	$1 573	$1 580	$1 618
11	$300 000	$1 905	$1 887	$1 896	$1 942
12	$350 000	$2 223	$2 202	$2 212	$2 266
13	$400 000	$2 541	$2 516	$2 529	$2 589
14					

Sheet 1 / Sheet 2 /

i What is the change in monthly repayments on a $250 000 loan from February 2009 to April 2009? *(1 mark)* Easy

ii Xiang wants to borrow $307 000 to buy a house.
Xiang's bank approves loans for customers if their loan repayments are no more than 30% of their monthly gross salary.
Xiang's monthly gross salary is $6500.
If she had applied for the loan in October 2009, would her bank have approved her loan?
Justify your answer with suitable calculations. *(3 marks)* Medium

iii Jack took out a loan at the same time and for the same amount as Xiang. Graphs of their loan balances are shown.

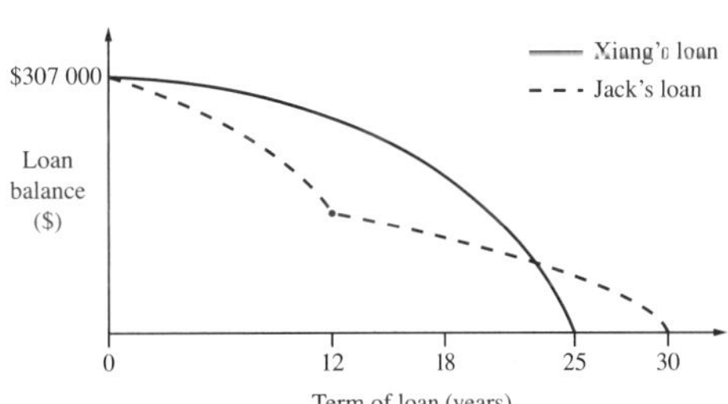

Identify TWO differences between the graphs and provide a possible explanation for each difference, making reference to interest rates and/or loan repayments. *(2 marks)* Medium

(Q28a, **2010 HSC**)

---

20 Jay bought a computer for $3600. His friend Julie said that all computers are worth nothing (i.e. the value is $0) after 3 years.

i Find the amount that the computer would depreciate each year to be worth nothing after 3 years, if the straight line method of depreciation is used. *(1 mark)* Easy

ii Explain why the computer would never be worth nothing if the declining balance method of depreciation is used, with 30% per annum rate of depreciation.

Use suitable calculations to support your answer. *(2 marks)* **Medium**

(Q24e, **2009 HSC**)

21 A plasma TV depreciated in value by 15% per annum. Two years after it was purchased it had depreciated to a value of $2023, using the declining balance method.

What was the purchase price of the plasma TV? *(2 marks)*

(Q27c, **2008 HSC**) **Medium**

22 The value of a car is depreciated using the declining balance method.

Which graph best illustrates the value of the car over time?

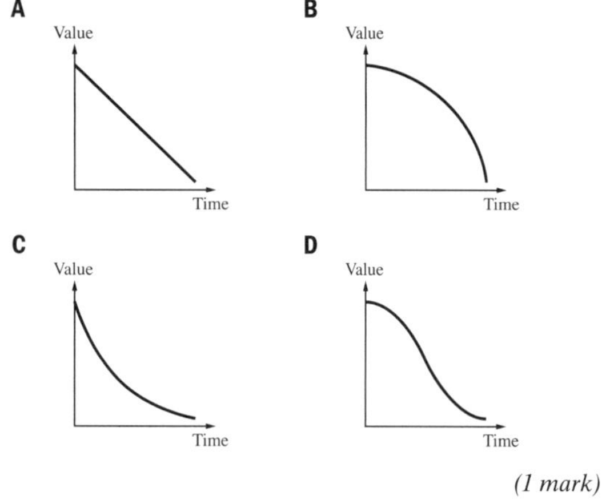

(1 mark)

(Q12, **2007 HSC**) **Easy**

23 Two families borrow different amounts of money on the same day.

The Wang family has a flat rate loan. The Salama family has a reducing balance loan and repays the loan earlier than the Wang family.

Which graph best represents this situation?

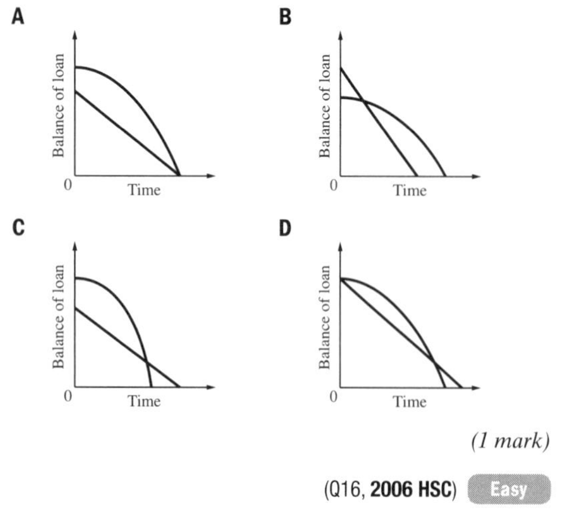

(1 mark)

(Q16, **2006 HSC**) **Easy**

24 Bill borrows $420 000 to buy a house. Interest is charged at 7.2% per annum, compounded monthly.

How much does he owe at the end of the first month, after he has made a $4000 repayment?

A $418 496 **B** $418 520
C $445 952 **D** $446 240 *(1 mark)*

(Q21, **2006 HSC**) **Easy**

25 Paul invested money in a bank for 4 years. The stated interest rate on the account was 6.1% per annum *compounded annually*. This is equivalent to an effective simple interest rate of 6.68% per annum.

The formula Paul used to calculate the effective simple interest rate was:

$$E=\frac{(1+r)^n-1}{n}$$

where r is the stated interest rate per period (expressed as a decimal)

E is the effective simple interest rate per period (expressed as a decimal)

n is the number of periods

Martha invested money in a different bank for 4 years. The stated interest rate on her investment was 6% per annum *compounded monthly*.

Martha thinks that she has a better deal than Paul. Do you agree? Justify your answer by comparing their effective simple interest rates. *(4 marks)*

(Q25d, **2006 HSC**) **Hard**

26 Liliana wants to borrow money to buy a house. The bank sent her an email with the following table attached.

Monthly repayments

Amount borrowed	*Term of loan*				
	10 years 120 months	15 years 180 months	20 years 240 months	25 years 300 months	30 years 360 months
$80 000	$970.62	$764.52	$669.15	$617.45	$587.01
$90 000	$1091.95	$860.09	$752.80	$694.63	$660.39
$100 000	$1213.28	$955.65	$836.44	$771.82	$733.76
$110 000	$1334.60	$1051.22	$920.08	$849.00	$807.14
$120 000	$1455.93	$1146.78	$1003.73	$926.18	$880.52
$130 000	$1577.26	$1242.35	$1087.37	$1003.36	$953.89
$140 000	$1698.59	$1337.91	$1171.02	$1080.54	$1027.27
$150 000	$1819.91	$1433.48	$1254.66	$1157.72	$1100.65
$160 000	$1941.24	$1529.04	$1338.30	$1234.91	$1174.02

i Liliana decides that she can afford $1000 per month on repayments.
What is the maximum amount she can borrow, and how many years will she have to repay the loan? *(1 mark)* **Medium**

ii Zali intends to borrow $160 000 over 15 years from the same bank.
If she chooses to borrow $160 000 over 20 years instead, how much more interest will she pay? *(2 marks)* **Hard**

(Q27a, **2006 HSC**)

27 Kai purchased a new car for \$30 000. It depreciated in value by \$2000 per year for the first three years.

After the end of the third year, Kai changed the method of depreciation to the declining balance method at the rate of 25% per annum.

- **i** Calculate the value of the car at the end of the third year. *(1 mark)* Easy
- **ii** Calculate the value of the car seven years after it was purchased. *(2 marks)* Easy
- **iii** Without further calculations, sketch a graph to show the value of the car over the seven years.
  Use the horizontal axis to represent time and the vertical axis to represent the value of the car. *(3 marks)* Medium

(Q27c, **2006 HSC**)

28 The table is used to calculate monthly loan repayments.

Monthly loan repayments (in dollars) per \$1000 borrowed

Interest rate % pa	*5 years*	*10 years*	*15 years*	*20 years*
5%	18.87	10.61	7.91	6.60
6%	19.33	11.10	8.44	7.16
7%	19.80	11.61	8.99	7.75
8%	20.28	12.13	9.56	8.36
9%	20.76	12.67	10.14	9.00

Samantha has borrowed \$70 000 at 8% per annum for 15 years.

What is her monthly loan repayment?

A \$143.40 **B** \$669.20
C \$8030.40 **D** \$10 038.00 *(1 mark)*

(Q10, **2005 HSC**) Easy

29 A car bought for \$50 000 is depreciated using the declining balance method.

Which graph best represents the salvage value of the car over time?

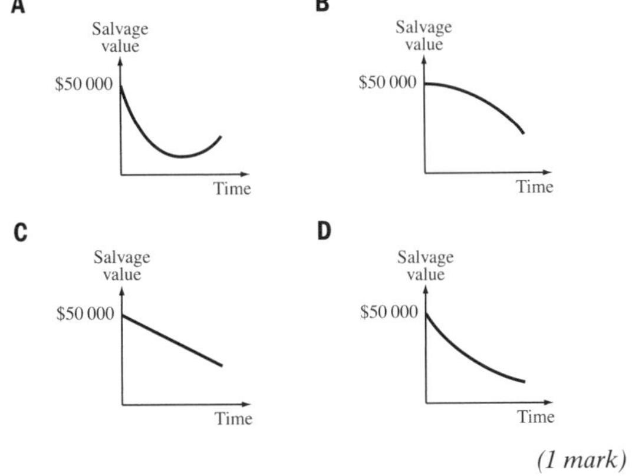

(1 mark)

(Q15, **2005 HSC**) Easy

30 A printing machine worth \$150 000 is bought in December 2005.

In December each year, beginning in 2006, the value of the printing machine is depreciated by 10% using the declining balance method of depreciation.

In which year will the depreciated value first fall below \$120 000? *(2 marks)*

(Q26a, **2005 HSC**) Medium

31 Kerry has a credit card. She is charged 0.05% compound interest per day on outstanding balances.

How much interest is Kerry charged on an amount of \$250, which is outstanding on her credit card for 30 days?

A \$3.75 **B** \$3.78
C \$253.75 **D** \$253.78 *(1 mark)*

(Q19, **2004 HSC**) Medium

32 Tai uses the declining balance method of depreciation to calculate tax deductions for her business. Tai's computer is valued at \$6500 at the start of the 2003 financial year. The rate of depreciation is 40% per annum.

- **i** Calculate the value of her tax deduction for the 2003 financial year. *(1 mark)* Medium
- **ii** What is the value of her computer at the start of the 2006 financial year? *(2 marks)* Medium

(Q25a, **2004 HSC**)

33 Aaron decides to borrow \$150 000 over a period of 20 years at a rate of 7.0% per annum.

MONTHLY REPAYMENT TABLE

Principal and interest per \$1000 borrowed

Interest rate (pa)	Term of loan – years					
	5	10	15	20	25	30
6.5%	19.57	11.35	8.71	7.46	6.75	6.32
7.0%	19.80	11.61	8.99	7.75	7.07	6.65
7.5%	20.04	11.87	9.27	8.06	7.39	6.99
8.0%	20.28	12.13	9.56	8.36	7.72	7.34

- **i** Using the Monthly Repayment Table, calculate Aaron's monthly repayment. *(2 marks)* Easy
- **ii** How much interest does he pay over the 20 years? *(2 marks)* Medium
- **iii** Aaron calculates that if he repays the loan over 15 years, his total repayments would be \$242 730.
  How much interest would he save by repaying the loan over 15 years instead of 20 years? *(2 marks)* Hard

(Q27a, **2004 HSC**)

34 Jim bought a new car at the beginning of 2001 for $40 000. At the end of 2001 the value of the car had depreciated by 30%. In 2002 the value of the car depreciated by 25% of the value it had at the end of 2001.

What was the value of the car at the end of 2002?

A $18 000 **B** $19 600
C $21 000 **D** $22 000 *(1 mark)*

(Q5, **2003 HSC**) Easy

35 While Sandra is on holidays she buys a digital camera for $800. Using the straight-line method of depreciation, the salvage value, S, of the camera will be zero in four years time, as shown on the graph below.

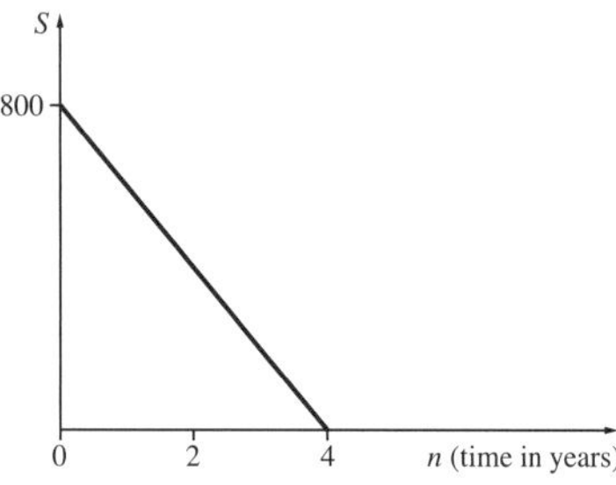

Using the declining balance method of depreciation, at a rate of R% per annum, the salvage value is represented by the dotted curve on the graph below.

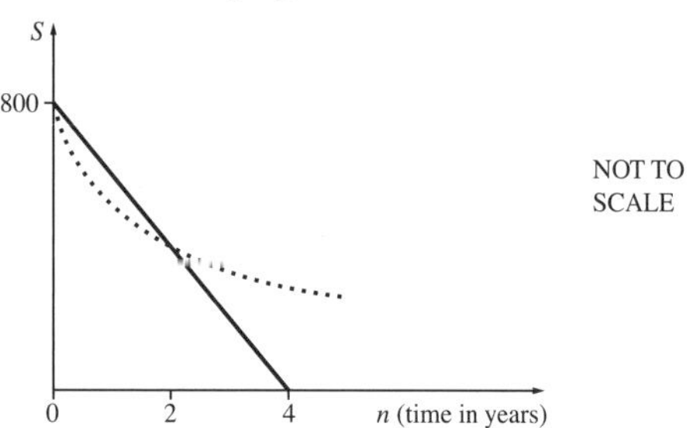

Both methods give the same salvage value after two years. Find the value of R. *(4 marks)*

(Q28d, **2003 HSC**) Hard

36 The table shows monthly repayments for loans over 30 years.

Interest rate per annum	*Loan amount*			
	$100 000	**$150 000**	**$200 000**	**$250 000**
5.0%	$537	$806	$1074	$1343
5.5%	$568	$852	$1136	$1420
6.0%	$600	$900	$1200	$1499
6.5%	$633	$949	$1265	$1581
7.0%	$666	$998	$1331	$1664
7.5%	$700	$1049	$1399	$1749

James borrows $200 000 over a period of 30 years at 6.5% per annum. Repayments are to be made monthly according to the table.

How much would James repay over 30 years if the interest rate were to remain the same?

A $1265 **B** $37 950
C $390 000 **D** $455 400 *(1 mark)*

(Q9, **2002 HSC**) Medium

37 At the end of 2000, Zara purchased a new computer for $4999. Use the declining balance method to determine the value of the computer at the end of 2002, assuming a depreciation rate of 40% per annum. (Answer to the nearest dollar.)

A $800 **B** $1000
C $1800 **D** $2000 *(1 mark)*

(Q11, **2002 HSC**) Easy

38 Minh wants to buy a hi-fi system from Advanced Sound Systems for $5000.

i Minh considers buying the system on hire purchase.
What would Minh repay monthly if he were to buy the hi-fi system on the following hire purchase terms?

> ADVANCED SOUND SYSTEMS
> Hire Purchase Terms:
> **10% deposit**
> 15% pa simple interest on the balance
> *Equal monthly repayments over 3 years*

(3 marks) Hard

ii Minh decides instead to borrow the required $5000 from his local bank. The loan is a reducing balance loan over 5 years, with monthly repayments.
The graph shows the balance owing on the loan over time.

1 Use the graph to determine the balance owing after 2 years. *(1 mark)* Easy

2 Use the graph to determine when the loan is half-paid. *(1 mark)* Easy

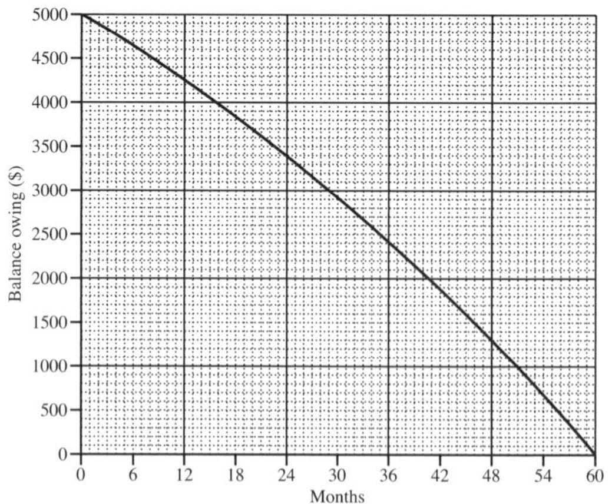

(Q23c, **2002 HSC**)

39 Frank has a credit card with an interest rate of 0.05% per day and no interest-free period.

Frank used the credit card to pay for car repairs costing \$480. He paid the credit card account 16 days later. What is the total amount (including interest) that he paid for the repairs?

A \$480.24 **B** \$483.84
C \$504.00 **D** \$864.00 *(1 mark)*

(Q4, **2001 HSC**) **Medium**

---

40 A car is purchased for \$42 000.

Use the declining balance method to calculate the salvage value of the car after 4 years at a depreciation rate of 15% per annum. *(2 marks)*

(Q27b, **2001 HSC**) 

Year 12 Depreciation and loans—Worked answers

1 $V_0 = \$2467, r = 0.15, n = 3$

$S = V_0(1-r)^n$

$= \$2467 \times (1-0.15)^3$

$= \$1515.04637\ldots$

$= \$1515$ (nearest dollar)

Answer D

2 $V_0 = \$10000, r = 0.08,$

$n = 5 \times 2 = 10$

$S = V_0(1-r)^n$

$= \$10000(1-0.08)^{10}$

$= \$4343.88454\ldots$

$= \$4343.88$ (nearest cent)

Answer C

3 $PV = \$500, r = \dfrac{0.17}{365}, n = 15$

$FV = PV(1+r)^n$

$= \$500\left(1+\dfrac{0.17}{365}\right)^{15}$ ✓

$= \$503.504562\ldots$

$= \$503.50$ (nearest cent) ✓

Amount owing $= \$503.50 - \$250$

$= \$253.50$ ✓

(3 marks)

4 Number of days where interest is paid is 11. ✓

$FV = PV(1+r)^n$

$= 3700\left(1+\dfrac{0.1825}{365}\right)^{11}$

$= 3720.40095\ldots$ ✓

So the closing balance is \$3720.40

Minimum payment

$= 2\%$ of \$3720.40

$= \$74.41$ (nearest cent) ✓

(3 marks)

5 Loss in value in first year

$= \$24\,950 - \$21\,457$

$= \$3493$ ✓

Rate of depreciation

$= \dfrac{3493}{24950} \times 100\%$

$= 14\%$ ✓

$S = V_0(1-r)^n$

$= \$24950(1-0.14)^{10}$

$= \$5521.47439\ldots$

$= \$5521$ (nearest dollar)

The value of the car at the end of 10 years is \$5521 to the nearest dollar. ✓

(3 marks)

6 $V_0 = \$23900, r = 0.115, n = 3$

$S = V_0(1-r)^n$

$= \$23900(1-0.115)^3$ ✓

$= \$16566.3835\ldots$

$= \$16566$ (nearest dollar)

The value of the car after three years is \$16566 to the nearest dollar. ✓

(2 marks)

7 Continuing to repay the original amount, Andrew would pay another 6 years at \$243 per month.

Total $= 6 \times 12 \times \$243$

$= \$17496$ ✓

At the higher amount he pays \$281 per month for 5 years.

New total $= 5 \times 12 \times \$281$

$= \$16860$

Difference $= \$17496 - \$16860$

$= \$636$

Andrew repaid \$636 less altogether by making the change. ✓

(2 marks)

8 Depreciation $= \$19900 - \$16300$

$= \$3600$

% depreciation

$= \dfrac{\$3600}{\$19\,900} \times 100\%$

$= 18.090452\ldots\%$

The approximate depreciation is 18%.

Answer A

9 Interest rate = 12% p.a.

= 1% per month

First amount of interest

$= 0.01 \times \$100000$

$= \$1000$ ✓

Amount owing after first repayment

$= \$100000 + \$1000 - \$1029$

$= \$99971$ ✓

Second amount of interest

$= 0.01 \times \$99971$

$= \$999.71$

Amount owing after second repayment

$= \$99971 + \$999.71 - \$1029$

$= \$99941.71$ ✓

(3 marks)

10 $A = P(1+r)^n$

$= \$780(1+0.00036)^{24}$

$= \$786.767174\ldots$

$= \$786.77$ (nearest cent)

$CI = \$786.77 - \$780$

$= \$6.77$

Answer B

11 $S = V_0(1-r)^n$

$= 56\,000(1-0.08)^n$

When $n = 30, S \approx 4590$

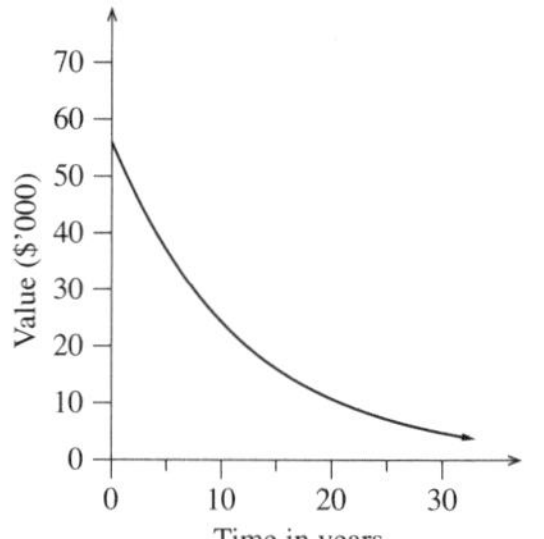

Answer A

12 From the graph, Jamal pays off the loan 8 years earlier by making the lump sum payment of \$80 000.

Saving in repayments

$= 8 \times 12 \times \$1880$

$= \$180\,480$ ✓

Amount actually saved

$= \$180480 - \$80000$

$= \$100480$

So Jamal paid \$100480 less by making the lump sum payment. ✓

(2 marks)

13 $V_0 = \$19990, r = 0.18, n = 3$

$S = V_0(1-r)^n$

$= \$19990(1-0.18)^3$

$= \$11021.846\ldots$

$= \$11022$ (nearest dollar)

Answer C

14 Method 1:

Depreciation

= \$1250 per year for 2 years

= \$2500 ✓

Method 2:

$V_0 = \$5000; r = 0.35; n = 2$

$S = V_0(1-r)^n$

$= \$5000(1-0.35)^2$

$= \$2112.50$ ✓

Amount of depreciation

$= \$5000 - \$2112.50$

$= \$2887.50$

So Method 2 gives the greater amount of depreciation. ✓

(3 marks)

15 [The declining balance method of depreciation is used, not the straight-line method, so options B and D are not correct. The machine will still have some value after 10 years so option C is not correct.]

The graph which best represents the salvage value of the machine is this graph.

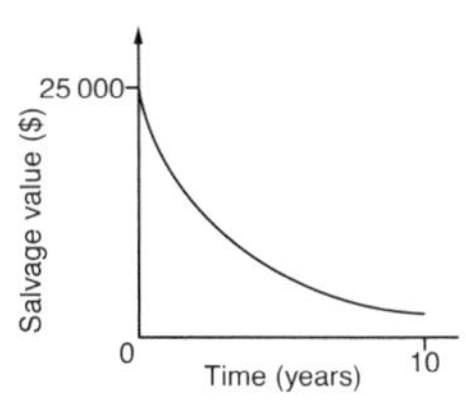

Answer A

16 Monthly repayments:

\$2796.86 over 30 years

Total repaid

= \$2796.86 × 30 × 12

= \$1 006 869.60

Fortnightly repayments:

\$1404.76 over 23 years

Total repaid

= \$1404.76 × 23 × 26

= \$840 046.48

Difference

= \$1 006 869.60 – \$840 046.48

= \$166 823.12

= \$166 823 [nearest dollar]

Answer B

17 $V_0 = 22000 \quad r = 0.15 \quad n = 3$

$S = V_0(1 - r)^n$

$= 22000(1 - 0.15)^3$ ✓

$= 13510.75$

The value after 3 years is \$13 510.75. ✓

(2 marks)

18 i Gradient $= \dfrac{-60000}{15}$

$= -4000$

(1 mark)

ii The gradient represents the change in value of the tractor (in dollars) each year.

(1 mark)

iii $S = 60000 - 4000n$

[Or: $S = -4000n + 60000$]

(1 mark)

iv Norman cannot use any negative values of n because they would represent years ✓ before he bought the tractor.

Norman cannot use values of n that are greater than 15 because they give his tractor a negative value which doesn't make sense. ✓

(2 marks)

v $V_0 = \$60\,000, r = 0.2, n = 14$

$S = V_0(1 - r)^n$

$= \$60\,000(1 - 0.2)^{14}$ ✓

$= \$2638.827\,907\ldots$

$= \$2639$ [nearest dollar] ✓

(2 marks)

vi When n is greater than 15, the salvage value gets smaller and smaller each year, and is closer and closer to \$0. The amount by which the value decreases each year also gets smaller.

(1 mark)

19 i February repayments = \$1588

April repayments = \$1573

Difference = \$1588 – \$1573

= \$15

The change is a decrease of \$15.

(1 mark)

ii 30% of Xiang's salary

= 0.3 × \$6500

= \$1950 ✓

In October:

Repayments on \$1000 = \$6.47

Repayments on \$307 000

= 307 × \$6.47

= \$1986.29 ✓

The repayments are more than 30% of Xiang's monthly salary.

The bank would not have approved the loan. ✓

(3 marks)

iii Possible answers are:

- Xiang's graph is smooth while Jack's changes after 12 years. There must have been a change to Jack's interest rate and/or repayments at that time. ✓
- Jack's loan balance decreased much more quickly than Xiang's over the first 12 years. He could have been making higher repayments or have been charged a lower interest rate for the first 12 years. ✓
- After 12 years Jack's balance decreases much more slowly than Xiang's. Jack's interest rate could have increased or the amount of his repayments may have decreased.
- Xiang's loan is paid off 5 years earlier than Jack's. Although she might have been paying a higher interest rate at first she paid less interest in total than Jack.

(2 marks)

20 i Depreciation over 3 years is \$3600

∴ Depreciation per year

= \$3600 ÷ 3

= \$1200

(1 mark)

ii Declining balance formula is

$S = V_0(1 - r)^n$

where $V_0 = \$3600$ and $r = 0.3$

$\therefore S = 3600(1 - 0.3)^n$ ✓

$= 3600(0.7)^n$

This can never be zero, since $0.7^n \neq 0$ for any value of n. ✓

(2 marks)

21 $S = V_0(1 - r)^n$

$r = 15\% = 0.15$

$n = 2$ years

$S = \$2023$

$\therefore 2023 = V_0(1 - 0.15)^2$

$2023 = V_0 \times 0.7225$ ✓

$V_0 = 2023 \div 0.7225$

$= 2800$

∴ The purchase price is \$2800. ✓

(2 marks)

22 Formulae for declining balance method is $S = V_0(1 - r)^n$

This is not a linear graph as in *A*.

It is an exponential graph as in *C*.

or

Check graph on graphics calculator substituting values for V_0 and r.

Answer C

23 Flat rate loan is a straight-line graph. Reducing balance loan is a curved graph.

Both graphs C and D show that the time taken to repay the loan is less for a reducing balance loan, but graph C starts with different loan amounts.

Answer C

24 $r = 7.2\%$ per annum

$= (7.2 \div 12)\%$ per month

$= 0.6\%$

Interest for 1 month

$= \dfrac{0.6}{100} \times \$420\,000$

= \$2520 (by calc.)

∴ Amount owing after 1 month

= \$420 000 + \$2520 – \$4000

= \$418 520

Answer B

25 $E = \dfrac{(1+r)^n - 1}{n}$

$r = 6\%$ p.a. compounded monthly
$= 0.5\%$ per month
$= 0.005$

$n = 4$ years ✓
$= 48$ months

$\therefore E = \dfrac{(1+0.005)^{48} - 1}{48}$ ✓

$= 0.005\,635\ldots$ (by calc.)
$= 0.5635\ldots\%$ per month
$= 6.7622\ldots\%$ per annum ✓

∴ Martha does have a better deal than Paul as her effective interest rate is higher. ✓

(4 marks)

26 i From the table:
Over 10 years, Liliana can borrow $80 000.
Over 15 years, Liliana can borrow $100 000.
Continuing to look at the table, the maximum she can borrow is $130 000 over 30 years.

(1 mark)

ii Interest over 15 years
(= 180 months)
= ($1529.04 × 180) – $160 000
= $275 227.20 – $160 000
= $115 227.20 ✓

Interest over 20 years
(= 240 months)
= ($1338.30 × 240) – $160 000
= $321 192 – $160 000
= $161 192

∴ Additional interest
= $161 192 – $115 227.20
= $45 964.80 ✓

(2 marks)

27 i Value at end of 3rd year
= $30 000 – $2000 × 3
= $30 000 – $6000
= $24 000

(1 mark)

ii For the next 4 years, the car is depreciated using:
$S = V_0(1 - r)^n$
$r = 0.25$
$V_0 = 24\,000$
$n = 4$

$\therefore S = 24\,000(1 - 0.25)^4$ ✓
$\therefore S = \$7593.75$

∴ Value of car seven years after purchase is $7593.75. ✓

(2 marks)

iii

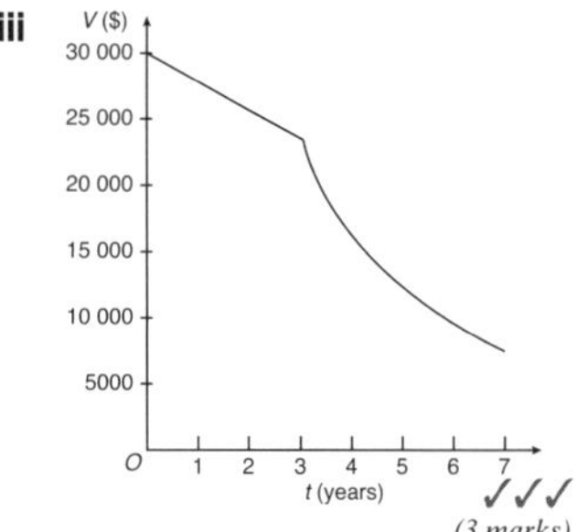

✓✓✓
(3 marks)

28 From the table, each $1000 borrowed for 15 years at 8% p.a. is repaid at $9.56 per month.
∴ Samantha's monthly loan repayment
= 70 × $9.56
= $669.20
Answer B

29 Using declining balance formula,
$S = 50\,000(1 - r)^n$

This is an exponential graph.

Answer D

OR

Substituting a value or r, say 20% into the formula,

$S = 50\,000(1 - 0.2)^n$
$= 50\,000(0.8)^n$

Plotting points:

n	0	1	2
S	50 000	40 000	32 000

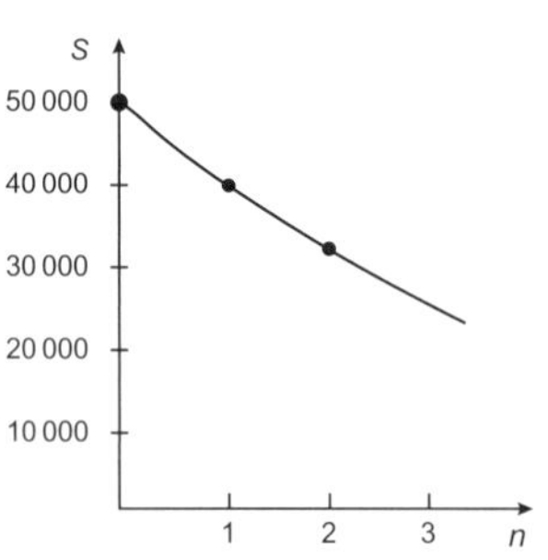

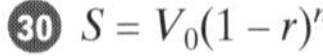
Answer D

30 $S = V_0(1 - r)^n$
$V_0 = \$150\,000$
$r = 10\% = 0.1$
$S = \$120\,000$
$n = ?$
Since 10% of $150 000 = $15 000, it would be reasonable to assume that it will take between 2 and 3 years for the depreciated value to reach $120 000.

By trial and error,
when $n = 2$,
$S = 150\,000(1 - 0.1)^2$
$= \$121\,500$ ✓
when $n = 3$,
$S = 150\,000(1 - 0.1)^3$
$= \$109\,350$
∴ The value will first fall below $120 000 after 2 years, (i.e. after December 2007) in 2008. ✓

(2 marks)

31 $A = P(1 + r)^n$
$= 250\left(1 + \dfrac{0.05}{100}\right)^{30}$
$= 253.777\ldots$
$= \$253.78$
∴ Interest = $253.78 – $250.00
= $3.78

Answer B

32 i Tax deduction
= 40% of $6500
= $2600

(1 mark)

ii $S = V_0(1 - r)^n$
$= \$6500(1 - 40\%)^3$ ✓
$= \$6500(1 - 0.4)^3$
$= \$1404$ ✓

(2 marks)

33 i Monthly repayment
= 7.75 × 150 ✓
= $1162.50 ✓

(2 marks)

ii Total repayments
= $1162.50 × 12 × 20
= $279 000 ✓

∴ Interest
= $279 000 – $150 000
= $129 000 ✓

(2 marks)

iii Interest over 15 years
= $242 730 – $150 000
= $92 730 ✓

∴ Saving in interest
= $129 000 – $92 730
= $36 270 ✓

(2 marks)

34 Value of car at end of 2001
= (100% – 30%) × $40 000
= 70% × $40 000
= 0.7 × $40 000
= $28 000

∴ Value of car at end of 2002

= 0.75 × $28 000
= $21 000

Answer C

35 *Straight line method:*

$S = V_0 - Dn$

$\therefore S = 800 - Dn$

When $n = 4$, $S = 0$

$\therefore 0 = 800 - 4D$

$4D = 800$

$\therefore D = 200$ ✓

$\therefore S = 800 - 200n$

When $n = 2$,

$S = 800 - 200 \times 2$

$S = 800 - 400$

$S = 400$ ✓

Declining balance method:

$S = V_0(1 - r)^n$

$\therefore S = 800(1 - r)^n$

When $n = 2$,

$S = 800(1 - r)^2$ ✓

Since after 2 years, the salvage value (S) is the same using either method:

$400 = 800(1 - r)^2$

$(1 - r)^2 = \frac{400}{800}$

$(1 - r)^2 = \frac{1}{2}$

$1 - r = \sqrt{\frac{1}{2}}$

$-r = \sqrt{\frac{1}{2}} - 1$

$-r = -0.292\ldots$

$r = 0.292\ldots$

$\therefore R = 0.292\ldots \times 100$ ✓

$\doteqdot 29\%$

(4 marks)

36 Total repayments

$= 30 \times 12 \times \$1265$

$= \$455\,400$

Answer D

37 $S = V_0\,(1 - r^n)$

$= \$4999\,(1 - 0.4)^2$

$= \$1799.64$

$\doteqdot \$1800$

Answer C

38 **i** Deposit = 10% of \$5000

$= \frac{10}{100} \times \$5000$

$= \$500$

Balance = \$5000 – \$500

= \$4500 ✓

Interest = 15% of \$4500 × 3

$= \frac{15}{100} \times \$4500 \times 3$

$= \$675 \times 3$

$= \$2025$ ✓

$\therefore$ Total payable

= \$4500 + \$2025

= \$6525

Monthly repayments

= \$6525 ÷ 36

= \$181.25 ✓

(3 marks)

ii **1** \$3400

(1 mark)

2 \$2500 has been paid by the end of 35 months.

(1 mark)

39 Interest $= \frac{0.05}{100} \times \$480 \times 16$

$= \$3.84$

$\therefore$ Amount owing = \$480 + \$3.84

= \$483.84

Answer B

40 $S = V_0(1 - r)^n$

$= 42\,000(1 - 0.15)^4$ ✓

$= \$21\,924.26$ ✓

(2 marks)

1 Julie invests \$12 500 in a savings account. Interest is paid at a fixed monthly rate. At the end of each month, after the monthly interest is added, Julie makes a deposit of \$500.

Julie has created a spreadsheet to show the activity in her savings account. The details for the first 6 months are shown.

Month	*Amount in account at beginning of month*	*Monthly interest*	*Deposit*	*Amount in account at end of month*
1	12 500.00	18.75	500	13 018.75
2	13 018.75	19.53	500	13 538.28
3	13 538.28	20.31	500	14 058.59
4	14 058.59	21.09	500	14 579.68
5	14 579.68	21.87	500	15 101.55
6	15 101.55	22.65	500	15 624.20
7			500	

By finding the monthly rate of interest, complete the final row above for the 7th month. *(3 marks)*

(Q21, **2021 HSC**) **Medium**

2 Present value interest factors for an annuity of \$1 for various interest rates (r) and numbers of periods (N) are given in the table.

Table of present value interest factors

N \ r	*Interest rate per period as a decimal*			
	0.001	0.00125	0.0015	0.00175
300	259.07072	250.03980	241.43789	233.24180
330	280.95771	270.26900	260.13532	250.52386
360	302.19816	289.75411	278.01062	266.92278

A bank lends Martina \$500 000 to purchase a home, with interest charged at 1.5% per annum compounding monthly. She agrees to repay the loan by making equal monthly repayments over a 30-year period.

How much should the monthly payment be in order to pay off the loan in 30 years? Give your answer correct to the nearest cent. *(2 marks)*

(Q31, **2021 HSC**) **Medium**

3 A table of future value interest factors for an annuity of \$1 is shown.

Table of future value interest factors

Number of periods	*Interest rate per period*				
	0.25%	0.5%	0.75%	1%	1.25%
2	2.0025	2.0050	2.0075	2.0100	2.0125
4	4.0150	4.0301	4.0452	4.0604	4.0756
6	6.0376	6.0755	6.1136	6.1520	6.1907
8	8.0704	8.1414	8.2132	8.2857	8.3589
10	10.1133	10.2280	10.3443	10.4622	10.5817

Simone deposits \$1000 into a savings account at the end of each year for 8 years. The interest rate for these 8 years is 0.75% per annum, compounded annually.

After the 8th deposit, Simone stops making deposits but leaves the money in the savings account. The money in her savings account then earns interest at 1.25% per annum, compounded annually, for a further two years.

Find the amount of money in Simone's savings account at the end of ten years. *(3 marks)*

(Q40, **2021 HSC**) **Hard**

4 An annuity consists of ten payments, each equal to \$1000. Each payment is made on 30 June each year from 2021 through to 2030 inclusive.

The rate of compound interest is 5% per annum.

The present value of the annuity is calculated at 30 June 2020.

The future value of the annuity is calculated at 30 June 2030.

Without performing any calculations, which of the following statements is true?

A Present value of the annuity < \$10 000 < future value of the annuity

B \$10 000 < present value of the annuity < future value of the annuity

C Future value of the annuity < \$10 000 < present value of the annuity

D \$10 000 < future value of the annuity < present value of the annuity *(1 mark)*

(Q14, **2020 HSC**) **Medium**

5 Tina inherits \$60 000 and invests it in an account earning interest at a rate of 0.5% per month. Each month, immediately after the interest has been paid, Tina withdraws \$800.

The amount in the account immediately after the nth withdrawal can be determined using the recurrence relation

$$A_n = A_{n-1}(1.005) - 800,$$

where $n = 1, 2, 3, \ldots$ and $A_0 = 60000$.

a Use the recurrence relation to find the amount of money in the account immediately after the third withdrawal. *(2 marks)* **Medium**

b Calculate the amount of interest earned in the first three months. *(2 marks)* **Hard**

6 Wilma deposited a lump sum into a new bank account which earns 2% per annum compound interest.

Present value interest factors for an annuity of \$1 for various interest rates (r) and numbers of periods (N) are given in the table.

Table of present value interest factors

N \ r	*Interest rate per period as a decimal*			
	0.01	0.015	0.02	0.025
10	9.471	9.222	8.983	8.752
20	18.046	17.169	16.351	15.589
30	25.808	24.016	22.396	20.930

Wilma was able to make the following withdrawals from this account.

- \$1000 at the end of each year for twenty years (starting one year after the account is opened)
- \$3000 each year for ten years starting 21 years after the account is opened.

Calculate the minimum lump sum Wilma must have deposited when she opened the new account. *(3 marks)*

(Q37, **2020 HSC**) **Hard**

7 The table shows the future values of an annuity of \$1 for different interest rates for 4, 5 and 6 years. The contributions are made at the end of each year.

Future value of an annuity of \$1

Years	*Interest rate per annum*			
	1%	*2%*	*3%*	*4%*
4	4.060	4.122	4.184	4.246
5	5.101	5.204	5.309	5.416
6	6.152	6.308	6.468	6.633

An annuity account is opened and contributions of \$2000 are made at the end of each year for 7 years.

For the first 6 years, the interest rate is 4% per annum, compounding annually. For the 7th year, the interest rate increases to 5% per annum, compounding annually.

Calculate the amount in the account immediately after the 7th contribution is made. *(3 marks)*

(Q42, **2019 HSC**) **Hard**

8 Sally is currently aged 50 and is planning to retire at the end of the year in which she turns 65. Her current self-managed super fund has a balance of \$530 000 and is delivering 3.2% compound interest. At the start of the year she makes a payment of \$8600 into the fund. However, Sally plans to increase this annual deposit by n% each year. A table for the annuity is shown below including the balance of the annuity at the end of each year.

Payment number	Payment received	Interest earned	Principal increase	Balance of annuity
0	\$0.00	\$0.00	\$0.00	\$530 000.00
1	\$8600.00	\$17 235.20	\$25 835.20	\$555 835.20
2	\$8772.00	\$18 067.43	\$26 839.43	\$582 674.63
3	\$8947.44	\$18 931.91	\$27 879.35	\$610 553.98
4	\$9126.39	\$19 829.77	\$28 956.16	\$639 510.14
5	\$9308.92	\$20 762.21	\$30 071.13	\$669 581.26
6	\$9495.09	\$21 730.44	\$31 225.54	\$700 806.80
7	\$9685.00	\$22 735.74	\$32 420.73	\$733 227.54
8	\$9878.70	\$23 779.40	\$33 658.10	\$766 885.63
9	\$10 076.27	\$24 862.78	\$34 939.05	\$801 824.68
10	\$10 277.80	\$25 987.28	\$36 265.08	\$838 089.76
11	\$10 483.35	\$27 154.34	\$37 637.69	\$875 727.45
12	\$10 693.02	\$28 365.46	\$39 058.47	\$914 785.93
13	\$10 906.88	***A***		***B***
14	\$11 125.02	\$30 926.08	\$42 051.10	\$997 366.07
15	\$11 347.52	\$32 278.83	\$43 626.35	\$1 040 992.42

a Find the value of n. *(1 mark)* **Easy**

b Find ***A*** and ***B***. *(2 marks)* **Medium**

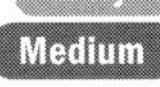

c Sally has decided to delay retirement for one year.

i What will be her contribution for payment 16? *(1 mark)* **Medium**

ii What will be the *new* balance for the annuity at the end of her 66th year? *(1 mark)* **Medium**

Bonus question (see page iv)

9 Mark needs $8000 to go on a holiday in three years time. He has a holiday savings account with a balance of $600. He arranges to deposit $150 into this account at the end of each month for the next 3 years. He earns 6% per annum interest on the money in the account, compounded monthly. The table below shows future values of an annuity of $1 and can be used in your calculations.

Period	Table of future value interest factors							
	Interest rate per period							
	0.50%	**1.00%**	**1.50%**	**2.00%**	**3.00%**	**4.00%**	**5.00%**	**6.00%**
5	5.0503	5.1010	5.1523	5.2040	5.3091	5.4163	5.5256	5.6371
6	6.0755	6.1520	6.2296	6.3081	6.4684	6.6330	6.8019	6.9753
7	7.1059	7.2135	7.3230	7.4343	7.6625	7.8983	8.1420	8.3938
8	8.1414	8.2857	8.4328	8.5830	8.8923	9.2142	9.5491	9.8975
12	12.3356	12.6825	13.0412	13.4121	14.1920	15.0258	15.9171	16.8699
24	25.4320	26.9735	28.6335	30.4219	34.4265	39.0826	44.5020	50.8156
36	39.3361	43.0769	47.2760	51.9944	63.2759	77.5983	95.8363	119.1209
48	54.0978	61.2226	69.5652	79.3535	104.4084	139.2632	188.0254	256.5645
60	69.7700	81.6697	96.2147	114.0515	163.0534	237.9907	353.5837	533.1282
72	86.4089	104.7099	128.0772	158.0570	246.6672	396.0566	650.9027	1089.6286

Will Mark have enough money for his trip at the end of the 3 years? Justify your answer with calculations. *(3 marks)*

Bonus question **Hard**

---

10 The table shows the present value interest factors for various interest rates and periods of time.

Table of present value interest factors							
	Interest rate per period						
Period	**0.50%**	**1.00%**	**1.25%**	**1.50%**	**2.00%**	**3.00%**	**4.00%**
20	18.9874	18.0456	17.5993	17.1686	16.3514	14.8775	13.5903
30	27.7941	25.8077	24.8889	24.0158	22.3965	19.6004	17.2920
40	36.1722	32.8347	31.3269	29.9158	27.3555	23.1148	19.7928
50	44.1428	39.1961	37.0129	34.9997	31.4236	25.7298	21.4822
120	90.0735	69.7005	61.9828	55.4985	45.3554	32.3730	24.7741
180	118.5035	83.3217	71.4496	62.0956	48.5844	33.1703	24.9785
240	139.5808	90.8194	75.9423	64.7957	49.5686	33.3057	24.9980

Mel is 40 years old today and plans to retire when she turns 60 years old. In retirement, she would like to receive an annual annuity of $60 000 until she turns 90. What single amount should she deposit today in a savings account in order to have an investment large enough to receive her annual annuity in retirement, assuming an interest rate of 4% per annum for the entire 50 years? Justify your answers with calculations. *(4 marks)*

Bonus question **Hard**

---

11 Tracey wants an annuity of $15 000 every 3 months for the next 30 years. Her financial institution can guarantee an interest rate of 2% p.a. compounded quarterly for the life of the annuity. Use the table below to find the amount Tracey will need to invest today to provide the annuity.

Table of present value interest factors							
	Interest rate per period						
Period	**0.50%**	**1.00%**	**1.25%**	**1.50%**	**2.00%**	**3.00%**	**4.00%**
20	18.9874	18.0456	17.5993	17.1686	16.3514	14.8775	13.5903
30	27.7941	25.8077	24.8889	24.0158	22.3965	19.6004	17.2920
40	36.1722	32.8347	31.3269	29.9158	27.3555	23.1148	19.7928
50	44.1428	39.1961	37.0129	34.9997	31.4236	25.7298	21.4822
120	90.0735	69.7005	61.9828	55.4985	45.3554	32.3730	24.7741
180	118.5035	83.3217	71.4496	62.0956	48.5844	33.1703	24.9785
240	139.5808	90.8194	75.9423	64.7957	49.5686	33.3057	24.9980

(2 marks)

Bonus question 

12 A table of future value interest factors for an annuity of $1 is shown.

Table of future value interest factors					
Period	*Interest rate per period*				
	1%	2%	3%	4%	5%
3	3.0301	3.0604	3.0909	3.1216	3.1525
4	4.0604	4.1216	4.1836	4.2465	4.3101
5	5.1010	5.2040	5.3091	5.4163	5.5256
6	6.1520	6.3081	6.4684	6.6330	6.8019

An annuity involves contributions of $12 000 per annum for 5 years. The interest rate is 4% per annum, compounded annually.

i Calculate the future value of this annuity. *(1 mark)* **Easy**

ii Calculate the interest earned on this annuity. *(1 mark)* **Hard**

(Q27c, **2017 HSC**)

13 The table gives the contribution per period for an annuity with a future value of $1 at different interest rates and different periods of time.

Contribution per period for an annuity with a future value of $1

Number of periods	*Interest rate (% per period)*					
	0.25%	0.5%	0.75%	1%	1.25%	1.5%
6	0.1656	0.1646	0.1636	0.1625	0.1615	0.1605
12	0.0822	0.0811	0.0800	0.0788	0.0778	0.0767
18	0.0544	0.0532	0.0521	0.0510	0.0499	0.0488
24	0.0405	0.0393	0.0382	0.0371	0.0360	0.0349
30	0.0321	0.0310	0.0298	0.0287	0.0277	0.0266
36	0.0266	0.0254	0.0243	0.0232	0.0222	0.0212

Margaret needs to save $75 000 over 6 years for a deposit on a new apartment. She makes regular quarterly contributions into an investment account which pays interest at 3% pa.

How much will Margaret need to contribute each quarter to reach her savings goal? *(2 marks)*

(Q28d, **2016 HSC**) **Medium**

14 The table gives the present value interest factors for an annuity of $1 per period, for various interest rates (r) and numbers of periods (N).

Table of present value interest factors					
r \ N	Interest rate per period (as a decimal)				
	0.0075	0.0080	0.0085	0.0090	0.0095
70	54.30462	53.43960	52.59397	51.76724	50.95891
71	54.89293	54.00754	53.14226	52.29657	51.46995
72	55.47685	54.57097	53.68593	52.82118	51.97618
73	56.05643	55.12993	54.22502	53.34111	52.47764
74	56.63169	55.68446	54.75957	53.85641	52.97438

i Oscar plans to invest $200 each month for 74 months. His investment will earn interest at the rate of 0.0080 (as a decimal) per month. Use the information in the table to calculate the present value of this annuity. *(1 mark)* **Medium**

ii Lucy is using the same table to calculate the loan repayment for her car loan. Her loan is $21 500 and will be repaid in equal monthly repayments over 6 years. The interest rate on her loan is 10.8% per annum. Calculate the amount of each monthly repayment, correct to the nearest dollar. *(2 marks)* **Hard**

(Q30c, **2015 HSC**)

15 A table of future value interest factors is shown.

Table of future value interest factors

Period	*Interest rate per period*				
	1%	*2%*	*3%*	*4%*	*5%*
1	1.0000	1.0000	1.0000	1.0000	1.0000
2	2.0100	2.0200	2.0300	2.0400	2.0500
3	3.0301	3.0604	3.0909	3.1216	3.1525
4	4.0604	4.1216	4.1836	4.2465	4.3101

A certain annuity involves making equal contributions of $25 000 into an account every 6 months for 2 years at an interest rate of 4% per annum.

Based on the information provided, what is the future value of this annuity?

A $50 500 **B** $51 000
C $103 040 **D** $106 162 *(1 mark)*

(Q21, **2014 HSC**) **Medium**

16 Zina opened an account to save for a new car. Six months after opening the account, she made her first deposit of $1200 and continued depositing $1200 at the end of each six month period. Interest was paid at 3% per annum, compounded half-yearly.

How much was in Zina's account two years after first opening it?

A $4909.08 **B** $4982.72
C $5018.16 **D** $5094.55 *(1 mark)*

(Q23, **2013 HSC**) **Medium**

17 An amount of $5000 is invested at 10% per annum, compounded six-monthly.

Compounded values of $1

Period	**Interest rate per period**				
	1%	*5%*	*10%*	*15%*	*20%*
1	1.010	1.050	1.100	1.150	1.200
2	1.020	1.103	1.210	1.323	1.440
3	1.030	1.158	1.331	1.521	1.728
4	1.041	1.216	1.464	1.750	2.074
5	1.051	1.276	1.611	2.011	2.488
6	1.062	1.340	1.772	2.313	2.986

Use the table to find the value of this investment at the end of three years. *(2 marks)*

(Q23c, **2011 HSC**) **Hard**

18 The table shows the future value of a $1 annuity at different interest rates over different numbers of time periods.

Future values of a $1 annuity

Time Period	*Interest rate*				
	1%	*2%*	*3%*	*4%*	*5%*
1	1.0000	1.0000	1.0000	1.0000	1.0000
2	2.0100	2.0200	2.0300	2.0400	2.0500
3	3.0301	3.0604	3.0909	3.1216	3.1525
4	4.0604	4.1216	4.1836	4.2465	4.3101
5	5.1010	5.2040	5.3091	5.4163	5.5256
6	6.1520	6.3081	6.4684	6.6330	6.8019
7	7.2135	7.4343	7.6625	7.8983	8.1420
8	8.2857	8.5830	8.8923	9.2142	9.5491

i What would be the future value of a $5000 per year annuity at 3% per annum for 6 years, with interest compounding yearly? *(1 mark)* **Easy**

ii What is the value of an annuity that would provide a future value of $407 100 after 7 years at 5% per annum compound interest? *(1 mark)* **Medium**

iii An annuity of $1000 per quarter is invested at 4% per annum, compounded quarterly for 2 years. What will be the amount of interest earned? *(3 marks)* **Hard**

(Q27a, **2009 HSC**)

---

19 Rod is saving for a holiday. He deposits $3600 into an account at the end of every year for four years. The account pays 5% per annum interest, compounding annually.

The table shows future values of an annuity of $1.

Future values of an annuity of $1

End of year	*Interest rate*				
	1%	2%	3%	4%	5%
1	1.0000	1.0000	1.0000	1.0000	1.0000
2	2.0100	2.0200	2.0300	2.0400	2.0500
3	3.0301	3.0604	3.0909	3.1216	3.1525
4	4.0604	4.1216	4.1836	4.2465	4.3101
5	5.1010	5.2040	5.3091	5.4163	5.5256
6	6.1520	6.3081	6.4684	6.6330	6.8019
7	7.2135	7.4343	7.6625	7.8983	8.1420
8	8.2857	8.5830	8.8923	9.2142	9.5491

i Use the table to find the value of Rod's investment at the end of four years. *(2 marks)* **Easy**

ii How much interest does Rod earn on his investment over the four years? *(2 marks)* **Medium**

(Q26b, **2005 HSC**)

Year 12 Annuities—Worked answers

1 Monthly interest rate
$= \frac{18.75}{12\,500.00} \times 100\%$
$= 0.15\%$ ✓
The amount at the beginning of month 7 is 15 624.20.
Interest $= 0.0015 \times 15\,624.20$
$= 23.4363$
$= 23.44$ (2 d.p.) ✓
Amount at the end of the month
$= 15\,624.20 + 23.44 + 500$
$= 16\,147.64$ ✓

7	15 624.20	23.44	500	16 147.64

(3 marks)

2 30 years $= 30 \times 12$ months
$= 360$ months
So $N = 360$.
Interest rate per period $= \frac{0.015}{12}$
$= 0.00125$
Present value factor is 289.754 11 ✓
Monthly payment
$= \$500\,000 \div 289.754\,11$
$= \$1725.601\,06\ldots$
$= \$1725.60$ (nearest cent) ✓
(2 marks)

3 Interest rate $= 0.75\%$, 8 periods
Future value factor $= 8.2132$ ✓
After the 8th deposit,
$FV = \$1000 \times 8.2132$
$= \$8213.20$
For final 2 years, $PV = \$8213.20$, $r = 0.0125$ ✓
$FV = PV(1 + r)^n$
$= \$8213.20 \times (1 + 0.0125)^2$
$= \$8419.81$ (nearest cent)
The amount of money in the account at the end of 10 years is $8419.81. ✓
(3 marks)

4 Total payments = $10 000
The future value must be more than that as interest is earned.
The present value must be less than $10 000 as it is the single amount that could be invested at the same rate.
Answer A

5 **a** $A_n = A_{n-1}(1.005) - 800$
$A_0 = 60\,000$
$A_1 = 60\,000(1.005) - 800$
$= 59\,500$ ✓
$A_2 = 59\,500(1.005) - 800$
$= 58\,997.5$
$A_3 = 58\,997.5(1.005) - 800$
$= 58\,492.4875$
$= 58\,492.49$ (2 d.p.)
The amount of money in the account immediately after the third withdrawal is $58 492.49. ✓
(2 marks)

b Total amount withdrawn
$= 3 \times \$800$
$= \$2400$ ✓
Interest earned
$= \$58\,492.49 + \$2400 - \$60\,000$
$= \$892.49$ ✓
(2 marks)

6 $1000 each year for 20 years:
$r = 0.02, N = 20$
Present value factor = 16.351
Present value $= \$1000 \times 16.351$
$= \$16\,351$ ✓
So Wilma will need to deposit $16 351 to withdraw $1000 per year for the next 20 years.
$3000 each year for 10 years
$r = 0.02, N = 10$
Present value factor = 8.983
Present value $= \$3000 \times 8.983$
$= \$26\,949$
So Wilma will need to have $26 949 in the account at the end of 20 years to be able to withdraw $3000 each year for the following 10 years.
Amount required at the beginning of the 20 years
$= \frac{\$26\,949}{(1+0.02)^{20}}$ ✓
$= \$18\,135.9044\ldots$
$= \$18\,136$ (nearest dollar)
Minimum deposit
$= \$16\,351 + \$18\,136$
$= \$34\,487$ ✓
(3 marks)

7 Future value at end of 6 years
$= 6.633 \times \$2000$
$= \$13\,266$ ✓
Interest for 7th year
$= 0.05 \times \$13\,266$
$= \$663.30$ ✓
Total at end of 7 years
$= \$13\,266 + \$663.30 + \$2000$
$= \$15\,929.30$ ✓
(3 marks)

8 **a** Difference $= 8772 - 8600$
$= 172$
Percentage $= \frac{172}{8600} \times 100\%$
$= 2$
$\therefore n = 2\%$
(1 mark)

b $A: 0.032 \times (914\,785.93 + 10\,906.88)$
$= 29\,622.17$
$\therefore A = \$29\,622.17$ ✓
$B: 914\,785.93 + 10\,906.88 + 29\,622.17$
$= 955\,314.98$
$\therefore B = \$955\,314.98$ ✓
(2 marks)

c **i** $11\,347.52 \times 1.02$
$= 11\,574.47$
$\therefore$ contribution of $11 574.47.
(1 mark)

ii $(1\,040\,992.42 + 11\,574.47) \times 1.032$
$= 1\,086\,249.03$
$\therefore$ new balance of $1 086 249.03.
(1 mark)

9 6% per annum
$= 0.06$ per annum
$= \frac{0.06}{12}$ per month
$= 0.005$
3 years $= 3 \times 12$ months
$= 36$ months
Mark's account has two parts:
$600 earning 6% p.a. compounded monthly for 3 years:
$FV = PV(1 + r)^n$
$= 600(1.005)^{36}$
$= 718.008\,3149\ldots$ ✓
$\therefore \$718.01$
Also, $150 deposited each month. Using the table …
$FV = 150 \times 39.336$
$= 5900.4$
$\therefore \$5900.40$ ✓
Total $= \$718.01 + \$5900.40$
$= \$6618.41$
$\therefore$ Mark will not have enough money. ✓
(3 marks)

10 From 60 years old to 90 years old is 30 years. Also, 4%. ✓
$PV = 60\,000 \times 17.2920$
$= 1\,037\,520$
$\therefore$ Mel needs $1 037 520 in her savings account when she turns 60. ✓
Using $FV = 1\,037\,520$, $r = 0.04$ and $n = 20$:
$FV = PV(1 + r)^n$
$1\,037\,520 = PV(1.04)^{20}$ ✓
$PV = \frac{1\,037\,520}{1.04^{20}}$
$= 473\,510.5844\ldots$
$= 473\,511$ (nearest whole)
$\therefore$ Mel needs to deposit $473 511 in her savings account today. ✓
(4 marks)

11 2% p.a. = 0.5% per quarter
30 years = 120 quarters
$PV = 15\,000 \times 90.0735$ ✓
$= 1\,351\,102.50$
∴ Tracey should invest $1 351 102.50. ✓
(2 marks)

12 **i** Period 5 years, interest rate = 4%
Interest rate factor is 5.4163.
Future value
= 5.4163 × $12 000
= $64 995.60
(1 mark)

ii Total contributions
= 5 × $12 000
= $60 000
Interest
= $64 995.60 – $60 000
= $4995.60
(1 mark)

13 Number of quarters = 6 × 4 = 24
Interest rate per quarter
= 3% ÷ 4 = 0.75% ✓
Contribution required
= 0.0382 × $75 000
= $2865 ✓
(2 marks)

14 **i** For 74 months at 0.0080 per month the interest rate factor is 55.684 46
Present value
= $200 × 55.684 46
= $11 136.892
= $11 137 (nearest dollar)
The present value of the annuity is $11 137 to the nearest dollar.
(1 mark)

ii rate = 10.8% pa = 0.9% per month
So $r = 0.0090$
6 years
= 6 × 12 months
= 72 months
So $N = 72$
The required interest factor is 52.821 18 ✓
Amount of each repayment
= $21 500 ÷ 52.821 18
= $407.033 693….
= $407 (nearest dollar) ✓
(2 marks)

15 Number of periods = 2 × 2 = 4
Interest rate per period
= 4% ÷ 2 = 2%
Interest rate factor = 4.1216
FV = $25 000 × 4.1216
= $103 040
Answer C

16 Using 3% per annum
= 1.5% per half-year:
Total amount
$= 1200 \times 1.015^3 + 1200 \times 1.015^2 + 1200 \times 1.015 + 1200$
= 4909.084 05 …
= 4909.08
∴ there is $4909.08 in her account
Answer A

17 Rate = 10% pa
= 5% per six-months

Period = 3 years
= 6 six-months ✓

Value of investment
= $5000 × 1.340
= $6700 ✓
(2 marks)

18 **i** Future value of $1 = $6.4684

∴ Future value of $5000
= $6.4684 × 5000
= $32 342
(1 mark)

ii Future value of $1 = $8.1420

Annuity which yields $407 100
= $407 100 ÷ 8.1420
= $50 000
(1 mark)

iii $1000 invested at 4% per annum compounded quarterly for 2 years is equal to $1000 invested for 8 periods at 1% per period. ✓
∴ Future value
= 1000 × $8.2857
= $8285.70 ✓
∴ Interest
= $8285.70 – 8 × $1000
= $285.70 ✓
(3 marks)

19 **i** $1 invested at the end of every year for 4 years at 5% p.a. compounded annually becomes $4.3101. ✓
∴ Rod's investment will become $4.3101 × 3600
= $15 516.36 ✓
(2 marks)

ii Interest
= $15 516.36 – 4 × $3600 ✓
= $15 516.36 – $14 400
= $1116.36 ✓
(2 marks)

1 A salesperson is interested in the relationship between the number of bottles of lemonade sold per day and the number of hours of sunshine in the day.

The diagram shows the dataset used in the investigation and the least-squares regression line.

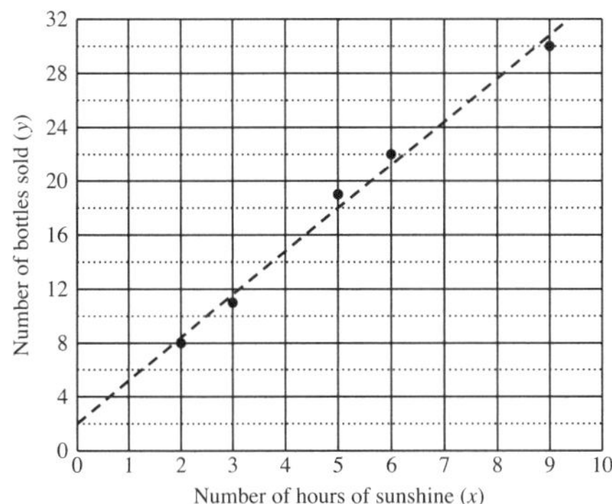

a Find the equation of the least-squares regression line relating to the dataset. *(2 marks)* **Hard**

b Suppose a sixth data point was collected on a day which had 10 hours of sunshine. On that day 45 bottles of lemonade were sold.
What would happen to the gradient found in part **a**? *(1 mark)* **Medium**

(Q28, **2021 HSC**)

2 For a sample of 17 inland towns in Australia, the height above sea level, x (metres), and the average maximum daily temperature, y (°C), were recorded.

The graph shows the data as well as a regression line.

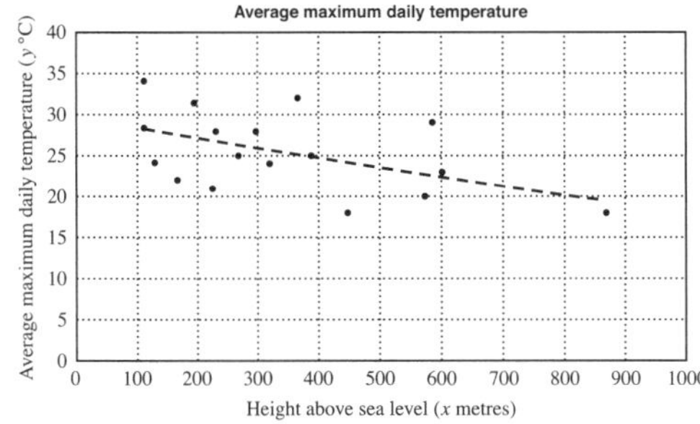

The equation of the regression line is $y = 29.2 - 0.011x$.

The correlation coefficient is $r = -0.494$.

a **i** By using the equation of the regression line, predict the average maximum daily temperature, in degrees Celsius, for a town that is 540 m above sea level. Give your answer correct to one decimal place. *(1 mark)* **Medium**

ii The gradient of the regression line is -0.011. Interpret the value of this gradient in the given context. *(2 marks)* **Hard**

b The graph below shows the relationship between the latitude, x (degrees south), and the average maximum daily temperature, y (°C), for the same 17 towns, as well as a regression line.

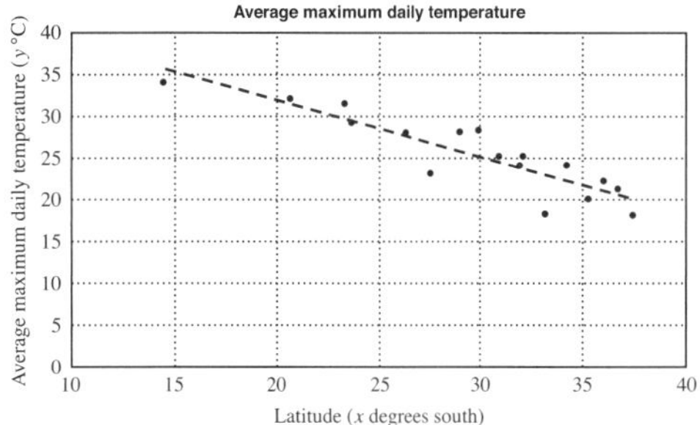

The equation of the regression line is $y = 45.6 - 0.683x$.

The correlation coefficient is $r = -0.897$.

Another inland town in Australia is 540 m above sea level. Its latitude is 28 degrees south.

Which measurement, height above sea level or latitude, would be better to use to predict this town's average maximum daily temperature? Give a reason for your answer. *(1 mark)* **Hard**

CQ (Q33, **2021 HSC**)

3 For a set of bivariate data, Pearson's correlation coefficient is –1.

Which graph could best represent this set of bivariate data?

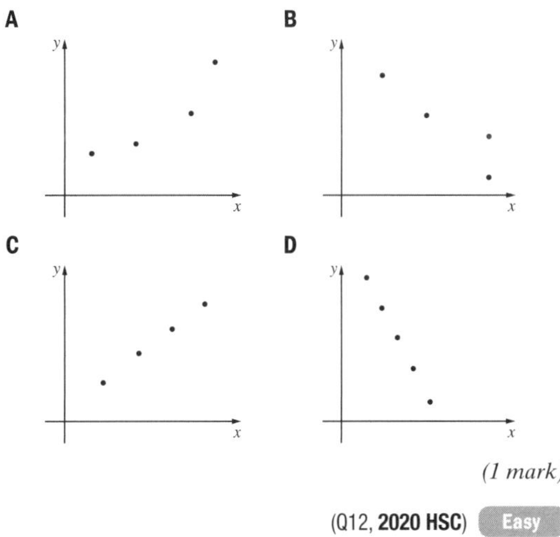

(1 mark)

(Q12, **2020 HSC**) **Easy**

4 A cricket is an insect. The male cricket produces a chirping sound.

A scientist wants to explore the relationship between the temperature in degrees Celsius and the number of cricket chirps heard in a 15-second time interval.

Once a day for 20 days, the scientist collects data. Based on the 20 data points, the scientist provides the information below.

- A box-plot of the temperature data is shown.

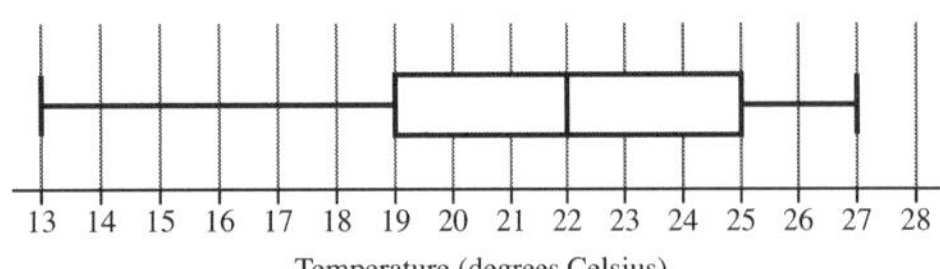

- The mean temperature in the dataset is 0.525°C below the median temperature in the dataset.
- A total of 684 chirps was counted when collecting the 20 data points.

The scientist fits a least-squares regression line using the data (x, y), where x is the temperature in degrees Celsius and y is the number of chirps heard in a 15-second time interval. The equation of this line is

$$y = -10.6063 + bx,$$

where b is the slope of the regression line.

The least-squares regression line passes through the point $(\bar{x}, \bar{y})$ where $\bar{x}$ is the sample mean of the temperature data and $\bar{y}$ is the sample mean of the chirp data.

Calculate the number of chirps expected in a 15-second interval when the temperature is 19° Celsius. Give your answer correct to the nearest whole number. *(5 marks)*

CQ (Q36, **2020 HSC**) **Hard**

5 A set of bivariate data is collected by measuring the height and arm span of seven children. The graph shows a scatterplot of these measurements.

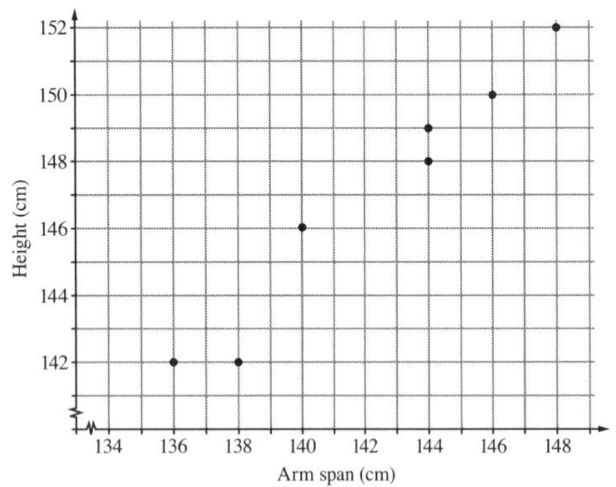

a Calculate Pearson's correlation coefficient for the data, correct to two decimal places. *(1 mark)* **Medium**

b Identify the direction and the strength of the linear association between height and arm span. *(1 mark)* **Medium**

c The equation of the least-squares regression line is shown.

Height = 0.866 × (arm span) + 23.7

A child has an arm span of 143 cm. Calculate the predicted height for this child using the equation of the least-squares regression line. *(1 mark)* **Easy**

(Q23, **2019 HSC**)

6 Data for life expectancy (expected remaining years of life) for females at selected ages are given in the table.

Age (x years)	*Life expectancy* (y years) *for females*
0	84.6
5	79.9
10	74.9
15	69.9
20	65.0
25	60.1
30	55.1
35	50.2
40	45.4

Find the equation of the regression line for this data, using accuracy to three decimal places. *(2 marks)*

Bonus question (see page iv) **Hard**

7 The table shows the number of hours that six students spent studying (n) and their marks scored in an exam (M).

n	5	6	8	8	11	14
M	71	79	84	76	94	89

Find the equation of the regression line for this data, using accuracy to two decimal places. *(2 marks)*

Bonus question **Hard**

8 Ciara surveyed a group of students comparing their age (in years) to their heights (in centimetres). The results are listed in the table below.

Age	Height
11	141
12	139
12	145
13	153
13	149
14	160
14	155
15	154
16	157
17	168

a Calculate Pearson's correlation for the data, correct to three decimal places. *(1 mark)* **Medium**

b Ciara plotted the data on the graph below and drew a line of best fit by eye.

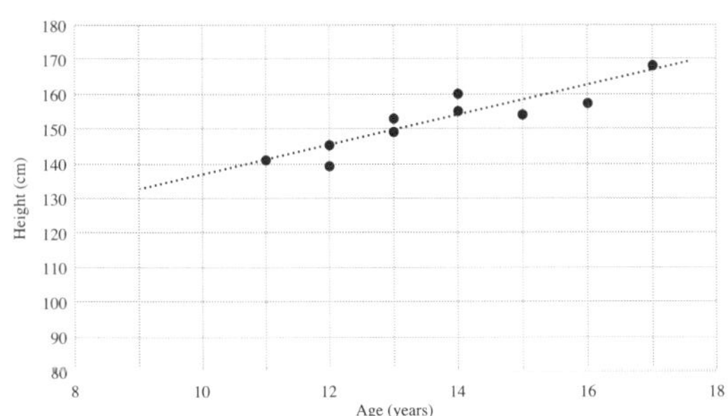

i Write down the gradient of Ciara's line of best fit. *(1 mark)* **Medium**

ii Using height = h and age = n, find the equation of Ciara's line. *(2 marks)* **Medium**

c Ciara wanted to use the equation determined in part **ii** to find the height of her 25-year-old sister. Explain the limitations of the equation. *(1 mark)* **Medium**

Bonus question

9 The table shows the relationship between the number of hours slept (t) and the number of mistakes (n) made by 10 students in a simple test.

Hours of sleep (t)	Mistakes (n)
9	1
8.5	3
7	5
6.5	6
7.5	4
8	5
7.5	5
7	4
9.5	1
8.5	2

a Calculate Pearson's correlation (r) for the data. Answer correct to two decimal places. *(2 marks)* **Medium**

b Describe the relationship between the number of hours slept and the mistakes made in the test. *(1 mark)* **Easy**

c The data has been plotted on the graph below. Plot a trendline by eye for the data. *(1 mark)* **Medium**

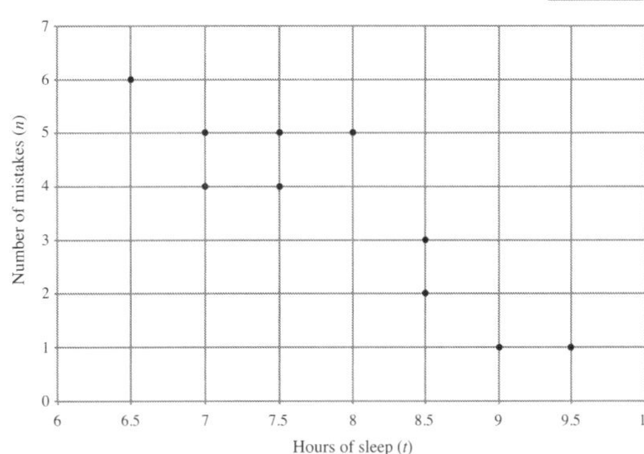

Bonus question

10 Which of the data sets graphed below has the largest positive correlation coefficient value?

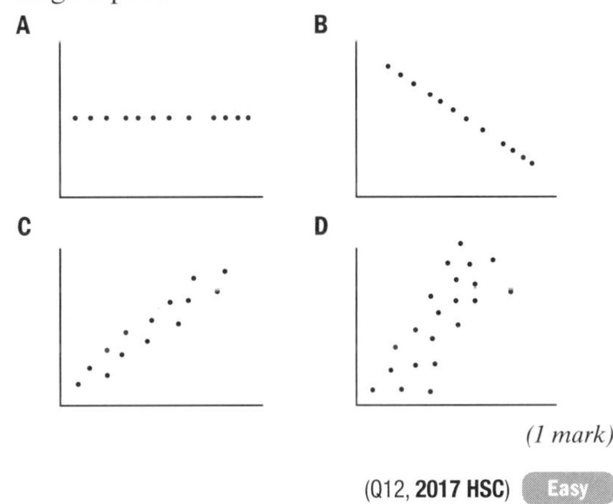

(1 mark)

(Q12, **2017 HSC**) **Easy**

11 The graph shows a scatterplot for a set of data.

Which of the following is the best approximation for the correlation coefficient of this set of data?

A −1 **B** − 0.3

C 0.3 **D** 1 *(1 mark)*

(Q3, **2016 HSC**) **Easy**

12 The shoe size and height of ten students were recorded.

Shoe size	6	7	7	8	8.5	9.5	10	11	12	12
Height	155	150	165	175	170	170	190	185	200	195

i Complete the scatter plot AND draw a line of fit by eye. *(2 marks)* **Medium**

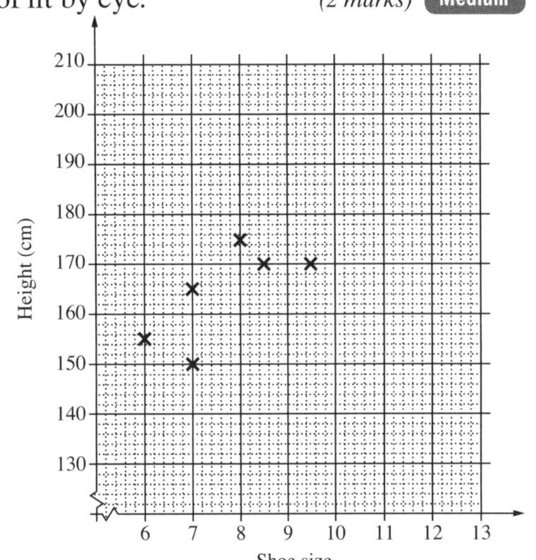

ii Use the line of fit to estimate the height difference between a student who wears a size 7.5 shoe and one who wears a size 9 shoe. *(1 mark)* **Easy**

iii A student calculated the correlation coefficient to be 1 for this set of data. Explain why this cannot be correct. *(1 mark)* **Medium**

(Q28e, **2015 HSC**)

13 The scatterplot shows the relationship between expenditure per primary school student, as a percentage of a country's Gross Domestic Product (GDP per capita), and the life expectancy in years for 15 countries.

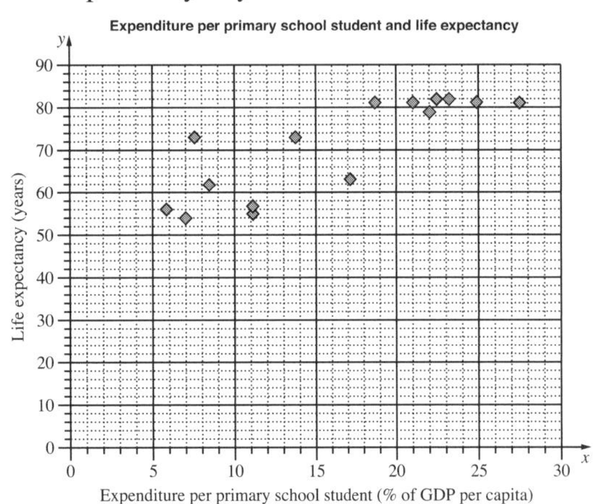

i For the given data, the correlation coefficient, r, is 0.83. What does this indicate about the relationship between expenditure per primary school student and life expectancy for the 15 countries? *(1 mark)* **Medium**

ii For the data representing expenditure per primary school student, Q_L is 8.4 and Q_U is 22.5. What is the interquartile range? *(1 mark)* **Easy**

iii Another country has an expenditure per primary school student of 47.6% of its GDP per capita. Would this country be an outlier for this set of data? Justify your answer with calculations. *(2 marks)* **Medium**

[Parts **iv** and **v** not in syllabus]

vi On the scatterplot, draw the least-squares line of best fit, $y = 1.29x + 49.9$. *(2 marks)* **Medium**

vii Using this line, or otherwise, estimate the life expectancy in a country which has an expenditure per primary school student of 18% of its GDP per capita. *(1 mark)* **Medium**

viii Why is this line NOT useful for predicting life expectancy in a country which has expenditure per primary school student of 60% of its GDP per capita? *(1 mark)* **Hard**

(Q30b, **2014 HSC**)

14 Which graph best shows data with a correlation closest to 0.3?

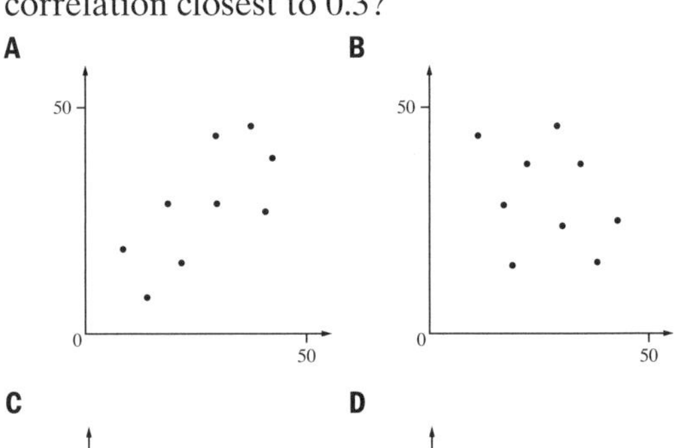

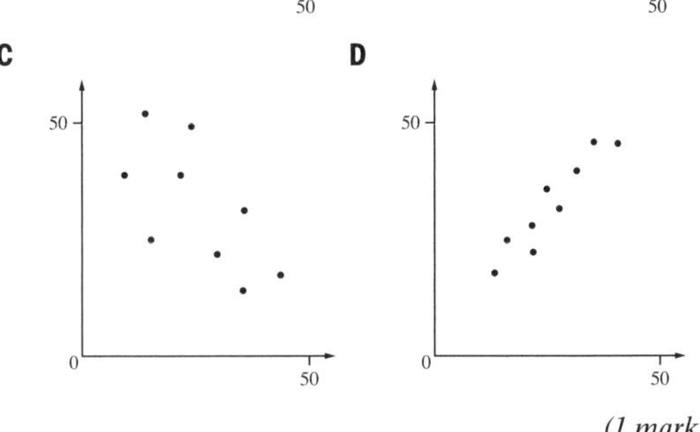

(1 mark)

(Q2, **2013 HSC**) **Medium**

15 Ahmed collected data on the age (a) and height (h) of males aged 11 to 16 years. He created a scatterplot of the data and constructed a line of best fit to model the relationship between the age and height of males.

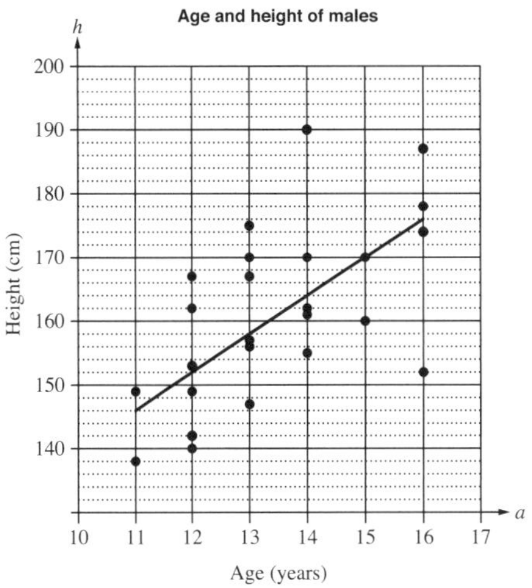

i Determine the gradient of the line of best fit shown on the graph. *(1 mark)* **Medium**

ii Explain the meaning of the gradient in the context of the data. *(1 mark)* **Hard**

iii Determine the equation of the line of best fit shown on the graph. *(2 marks)* **Hard**

iv Use the line of best fit to predict the height of a typical 17-year-old male. *(1 mark)* **Easy**

v Why would this model not be useful for predicting the height of a typical 45-year-old male? *(1 mark)* **Easy**

(Q28b, **2013 HSC**)

16 Which of the following relationships would most likely show a negative correlation?

A The population of a town and the number of hospitals in that town.

B The hours spent training for a race and the time taken to complete the race.

C The price per litre of petrol and the number of people riding bicycles to work.

D The number of pets per household and the number of computers per household. *(1 mark)*

(Q11, **2012 HSC**) **Medium**

17 Tourists visit a park where steam erupts from a particular geyser.

The brochure for the park has a graph of the data collected for this geyser over a period of time.

The graph shows the duration of an eruption and the time until the next eruption, timed from the end of one eruption to the beginning of the next.

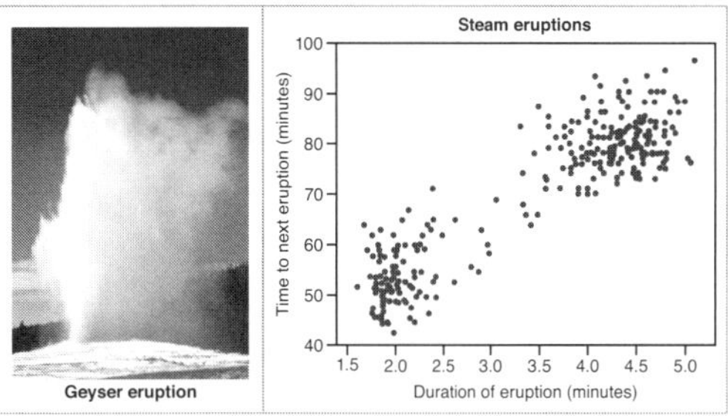

i Tony sees an eruption that lasts 4 minutes. Based on the data in the graph, what is the minimum time that he can expect to wait for the next eruption? *(1 mark)* **Easy**

ii Julia saw two consecutive eruptions, one hour apart. Based on the data in the graph, what was the longest possible duration of the first eruption that she saw? *(1 mark)* **Medium**

iii What does the graph suggest about the relationship between the duration of an eruption and the time to the next eruption? *(1 mark)* **Easy**

(Q29a, **2012 HSC**)

18 In which graph would the data have a correlation coefficient closest to –0.9?

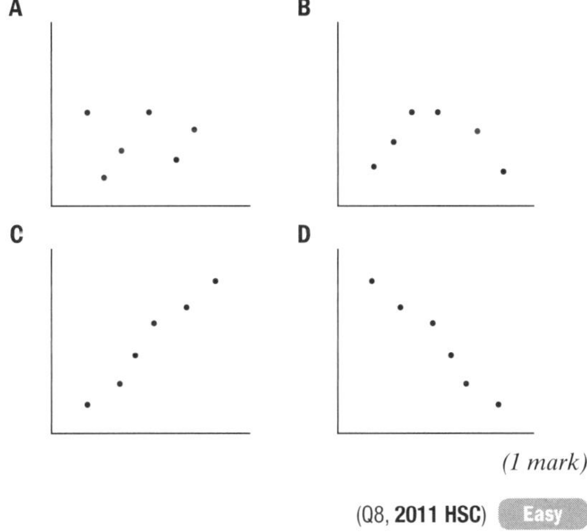

(1 mark)

(Q8, **2011 HSC**) **Easy**

19 A survey of Year 7 students found a number of relationships with a high degree of correlation.

Which of the following relationships also demonstrates causality?

A Students' height and the length of their arm span

B The size of students' left feet and the size of their right feet

C Students' test scores in Mathematics and their test scores in Music

D The number of hours students spent studying for a test and their results in that test *(1 mark)*

(Q6, **2010 HSC**) **Medium**

20 The height and mass of a child are measured and recorded over its first two years.

Height (cm), H	45	50	55	60	65	70	75	80
Mass (kg), M	2.3	3.8	4.7	6.2	7.1	7.8	8.8	10.2

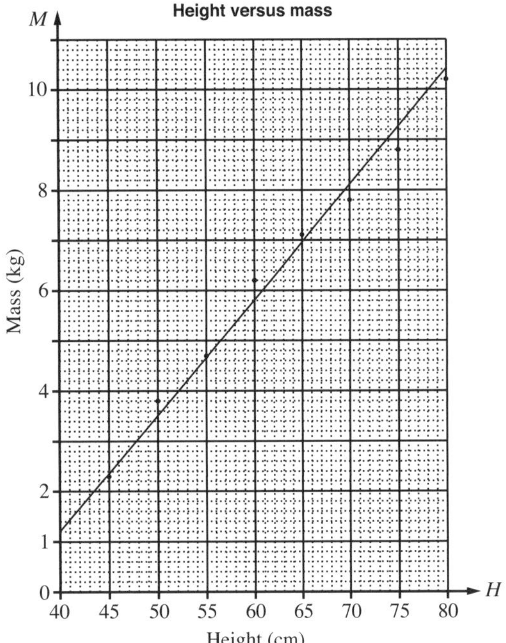

This information is displayed in a scatter graph.

i Describe the correlation between the height and mass of this child, as shown in the graph. *(1 mark)* **Medium**

ii A line of best fit has been drawn on the graph. Find the equation of this line. *(2 marks)* **Hard**

(Q28b, **2009 HSC**)

21 A scatterplot is shown.

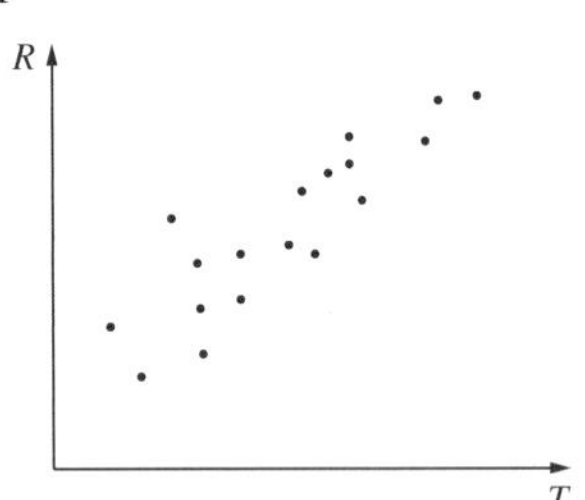

Which of the following best describes the correlation between R and T?

A Positive **B** Negative

C Positively skewed **D** Negatively skewed

(1 mark)

(Q12, **2008 HSC**) **Medium**

22 Which of the following would be most likely to have a positive correlation?

A The population of a town and the number of schools in that town

B The price of petrol per litre and the number of litres of petrol sold

C The hours training for a marathon and the time taken to complete the marathon

D The number of dogs per household and the number of televisions per household *(1 mark)*

(Q9, **2007 HSC**) **Medium**

23 Each member of a group of males had his height and foot length measured and recorded. The results were graphed and a line of fit drawn.

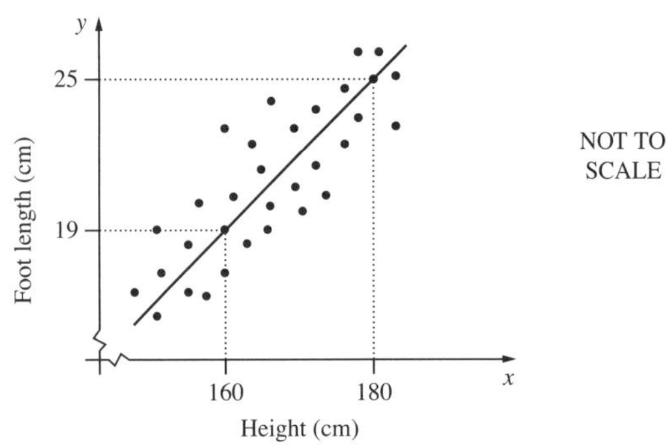

i Why does the value of the y-intercept have no meaning in this situation? *(1 mark)* **Easy**

ii George is 10 cm taller than his brother Harry. Use the line of fit to estimate the difference in their foot lengths. *(1 mark)* **Easy**

iii Sam calculated a correlation coefficient of −1.2 for the data. Give TWO reasons why Sam must be incorrect. *(2 marks)* **Easy**

(Q27b, **2006 HSC**)

24 Which scatterplot shows a low (weak) positive correlation?

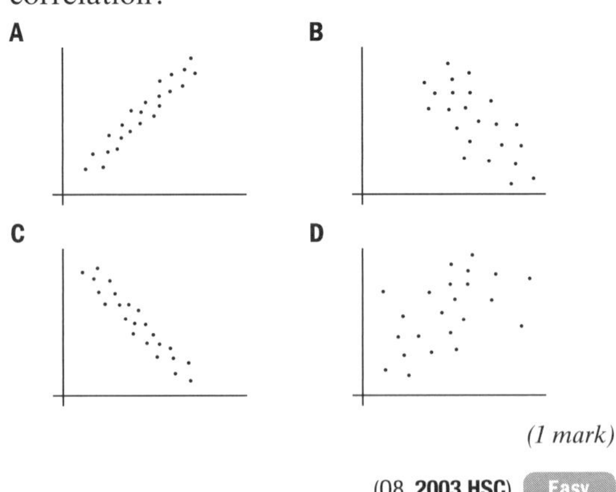

(1 mark)

(Q8, **2003 HSC**) **Easy**

25 A class of 30 students sat for an algebra test and a geometry test. The results were displayed in a scatterplot, and a line of fit was drawn, as shown.

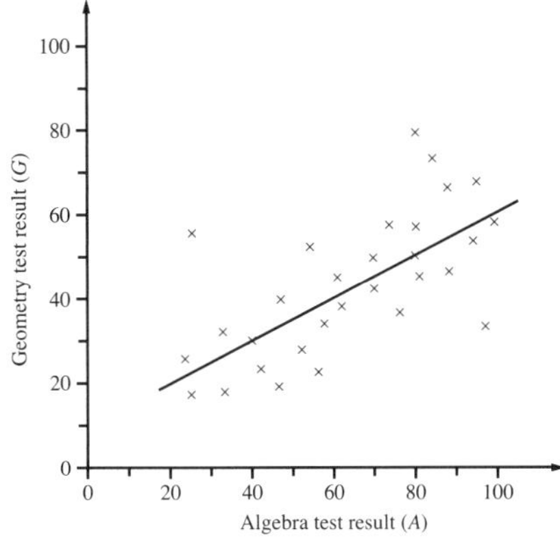

i How many students scored less than 30 on the algebra test? *(1 mark)* Easy

ii Calculate the gradient of the line of fit drawn. *(1 mark)* Easy

iii What is the equation of the line of fit drawn? *(2 marks)* Easy

iv Describe the correlation between geometry test results and algebra test results. *(1 mark)* Easy

v Mitchell looked at the scatterplot and said: 'In this class, all students who are near the top in algebra are also near the top in geometry'. Explain why his statement is incorrect. *(1 mark)* Medium

(Q26c, **2002 HSC**)

26 The 11 people in Sam's cricket team always bat in the same order. Sam recorded the batting order and the average number of runs scored by each player during the season.

Batting order	*Average number of runs*
1	16
2	10
3	11
4	8
5	7
6	4
7	4
8	5
9	3
10	1
11	1

i Display the data as a scatterplot on the graph paper [provided following this question]. Make sure that you have labelled the axes. *(2 marks)* Easy

ii Draw a line of fit on your scatterplot on graph paper. (No calculations are necessary.) *(1 mark)* Easy

iii Using your scatterplot, describe the correlation between the batting order and the average number of runs. *(1 mark)* Medium

(Q23a, **2001 HSC**)

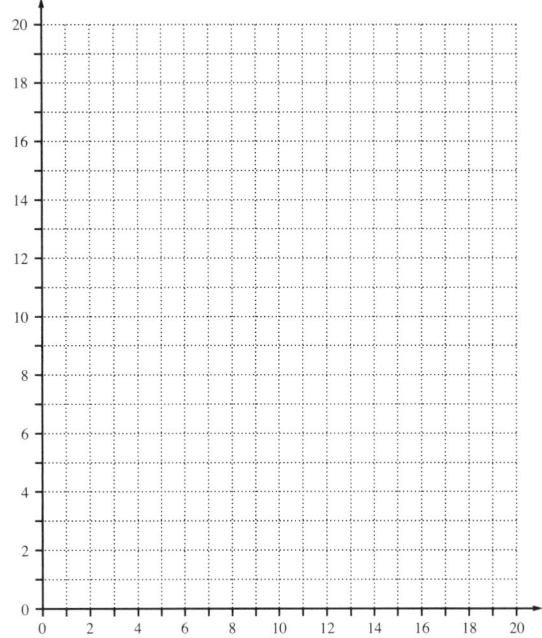

Year 12 Bivariate data analysis—Worked answers

1 **a** Vertical intercept = 2 ✓
The line passes through (0, 2) and (5, 18).
$\text{Gradient} = \frac{18-2}{5}$
$= 3.2$
The equation is $y = 3.2x + 2$. ✓
(2 marks)

b Assuming that a new least-squares regression line is drawn for the new data, the gradient will increase.
(1 mark)

2 **a** **i** $y = 29.2 - 0.011x$
When $x = 540$,
$y = 29.2 - 0.011 \times 540$
$= 23.26$
The average maximum daily temperature would be 23.3° to one decimal place. *(1 mark)*

ii For every 100 m increase in height above sea level, there is a 1.1° drop in average temperature. ✓✓
(2 marks)

b The latitude measurement would be better to predict the town's average temperature. Both correlation coefficients are negative but because the latitude correlation is closer to –1 it means there is a closer relationship between latitude and temperature than between sea level and temperature. *(1 mark)*

3 A correlation coefficient of –1 means perfect negative correlation.

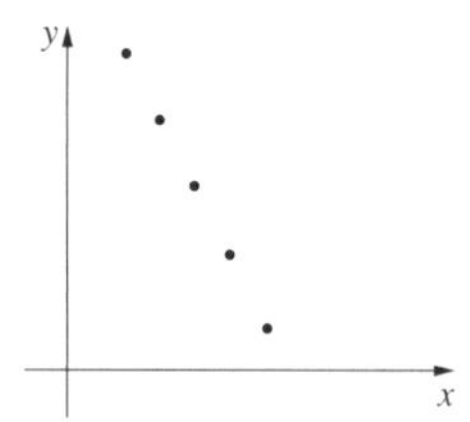

Answer D

4 $y = -10.6063 + bx$
Median temperature is 22° from the box-plot.
Mean temperature = 22° – 0.525°
So $\bar{x} = 21.475$ ✓
$\bar{y} = 684 \div 20$
$= 34.2$ ✓
The point (21.475, 34.2) lies on the regression line.
$34.2 = -10.6063 + b \times 21.475$ ✓
$44.8063 = 21.475b$
$b = 2.08644004\ldots$ ✓

When $x = 19$,
$y = -10.6063 + 2.08644004\ldots \times 19$
$= 29.0360608\ldots$
$= 29$ (nearest whole number)
The number of chirps expected when the temperature is 19 °C is 29 to the nearest whole number. ✓
(5 marks)

5 **a** $r = 0.98$ (2 decimal places; by calculator)
(1 mark)

b Very strong positive association
(1 mark)

c Arm span = 143 cm
Height = $0.866 \times 143 + 23.7$
$= 147.538$ ✓
So the predicted height is 148 cm to the nearest centimetre.
(1 mark)

6 From the calculator, $a = 84.698$ (3 dec. pl.) and $b = -0.984$ (2 dec. pl.). ✓
Using $y = bx + a$:
$y = -0.984x + 84.698$ ✓
(2 marks)

7 From the calculator, $a = 63.95$ (2 dec. pl.) and $b = 2.10$ (2 dec. pl.) ✓
Using $M = bn + a$:
$M = 2.10n + 63.95$ ✓
(2 marks)

8 **a** $r = 0.898186016\ldots$
$= 0.898$ (3 dec. pl.)
(1 mark)

b **i** Line passes through the points (11, 141) and (12, 145).
$\text{Gradient} = \frac{145-141}{12-11}$
$= 4$
(1 mark)

ii

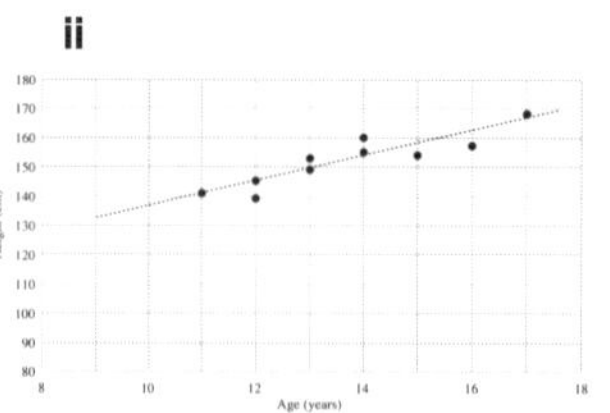

As h = gradient × n + vertical intercept, then $h = 4n + c$.
Substitute (11, 141):
$141 = 4(11) + c$
$c = 97$
$\therefore h = 4n + 97$ ✓✓
(2 marks)

c The girls involved in the survey are at an age where they have growth spurts. This does not continue after adolescence. The same growth rate would not be sustained for a 25-year-old and so the model is not relevant. *(1 mark)*

9 **a** $r = -0.899945038\ldots$ ✓
$= -0.90$ (2 dec. pl.) ✓
(2 marks)

b There is a strong negative correlation. The more students sleep, the fewer mistakes made. *(1 mark)*

c

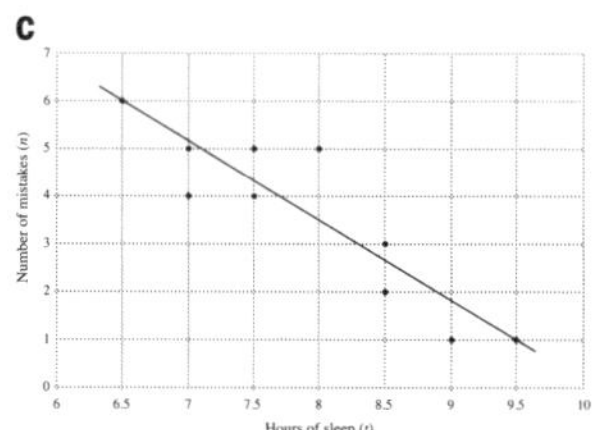

(1 mark)

10 Only graphs C and D show positive correlation.
The dots in C more closely form a straight line, so that graph has the largest positive correlation coefficient.

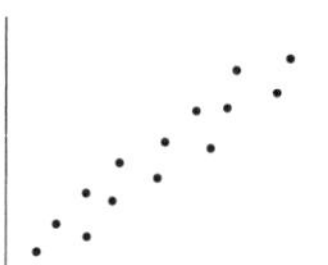

Answer C

11 [The correlation is negative, but weak, so the best approximation is –0.3.]

Answer B

12 **i** [One possible line of fit is shown.]

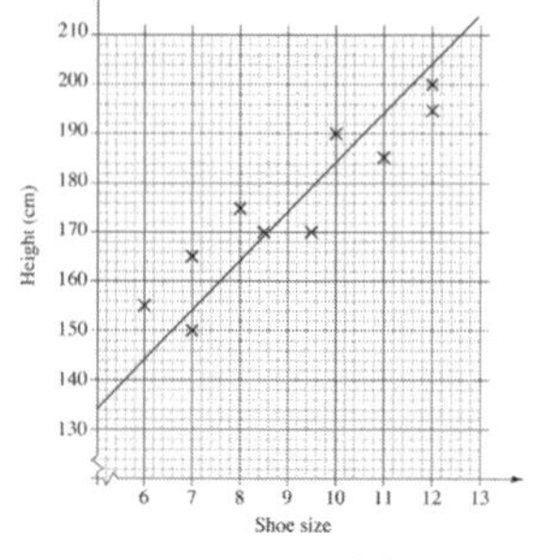

✓✓ *(2 marks)*

ii Using the line of fit, a student who wears size 7.5 shoes would be 159 cm tall and one who wears size 9 shoes would be 174 cm tall.

Difference in height
= (174 – 159) cm
= 15 cm

(1 mark)

iii A correlation coefficient of 1 means perfect correlation. The points would all need to form a straight line. As they do not do this, the coefficient cannot be correct.

(1 mark)

13 **i** There is a strong positive correlation between expenditure per primary school student and life expectancy. In general, the more that is spent on students in primary school the greater the life expectancy.

(1 mark)

ii $IQR = Q_U - Q_L$
$= 22.5 - 8.4$
$= 14.1$

(1 mark)

iii $Q_U + 1.5 \times IQR$
$= 22.5 + 1.5 \times 14.1$
$= 43.65$ ✓

Now 47.6 > 43.65

So this country would be an outlier for this set of data. ✓

(2 marks)

vi

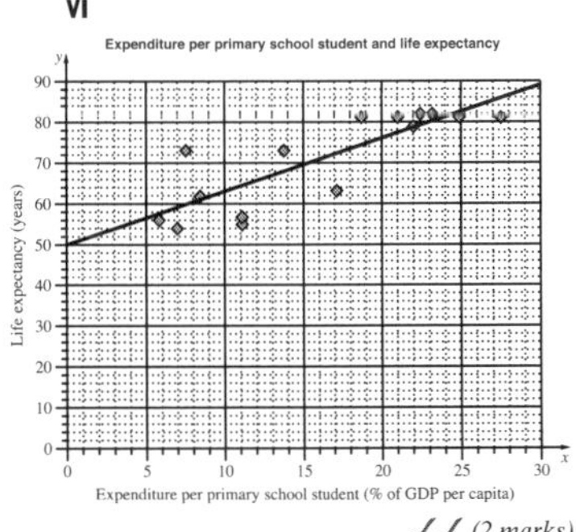

✓✓ *(2 marks)*

vii Using the line, the estimated life expectancy is about 73 years.
[Or using the equation:

When $x = 18$,
$y = 1.29 \times 18 + 49.9$
$= 73.12$
≈ 73 years]

(1 mark)

viii 60% of GDP is outside the limits of this data set and predictions should not be made beyond the bounds of the data set. [With such a high expenditure on education, other influences on life expectancy such as health spending might be neglected.]

(1 mark)

14 Correlation of 0.3 means that there is a weak positive relationship. This is shown in option A.

Answer A

15 **i** $m = \dfrac{\text{vertical change in position}}{\text{horizontal change in position}}$
$= \dfrac{6}{1}$
$= 6$

(1 mark)

ii The gradient gives the number of centimetres that a male grows per year according to the data.

(1 mark)

iii The equation is of the form $h = ma + b$

But, $m = 6$

So $h = 6a + b$ ✓

When $a = 15, h = 170$

$170 = 6 \times 15 + b$
$= 90 + b$
$b = 170 - 90$
$= 80$ ✓

So the equation is $h = 6a + 80$

(2 marks)

iv When $a = 17$,
$h = 6 \times 17 + 80$
$= 182$

The height of a typical 17-year-old male would be 182 cm.

(1 mark)

v Once reaching a certain age boys stop growing. Assuming that males continue to grow at the same rate is wrong. Using a model based on this wrong assumption would not be useful.

(1 mark)

16 For a relationship to show negative correlation one part should increase as the other decreases. This happens in option B. As the hours spent training increase, the time to complete the race should decrease. This relationship is most likely to show negative correlation.

Answer B

17 **i** Using the graph, for an eruption of 4 minutes, the minimum time to wait for the next eruption is 70 minutes.

(1 mark)

ii For eruptions one hour apart, the longest possible duration of the first eruption, based on the data, is 3 minutes.

(1 mark)

iii The graph suggests a strong positive relationship between the duration of an eruption and the length of time until the next eruption. The longer the duration of the first eruption, the longer the wait until the next eruption.

(1 mark)

18 A correlation coefficient of –0.9 means there is a very strong negative association between the two variables. This is shown in [D].

Answer D

19 More hours spent studying could cause better test results, and poor results could be caused by lack of time spent studying.

Answer D

20 **i** As the height increases, the mass increases. There is a strong positive correlation between the height and the mass.

(1 mark)

ii $y = mx + b$

m is the gradient

Using the points (60, 5.8) and (40, 1.2),

$m = \dfrac{5.8 - 1.2}{60 - 40}$ ✓
$= \dfrac{4.6}{20}$
$= 0.23$

$\therefore M = 0.23H + C$

using (40, 1.2)
$1.2 = 0.23(40) + C$
$C = -8$
i.e. $M = 0.23H - 8$ ✓

(2 marks)

21 If a line was drawn through the points, its gradient would be positive.

Answer A

22 As the population of a town increases, so will the number of schools, indicating a positive correlation.

Answer A

23 **i** The foot length of a person with zero height is meaningless. (Also note that the scale on the y-axis does not start at zero.)

(1 mark)

ii If Harry is 160 cm, his foot length is 19 cm. George is 170 cm and his foot length is approximately 22 cm from the graph.

$\therefore$ Difference in foot lengths is approximately 3 cm.

(1 mark)

iii
- The correlation shown is positive (positive slope). ✓
- The correlation coefficient is always between –1 and 1, $\therefore$ it cannot be –1.2. ✓

(2 marks)

24 Answer D

25 **i** 3 students (If a vertical line is drawn through 30 on the algebra axis, only 3 scores are to the left of it.)

(1 mark)

ii Using the points (40, 30) and (80, 50) on the line,

$$\text{gradient} = \frac{50-30}{80-40} = \frac{20}{40} = \frac{1}{2}$$

(1 mark)

iii $y = mx + b$

where m is the gradient and b is where the line crosses the vertical axis.

$\therefore G = \frac{1}{2}A + 10$ ✓✓

(2 marks)

iv There is a strong positive correlation between algebra and geometry. (However, not a correlation of 1.)

(1 mark)

v Of the top 11 algebra students, only 5 were also in the top 11 of geometry, therefore his statement is incorrect.

(1 mark)

26 **i** **ii**

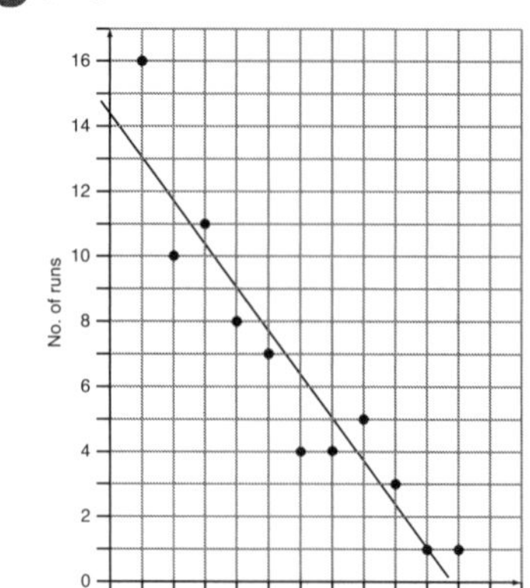

✓✓✓

(3 marks)

iii Since the correlation is negative (about –1), then the batting order and number of runs are inversely related, i.e. the earlier you bat, the more runs you score.

(1 mark)

1 On a test, Zac's mark corresponded to a z-score of 2. The test scores had a mean of 63 and a standard deviation of 8.

What was Zac's actual mark on the test?

A 65 **B** 67
C 73 **D** 79 *(1 mark)*

(Q8, **2021 HSC**) Medium

2 A random variable is normally distributed with mean 0 and standard deviation 1. The table gives the probability that this random variable lies between 0 and z for different values of z.

z	0.1	0.2	0.3	0.4	0.5	0.6
Probability	0.0398	0.0793	0.1179	0.1554	0.1915	0.2257

The probability values given in the table for different values of z are represented by the shaded area in the following diagram.

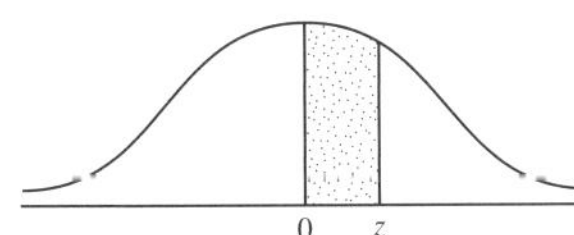

a Using the table, show that the probability that a value from a random variable that is normally distributed with mean 0 and standard deviation 1 is greater than 0.3 is equal to 0.3821. *(1 mark)* Hard

b Birth weights are normally distributed with a mean of 3300 grams and a standard deviation of 570 grams. By first calculating a z-score, find how many babies, out of 1000 born, are expected to have a birth weight greater than 3471 grams. *(3 marks)*

CQ (Q38, **2021 HSC**) Hard

3 In a particular city, the heights of adult females and the heights of adult males are each normally distributed.

Information relating to two females from that city is given in Table 1.

Table 1

Height	*Gender*	*Percentage of females in this city shorter than this person*
175 cm	Female	97.5%
160.6 cm	Female	16%

The means and standard deviations of adult females and males, in centimetres, are given in Table 2.

Table 2

	Mean	*Standard deviation*
Females	μ	σ
Males	1.05μ	1.1σ

A selected male is taller than 84% of the population of adult males in this city.

By first labelling the normal distribution curve below with the heights of the two females given in Table 1, calculate the height of the selected male, in centimetres, correct to two decimal places.

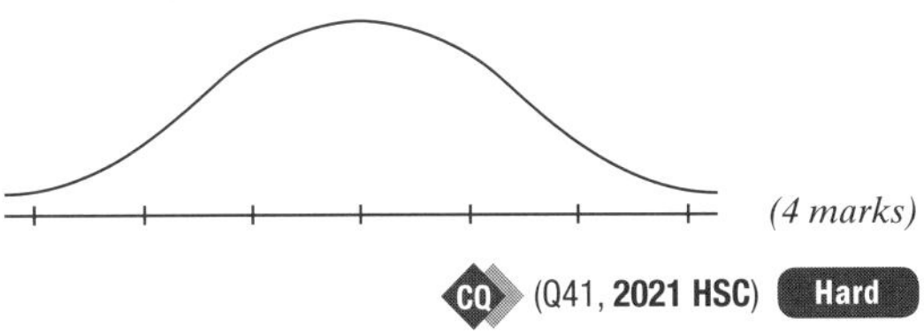

(4 marks)

CQ (Q41, **2021 HSC**) Hard

4 John recently did a class test in each of three subjects. The class scores on each test were normally distributed.

The table shows the subjects and John's scores as well as the mean and standard deviation of the class scores on each test.

Subject	*John's score*	*Mean*	*Standard deviation*
French	82	70	8
Commerce	80	65	5
Music	74	50	12

Relative to the rest of the class, which row of the table below shows John's strongest subject and his weakest subject?

	Strongest subject	*Weakest subject*
A	Commerce	French
B	French	Music
C	Music	French
D	Commerce	Music

(1 mark)

(Q8, **2020 HSC**)

5 The Intelligence Quotient (IQ) scores for adults in City A are normally distributed with a mean of 108 and a standard deviation of 10. The IQ scores for adults in City B are normally distributed with a mean of 112 and a standard deviation of 16.

a Yin is an adult who lives in City A and has an IQ score of 128. What percentage of the adults in this city have an IQ score higher than Yin's? *(2 marks)* Medium

b There are 1 000 000 adults living in City B. Calculate the number of adults in City B that would be expected to have an IQ score lower than Yin's. *(2 marks)* Medium

c Simon, an adult who lives in City A, moves to City B. The z-score corresponding to his IQ score in City A is the same as the z-score corresponding to his IQ score in City B. By first forming an equation, calculate Simon's IQ score.
Give your answer correct to one decimal place. *(3 marks)* **Hard**

(Q35, **2020 HSC**)

---

6 The scores on an examination are normally distributed with a mean of 70 and a standard deviation of 6. Michael received a score on the examination between the lower quartile and the upper quartile of the scores.
Which shaded region most accurately represents where Michael's score lies?

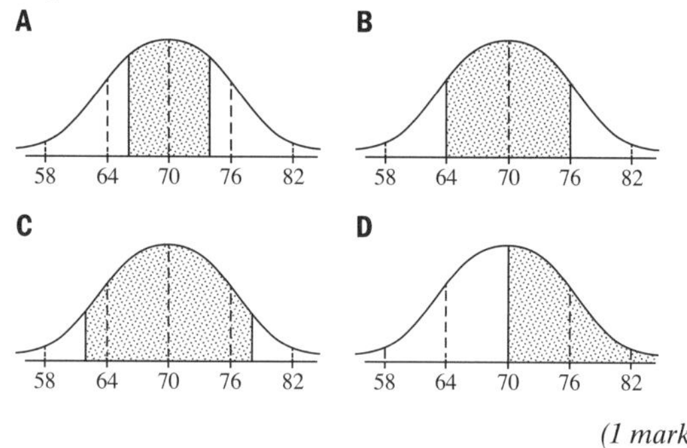

(1 mark)

(Q15, **2019 HSC**) **Hard**

---

7 In a particular country, the birth weight of babies is normally distributed with a mean of 3000 grams. It is known that 95% of these babies have a birth weight between 1600 grams and 4400 grams.

One of these babies has a birth weight of 3497 grams. What is the z-score of this baby's birth weight? *(2 marks)*

(Q38, **2019 HSC**) **Medium**

---

8 A set of data is normally distributed with a mean of 48 and a standard deviation of 3. Approximately what percentage of the scores lies between 39 and 45?

A 15.85% **B** 31.7%
C 47.5% **D** 49.85% *(1 mark)*

(Q23, **2018 HSC**) **Medium**

---

9 Joanna sits a Physics test and a Biology test.

i Joanna's mark in the Physics test is 70. The mean mark for this test is 58 and the standard deviation is 8.
Calculate the z-score for Joanna's mark in this test. *(1 mark)* **Easy**

ii In the Biology test, the mean mark is 64 and the standard deviation is 10. Joanna's z-score is the same in both the Physics test and the Biology test. What is her mark in the Biology test? *(2 marks)* **Easy**

(Q27e, **2018 HSC**)

---

10 The heights of Year 12 girls are normally distributed with a mean of 165 cm and a standard deviation of 5.5 cm.

What is the z-score for a height of 154 cm?

A −2 **B** −0.5
C 0.5 **D** 2 *(1 mark)*

(Q13, **2017 HSC**) **Easy**

---

11 All the students in a class of 30 did a test.

The marks, out of 10, are shown in the dot plot.

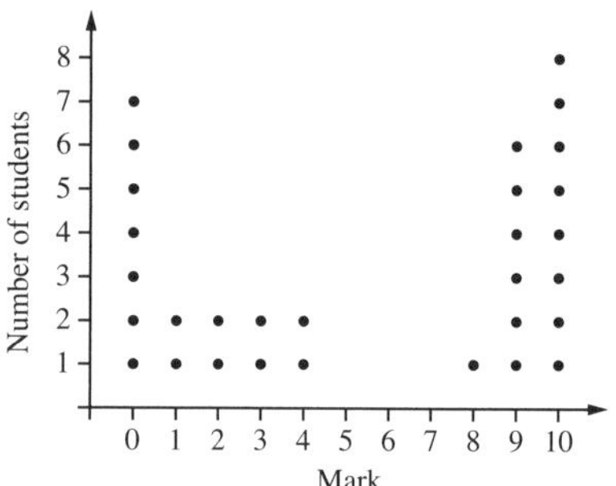

i Find the median test mark. *(1 mark)* **Medium**

ii The mean test mark is 5.4. The standard deviation of the test marks is 4.22.
Using the dot plot, calculate the percentage of the marks which lie within one standard deviation of the mean. *(2 marks)* **Hard**

iii A student states that for any data set, 68% of the scores should lie within one standard deviation of the mean.
With reference to the dot plot, explain why the student's statement is NOT relevant in this context. *(1 mark)* **Hard**

(Q29d, **2017 HSC**)

---

12 The speed limit outside a school is 40 km/h. Year 11 students measured the speed of passing vehicles over a period of time. They found the set of data to be normally distributed with a mean speed of 36 km/h and a standard deviation of 2 km/h.

What percentage of the vehicles passed the school at a speed greater than 40 km/h?

A 2.5% **B** 5%
C 47.5% **D** 95% *(1 mark)*

(Q13, **2016 HSC**) 

---

13 A machine produces cylindrical pipes. The mean of the diameters of the pipes is 8 cm and the standard deviation is 0.04 cm.

Assuming a normal distribution, what percentage of cylindrical pipes produced will have a diameter less than 7.96 cm?

A 16% **B** 32%
C 34% **D** 68% *(1 mark)*

(Q20, **2015 HSC**) Medium

14 The results of two tests are normally distributed. The mean and standard deviation for each test are displayed in the table.

	Mathematics	English
$\bar{x}$	70	75
s	6.5	8

Kristoff scored 74 in Mathematics and 80 in English. He claims that he has performed better in English.

Is Kristoff correct? Justify your answer using appropriate calculations. *(2 marks)*

(Q28b, **2015 HSC**) Medium

15 The weights of 10000 newborn babies in NSW are normally distributed. These weights have a mean of 3.1 kg and a standard deviation of 0.35 kg.

How many of these newborn babies have a weight between 2.75 kg and 4.15 kg?

A 4985 **B** 6570
C 8370 **D** 8385 *(1 mark)*

(Q24, **2014 HSC**) Medium

16 There are 60000 students sitting a state-wide examination. If the results form a normal distribution, how many students would be expected to score a result between 1 and 2 standard deviations above the mean?

You may assume for normally distributed data that:

- 68% of scores have z-scores between minus –1 and 1
- 95% of scores have z-scores between minus –2 and 2
- 99.7% of scores have z-scores between minus –3 and 3.

A 8100 **B** 16200
C 20400 **D** 28500 *(1 mark)*

(Q20, **2013 HSC**) Hard

17 A machine produces nails. When the machine is set correctly, the lengths of the nails are normally distributed with a mean of 6.000 cm and a standard deviation of 0.040 cm.

To confirm the setting of the machine, three nails are randomly selected. In one sample the lengths are 5.950, 5.983 and 6.140.

The setting of the machine needs to be checked when the lengths of two or more nails in a sample lie more than 1 standard deviation from the mean.

Does the setting on the machine need to be checked? Justify your answer with suitable calculations. *(2 marks)*

(Q29b, **2012 HSC**) Medium

18 Two brands of light bulbs are being compared. For each brand, the life of the light bulbs is normally distributed.

Life of light bulbs (in hours)

	Mean	*Standard deviation*
Brand A	450	25
Brand B	500	50

i One of the Brand B light bulbs has a life of 400 hours.
What is the z-score of the life of this light bulb? *(1 mark)* Easy

ii A light bulb is considered defective if it lasts less than 400 hours. The following claim is made:

'Brand A light bulbs are more likely to be defective than Brand B light bulbs.'

Is this claim correct? Justify your answer, with reference to z-scores or standard deviations or the normal distribution. *(2 marks)* Medium

(Q27c, **2011 HSC**)

19 Which of the following frequency histograms shows data that could be normally distributed?

A

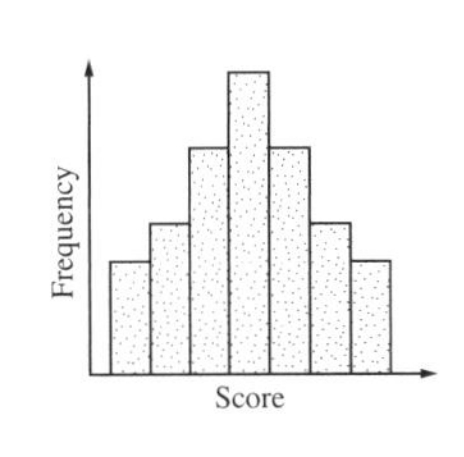

B

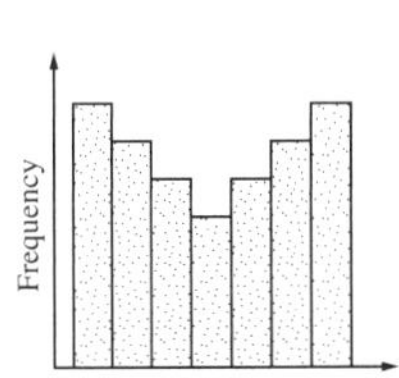

C

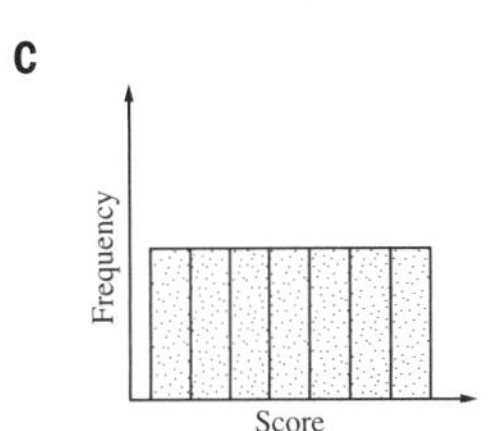

D

(1 mark)

(Q4, **2010 HSC**) Easy

20 The marks in a class test are normally distributed. The mean is 100 and the standard deviation is 10.

i Jason's mark is 115. What is his z-score? *(1 mark)* Easy

ii Mary has a z-score of 0. What mark did she achieve in the test? *(1 mark)* Easy

iii What percentage of marks lie between 80 and 110?

You may assume the following:

- 68% of marks have z-scores between –1 and 1
- 95% of marks have z-scores between –2 and 2
- 99.7% of marks have z-scores between –3 and 3. *(2 marks)* Medium

(Q24c, **2010 HSC**)

21 In Broken Hill, the maximum temperature for each day has been recorded. The mean of these maximum temperatures during spring is 25.8°C, and their standard deviation is 4.2°C.

i What temperature has a z-score of –1? *(1 mark)* Easy

ii What percentage of spring days in Broken Hill would have maximum temperatures between 21.6°C and 38.4°C?

You may assume that these maximum temperatures are normally distributed and that

- 68% of maximum temperatures have z-scores between –1 and 1
- 95% of maximum temperatures have z-scores between –2 and 2
- 99.7% of maximum temperatures have z-scores between –3 and 3. *(3 marks)* Medium

(Q25d, **2009 HSC**)

22 The following graph indicates z-scores of 'height-for-age' for girls aged 5–19 years.

i What is the z-score for a six year old girl of height 120 cm? *(1 mark)* Easy

ii Rachel is $10\frac{1}{2}$ years of age.

1 If 2.5% of girls of the same age are taller than Rachel, how tall is she? *(1 mark)* Medium

2 Rachel does not grow any taller. At age $15\frac{1}{2}$, what percentage of girls of the same age will be taller than Rachel? *(2 marks)* Medium

iii What is the average height of an 18 year old girl? *(1 mark)* Medium

iv For adults (18 years and older), the Body Mass Index is given by $b = \frac{m}{h^2}$ where

m = mass in kilograms and h = height in metres.

The medically accepted healthy range for B is $21 \le B \le 25$.

What is the minimum weight for an 18 year old girl of average height to be considered healthy? *(2 marks)* Medium

(Q28a, **2008 HSC**)

23 The results of two class tests are normally distributed. The means and standard deviations of the tests are displayed in the table.

	Test 1	*Test 2*
Mean	60	58
Standard deviation	6.2	6.0

i Stuart scored 63 in Test 1 and 62 in Test 2. He thinks that he has performed better in Test 1. Do you agree? Justify your answer using appropriate calculations. *(2 marks)* Medium

ii If 150 students sat for Test 2, how many students would you expect to have scored less than 64? *(2 marks)* Hard

(Q25d, **2007 HSC**)

24 A clubhouse uses four long-life light globes for five hours every night of the year. The purchase price of each light globe is \$6.00 and they each cost \$$d$ per hour to run.

The manufacturer's specifications state that the expected life of the light globes is normally distributed with a standard deviation of 170 hours. What is the mean life, in hours, of these light globes if 97.5% will last up to 5000 hours? *(1 mark)*

(Q27b iv, **2007 HSC**) Medium

25 In a normally distributed set of scores, the mean is 23 and the standard deviation is 5.

Approximately what percentage of the scores will lie between 18 and 33?

A 34% **B** 47.5%
C 68% **D** 81.5% *(1 mark)*

(Q17, **2006 HSC**) Medium

26 The weights of boxes of Brekky Bicks are normally distributed. The mean is 754 grams and the standard deviation is 2 grams.

i What is the z-score of a box of Brekky Bicks with a weight of 754 g? *(1 mark)* Easy

ii What is the weight of a box that has a z-score of –1? *(1 mark)* Medium

iii Brekky Bicks boxes are labelled as having a weight of 750 g.
What percentage of boxes will have a weight less than 750 g? *(2 marks)* Medium

(Q26c, **2005 HSC**)

27 The normal distribution shown has a mean of 170 and a standard deviation of 10.

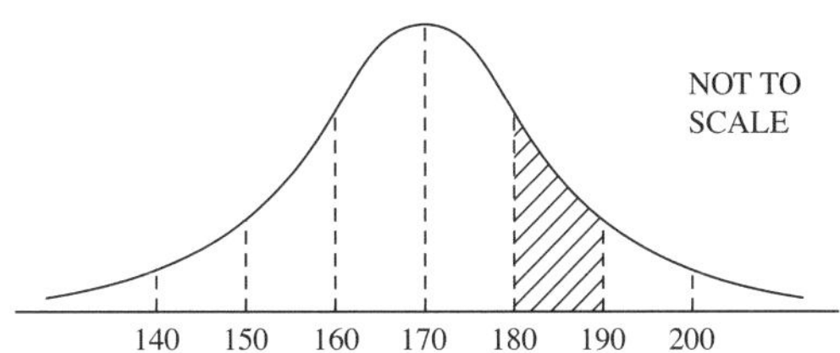

i Roberto has a raw score in the shaded region. What could his z-score be? *(1 mark)* Medium

ii What percentage of the data lies in the shaded region? *(2 marks)* Medium

(Q24c, **2004 HSC**)

28 Results for an aptitude test are given as z-scores. In this test, Hardev gains a z-score of 1.

i Interpret Hardev's score with reference to the mean and standard deviation of the test. *(2 marks)* Medium

zii The scores for the test are normally distributed. What proportion of people sitting the test obtain a higher score than Hardev? *(1 mark)* Medium

(Q25c, **2003 HSC**)

29 Results for an aptitude test are given as z-scores. In this test Di gained a z-score of 3. The test has a mean of 55 and a standard deviation of 6.

What was Di's actual mark in this test?

A 57 **B** 58
C 64 **D** 73 *(1 mark)*

(Q8, **2002 HSC**) Medium

30 In the town of Burrow the ages of the residents are normally distributed. The mean age is 40 years and the standard deviation is 12 years.

Approximately what percentage of the residents are younger than 52?

A 16% **B** 32%
C 68% **D** 84% *(1 mark)*

(Q12, **2002 HSC**) Medium

31 A factory produces bags of flour. The weights of the bags are normally distributed, with a mean of 900 g and a standard deviation of 50 g.

What is the best approximation for the percentage of bags that weigh more than 1000 g?

A 0% **B** 2.5%
C 5% **D** 16% *(1 mark)*

(Q19, **2001 HSC**) Medium

32 Results for a reading test are given as z-scores. In this test, Kim gained a z-score equal to –2.

i Interpret this z-score in terms of the mean and standard deviation of the test. *(2 marks)* Easy

ii If the test has a mean of 75 and a standard deviation of 5, calculate the actual mark scored by Kim. *(1 mark)* Medium

(Q23b, **2001 HSC**)

Year 12 The normal distribution—Worked answers

Since 2020 the formula provided on the Formula Sheet to calculate z-scores is $z = \frac{x - \mu}{\sigma}$. This formula has been used in all solutions in this topic.

1 Zac's mark was two standard deviations above the mean.

Zac's mark $= 63 + 2 \times 8$
$= 79$

Answer D

2 **a** From the table,

$P(0 < z < 0.3) = 0.1179$

Now $P(z > 0) = 0.5$

So $P(z > 0.3) = 0.5 - 0.1179$
$= 0.3821$

(1 mark)

b $\mu = 3300$, $\sigma = 570$,
$x = 3471$

$z = \frac{x - \mu}{\sigma}$

$= \frac{3471 - 3300}{570}$

$= 0.3$ ✓

$P(\text{birth weight} > 3471 \text{ g})$
$= 0.3821$ (from **a**)

Expected number in 1000 births
$= 0.3821 \times 1000$
$= 382.1$ ✓

So, out of 1000 babies born, 382 would be expected to have a birth weight greater than 3471 g. ✓ *(3 marks)*

3 68% of scores have z-scores between –1 and 1.

34% have z-scores between –1 and 0.

(50 – 34)% = 16% will have z-scores less than –1.

So a height of 160.6 cm for a female corresponds to a z-score of –1.

95% of scores have z-scores between –2 and 2.

2.5% have z-scores less than –2, meaning 97.5% have z-scores less than 2.

So a height of 175 cm for a female corresponds to a z-score of 2.

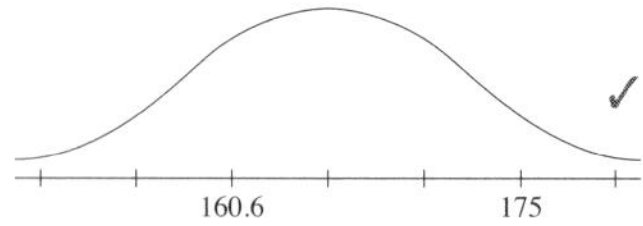

✓

There are three standard deviations between $z = -1$ and $z = 2$.

$3\sigma = 175 - 160.6$
$= 14.4$

$\sigma = 4.8$ ✓

$\mu = 160.6 + 4.8$
$= 165.4$

Now, $1.05\mu = 1.05 \times 165.4$
$= 173.67$

$1.1\sigma = 1.1 \times 4.8$
$= 5.28$

So the mean for males is 173.67 and the standard deviation is 5.28. ✓

84% of scores are less than a z-score of 1.

So the height of a male who is taller than 84% of the population corresponds to a z-score of 1.

Height = (173.67 + 5.28) cm
= 178.95 cm ✓

(4 marks)

4 French: $z = \frac{82 - 70}{8} = 1.5$

Commerce: $z = \frac{80 - 65}{5} = 3$

Music: $z = \frac{74 - 50}{12} = 2$

So Commerce is strongest and French weakest.

Answer A

5 **a** City A: $x = 128$, $\mu = 108$, $\sigma = 10$

$z = \frac{x - \mu}{\sigma}$

$= \frac{128 - 108}{10}$

$= 2$

Now 95% of scores are within two standard deviations of the mean.

So 5% are more than two standard deviations from the mean. ✓

Half of these are above two standard deviations.

So 2.5% of adults will have an IQ higher than Yin's. ✓

(2 marks)

b City B: $x = 128$, $\mu = 112$, $\sigma = 16$

$z = \frac{x - \mu}{\sigma}$

$= \frac{128 - 112}{16}$

$= 1$ ✓

Now 68% of scores are within one standard deviation of the mean.

Half of these, 34%, are between the mean and one standard deviation above the mean.

50% are below the mean.

So 84% of adults will have an IQ lower than Yin's.

Number of adults = 84% of 1 000 000

840 000 adults would be expected to have an IQ lower than Yin's. ✓

(2 marks)

c Let x be Simon's IQ.

Now $\frac{x - 108}{10} = \frac{x - 112}{16}$ ✓✓

$16(x - 108) = 10(x - 112)$

$16x - 1728 = 10x - 1120$

$6x = 608$

$x = 101.333\,333\ldots$
$= 101.3$ (1 d.p.)

So Simon's IQ would be 101.3 to one decimal place. ✓

(3 marks)

6 $\mu = 70$, $\sigma = 6$

So scores between 64 and 76 are within 1 standard deviation of the mean.

Now 68% of scores are within 1 standard deviation of the mean.

50% lie between the upper and lower quartiles.

So the shaded region can only be that in option A

Answer A

7 $\mu = 3000$ g

In normally distributed data 95% are within two standard deviations of the mean.

But 95% of these babies lie between 1600 g and 4400 g.

So two standard deviations is 1400 g. ✓

$\therefore \sigma = 700$ g

$z = \frac{x - \mu}{\sigma}$

$= \frac{3497 - 3000}{700}$

$= 0.71$ ✓

(2 marks)

8 $\mu = 48$ $\sigma = 3$

So 45 is one and 39 is three standard deviations below the mean.

Now 99.7% of scores lie within three standard deviations of the mean.

68% of scores lie within one standard deviation of the mean.

$\therefore$ 99.7% – 68% or 31.7% lie between one and three standard deviations.

Half of these are below the mean.

Required percentage
$= 0.5 \times 31.7\%$
$= 15.85\%$
Answer A

9 **i** Physics: $x = 70$, $\mu = 58$, $\sigma = 8$

$$z = \frac{x-\mu}{\sigma} = \frac{70-58}{8} = 1.5$$

(1 mark)

ii Biology: $z = 1.5$, $\mu = 64$, $\sigma = 10$

$$z = \frac{x-\mu}{\sigma}$$
$$1.5 = \frac{x-64}{10} \quad ✓$$
$$15 = x - 64$$
$$x = 79$$

Joanna's mark in Biology is 79. ✓

(2 marks)

10 $x = 154$, $\mu = 165$, $\sigma = 5.5$

$$z = \frac{x-\mu}{\sigma} = \frac{154-165}{5.5} = -2$$

Answer A

11 **i** $$\text{median} = \frac{4+8}{2} = 6$$

(1 mark)

ii $\mu = 5.4$, $\sigma = 4.22$
$\mu + \sigma = 5.4 + 4.22$
$= 9.62$
$\mu - \sigma = 5.4 - 4.22$
$= 1.18$ ✓

So scores within one standard deviation of the mean are scores greater than 1.18 but less than 9.62.

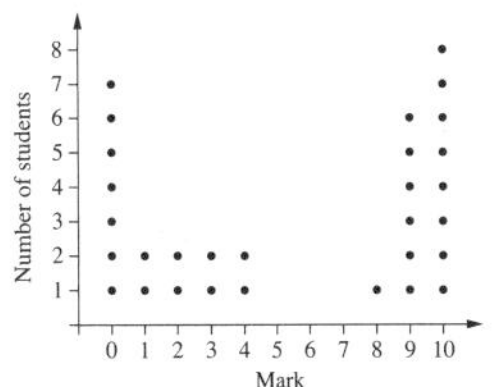

From the dot plot there are 13 such scores.

% within 1 standard deviation of the mean $= \frac{13}{30} \times 100\%$

$= 43\frac{1}{3}\%$ ✓

(2 marks)

iii 68% of scores lie within one standard deviation of the mean if the data is normally distributed. Normally distributed data has a bell-shaped graph. From the dot plot it is obvious that the distribution is not bell-shaped and the data is not normally distributed.

(1 mark)

12 40 km/h = 36 km/h + 2 × 2 km/h
So it is 2 standard deviations above the mean.
95% lie within 2 standard deviations of the mean so 5% lie outside that. Half of those are above the mean.
So 2.5% of the vehicles will have a speed greater than 40 km/h.
Answer A

13 $8 - 0.04 = 7.96$
So a diameter of 7.96 cm is one standard deviation below the mean.
Now 68% of scores are within one standard deviation of the mean.
Percentage more than one standard deviation from the mean = $(100 - 68)\%$
or 32%
Half of those will be more than one standard deviation below the mean.
So the percentage with a diameter less than 7.96 cm is $(32 \div 2)\%$ or 16%.
Answer A

14 Mathematics: $$z = \frac{x-\mu}{\sigma} = \frac{74-70}{6.5} = 0.615\,3846\ldots \quad ✓$$

English: $$z = \frac{x-\mu}{\sigma} = \frac{80-75}{8} = 0.625$$

So, yes Kristoff is correct. The z-score is slightly higher for English than for Mathematics so he performed slightly better in English. ✓

(2 marks)

15 $3.1 - 2.75 = 0.35$
So a weight of 2.75 kg is 1 standard deviation below the mean.
Half of 68%, or 34%, of babies will have weights between 2.75 kg and 3.1 kg.
$4.15 - 3.1 = 1.05$
$1.05 \div 0.35 = 3$
So a weight of 4.15 kg is 3 standard deviations above the mean.
Half of 99.7%, or 49.85%, of babies will have weights between 3.1 kg and 4.15 kg.
Now $34\% + 49.85\% = 83.85\%$
83.85% of 10 000 = 8385
Answer D

16 68% of scores have z-scores between –1 and 1.
Half of these, 34%, will have z-scores between 0 and 1.
95% have z-scores between –2 and 2.
So 47.5% will have z-scores between 0 and 2.
47.5% – 34% or 13.5% have z-scores between 1 and 2.
13.5% of 60 000 = 8100
So 8100 students would be expected to score a result between 1 and 2 standard deviations above the mean.
Answer A

17 $\mu = 6.000$ $\sigma = 0.040$
$\mu + \sigma = 6.000 + 0.040$
$= 6.040$
$\mu - \sigma = 6.000 - 0.040$
$= 5.960$ ✓

So the lengths of two or more of the 3 nails should be between 5.960 cm and 6.040 cm.
Now $5.950 < 5.960$
and $6.140 > 6.040$
So the lengths of two nails are more than one standard deviation from the mean.
Yes, the machine does need to be checked. ✓

(2 marks)

18 **i** $x = 400$, $\mu = 500$, $\sigma = 50$

$$z = \frac{x-\mu}{\sigma} = \frac{400-500}{50} = -2$$

(1 mark)

ii Brand A: $x = 400$, $\mu = 450$, $\sigma = 25$

$$z = \frac{x-\mu}{\sigma} = \frac{400-450}{25} = -2 \quad ✓$$

The z-scores for light bulbs with a life of 400 hours is the same for both brands. This means the same percentage of light bulbs are likely to be defective. The claim is not correct. ✓

(2 marks)

19 Normally distributed data has a bell-shaped curve.

Answer A

20 $\mu = 100,\ \sigma = 10$

i $x = 115$

$$z = \frac{x - \mu}{\sigma} = \frac{115 - 100}{10} = 1.5$$

Jason's z-score is 1.5.

(1 mark)

ii A z-score of 0 represents the mean. So Mary scored 100 in the test.

(1 mark)

iii 90 is 1 standard deviation below the mean and 110 is 1 standard deviation above the mean.

So 68% of marks are between 90 and 110.

80 is 2 standard deviations below the mean.

Percentage of marks with z-scores between 1 and 2 or between –1 and –2

= 95% – 68%

= 27% ✓

Half of these have z-scores between –1 and –2.

Percentage of marks between 80 and 90

= 27% ÷ 2

= 13.5%

Total percentage of marks between 80 and 110

= 13.5% + 68%

= 81.5% ✓ *(2 marks)*

21 **i** $z = \frac{x - \mu}{\sigma}$

$\therefore -1 = \frac{x - 25.8}{4.2}$

$-4.2 = x - 25.8$

$x = -4.2 + 25.8$

$= 21.6°C$

OR

Temperature with a z-score of –1 is one standard deviation below the mean, i.e. 25.8 – 4.2 = 21.6°C.

(1 mark)

ii

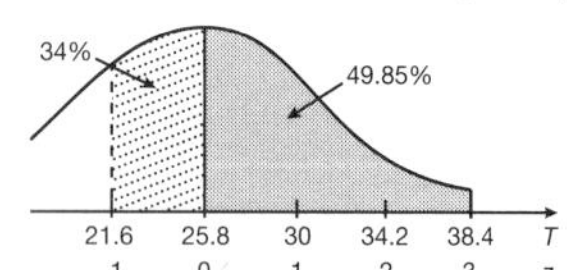

Shaded area from 25.8°C to 38.4°C is for z-scores between 0 and 3. ✓

This is 99.7% ÷ 2 = 49.85% of scores.

Dotted area from 21.6°C to 25.8°C is for z-scores between 0 and –1.

This is 68% ÷ 2 = 34% of scores. ✓

∴ Percentage of days with maximum temperatures between 21.6°C and 38.4°C = 49.85% + 34% = 83.85%. ✓

(3 marks)

22 **i** 1

(1 mark)

ii **1**

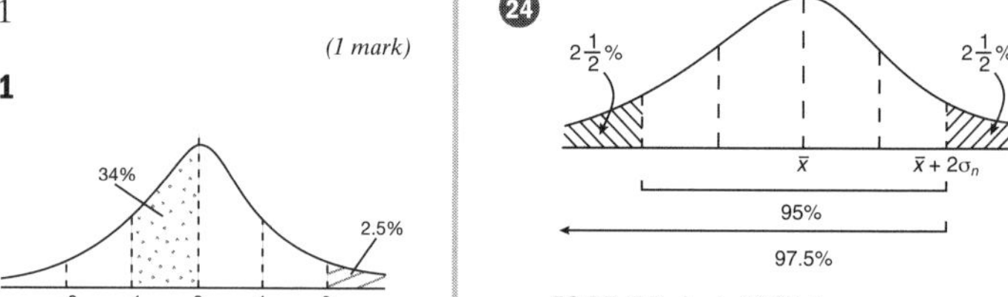

If 2.5% of girls are taller than Rachel, the corresponding z-score is 2.

From the graph, this is a height of 155 cm. Hence, Rachel is 155 cm tall.

(1 mark)

2 At age $15\frac{1}{2}$, 155 cm is a z-score of –1. ✓

∴ From a z-score of 0 to –1, the percentage of girls is 34%.

∴ (34 + 50)% = 84% of girls are taller than Rachel. ✓

(2 marks)

iii Average height is a z score of 0.

∴ Average height is 163 cm.

(1 mark)

iv $B = \frac{m}{h^2}$

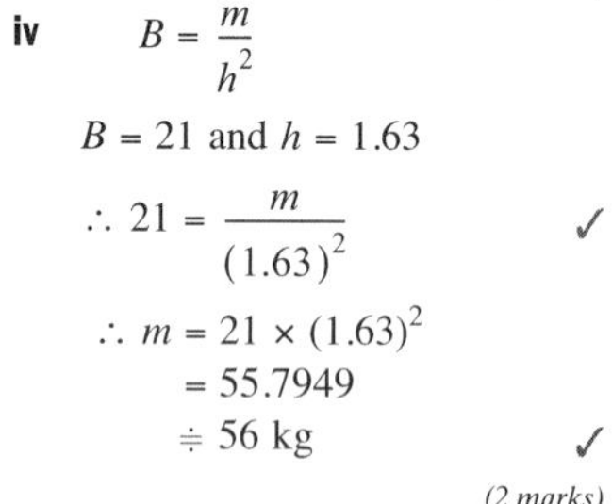

$B = 21$ and $h = 1.63$

$\therefore 21 = \frac{m}{(1.63)^2}$ ✓

$\therefore m = 21 \times (1.63)^2$

$= 55.7949$

$\doteqdot 56$ kg ✓

(2 marks)

23 **i** $z = \frac{x - \mu}{\sigma}$

Test 1: $z = \frac{63 - 60}{6.2} = 0.4838\ldots$ ✓

Test 2: $z = \frac{62 - 58}{6} = 0.6666\ldots$

Stuart performed better in Test 2 as his z-score is higher. ✓

(2 marks)

ii

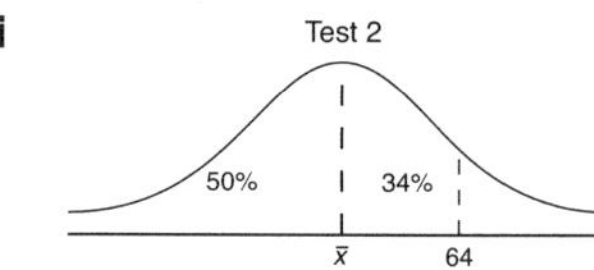

You would expect 84% of students to score less than 64. ✓

i.e. $\frac{84}{100} \times 150 = 126$ students. ✓

(2 marks)

24

If 97.5% last 5000 hours,

then $5000 = \mu + 2\sigma$

$5000 = \mu + 2 \times 170$

$\mu = 5000 - 2 \times 170$

$= 4660$ hours

(1 mark)

25

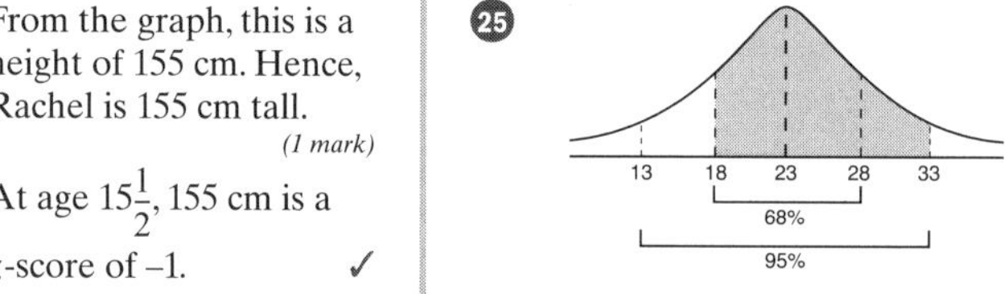

Now (95% – 68%) ÷ 2 = 13.5%

∴ Percentage of scores between 18 and 33

= 95% – 13.5%

= 81.5%

Answer D

26 **i** $\mu = 754$ g, $\sigma = 2$g

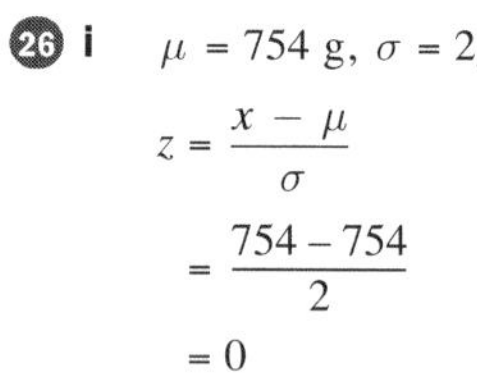

$$z = \frac{x - \mu}{\sigma} = \frac{754 - 754}{2} = 0$$

(This can be observed from the graph drawn in (iii)).

(1 mark)

ii $-1 = \frac{x - 754}{2}$

$-2 = x - 754$

$x = -2 + 754$

$= 752$

∴ Weight of box is 752 g.

(1 mark)

iii

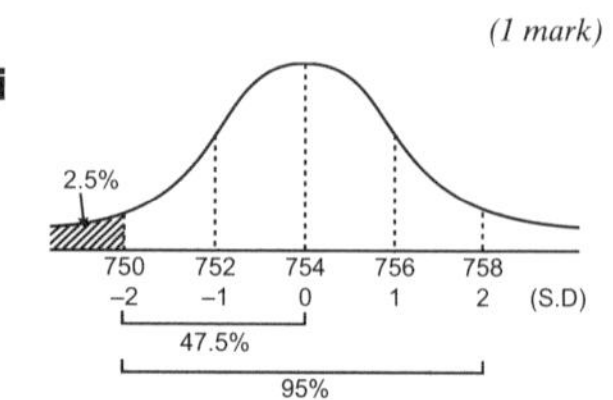

95% of the weights are within 2 standard deviations.

$\therefore (95 \div 2)\% = 47.5\%$ are between 750 g and 754 g. ✓

$\therefore 50\% + 47.5\% = 97.5\%$ are greater than 750 g.

$\therefore 2.5\%$ of boxes will have a weight less than 750 g. ✓

(2 marks)

27 **i** Roberto's z-score could be any number between 1 and 2 as this shaded region is the 2nd standard deviation from the mean. *(1 mark)*

ii

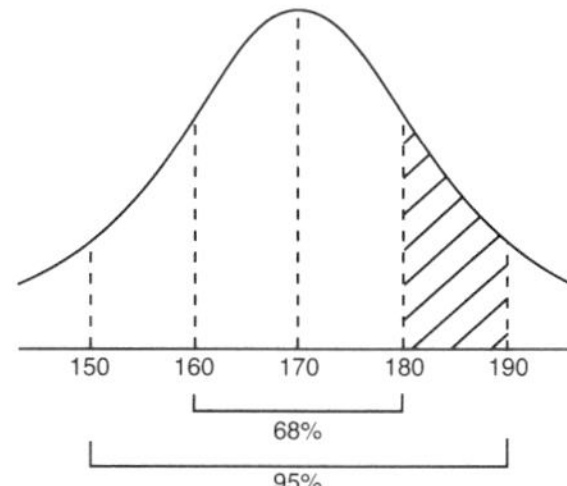

Percentage of data in shaded region

$\doteqdot (95 - 68) \div 2$ ✓

$\doteqdot 13.5\%$. ✓

(2 marks)

28 **i** $z = \dfrac{x - \mu}{\sigma}$

A z-score of 1 means that Hardev's score is exactly one standard deviation from the mean. For example, if Hardev's score was 80 and ✓ the mean was 60 and the standard deviation was 20 then,

$z = \dfrac{80 - 60}{20} = 1$ ✓

(2 marks)

ii About 68% of scores lie within a z-score range of –1 and 1.

$\therefore$ About 34% lie within a z-score range of 0 and 1.

$\therefore$ Proportion of people with higher score than Hardev

$\doteqdot 100\% - 50\% - 34\%$

$\doteqdot 16\%$

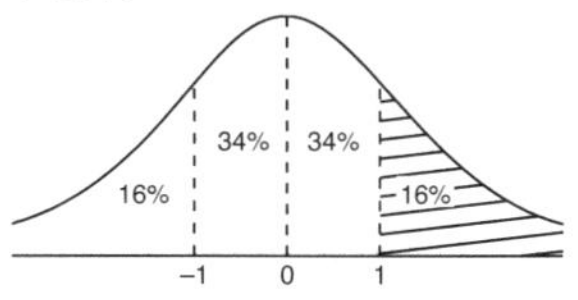

(1 mark)

29

$$z = \frac{x - \mu}{\sigma}$$

$$3 = \frac{x - 55}{6}$$

$$18 = x - 55$$

$$18 + 55 = x$$

$$\therefore x = 73$$

Answer D

30 $\mu = 40 \quad \sigma = 12$

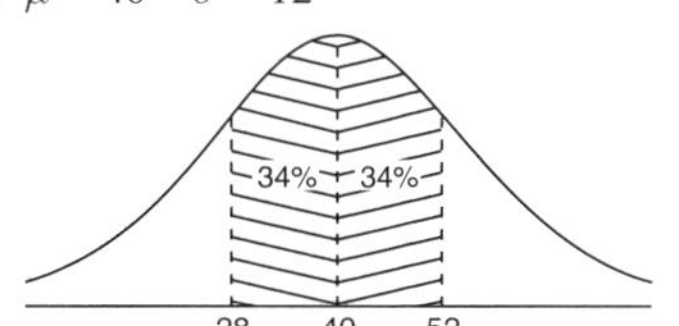

50% are younger than 40.

Appox. 68% lie within one standard deviation, i.e. between 28 and 52.

$\therefore$ 34% are between 40 and 52.

$\therefore$ Percentage younger than 52

$= 50\% + 34\%$

$= 84\%$

Answer D

31

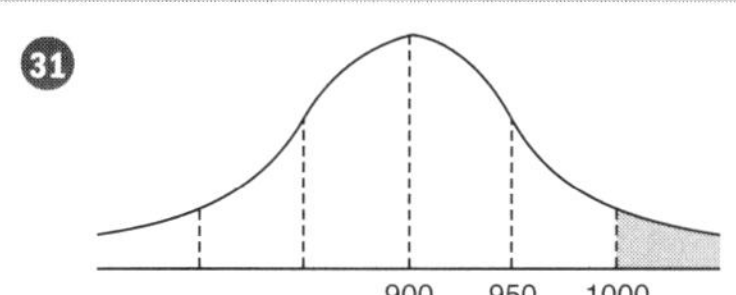

Shaded region $= (100\% - 95\%) \div 2$

$= 2.5\%$

Answer B

32 **i** z-score = –2

Kim's score was 2 standard deviations below the mean. ✓✓

(2 marks)

ii Actual mark

$= 75 - 2 \times 5$

$= 65$

or, using the formula:

$$z = \frac{x - \mu}{\sigma}$$

$$-2 = \frac{x - 75}{5}$$

$$-10 = x - 75$$

$$-10 + 75 = x$$

$$x = 65$$

(1 mark)

1 Consider the network diagram.

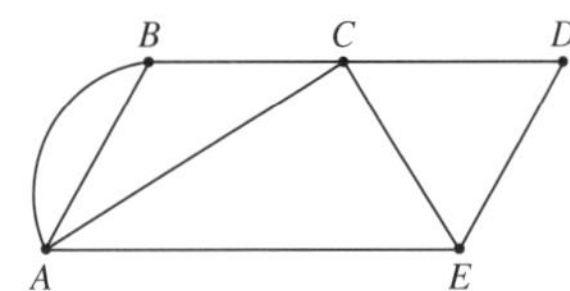

What is the sum of the degrees of all the vertices in this network?

A 5 **B** 8
C 14 **D** 16 *(1 mark)*

(Q2, **2021 HSC**) Medium

2 Team *A* and Team *B* have entered a chess competition.

Team *A* and Team *B* have three members each. Each member of Team *A* must play each member of Team *B* once.

Which of the following network diagrams could represent the chess games to be played?

A

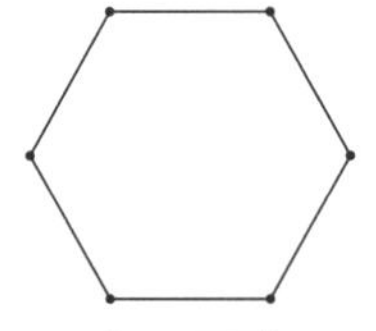

B

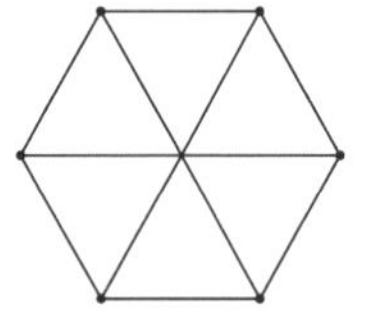

C

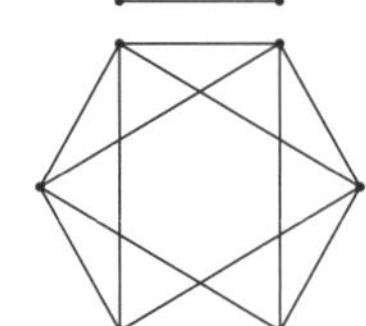

D

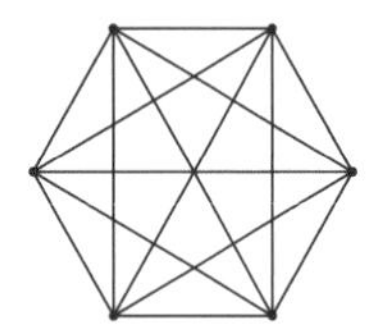

(1 mark)

(Q9, **2020 HSC**) Hard

3 A collection of bushwalking tracks is shown on the diagram below. There are six signposts labelled *P*, *Q*, *R*, *S*, *T* and *U*.

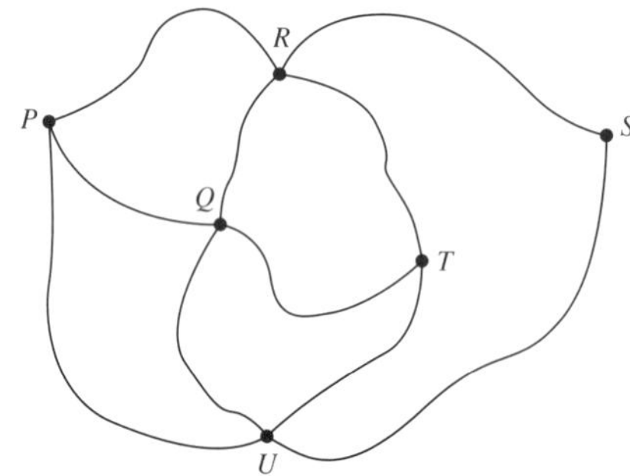

Which of the following edges should be removed so that a bushwalker could walk every track exactly once?

A *QR* and *QU* **B** *SR* and *SU*
C *PQ* and *PU* **D** *PQ* and *QT* *(1 mark)*

Bonus question (see page iv) Medium

4 Which of the following networks has more vertices than edges?

A / **C**

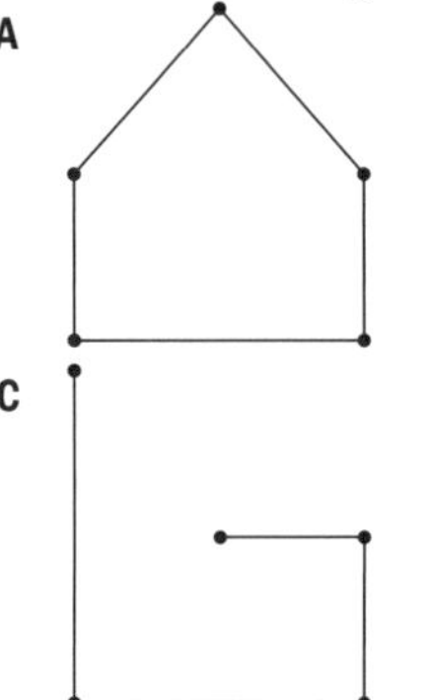

B / **D**

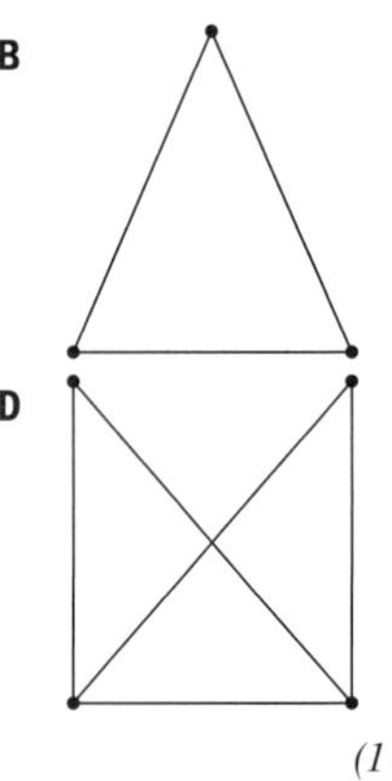

(1 mark)

Bonus question Easy

5 Here is a network diagram showing roads between five villages.

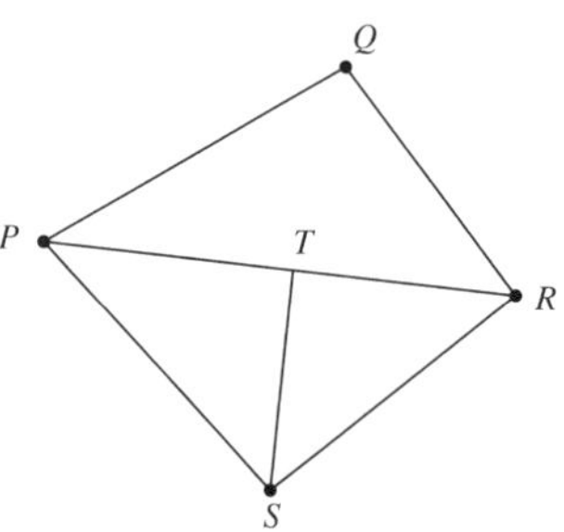

Which of these is a closed path?

A *PQRS* **B** *QTRSPQ*
C *PQRTS* **D** *TSRQPT* *(1 mark)*

Bonus question Easy

6 Use the table to add arrows and weights to the network diagram.

		To				
		A	**B**	**C**	**D**	**E**
From	**A**	–	6	–	–	7
	B	–	–	5	–	–
	C	–	–	–	9	4
	D	–	–	–	–	–
	E	–	8	–	3	–

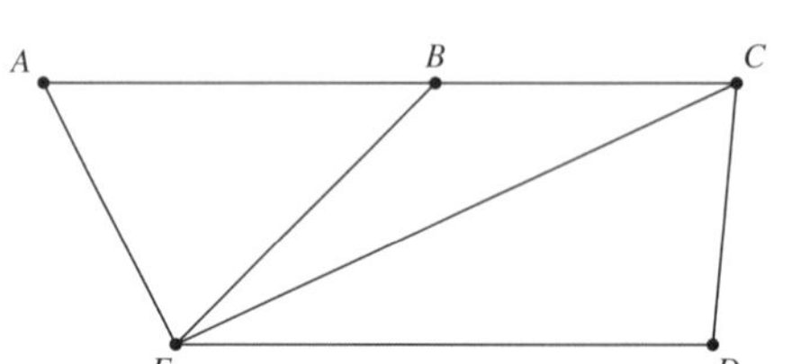

(2 marks)

Bonus question Easy

7 Draw a network diagram to represent the roads between towns A, B, C, D and E if:

- Town A is 21 km from B and 16 km from E
- Town B is 38 km from C and 24 km to D
- Town C is 29 km from E and 9 km from D.

(1 mark)

Bonus question Easy

8 Three airlines are named according to their base city and can only fly from their base airport to and from two other airports. Airline A flies to and from B and E, C flies to and from A, B and E, and D flies to and from C and E. The cost of a ticket for each airline is the same for any of its routes. Airline A charges \$120, C charges \$160 and D charges \$150.

Draw a weighted network diagram to represent the flight routes and costs.

(2 marks) Easy

Bonus question

9 Aleece drove from P, through Q and R, and returned to P. She had travelled a total of 36 km.

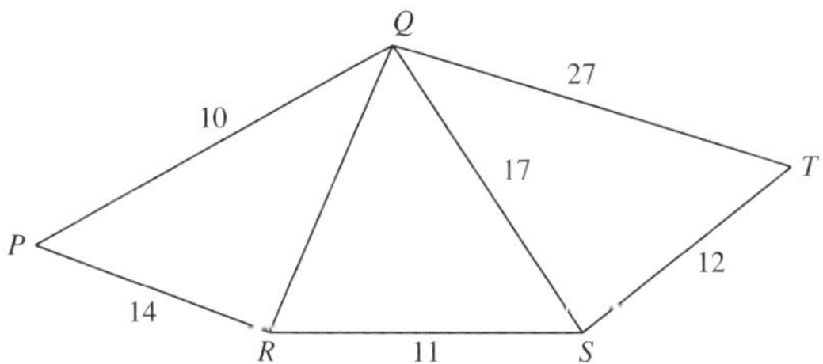

Her friend Gladys left R, drove through Q, T and S, and arrived back in R. How far did Gladys drive?

A 50 km **B** 60 km

C 62 km **D** 65 km *(1 mark)*

Bonus question Easy

10 Consider the diagram below.

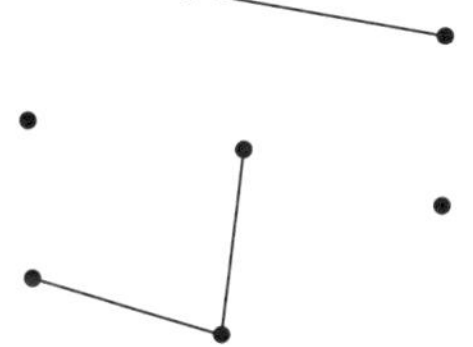

What is the smallest number of additional edges needed to make this a connected network?

A 1 **B** 2

C 3 **D** 4 *(1 mark)*

Bonus question Easy

11 An undirected connected network has five vertices. Three of these vertices are of even degree and the remaining of odd degree. Show the edges on the diagram below.

(1 mark)

Bonus question Easy

12 In the network below, how many vertices have an even degree?

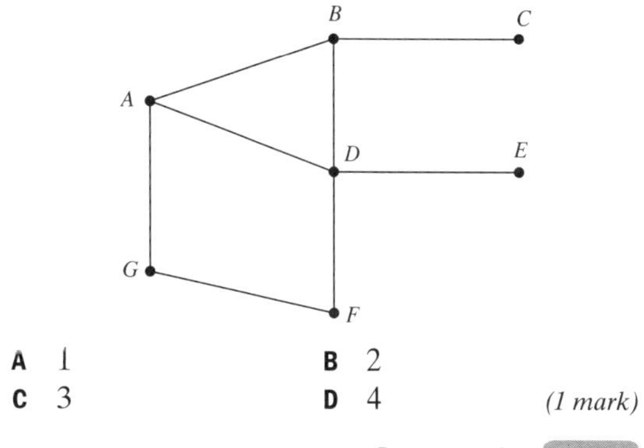

A 1 **B** 2

C 3 **D** 4 *(1 mark)*

Bonus question Easy

13 What is the most common degree of the vertices in the network?

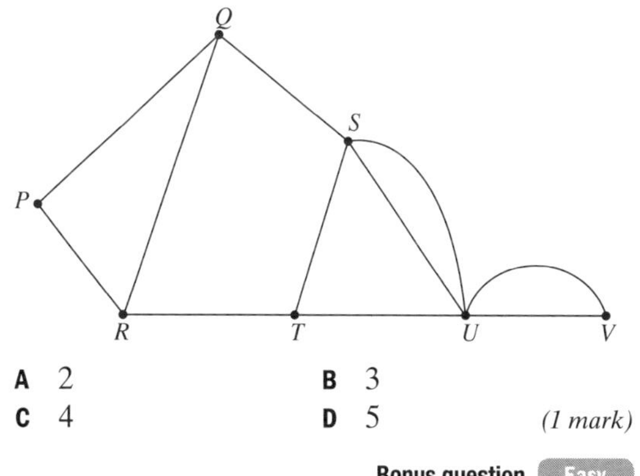

A 2 **B** 3

C 4 **D** 5 *(1 mark)*

Bonus question Easy

14 Brianna drew a network which has n edges. Which expression gives the sum of the degrees of the vertices in the network?

A n **B** $2n$

C $n + 1$ **D** $2n - 1$ *(1 mark)*

Bonus question Medium

15 The network diagram below shows the distances between six villages.

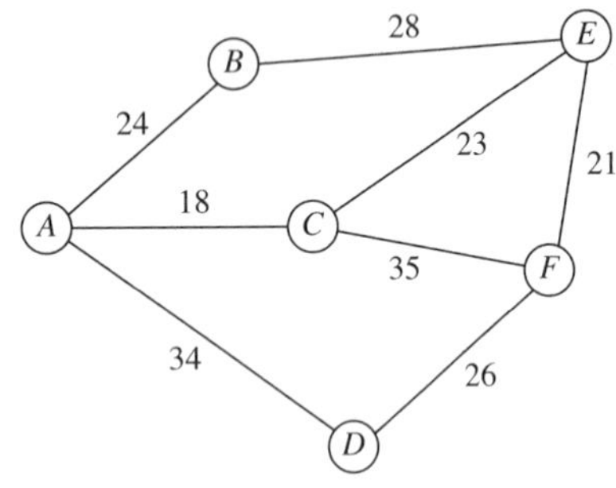

Complete the table representing this network.

	A	B	C	D	E	F
A	–					
B		–				
C			–			
D				–		
E					–	
F						–

(1 mark)

Bonus question Easy

16 Which of the following represents a cycle in the network diagram?

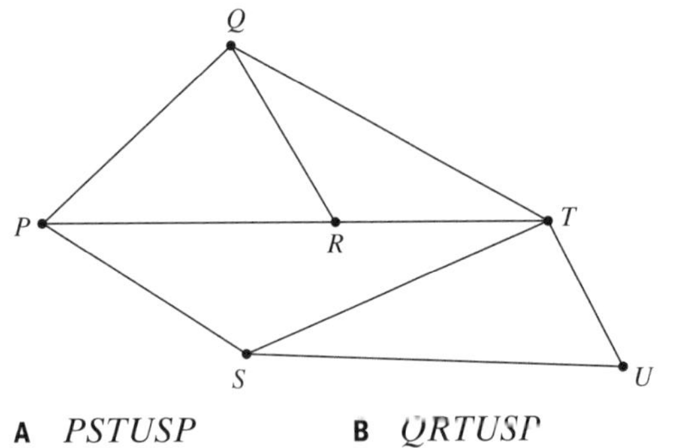

A *PSTUSP* **B** *QRTUSP*
C *TUSPQ* **D** *QPSTQ* *(1 mark)*

Bonus question Easy

17 The diagram shows a network of four points *P*, *Q*, *R* and *S*.

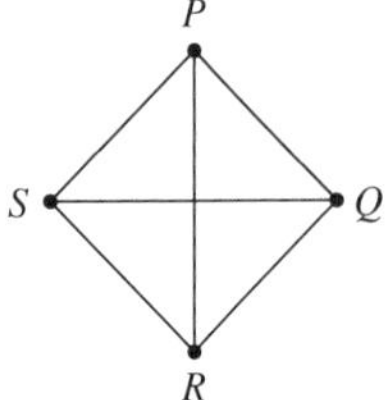

Which of the following network diagrams is **not** an equivalent graph of the network above?

A

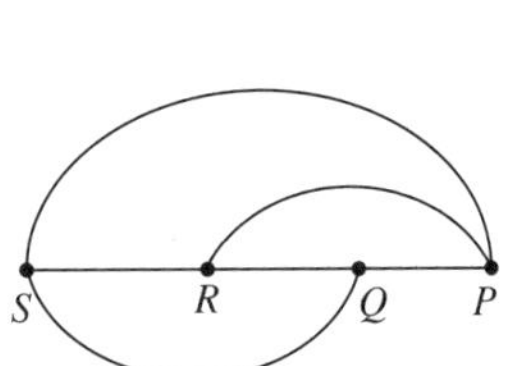

B

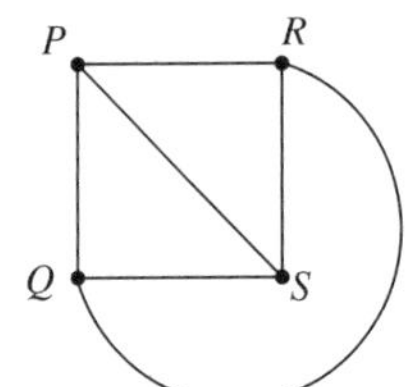

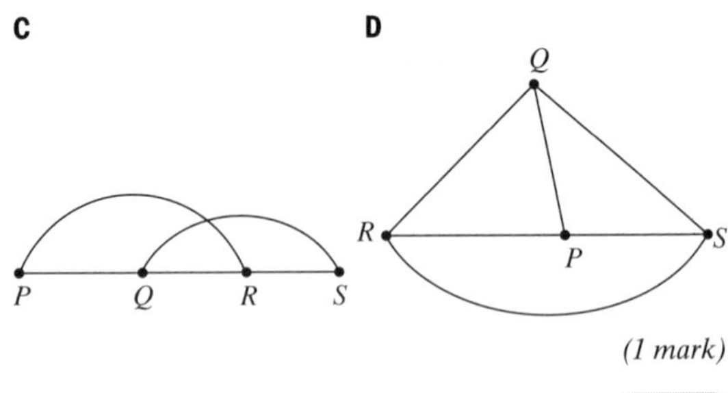

(1 mark)

Bonus question Easy

18 Which of the following walks is a path in the network diagram above?

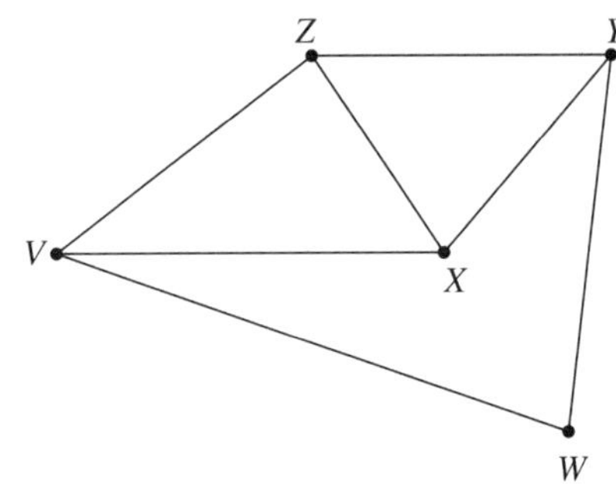

A *VZYW* **B** *VZYWV*
C *ZYXZV* **D** *XYXVZ* *(1 mark)*

Bonus question Easy

19 The table below represents a directed network.

		To			
		P	**Q**	**R**	**S**
From	**P**	–	6	–	4
	Q	2	–	–	7
	R	–	5	–	–
	S	–	–	3	–

Complete the network showing weights and arrows.

P • • *Q*

S • • *R*

(2 marks)

Bonus question Easy

20 Here is a list of people who invited friends to their birthday parties this year.

Person	Alex	Beck	Cael	Dana	Erin
Friends	Beck, Dana, Erin	Alex, Erin	Dana, Erin	Alex, Cael, Erin	Alex, Beck, Cael, Dana

Using the letters A, B, C, D and E, represent this information on a network.

(1 mark)

Bonus question Easy

21 A directed network is shown below.

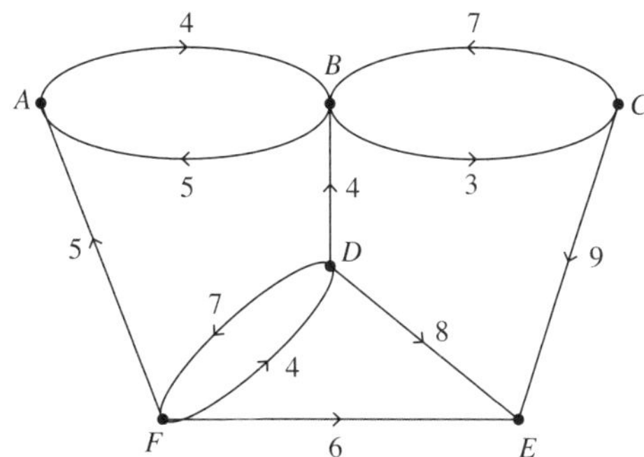

Complete the table network showing weights and arrows.

		To					
		A	**B**	**C**	**D**	**E**	**F**
From	**A**	–					
	B		–				
	C			–			
	D				–		
	E					–	
	F						–

(2 marks)

Bonus question Easy

22 The map shows a wide river with three islands B, C and D situated in the river. Bridges are shown connecting the banks of the river (A and E) to the islands.

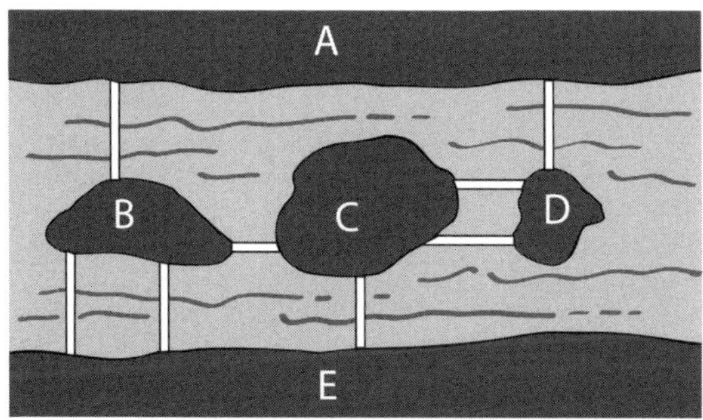

Draw a network diagram to show the road system linking the three islands to the riverbanks. (1 mark)

Bonus question Easy

23 The directed network represents a series of one-way streets. The vertices represent the intersection of these streets.

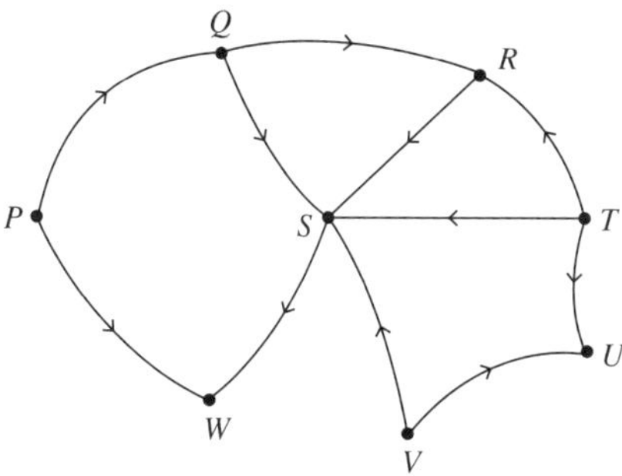

How many vertices can be reached from P?

A 7 **B** 4

C 3 **D** 5 (1 mark)

Bonus question Easy

24 A rugby league team welcomed five new players to the squad for the new season. The players were Joey, Kirk, Leon, Matt and Noah. The network diagram shows the players who had played together in other teams over previous seasons.

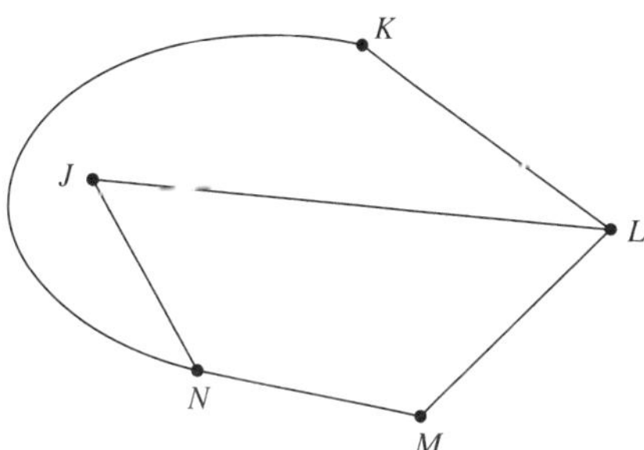

a How many of these players had Leon played with before joining the team? (1 mark) Easy

b Who had played with Noah before joining the team? (1 mark) Easy

c During the season Owen joined the team. He had played with Joey and Leon in previous seasons. Represent this information on the diagram above. (1 mark) Easy

Bonus question

25 The network diagram shows the directional traffic between seven rooms in an art gallery exhibition.

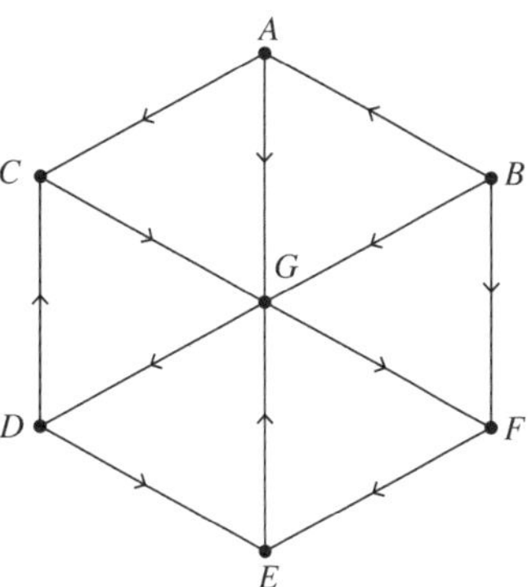

Between which two rooms is it not permitted to move?

A *A* to *E* **B** *B* to *E*

C *D* to *B* **D** *C* to *E* *(1 mark)*

Bonus question Easy

---

26 The table shows the distances in kilometres between towns ***A*** to ***E***.

	A	***B***	***C***	***D***	***E***
A	–	42	64	–	53
B	42	–	31	–	–
C	64	31	–	51	65
D	–	–	51	–	48
E	53	–	65	48	–

Draw a network diagram to represent the information in the table. *(2 marks)*

Bonus question Easy

Year 12 Network concepts—Worked answers

1 Sum of degrees
$= 4 + 3 + 4 + 2 + 3$
$= 16$
Answer D

2 Each member of Team A plays each member of team B.

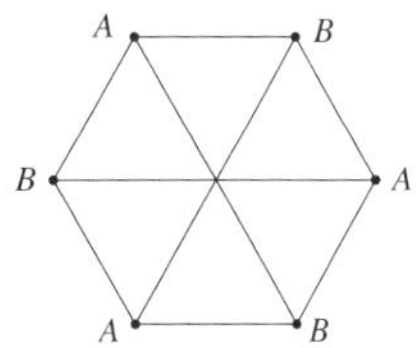

Answer B

3 Each vertex should have an even degree. Removing PQ and QT will mean P, Q and T are each degree 2.
Answer D

4 The network has 5 vertices and 4 edges.

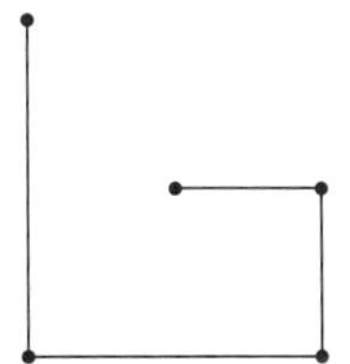

Answer C

5 The closed path is $TSRQPT$.
Answer D

6

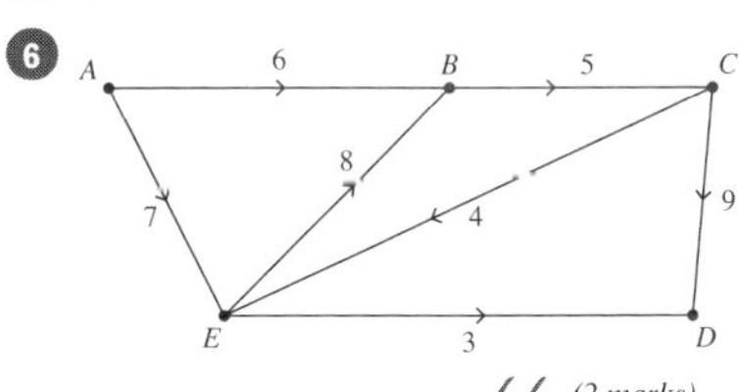

✓✓ *(2 marks)*

7

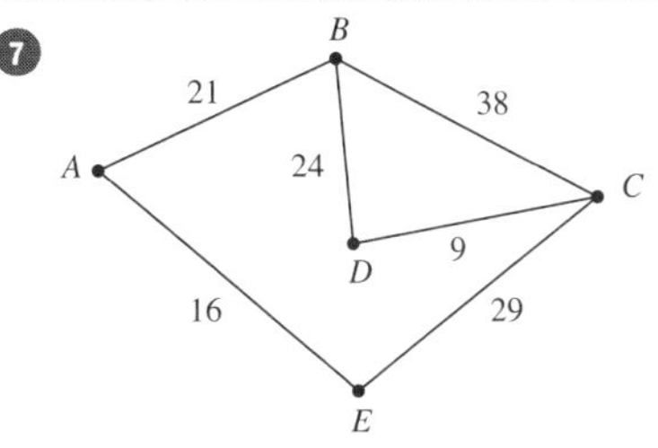

(1 mark)

8

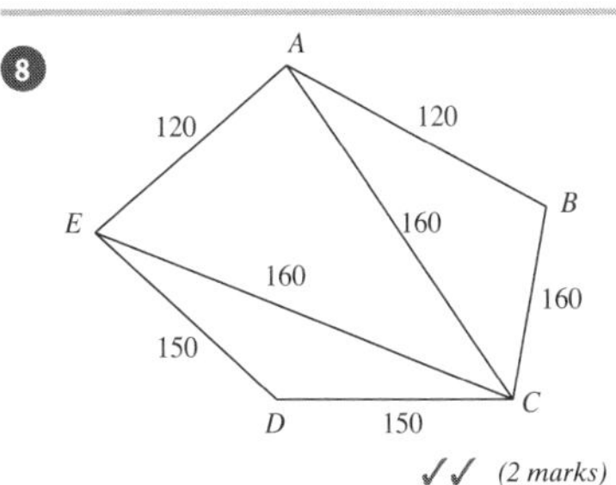

✓✓ *(2 marks)*

9 For $PQRP$:
$10 + QR + 14 = 36$
$QR + 24 = 36$
$QR = 12$
For $RQTSR$:
$12 + 27 + 12 + 11 = 62$
Gladys drove 62 km.
Answer C

10 Another 3 edges are required.

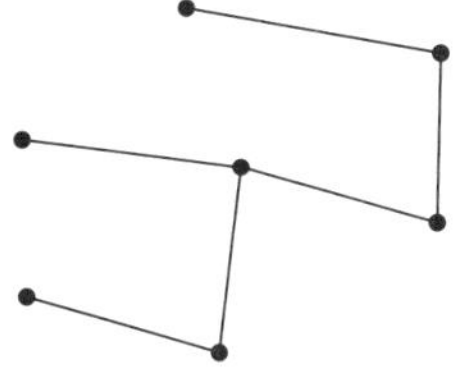

Answer C

11

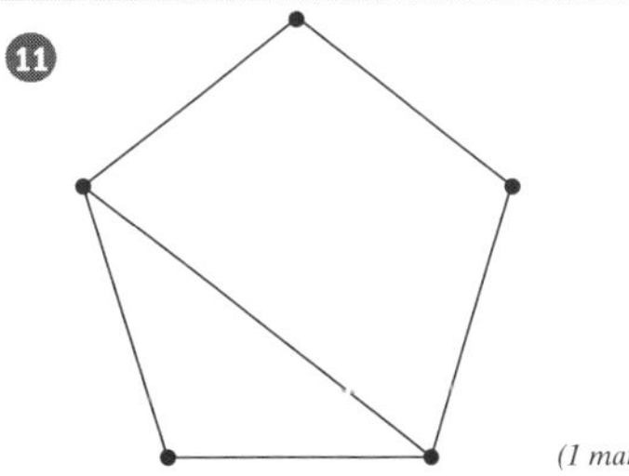

(1 mark)

12 A is degree 3, $B(3)$, $C(1)$, $D(4)$, $E(1)$, $F(2)$ and $G(2)$.
There are 3 vertices which have an even degree.
Answer C

13 P is degree 2, $Q(3)$, $R(3)$, $S(4)$, $T(3)$, $U(5)$ and $V(2)$.
The most common degree is 3.
Answer B

14 The sum of the degrees of the vertices of the network is twice the number of edges.
Answer B

15

	A	B	C	D	E	F
A	–	24	18	34	–	–
B		–	–	–	28	–
C			–	–	23	35
D				–	–	26
E					–	21
F						–

(1 mark)

16 A cycle is a walk where the initial and final vertices are the same, but no other vertices are repeated. $QPSTQ$ is a cycle. (A cycle is a closed path.)
Answer D

17

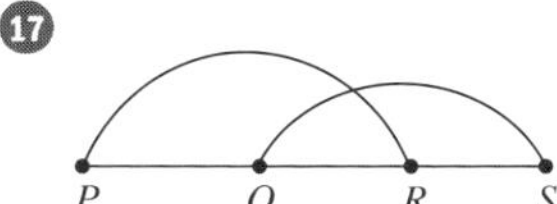

Answer C

18 A path is a walk which never visits the same vertex more than once. $VZYW$ is a path.
Answer A

19

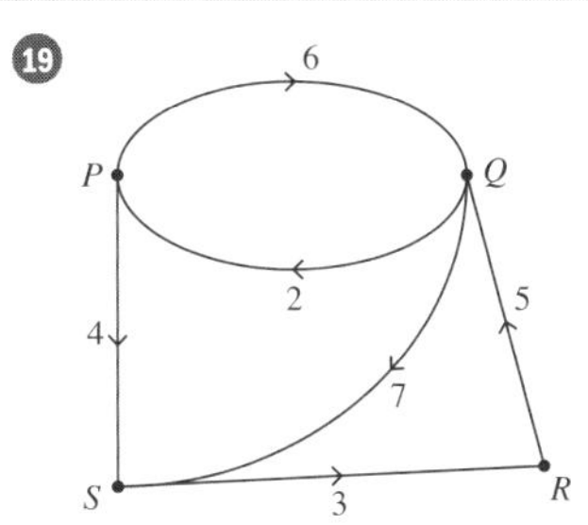

✓✓ *(2 marks)*

20

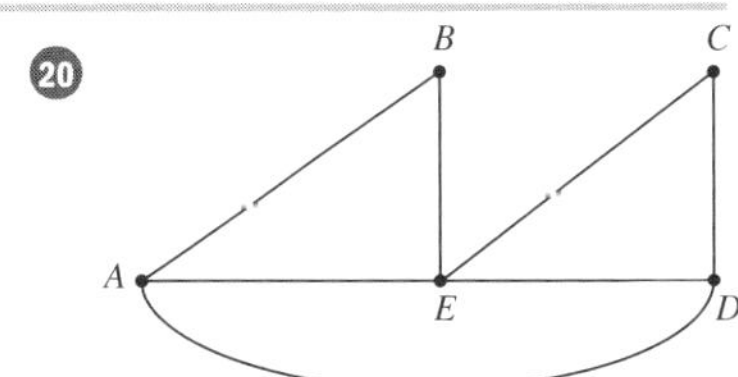

(1 mark)

21

		To					
		A	**B**	**C**	**D**	**E**	**F**
From	**A**	–	4	–	–	–	–
	B	5	–	3	–	–	–
	C	–	7	–	–	9	–
	D	–	4	–	–	8	7
	E	–	–	–	–	–	–
	F	5	–	–	4	6	–

✓✓ *(2 marks)*

22

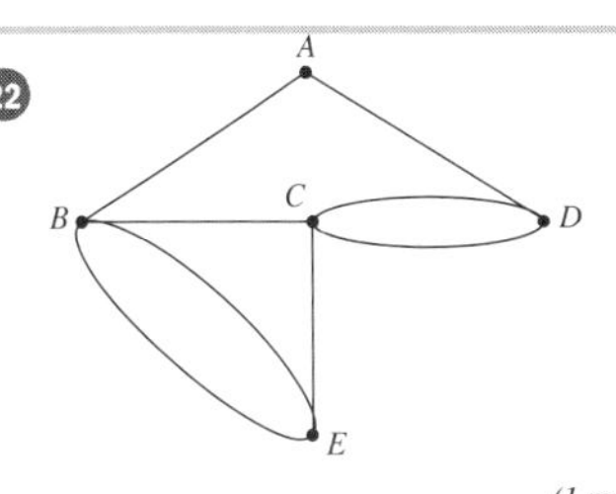

(1 mark)

23 From P, the vertices that can be reached are Q, S, W and R. There are 4 vertices.
Answer B

24 **a** 3 *(1 mark)*

b Joey, Kirk and Matt *(1 mark)*

c

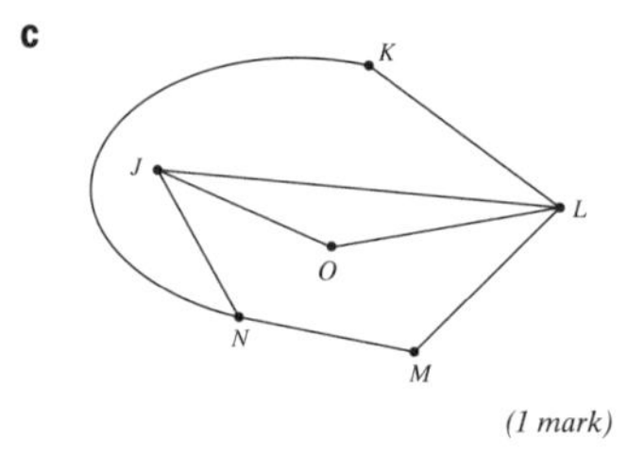

(1 mark)

25 You are not permitted to move from room D to room B.

Answer C

26

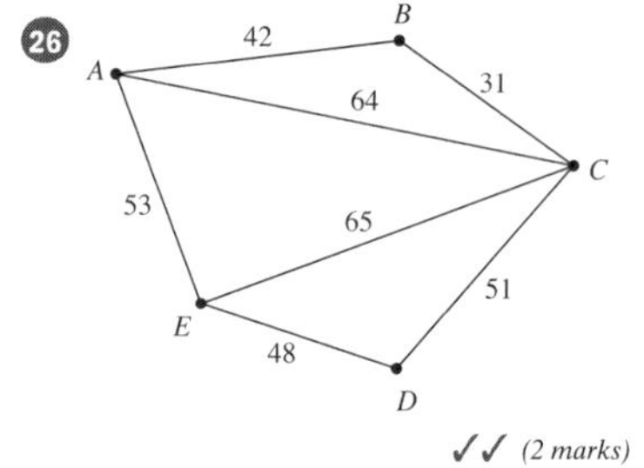

✓✓ *(2 marks)*

1 The network diagram shows the travel times in minutes along roads connecting a number of different towns.

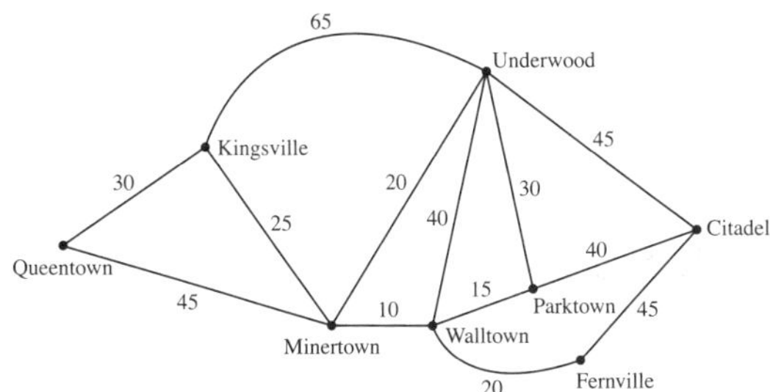

a Draw a minimum spanning tree for this network and determine its length. *(3 marks)* Easy

Length of minimum spanning tree = ...

b How long does it take to travel from Queentown to Underwood using the fastest route? *(1 mark)* Medium

(Q23, **2021 HSC**)

2 The diagram represents a network with weighted edges.

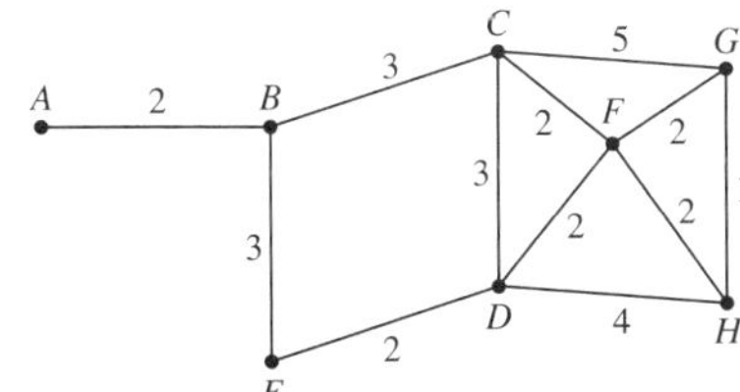

a Draw a minimum spanning tree for this network in the space below and determine its length. *(3 marks)* Medium

b The network is revised by adding another vertex, K. Edges AK and CK have weights of 12 and 10 respectively, as shown.

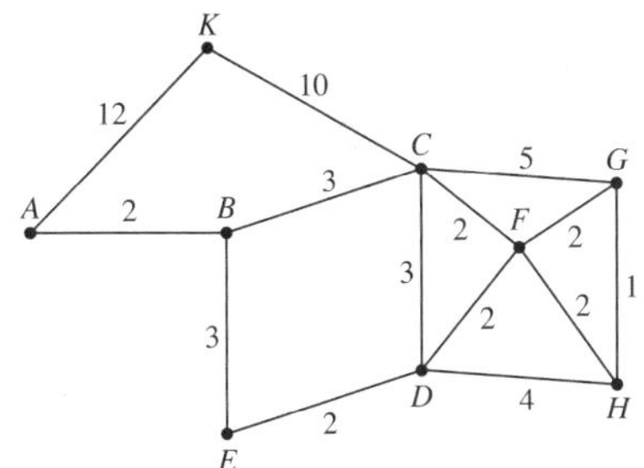

What is the length of the minimum spanning tree for this revised network? *(1 mark)* Medium

(Q18, **2020 HSC**)

3 The network diagram shows the tracks connecting 8 picnic sites in a nature park. The vertices A to H represent the picnic sites. The weights on the edges represent the distances along the tracks between the picnic sites, in kilometres.

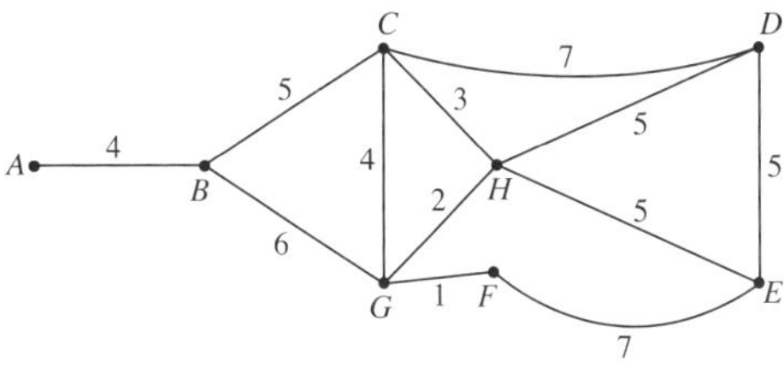

a Each picnic site needs to provide drinking water. The main water source is at site A. By drawing a minimum spanning tree in the space below, calculate the minimum length of water pipes required to supply water to all the sites if the water pipes can only be laid along the tracks. *(2 marks)* Medium

b One day, the track between C and H is closed. State the vertices that identify the shortest path from C to E that avoids the closed track. *(1 mark)* Medium

(Q30, **2019 HSC**)

4 A market gardener is expanding his cropping area and wants to place taps at locations A to K on the diagram below. Each tap is to be connected with a water pipe from the existing tap X. The distance, in metres, from each adjacent tap is recorded.

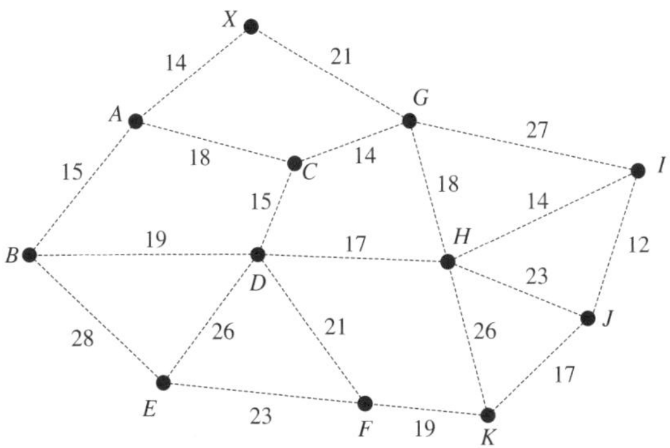

On the diagram draw the minimum spanning tree required to do this and determine the total cost if each new tap costs $22 and the cost of laying pipe is $4.50/metre, in metres. *(2 marks)*

Bonus question (see page iv) Hard

5 The diagram shows a directed network where the weights are in minutes.

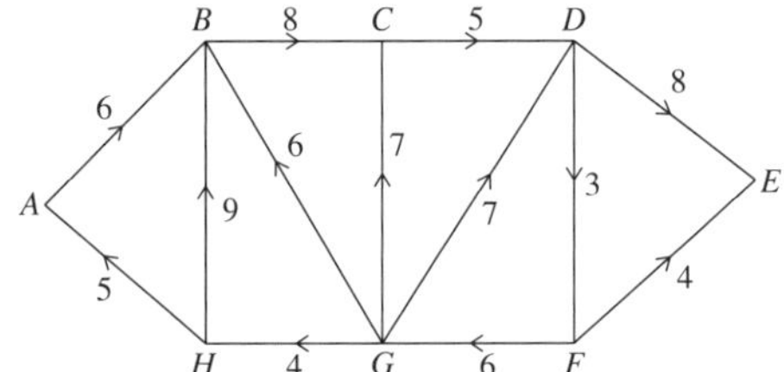

a From which vertex can vertex A not be reached? *(1 mark)* Easy

b What is the minimum time to travel from A to G? *(1 mark)* Medium

Bonus question

---

6 Once upon a time in the small village of Murgup the pathways between its 10 houses always became very muddy in wet weather. The villagers decided to pave some of the paths with cobblestones so that all householders could travel to everybody else's house on cobblestones without walking in mud.

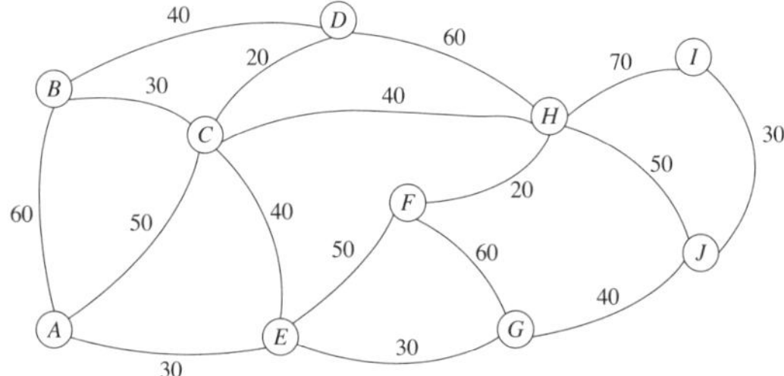

The distance between the houses is shown on the diagram, in metres, and 60 cobblestones are used for every metre of road. If every cobblestone has an average mass of 5 kg, find the minimum mass of cobblestones required to complete the project. *(3 marks)*

Bonus question Hard

---

7 A computer network is to be installed, consisting of nine computers which need to be connected to at least one other computer by cable. All dimensions are in metres and the cable and installation costs are $35 per metre.

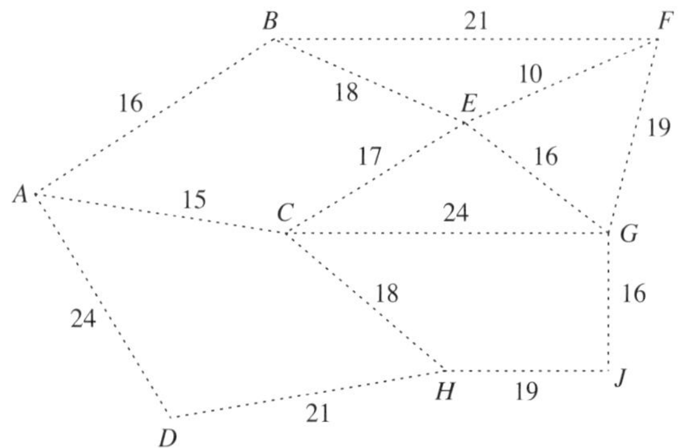

a Draw the minimum spanning tree on the network diagram and calculate the total cost of cable required. *(1 mark)* Medium

b Before work was to have commenced, it was decided that a computer should be removed from the network. What will be the savings if computer A is removed from the minimum spanning tree? *(2 marks)* Hard

Bonus question

---

8 The minimum spanning tree for the network below includes the edge with weight labelled m.

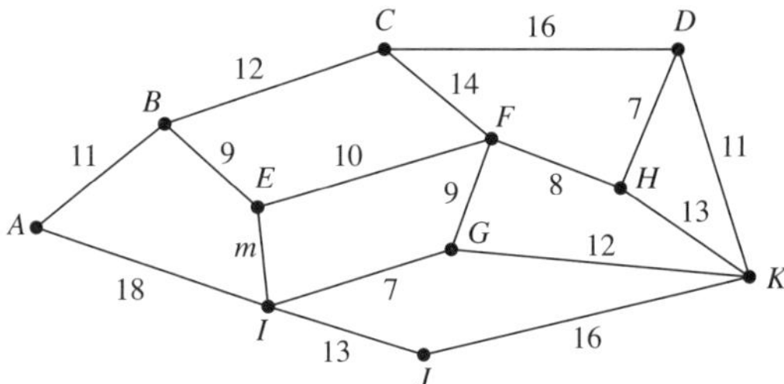

Find the value of m, if the total weight of the minimum spanning tree is 97. *(1 mark)*

Bonus question Medium

---

9 Andrew manages a holiday park and a contractor has quoted $42 per metre to provide connecting paths to each of the cabins as shown in the network diagram below. All measurements are in metres.

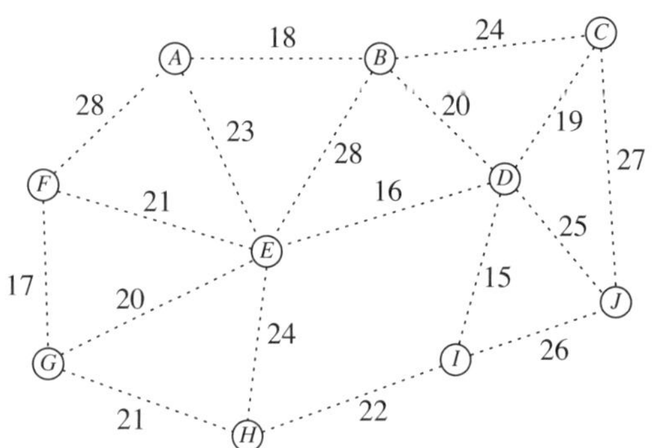

What will be the minimum cost of providing at least one path to each cabin in the network? *(2 marks)*

Bonus question Hard

---

10 The network diagram shows the locations of 10 rides at a theme park. The distance in metres between the rides is shown on the diagram. Draw the minimum spanning tree for the network on the diagram and write the length.

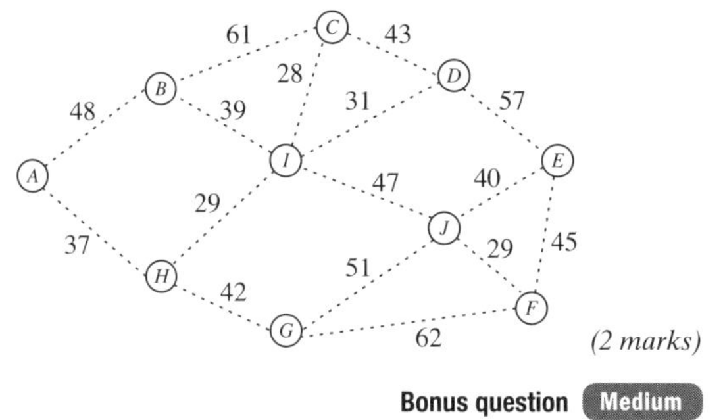

(2 marks)

Bonus question Medium

11 The network diagram shows the bus routes between stops in a certain town, and the time in minutes to travel between stops.

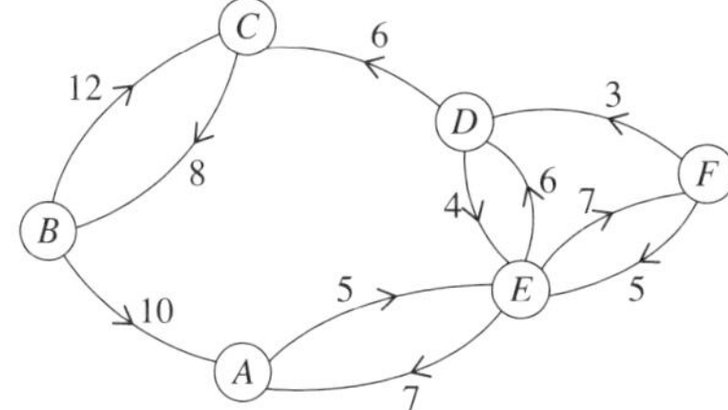

a What is the quickest time to travel between E and B? *(1 mark)* Medium

b In how many different ways is it possible for a person to travel from F to A without passing through a bus stop more than once? Explain your answer. *(1 mark)* Hard

Bonus question

12 The map shows the distances, in metres, between 11 new statues in a park shown by the letters A to K. As the statues are to be floodlit at night, electricity is to be provided to the base of each statue. The cost of laying the electricity cable is $42/metre.

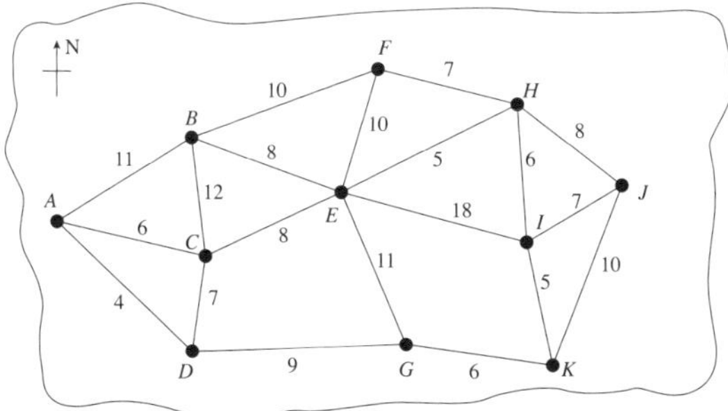

a What is the minimum cost of laying the cable that joins each of the statue locations? *(2 marks)* Medium

b Before work is to proceed, the project manager considers not running the cable to the statue at H to save money. Discuss the feasibility of the decision, showing calculations. *(2 marks)* Hard

Bonus question

13 Arnie has a small acreage and is planting trees across a paddock. He plans to lay water pipes to eight spots where he will locate taps. An existing tap is already located at C. The distance, in metres, between each tap is shown on the network diagram below.

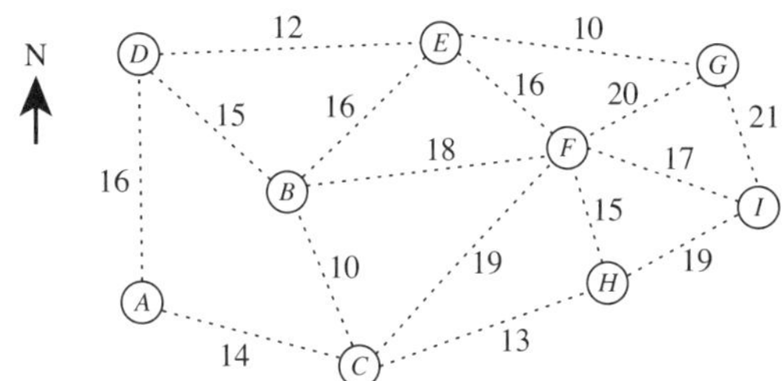

a On the diagram draw the minimum spanning tree required to do this and determine the total length of pipe required, in metres. *(2 marks)* Medium

b Before he starts, Arnie decides to include another tap at location J which is shown on the network diagram below, where $JC = 12$ m, $JH = 10$ m and $IJ = 16$ m.

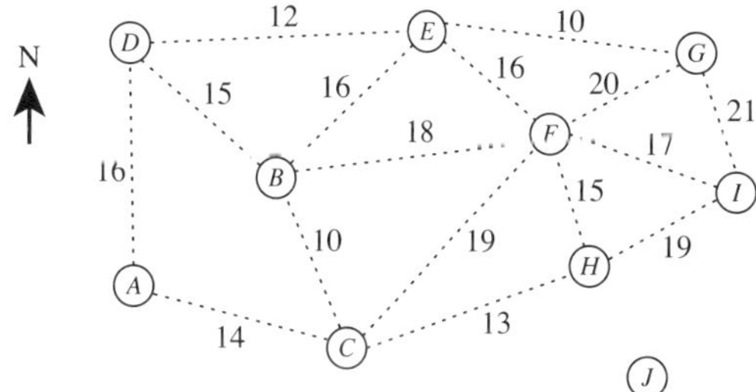

Determine whether the inclusion of the new tap will increase the total length of pipe required. Support your argument with calculations. *(2 marks)* Hard

Bonus question

14 Which of the following network diagrams is **not** a tree?

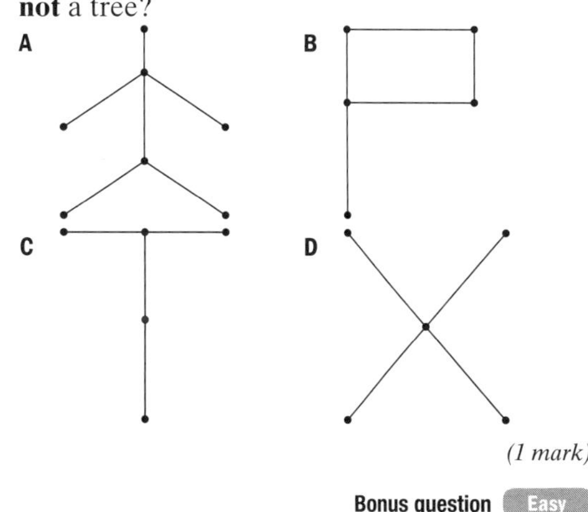

(1 mark)

Bonus question Easy

15

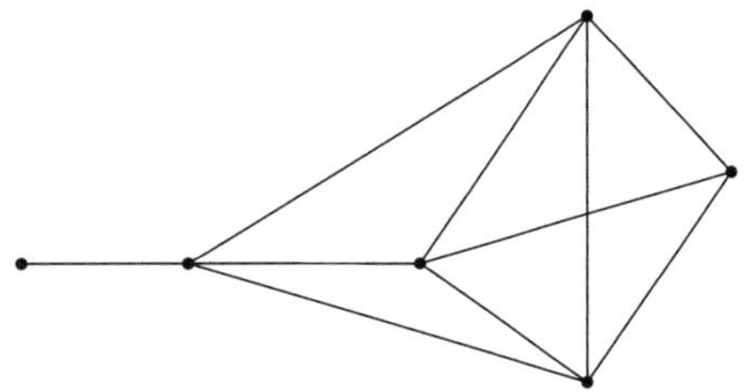

Which of the following is a spanning tree of the network above?

A

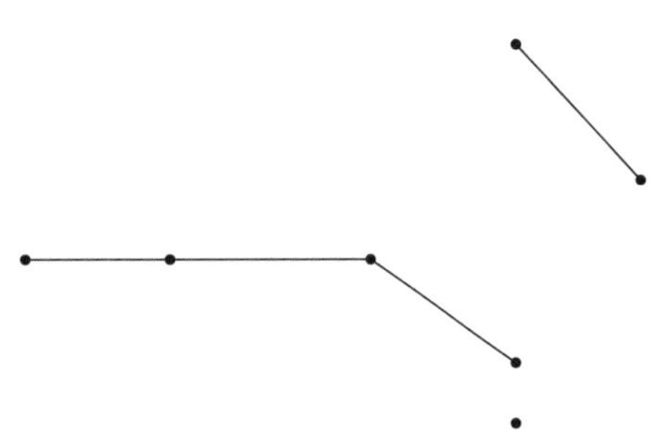

B

C

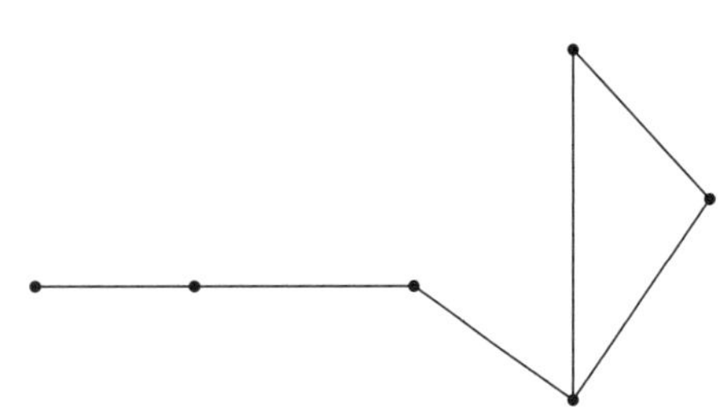

D

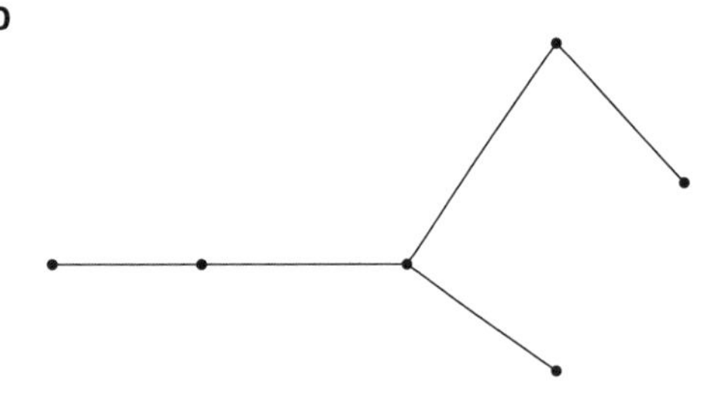

Bonus question Easy

16 The table below is used to represent a network.

	A	B	C	D	E
A	–	6	5	14	12
B	6	–	–	9	–
C	5	–	–	8	7
D	14	9	8	–	4
E	12	–	7	4	–

a Draw a weighted network that is represented by the given table.

A •   *C* •   • *E*

B •   • *D*

(2 marks) Easy

b Identify the minimum spanning tree on your network diagram above. What is the length of the minimum spanning tree? *(2 marks)* Easy

Bonus question

17 The network below shows the road distances in kilometres between seven towns labelled A, B, C, D, E, F and G. To establish a cable network of communications, it is proposed to lay the underground cable besides existing roads.

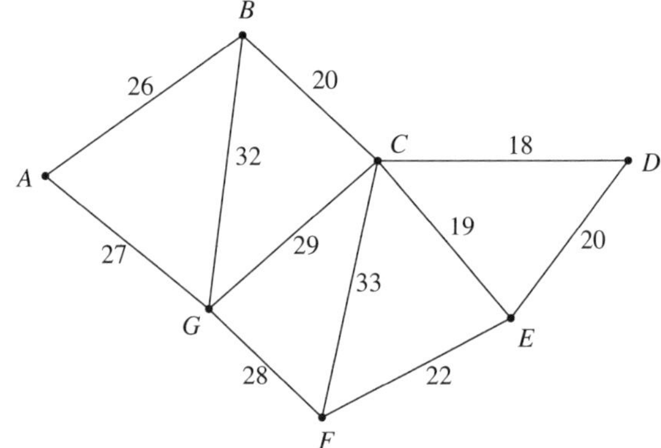

a By highlighting or emboldening show the minimum spanning tree on the network diagram that ensures all the towns are connected to the network and minimises the total amount of cable used. *(1 mark)* Medium

b A company quoted 4.1 million dollars to lay the cable. If it costs \$26.50/metre to lay the cable, what will be the amount of profit made by the company? *(2 marks)* Medium

Bonus question

18 The network diagram shows the roads joining six towns. All measurements are in kilometres.

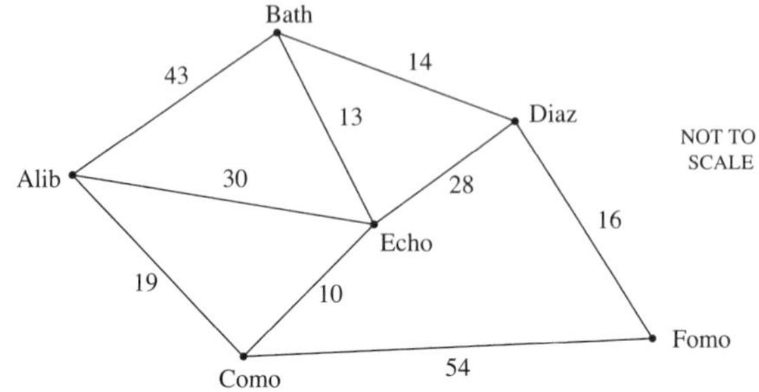

What is the shortest possible route between Alib and Fomo?

A 69 km **B** 70 km

C 72 km **D** 73 km *(1 mark)*

Bonus question Easy

19 A university residential college has completed a study to determine the cost of installing a new fibre-optic network on the campus. The cost of joining the five main buildings is provided in the table below, expressed in thousands of dollars. Buildings that cannot be joined are shown as a '–'. Using a minimum spanning tree, find the cost of connecting all five buildings.

	A	**B**	**C**	**D**	**E**
A	–	14	21	17	11
B		–	11	-	12
C			–	10	16
D				–	13
E					–

A B

E

D C *(4 marks)*

Bonus question Medium

20 The network diagram below shows a small park with six garden beds, labelled A, B, C, D, E and F. The park is accessed through a gate shown as G. All dimensions are in metres.

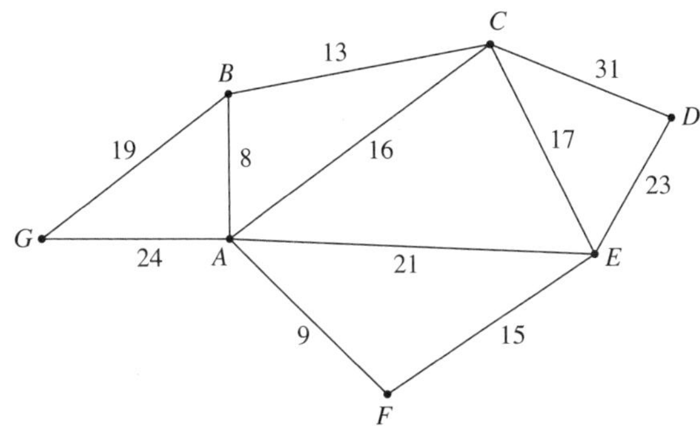

a What is the minimum distance a person needs to walk to visit each garden bed, starting and finishing at the gate? *(1 mark)* Easy

b Water is available at the entrance of the park. New pipes are to be laid along the paths to provide water to each of the garden beds. Find the minimum length of pipes required. Support your answer by drawing a minimum spanning tree on the diagram below.

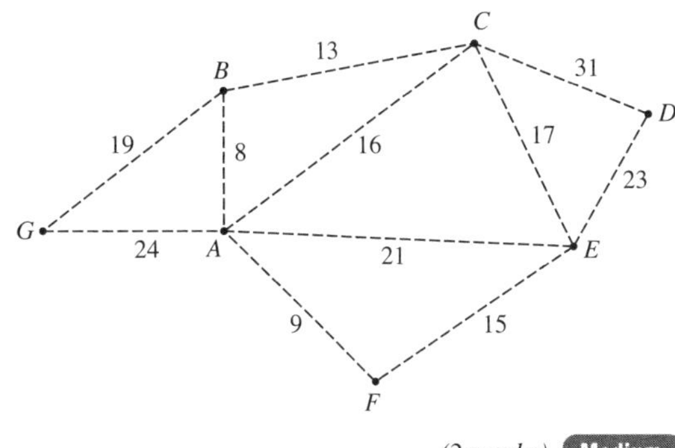

(2 marks) Medium

Bonus question

21 The network diagram shows the average time taken, in minutes, to walk between seven villages using existing tracks.

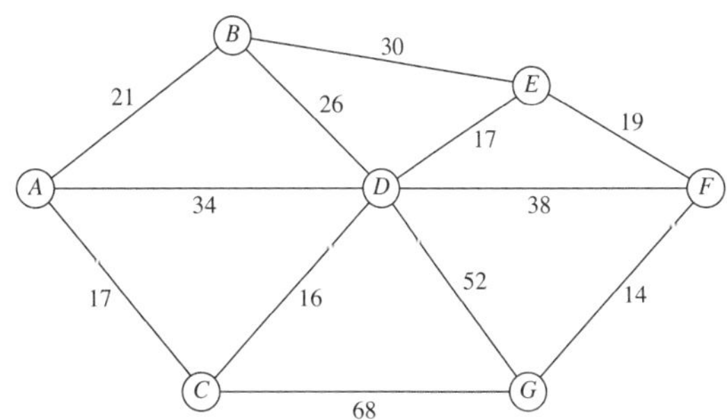

What is the shortest possible time to travel between A and G? Describe the path. *(2 marks)*

Bonus question Medium

22 A network contains n vertices. How many edges are required to form a spanning tree?

A n **B** $n-1$

C $n+1$ **D** $2n$ *(1 mark)*

Bonus question Easy

23 The diagram below shows a train network through six towns, Ashford, Bellbird, Cudal, Dungog, Euston and Falco.

The numbers on each edge represent the cost (in dollars) of an adult ticket between the two towns. Jannah and her eight-year-old daughter Ella live in Ashford and will be travelling by train to visit a friend who lives in Falco. The price of a child ticket is 75% of an adult ticket.

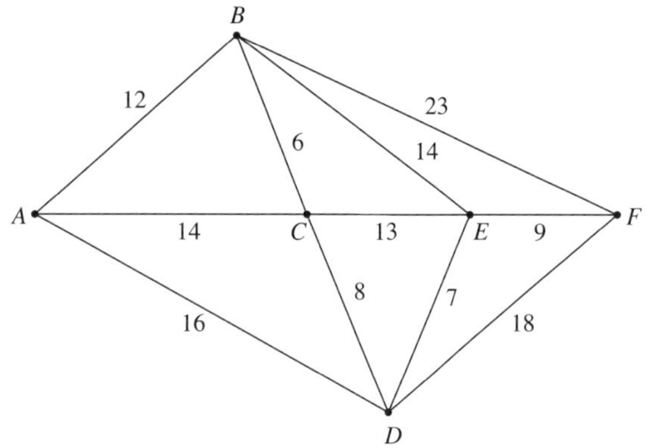

What is the minimum cost of the two tickets? *(2 marks)*

Bonus question **Medium**

---

24 The following network shows the length, in kilometres, of roads connecting nine towns. The local council plans to resurface some of the roads to ensure that people living in each town can access all other towns along a resurfaced road.

What length of road will connect all nine towns while keeping the length of road to be resurfaced to a minimum?

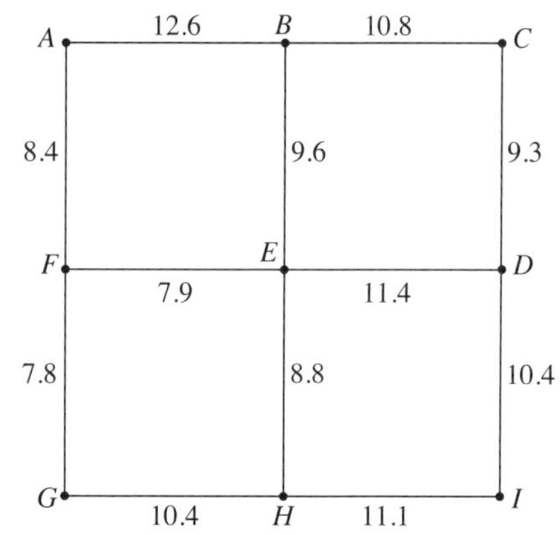

A 73.0 km **B** 74.6 km
C 75.0 km **D** 76.0 km *(1 mark)*

Bonus question **Medium**

---

25 Cabins in a national park camping ground are to be connected to a power source situated at P. The graph below shows the distances, in metres, between seven cabins and P.

The successful contractor has quoted a rate of $18.40 per metre. What would be the minimum cost of the project? *(3 marks)*

Bonus question **Medium**

---

26 The network diagram below represents a local fitness park which has 10 exercise stations named A to J. The edges on the graph represent the tracks between each station.

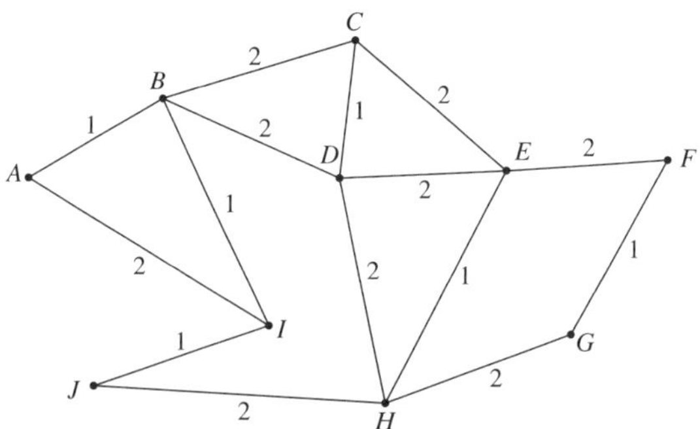

Logan starts at station C and runs every track exactly once. He rests for 5 minutes and then runs the shortest route back to C. What is the total distance covered? *(2 marks)*

Bonus question **Medium**

Year 12 Shortest paths—Worked answers

1 **a**

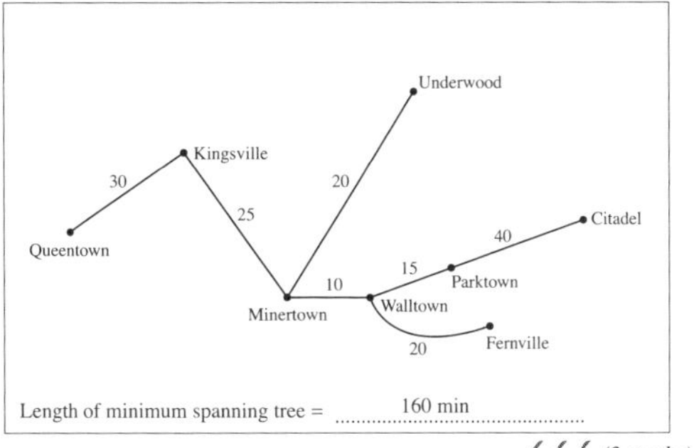

✓✓✓ *(3 marks)*

b Fastest route from Queentown to Underwood is through Minertown.

Time = (45 + 20) min
= 65 min
= 1 h 5 min

(1 mark)

2 **a**

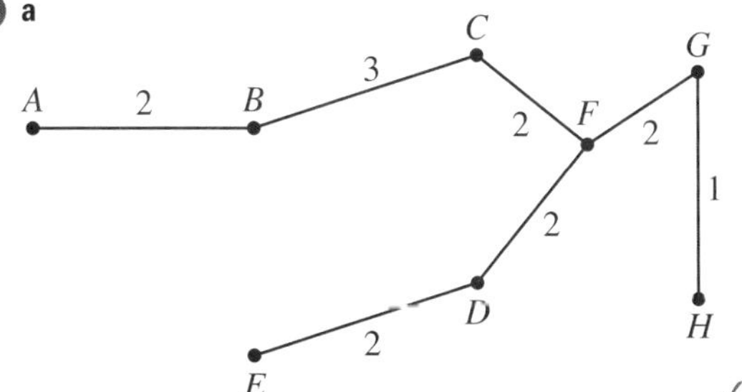

✓✓

Minimum length of spanning tree = 14 ✓

[BE could be joined instead of BC; FH instead of FG]

(3 marks)

b Edge CK would need to be added

Minimum length = 14 + 10 = 24

(1 mark)

3 **a** [HD could be joined instead of HE.]

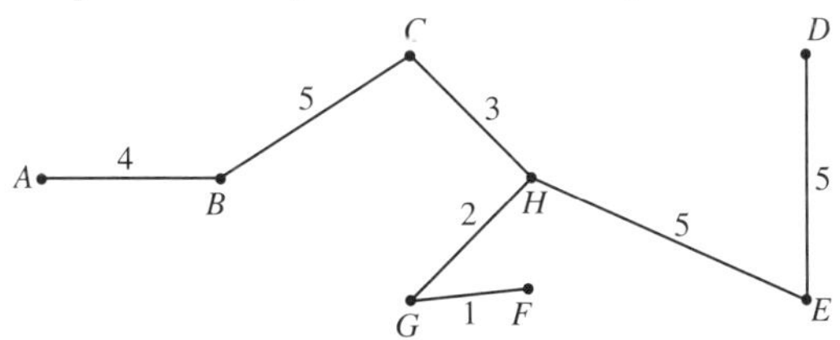

✓

Minimum length = 25 km ✓

(2 marks)

b The shortest path is $CGHE$.

(1 mark)

4

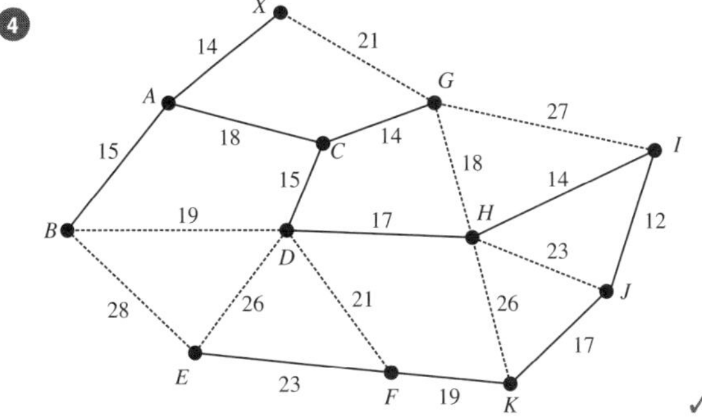

✓

14 + 15 + 18 + 14 + 15 + 17 + 14 + 12 + 17 + 19 + 23 = 178

∴ the length is 178 m.

∴ Cost = 178 × 4.50 + 11 × 22
= 1043 ✓

∴ the total cost is $1043

(2 marks)

5 **a** A cannot be reached from E.

(1 mark)

b The path is $ABCDFG$ which is 6 + 8 + 5 + 3 + 6 = 28 minutes.

(1 mark)

6 Distance = 30 + 30 + 20 + 40 + 40 + 30 + 20 + 40 + 30
= 280 ✓

Mass of cobblestones = 280 × 60 × 5
= 84 000

∴ the mass is 84 000 kg, or 84 tonnes. ✓✓

(3 marks)

7 **a**

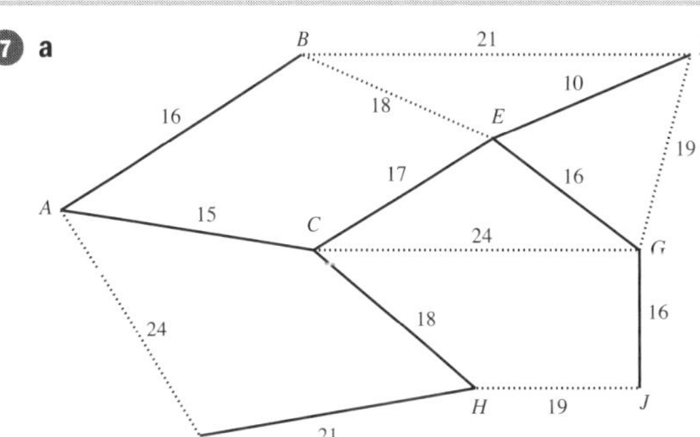

Length = 16 + 15 + 17 + 10 + 16 + 16 + 18 + 21
= 129

Cost = 129 × $35
= $4515

(1 mark)

b Reduction in length
= 15 + 16 − 18
= 13 ✓

Savings = 13 × $35
= $455 ✓

(2 marks)

8

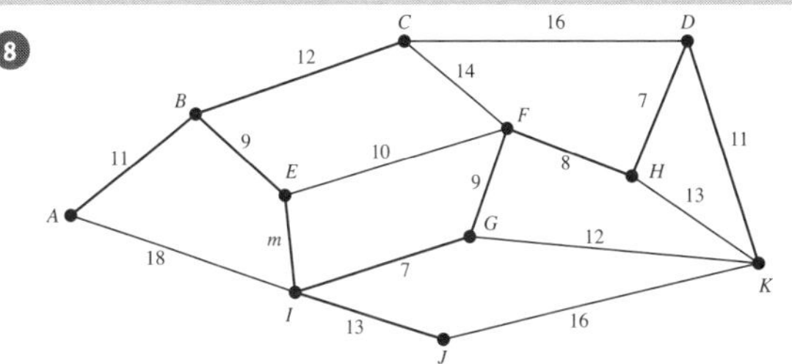

11 + 9 + m + 12 + 7 + 13 + 9 + 8 + 7 + 11 = 97

∴ m = 10

(1 mark)

9

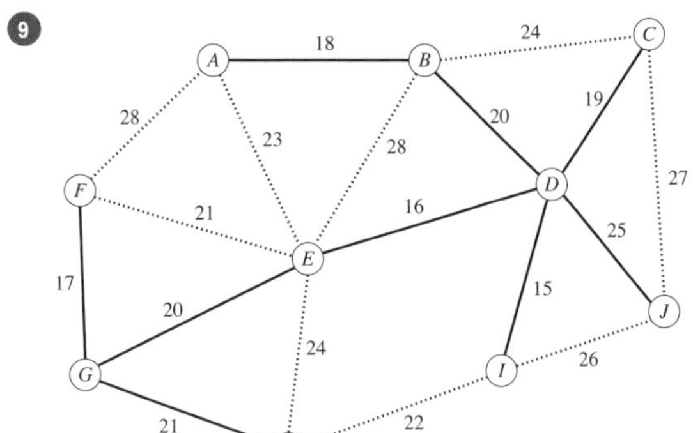

Length = 17 + 20 + 21 + 18 + 16 + 20 + 19 + 15 + 25
= 171 ✓

Cost = 171 × $42 = $7182

∴ the cost is $7182. ✓

(2 marks)

10

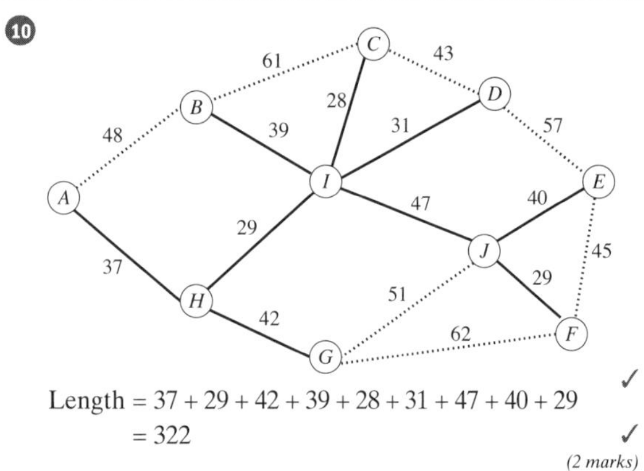

✓

Length = 37 + 29 + 42 + 39 + 28 + 31 + 47 + 40 + 29
= 322 ✓

(2 marks)

11 **a** *EDCB*: 6 + 6 + 8 = 20

∴ the quickest time is 20 minutes.

(1 mark)

b The paths are *FEA*, *FDEA*, *FDCBA*.

∴ there are 3 ways.

(1 mark)

12

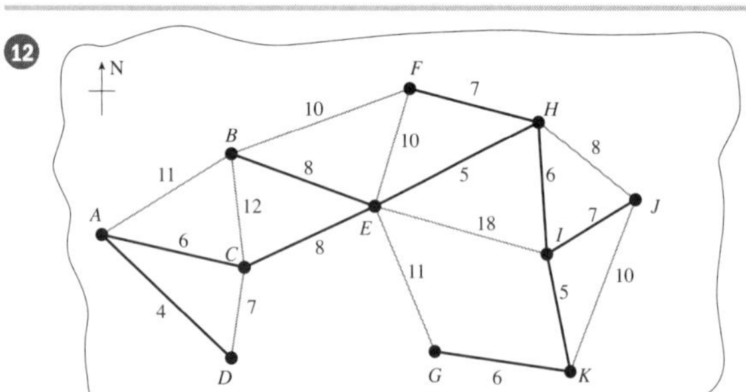

a Length of minimum spanning tree
= 6 + 4 + 8 + 6 + 5 + 7 + 6 + 5 + 7 + 8
= 62 ✓

Cost = 62 × 42
= 2604

∴ the cost is $2604. ✓

(2 marks)

b Length of new minimum spanning tree
= 6 + 4 + 9 + 6 + 5 + 7 + 8 + 8 + 10
= 63

Cost = 63 × 42
= 2646 ✓

∴ the cost is $2646.

Increase = $42

∴ even though there is one less statue involved, the price increases by $42. ✓

(2 marks)

13

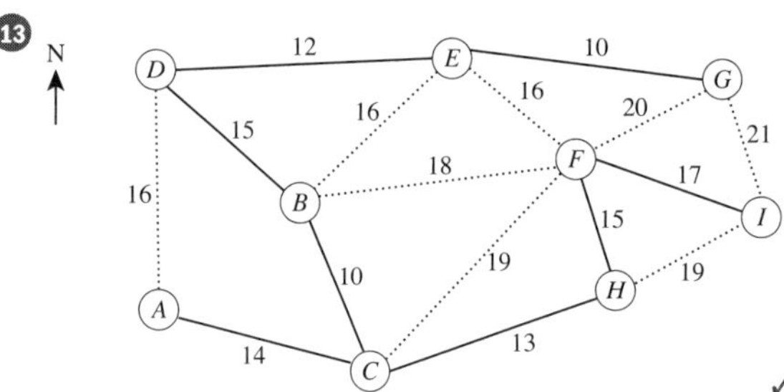

✓

a Length = 14 + 10 + 15 + 12 + 10 + 17 + 15 + 13
= 106

∴ 106 metres of water pipe ✓

(2 marks)

b

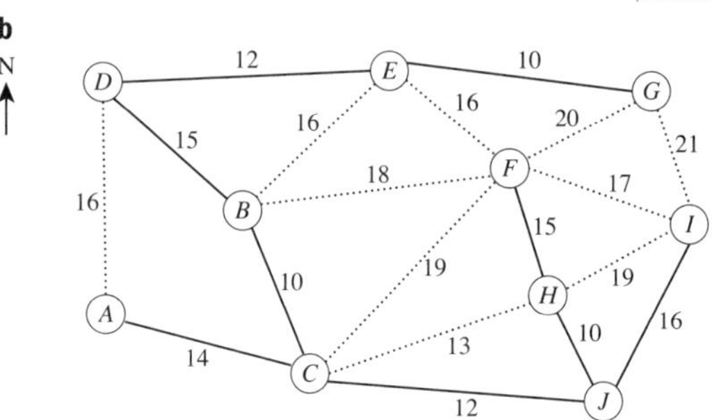

New length = 14 + 10 + 15 + 12 + 10 + 15 + 10 + 12 + 16
= 114 ✓

Difference = 114 − 106
= 8

∴ Yes, the length will increase. Arnie needs another 8 m of water pipe. ✓

(2 marks)

14 This network has vertices that are connected by more than one path. This means the network is not a tree.

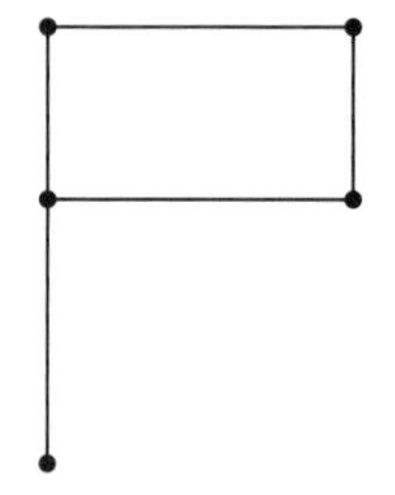

Answer B

15

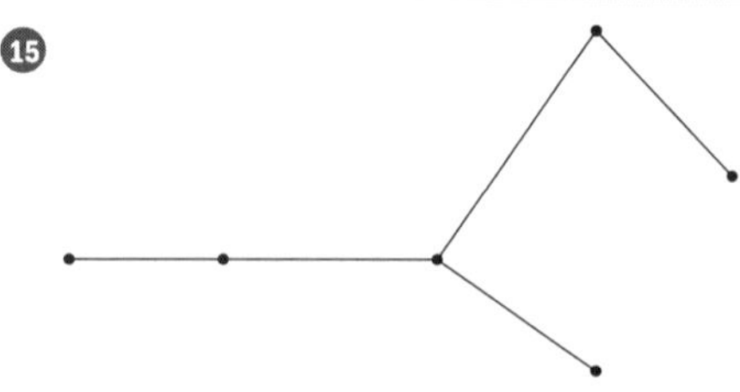

Answer D

16 **a** and **b**

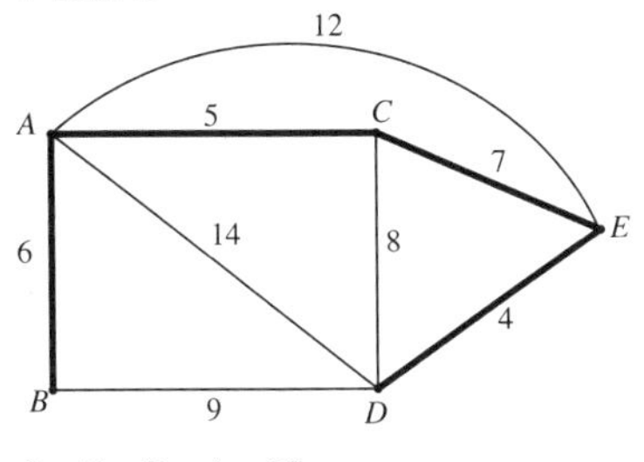

$6 + 5 + 7 + 4 = 22$ ✓✓✓ ✓

(4 marks)

17 **a**

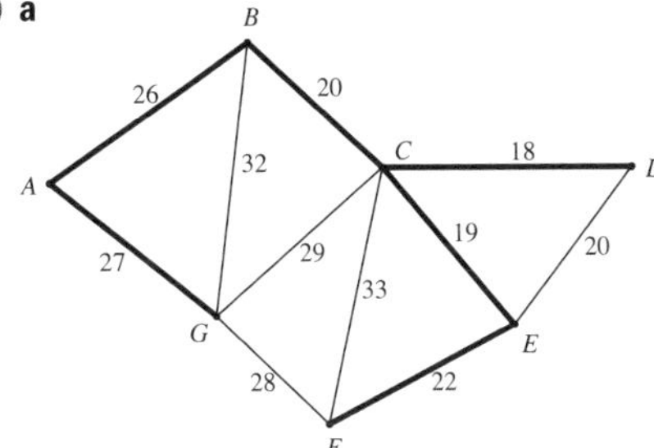

(1 mark)

b $27 + 26 + 20 + 18 + 19 + 22$
$= 132$
The length of the minimum spanning tree is 132 km. ✓
Total cost
$= \$26.50 \times 132 \times 1000$
$= \$3\,498\,000$
Profit
$= \$4\,100\,000 - \$3\,498\,000$
$= \$602\,000$
The company plans to make a profit of \$602 000. ✓

(2 marks)

18 $19 + 10 + 13 + 14 + 16 = 72$
The shortest distance is 72 km.
Answer C

19

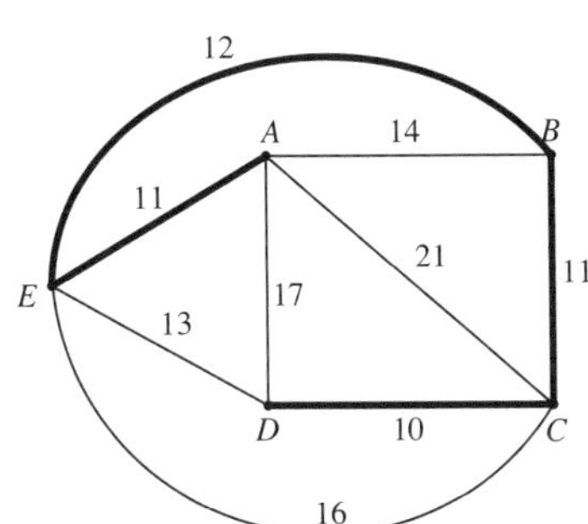

✓✓✓
$11 + 12 + 11 + 10 = 44$. The cost would be \$44 000. ✓

(4 marks)

20 **a** $19 + 13 + 31 + 23 + 15 + 9 + 24 = 134$.
The distance is 134 m.

(1 mark)

b

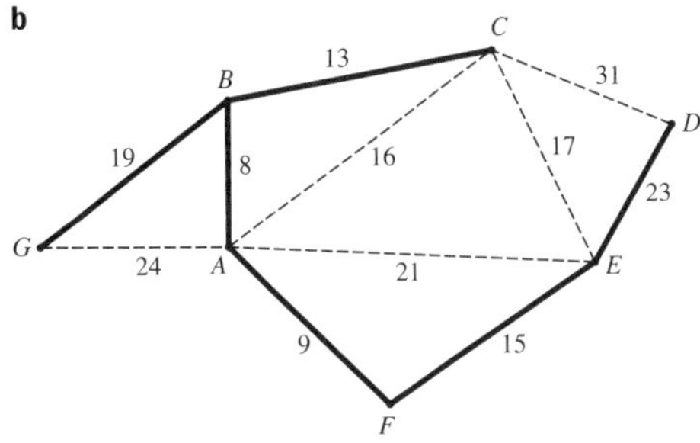

✓
$19 + 8 + 13 + 9 + 15 + 23 = 87$ The distance is 87 m. ✓

(2 marks)

21 The path is *ACDEFG*. ✓
As $17 + 16 + 17 + 19 + 14 = 83$, the minimum time is 83 minutes, or 1 h 23 min. ✓

(2 marks)

22 A spanning tree has one fewer edge than the number of vertices, which is $n - 1$.
Answer B

23 The 'shortest path' is *ADEF*. ✓
As $16 + 7 + 9 = 32$, the cost of an adult ticket is \$32.
Total cost $= 32 + 0.75 \times 32$
$= 56$
The cost would be \$56. ✓

(2 marks)

24 $7.8 + 7.9 + 8.4 + 8.8 + 9.6 + 10.8 + 9.3 + 10.4 = 73.0$.
The length of the minimum spanning tree is 73 km.

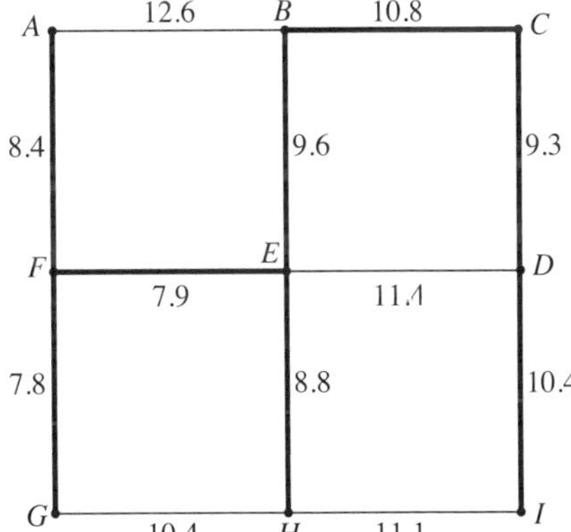

Answer A

25

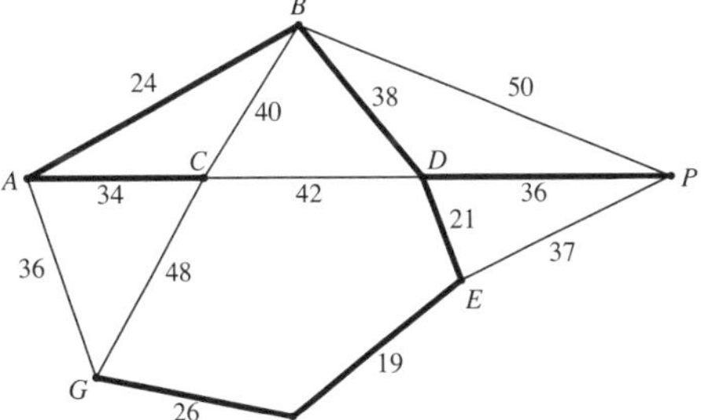

✓

Minimum length
$= 34 + 24 + 38 + 36 + 21 + 19 + 26$
$= 198$ ✓
Cost $= \$18.40 \times 198$
$= \$3643.20$ ✓

(3 marks)

26 Look for the vertices with odd degrees. He starts at C (degree 3) and finishes at I (degree 3). ✓

As $9 \times 2 + 6 = 24$, there is a total of 24 km of tracks.

From I to C is 3 km.

$$\begin{aligned}\text{Total distance} &= 24 + 3 \\ &= 27\end{aligned}$$

Logan ran 27 km. ✓

(2 marks)

1 A project requires completion of 11 tasks $A, B, C, \ldots, K$.

A network diagram for the project giving the completion time for each task, in minutes, is shown.

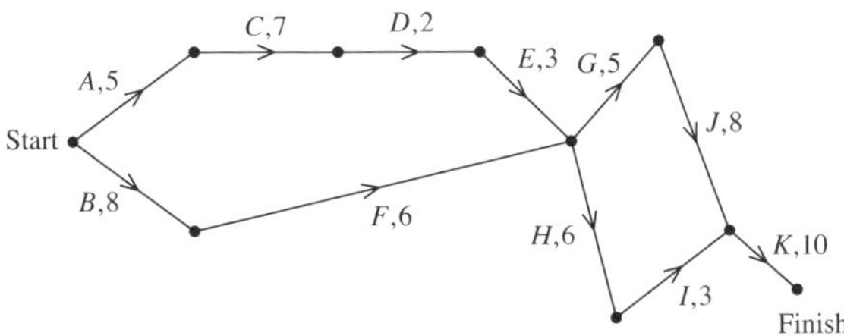

a Find the minimum time to complete the project. *(1 mark)* **Hard**

b State the critical path for this project. *(1 mark)* **Medium**

c A new task, X, is to be added to the project. The earliest starting time for X is 17 minutes, the latest starting time for X is 18 minutes and X has a completion time of 12 minutes.
Add task X to the given network diagram above AND state the float time for this task. *(2 marks)* **Hard**

(Q36, **2021 HSC**)

2 The preparation of a meal requires the completion of all ten activities A to J. The network diagram shows the activities and their completion times in minutes.

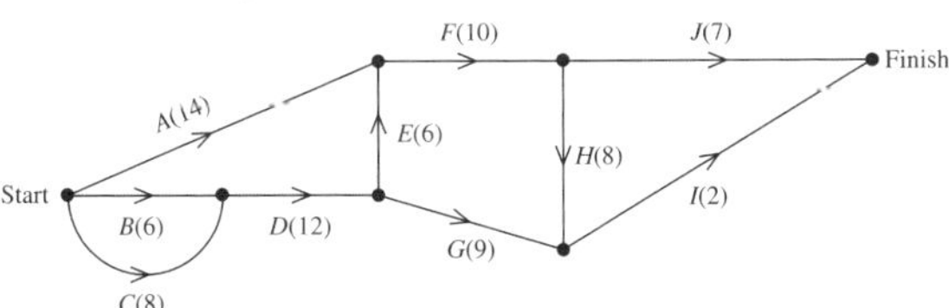

a What is the minimum time needed to prepare the meal? *(1 mark)* **Hard**

b List the activities which make up the critical path for this network. *(2 marks)* **Medium**

c Complete the table below, showing the earliest start time and float time for activities A and G. *(2 marks)* **Medium**

Activity	*Earliest start time* (minutes)	*Float time* (minutes)
A		
G		

(Q26, **2020 HSC**)

3 The network diagram shows a series of water channels and ponds in a garden. The vertices A, B, C, D, E and F represent six ponds. The edges represent the water channels which connect the ponds. The numbers on the edges indicate the maximum capacity of the channels.

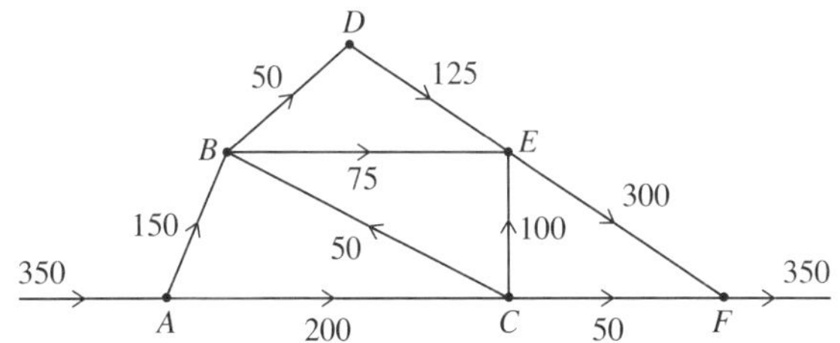

a Determine the maximum flow of the network. *(2 marks)* **Medium**

b A cut is added to the network, as shown.

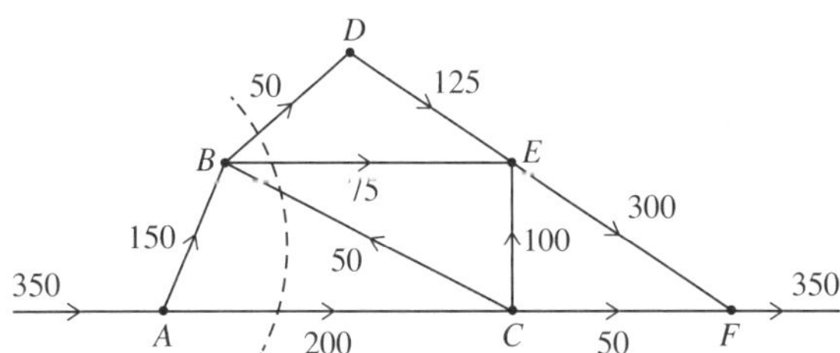

Is the cut shown a minimum cut? Give a reason for your answer. *(1 mark)* **Hard**

(Q30, **2020 HSC**)

4 A project requires activities A to F to be completed. The activity chart shows the immediate prerequisite(s) and duration for each activity.

Activity	*Immediate prerequisite(s)*	*Duration in hours*
A	–	2
B	A	6
C	A	5
D	B	2
E	C, D	4
F	E	1

a By drawing a network diagram, determine the minimum time for the project to be completed. *(3 marks)* **Medium**

b Determine the float time of the non-critical activity. *(1 mark)* **Hard**

(Q26, **2019 HSC**)

5 A museum is planning an exhibition using five rooms.

The museum manager draws a network to help plan the exhibition. The vertices A, B,

C, *D* and *E* represent the five rooms. The numbers on the edges represent the maximum number of people per hour who can pass through the security checkpoints between the rooms.

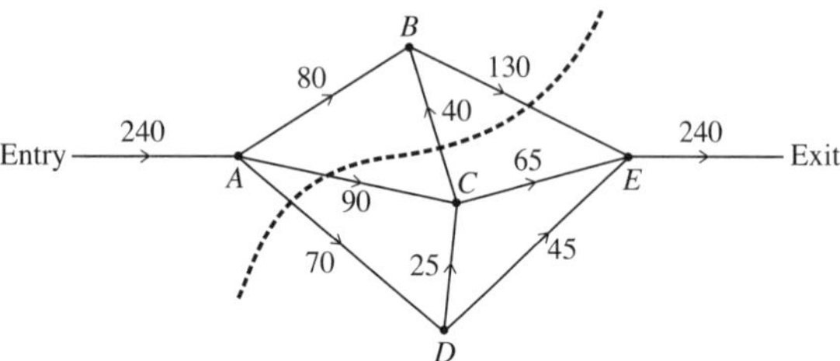

a What is the capacity of the cut shown? *(1 mark)* **Hard**

b The museum manager is planning for a maximum of 240 visitors to pass through the exhibition each hour. By using the 'minimum cut – maximum flow' theorem, the manager determines that the plan does not provide sufficient flow capacity. Draw the minimum cut onto the network below and recommend a change that the manager could make to one or more security checkpoints to increase the flow capacity to 240 visitors per hour.

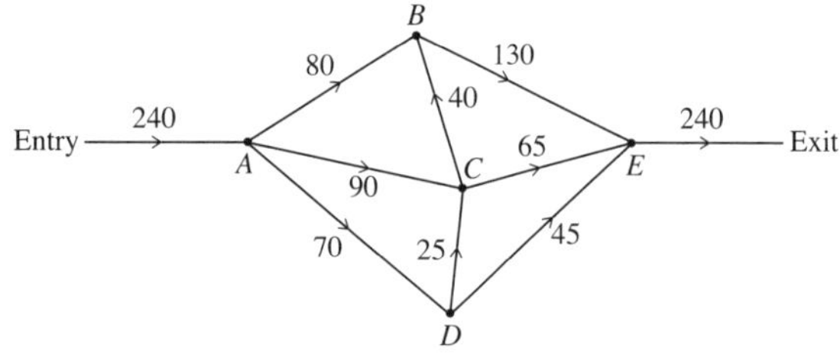

(2 marks) **Hard**

(Q40, **2019 HSC**)

6 In the network below, the values on the edges represent maximum possible flow between each pair of vertices. The arrows show the direction of flow. Cut 1 is shown on the network diagram.

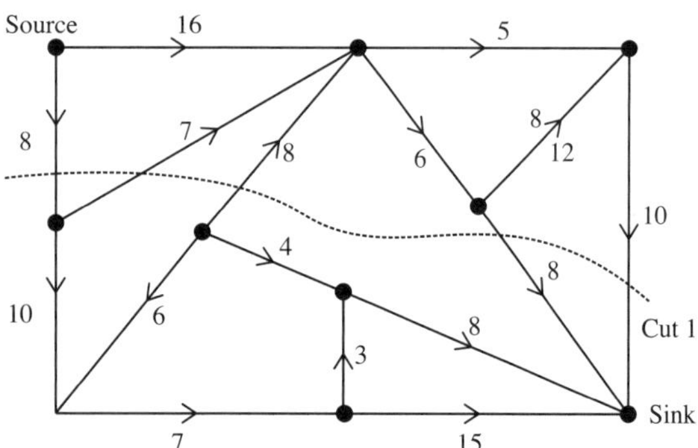

What is the capacity of Cut 1? *(1 mark)* **Easy**

Bonus question (see page iv)

7 To restore a vintage motor bike, a group of friends have determined there are 12 activities that need to be completed. The network below shows these 12 activities and their completion times in hours.

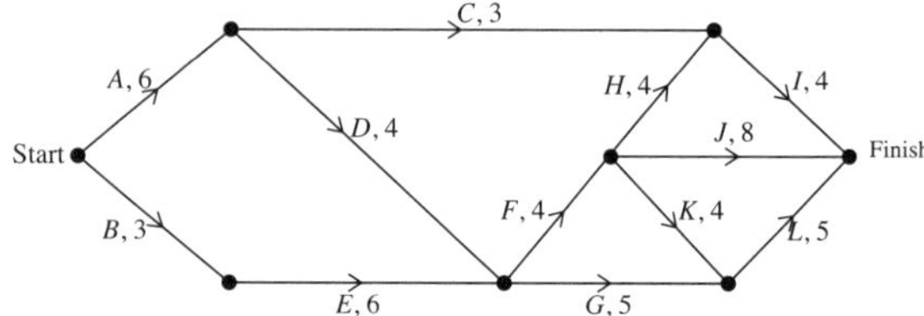

a Determine the earliest starting time of activity *G*. *(1 mark)* **Medium**

The minimum time in which all 12 activities can be completed is 23 hours.

b What is the latest starting time of activity *J*? *(1 mark)* **Medium**

c What is the float time of activity *E*? *(1 mark)* **Medium**

Just before the friends start the restoration they agree to add another activity, *P*, to the project. Activity *P* will take 3 hours to complete. Activity *P* will have no prerequisite activity, but will need to be completed before activity *F* commences.

d What is the latest starting time for activity *P* so as not to increase the minimum completion time of the project? *(1 mark)* **Hard**

The time allocation for activity *A* can be reduced by up to 4 hours at a cost of \$120 per hour. This may reduce the minimum completion time for the project, including activity *P*.

e Determine the least cost of this time reduction for activity *A* to give the greatest reduction in the minimum completion time of the project. *(1 mark)* **Hard**

Bonus question

8 The diagram below shows the flow capacity, in megalitres per hour (ML/h), of a network of gas pipes which connect a gas field (*X*) to a large storage facility (*Y*).

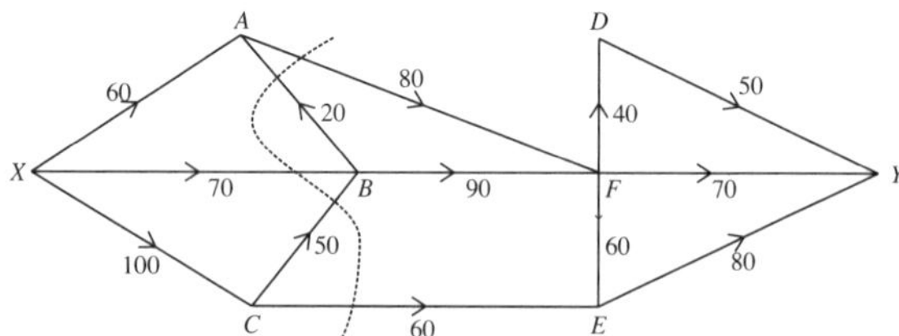

a What is the capacity of the cut shown on the diagram? *(1 mark)* **Easy**

b Find the maximum flow of gas across the network. *(1 mark)* **Medium**

c To increase the flow of gas, the energy company plans to build a new pipeline joining A to Y which will have a capacity of 70 ML/h.

i Draw the new pipeline on the diagram. *(1 mark)* **Medium**

ii What will be the new maximum flow capacity of the network? *(1 mark)* **Medium**

Bonus question

9 A project will be undertaken at a zoo. This project involves the 13 activities shown in the table below. The duration, in hours, and any prerequisite activities are also included in the table.

Activity	Duration	Prerequisites
A	4	–
B	3	F
C	2	H
D	5	J, M, B
E	6	G
F	5	A
G	2	–
H	4	F
I	5	H
J	7	G
K	4	E
L	3	C, D, K
M	3	–

Activity C is missing from the diagram for this project which is shown below.

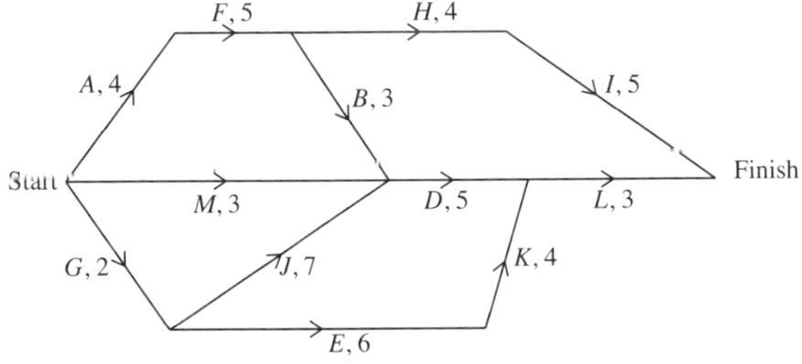

a Complete the activity diagram above by inserting activity C. *(1 mark)* **Medium**

b Determine the earliest starting time for activity K. *(1 mark)* **Medium**

c Given that activity C is not on the critical path:

i write down the activities that are on the critical path in the order they need to be completed. *(1 mark)* **Medium**

ii find the latest starting time for activity M. *(1 mark)* **Medium**

d James commented: 'If the time allocated to one activity is reduced by one hour then we could complete the project an hour earlier.' Under what circumstances would this occur? *(1 mark)* **Medium**

e The project manager can choose between two options to complete the project more quickly.

Option 1: The time allocated to activity D can be reduced by up to 3 hours at a cost of \$150 per hour.

Option 2: The time allocated to activity I can be reduced by up to 2 hours at a cost of \$120 per hour.

Which option should she choose? Support your conclusion with calculations. *(2 marks)* **Hard**

Bonus question

10 The network below represents the water flow, in kilolitres per minute, in a series of pipes.

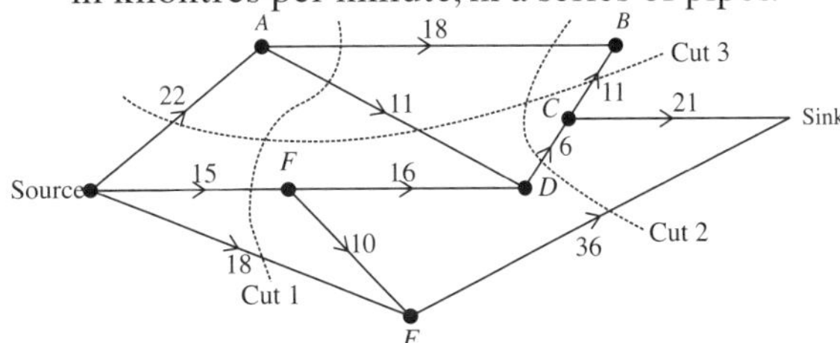

a What is the difference between the flow capacities of Cuts 1 and 2 shown on the diagram? *(1 mark)* **Medium**

b Explain why Cut 3 is NOT used to calculate the maximum flow. *(1 mark)* **Easy**

c What is the maximum flow of this network? *(1 mark)* **Medium**

Bonus question

11 The diagram shows a weighted network joining 10 vertices.

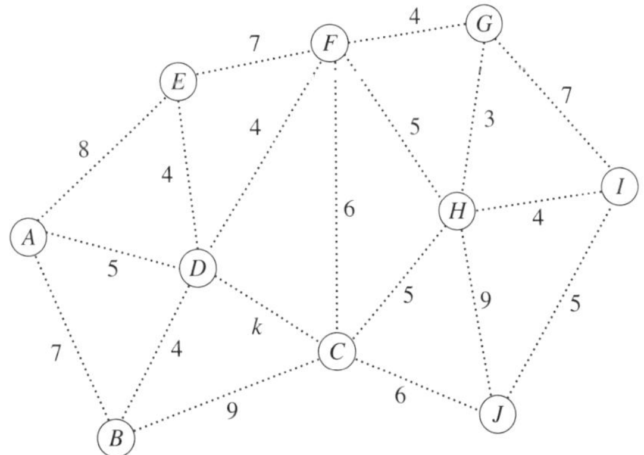

a If the minimum spanning tree of the network is 36, show that $k = 3$. *(1 mark)* **Medium**

b Another copy of the network is provided below for working for this question.

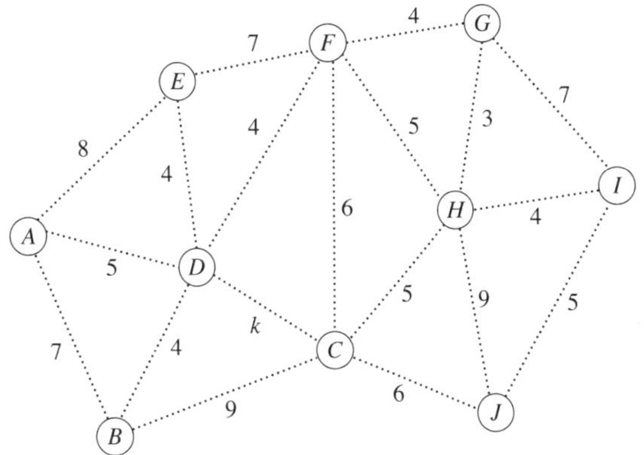

Vertex D is to be removed from the network. What is the weight of the new minimum spanning tree? *(2 marks)* **Medium**

Bonus question

12 A project is comprised of 13 activities and the duration of each activity is measured in days. The table below records the duration of each activity, any prerequisites that exist, the earliest starting time (EST) and the latest starting time (LST) of each activity.

Activity	Duration	Prerequisites	EST	LST
A	4	–	0	2
B	5	–	0	0
C	8	A	4	6
D	6	B	5	8
E	7	B	5	5
F	4	C, D	12	14
G	2	C, D	12	16
H	6	E	12	12
I	3	E	12	19
J	4	G, H	18	18
K	2	I, J	22	22
L	6	F	16	18
M	4	K, L	24	24

a What is the float time for activity I? *(1 mark)* **Medium**

b Name the critical path, and hence find the length of the project. *(2 marks)* **Medium**

c Use the table to complete the network diagram for the project, recording the duration of each activity. *(2 marks)* **Medium**

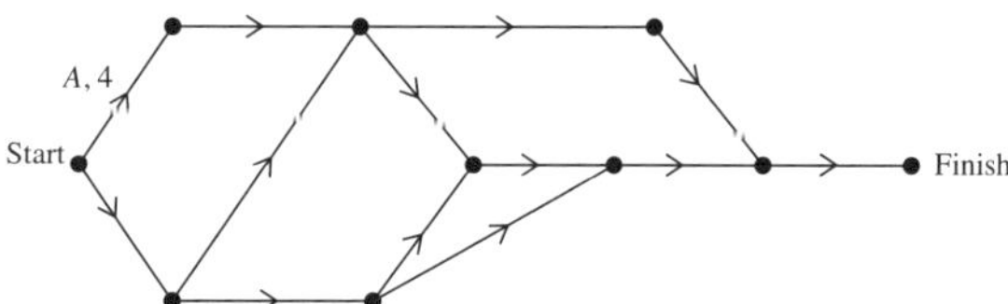

d Activity P is a new activity to be included in the project. It cannot start until after E is completed and must finish before activity M can commence. The duration of activity P is 18 days. What impact will this new activity have on the completion date of the project? *(2 marks)* **Hard**

Bonus question

13 The diagram shows the traffic flow of vehicles in vehicles/minute, using a road network travelling from a source, S, to a sink, T.

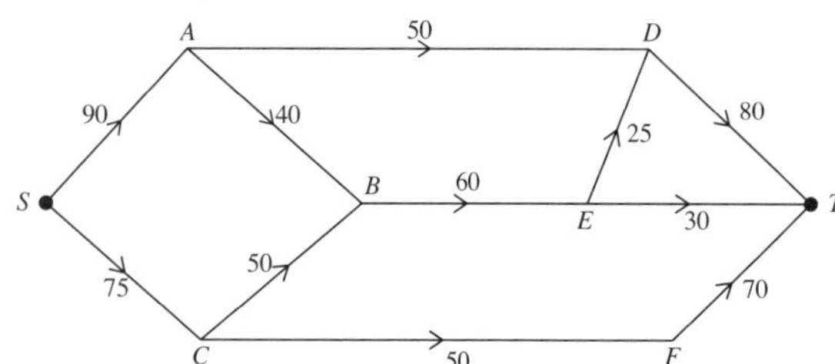

a Use a dotted line on the diagram to show the minimum cut and state the maximum flow of vehicles across the road network. *(1 mark)* **Medium**

b To increase the maximum flow in the network an engineer decides to upgrade the flow capacity of all roads to a minimum 40 vehicles/minute. The engineer expects this upgrade will improve the capacity of the network by at least 10%. Critically assess the effectiveness of this upgrade. *(2 marks)* **Hard**

Bonus question

14 A burst water main has flooded an underground carpark. The public will be stopped from using the carpark until repairs are made and the area cleaned up.

The table below shows the seven activities that need to be completed in order to repair the water pipe and open the carpark. The table also shows the duration for each activity, in minutes, the earliest start times (EST) and the prerequisites for activities.

Activity	Description	Duration	EST	Prerequisite(s)
A	Erect barriers to carpark		0	-
B	Turn off water	5	0	-
C	Pump water from carpark	150	15	A, B
D	Repair water pipe	60	165	C
E	Clean up entire area	45	165	C
F	Turn on water, check for leaks	15	225	D
G	Remove barriers	10		E, F

a What is the duration of activity A? *(1 mark)* **Medium**

b What is the earliest start time for activity G? *(1 mark)* **Hard**

c Use the table to draw a network diagram for the project. *(2 marks)* **Medium**

d What is the duration of the project? *(1 mark)* **Medium**

e What is the float time for activity E? *(1 mark)* **Medium**

f It is more complicated cleaning up the area (activity E) than expected as some additional damage was caused by the leak. It will now take twice the time originally planned.
Determine the shortest time in which activities A to G can now be completed. *(1 mark)* **Hard**

Bonus question

15 The network diagram shows the flow capacity in kL/min in water pipes connecting A to F.

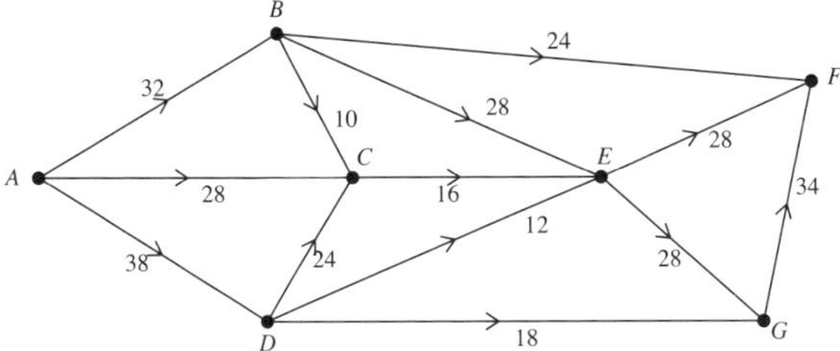

a Identify the sink of this network. *(1 mark)* Easy

b By drawing the minimum cut on the diagram, find the maximum flow of the network. *(2 marks)* Medium

c One pipe in the maximum flow network is to be replaced with a bigger pipe capable of a flow of 40 kL/min. Show that the maximum flow increases by only 8 kL/min. *(2 marks)* Hard

Bonus question

16 A new children's playground is to be built in Kahibah. The project involves 12 activities, A to L.

The directed network below shows these activities and their completion times in days.

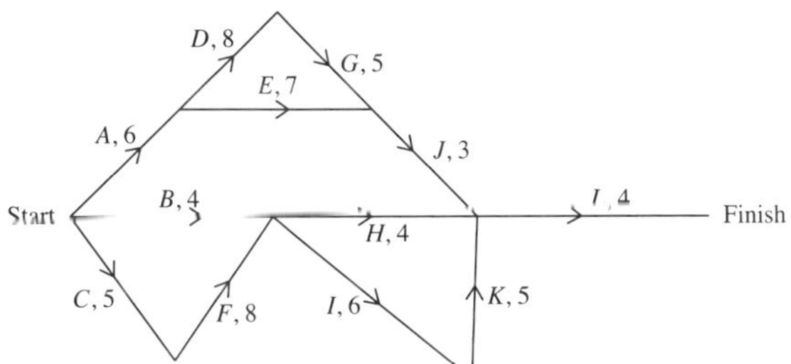

a Determine the earliest start time for activity J. *(1 mark)* Medium

b The minimum completion time for the playground is 28 days. Write down the critical path for this project. *(1 mark)* Medium

c Which activity has a float time of 7 days? *(1 mark)* Medium

d The completion times for activities D, F, H and K can be reduced by one day. The cost of reducing the completion time by one day for these activities is listed in the table below.

Activity	Cost ($)
D	4000
F	3000
H	2500
K	2800

What is the minimum additional cost to complete the project in 27 days? *(1 mark)* Hard

e The playground project will be repeated at Redhead, but with the addition of one extra activity, X. The inclusion of this new activity will not extend the project's duration. Activity X will commence once activity I has completed and will finish 2 days before the project is scheduled to be completed. If activity X has a float time of 2 days, what is the duration of the activity? *(2 marks)* Hard

Bonus question

17 A flow of oil through pipelines is represented by the network diagram where the direction of the flow (in litres per second) is shown by the arrows.

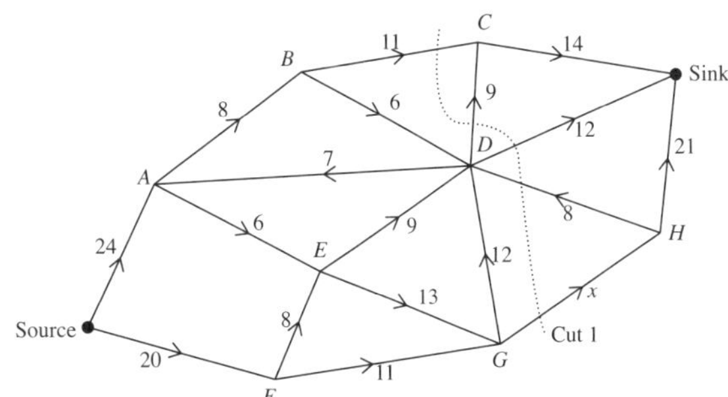

a The line shown as Cut 1 represents a flow of 43 L/second. What is the value of x? *(1 mark)* Easy

b What is the maximum flow of oil through the network from source to sink? Show working by drawing a line to represent the minimum cut through the network. *(1 mark)* Medium

Bonus question

18 The following table, consisting of 12 activities, contains information for a project in a small fabricating workshop.

Activity	Prerequisite(s)	Duration (in hours)
A	–	4
B	A, D	6
C	B, E	3
D	G	4
E	H	5
F	B, E	2
G	–	3
H	G	3
I	H	4
J	F, I, L	4
K	–	7
L	K	5

a Complete the project network below.

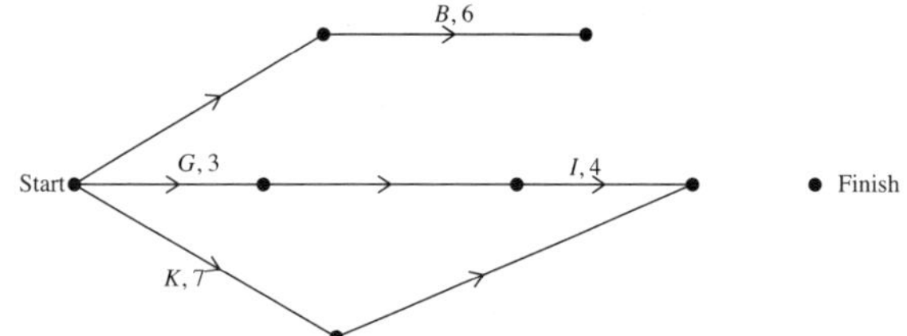

(2 marks) Medium

b State the critical path and the minimum completion time for this network. (2 marks) Hard

c Due to unforeseen circumstances, either activity C or activity H will take an extra 4 hours to complete. Which of these activities should be chosen if the completion time for the project is to be kept to a minimum? Justify your answer with calculations. (2 marks) Medium

Bonus question

19 The network diagram represents a series of roads connecting a shopping centre to a motorway. Traffic moves in the direction shown on the network. For example, the road from A to C indicates traffic can flow from A to C at a maximum flow of 40 vehicles per minute.

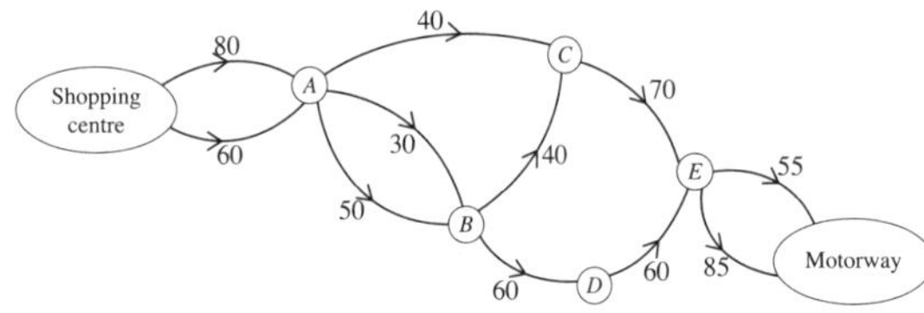

a What is the existing maximum flow for the road network? (1 mark) Medium

b It has been decided that one new road will be built to maximise the flow of traffic from the shopping centre to the motorway. This new road cannot start from the shopping centre, and also cannot join the motorway. Which is the best road to build and what will be the minimum capacity of this road to maximise the flow of vehicles across the network? (2 marks) Hard

Bonus question

20 A construction project involves 10 activities A to J.

The table shows the activities and their duration (in hours), earliest starting time (EST), latest starting time (LST) and prerequisites.

Activity	Duration	EST	LST	Prerequisite(s)
A	6	0	0	–
B	8			A
C	5			A
D	6			B, E
E	7			C
F	9			C
G	5			C
H	6			D, F
I	8			G
J	5			H, I

a What is the least number of activities that must be completed before activity H can commence? (1 mark) Medium

b Complete the network diagram for the project.

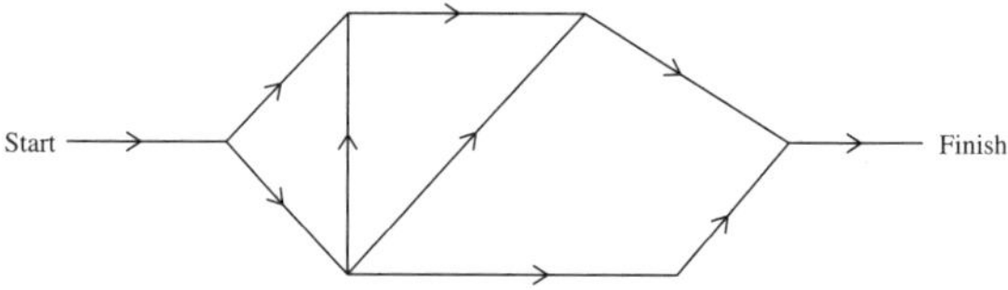

(1 mark) Hard

c Hence, or otherwise, complete the table. (2 marks) Hard

d What is the duration of the project? (1 mark) Medium

e A problem with the project means that activity G will be delayed for 12 hours after activity C has been completed. If any delay in the project costs \$1500 per hour, what will be the additional cost for the project? (1 mark) Hard

Bonus question

21 A project manager outlines a project, in days, using the network diagram below.

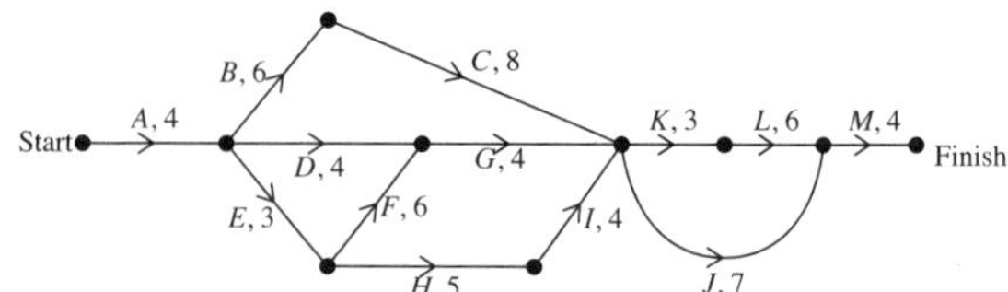

a Use the diagram to find the earliest and latest starting times for activity G. (2 marks) Medium

b What is the minimum time to complete the project? (1 mark) Medium

c The project manager decides to subcontract part of the project work. The subcontractors are able to speed up the project to enable an earlier completion time. The activities that can be reduced in time are C, J and L and the maximum reduction for each activity is 2 days. The cost of reducing the time of any activity is \$14 500 per day. Determine the minimum time, in days, for the project to be completed and its associated minimum additional cost. (3 marks) Hard

Bonus question

22 Stormwater enters a network of pipes at either Source 1 or Source 2 and flows into the ocean through either Outlet X or Outlet Y. The network diagram shows a series of pipes and the arrows represent the flow of stormwater. The numbers next to the arrows represent the maximum rate of stormwater, in kL/minute.

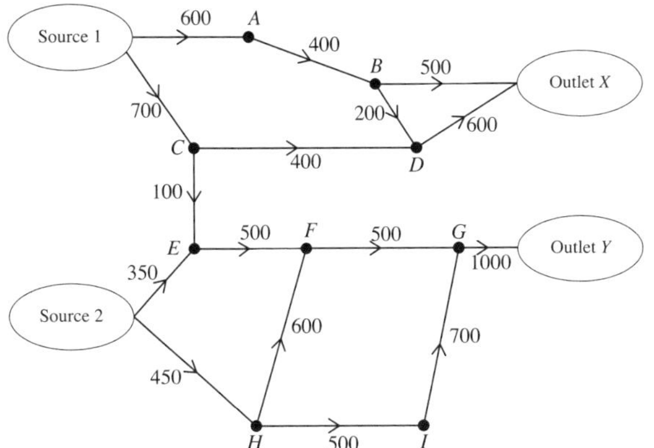

a Explain why water from Source 2 cannot flow to Outlet X. *(1 mark)* Easy

b What is the maximum flow into Outlet Y? *(1 mark)* Medium

c The pipe joining C to E is to be replaced with a larger pipe to increase the flow of stormwater from Source 1. The cost of the new pipe rises with an increase in flow capacity. What should be the flow capacity of the new pipe to minimise cost but maximise flow through Outlet Y? *(2 marks)* Hard

Bonus question

23 The table lists the 12 activities in a project.

Activity	Time (in days)	Prerequisite(s)
A	3	-
B	3	-
C	1	B
D	8	A, C
E	4	A, C
F	6	B
G	11	B
H	4	E, F
I	7	D, H
J	11	D, H
K	9	G, I
L	9	J, K

a Use the table to complete the network diagram below.

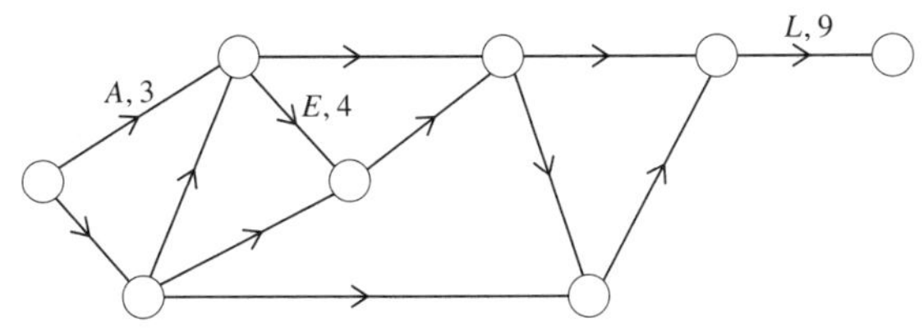

(2 marks) Medium

b What is the float time for activity J? *(1 mark)* Medium

c Name the critical path. *(1 mark)* Medium

d If the earliest date that activity F can start is 17 July, what is the latest starting date for activty I? *(1 mark)* Medium

Bonus question

24 The diagram shows activities in a project.

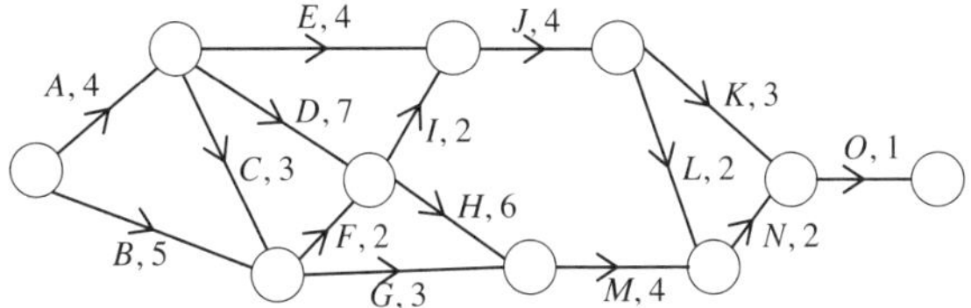

a Complete the activity chart for the network diagram above. *(2 marks)* Medium

Activity	Time (in days)	Prerequisite(s)
A	4	–
B	5	–
C	3	A
D	7	A
E	4	A
F	2	
G	3	B, C
H		
I	2	D, F
J	4	F, I
K	3	J
L	2	J
M	4	G, H
N	2	L, M
O	1	

b Identify the critical path and find its length. *(2 marks)* Hard

c What is the slack time for activity L? *(1 mark)* Medium

d The time for activity H can be reduced by up to 4 days at a cost of \$6000 per day. What is the total cost of minimising the project's duration? Support your explanation with calculations. There is another network diagram below to assist with your solution. *(2 marks)* Hard

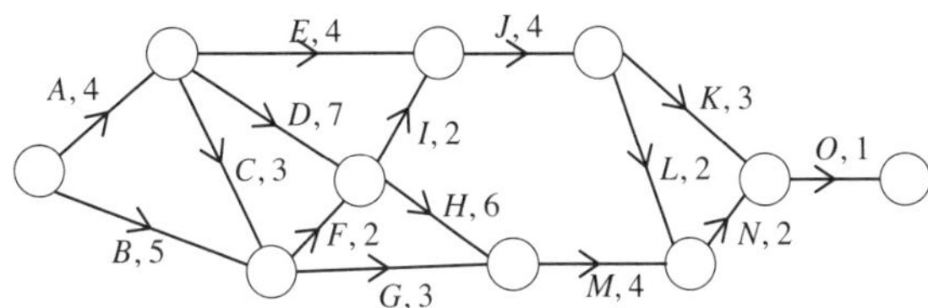

Bonus question

25 The diagram shows a network of water pipes where the rate of flow is represented on each length of pipe, in kL/minute. The direction of flow is shown by the direction of the arrow.

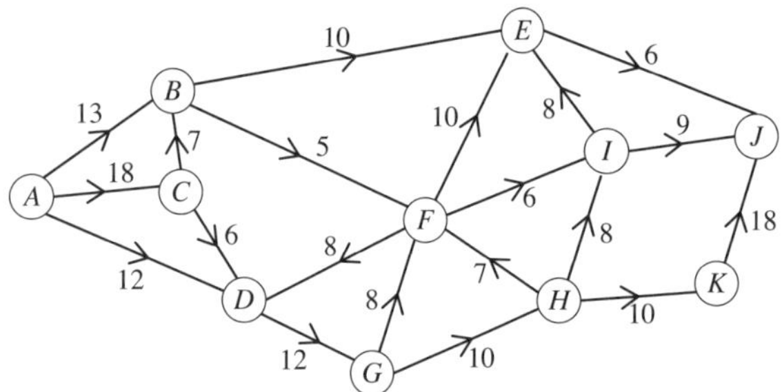

a Draw the line of minimum cut on the diagram. What is the maximum flow of water from A to J? *(2 marks)* **Medium**

b An additional pipeline is to be constructed from B to J. Water will flow through this pipe to J at the rate of 20 kL/minute. What is the increase in the maximum flow through the network? *(1 mark)* **Hard**

Bonus question

26 The diagram represents a network of sewer pipes where capacity is measured in litres per second and the direction of flow is represented by arrows.

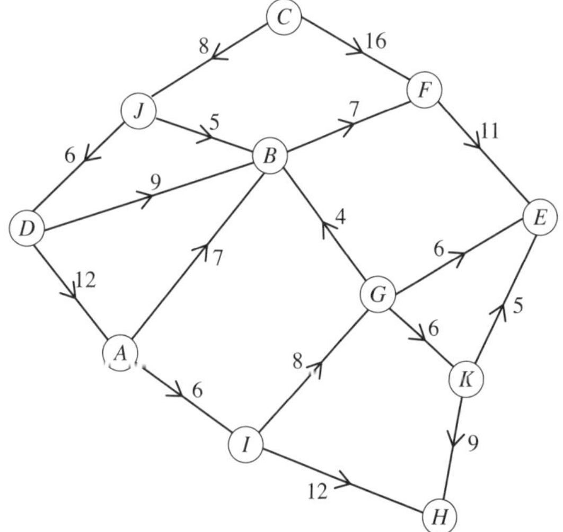

Supposing D is the source and E is the sink for the flow across the network, draw a minimum cut and calculate the maximum flow across the network. *(1 mark)*

Bonus question **Medium**

27 To revitalise a children's playground area, a project officer has determined 15 activities that need to be completed. The network diagram below shows these activities and their completion times in days. The minimum time in which all activities can be completed is 36 days.

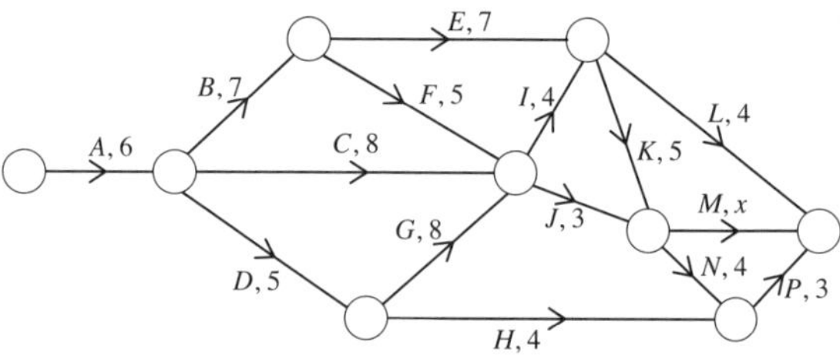

a The critical path is $ADGIKM$. Explain why $x = 8$. *(1 mark)* **Medium**

b What is the latest starting time for activity I? *(1 mark)* **Medium**

c What is the float time of activity H? *(1 mark)* **Medium**

d The time allocation for activity M can be reduced by up to 3 days at a cost of $2300 per day. Determine the least cost of this time reduction for activity M to give the greatest reduction in the minimum completion time of the project. *(1 mark)* **Hard**

Bonus question

Year 12 Critical path analysis—Worked answers

1 a Minimum time
$= (5 + 7 + 2 + 3 + 5 + 8 + 10)$ min
$= 40$ min

(1 mark)

b The critical path is A, C, D, E, G, J, K.

(1 mark)

c EST = 17 min
So X should start after task E.
LST = 18 min
Completion time = 12 min
Latest finish time = $(18 + 12)$ min
$= 30$ min
So X should finish before task K.

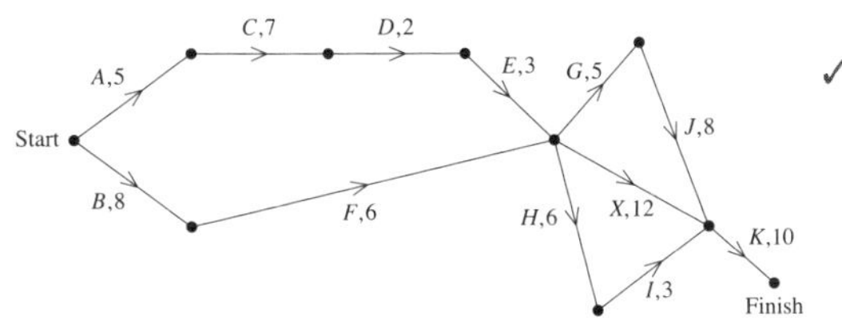

✓

The float time is 1 minute. ✓

(2 marks)

2 a $8 + 12 + 6 + 10 + 8 + 2 = 46$
Minimum time = 46 minutes

(1 mark)

b Critical path has activities C, D, E, F, H and I. ✓✓

(2 marks)

c

Activity	*Earliest start time* (minutes)	*Float time* (minutes)
A	0	12
G	20	15

✓✓ *(2 marks)*

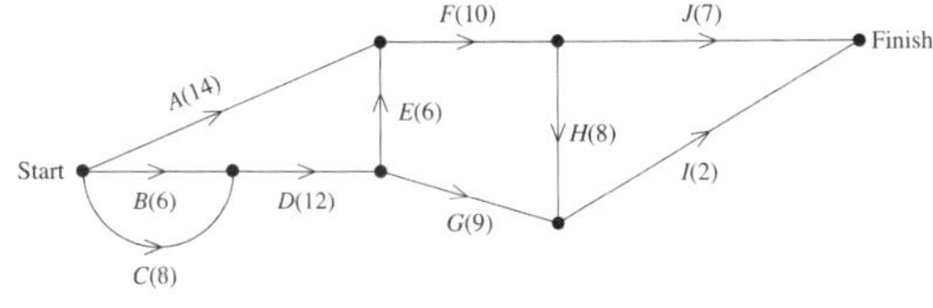

3 a Maximum flow = 275
Vertex B has input of 200 but output of 125. ✓
Vertex D has input of 50 so can only have output of 50, not 125. ✓
So vertex E has input and output of 225.

(2 marks)

b No, the value of the cut is $50 + 75 + 200 = 325$.
As this is more than the maximum flow it cannot be the minimum cut.
The minimum cut is:

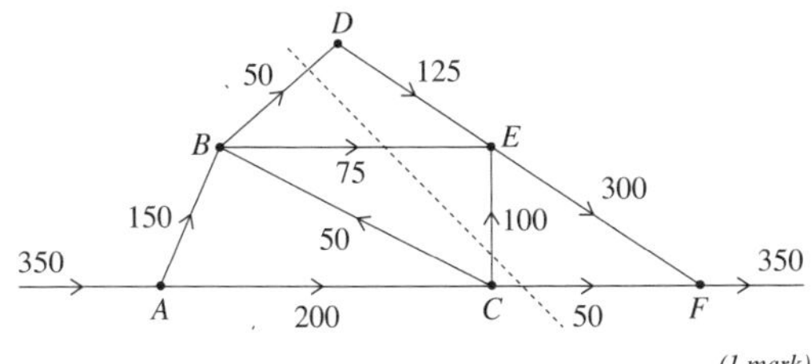

(1 mark)

4 a

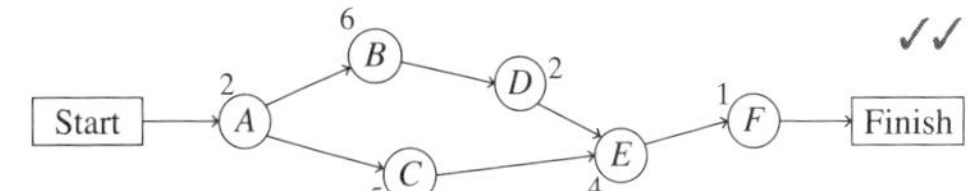

✓✓

Minimum time = 15 hours ✓

(3 marks)

b Float time of activity C is 3 hours.

(1 mark)

5 a Capacity $= 70 + 90 + 130$
$= 290$

(1 mark)

b

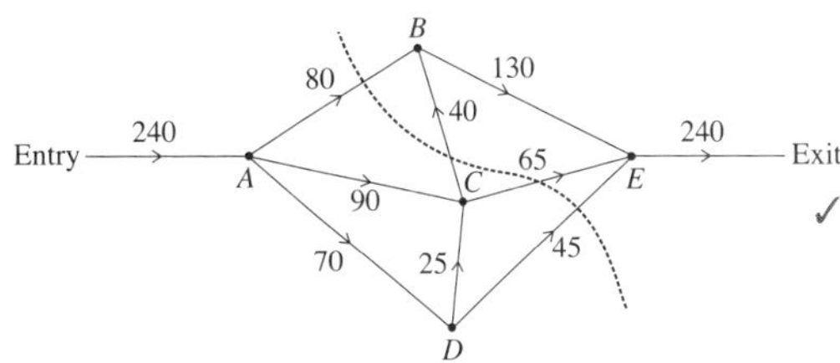

✓

Increasing the capacity of edge CB to 50 would increase the flow capacity to 240 visitors per hour. ✓

(2 marks)

6 $8 + 8 + 10 = 26$

(1 mark)

7 a $6 + 4 = 10$
$\therefore$ 10 hours

(1 mark)

b $23 - 8 = 15$
$\therefore$ 15 hours

(1 mark)

c $10 - 6 - 3 = 1$
$\therefore$ 1 hour

(1 mark)

d $6 + 4 - 3 = 7$
$\therefore$ 7 hours

(1 mark)

e Path $BEFKL$ is already 22 hours, which means the least cost is $120.

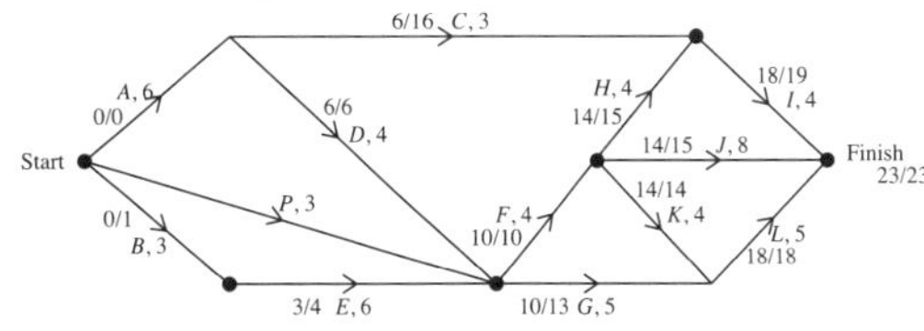

(1 mark)

8 a $80 + 70 + 50 + 60 = 260$
$\therefore$ 260 ML/h

(1 mark)

b $40 + 70 + 80 = 190$ (cutting DF, FY and EY)
$\therefore$ 190 ML/h

(1 mark)

c i

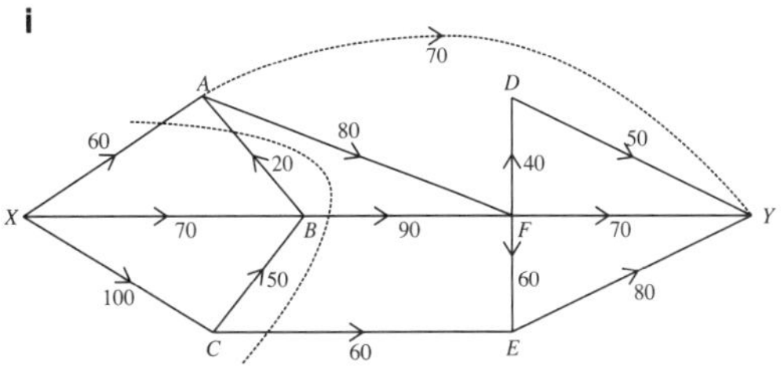

(1 mark)

ii The new minimum cut is $60 + 90 + 60 = 210$ (cutting AX, AB, BF and CE).
$\therefore$ 210 ML/h. (Note: '20' on AB is not used as the arrow is in the reverse direction, from 'sink' to 'source'.)

(1 mark)

---

9 **a**

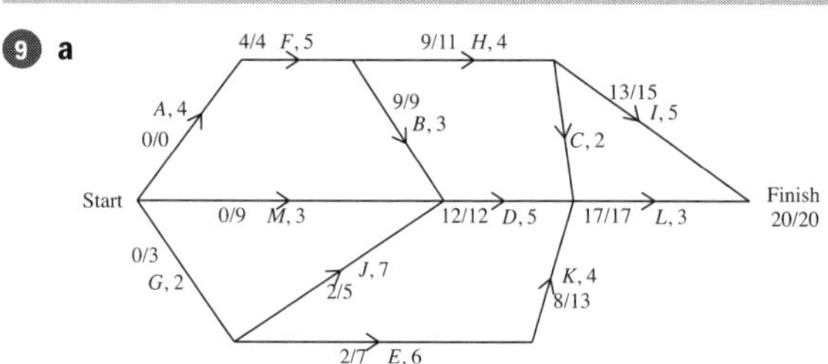

(1 mark)

b $2 + 6 = 8$
$\therefore$ 8 hours

(1 mark)

c **i** A, F, B, D, L

(1 mark)

ii $12 - 3 = 9$
$\therefore$ 9 hours

(1 mark)

d This would only occur if the activity was on the critical path: A, F, B, D, L.

(1 mark)

e $AFHI$ presently has a length of 18 hours.
Option 1: As $2 \times \$150 = \$300$, the project time is $20 - 2 = 18$ hours for critical path $AFBDL$ ✓
Reducing activity D by 2 hours means the two critical paths are now 18 hours each. ✓

(2 marks)

---

10 **a** Cut 1 is 62 kL/min and Cut 2 is 60 kL/min. The difference is 2 kL/min.

(1 mark)

b Cut 3 does not separate sink and source.

(1 mark)

c $18 + 6 + 10 = 34$
(the cut goes through Source: E, EF, DC, CB)
$\therefore$ maximum flow is 34 kL/min.

(1 mark)

---

11 **a** $5 + 4 + 4 + k + 4 + 4 + 3 + 4 + 5$
$= 36$
$33 + k = 36$
$k = 3$

(1 mark)

b New weight $= 7 + 8 + 7 + 4 + 3 + 5 + 4 + 5$
$= 43$ ✓

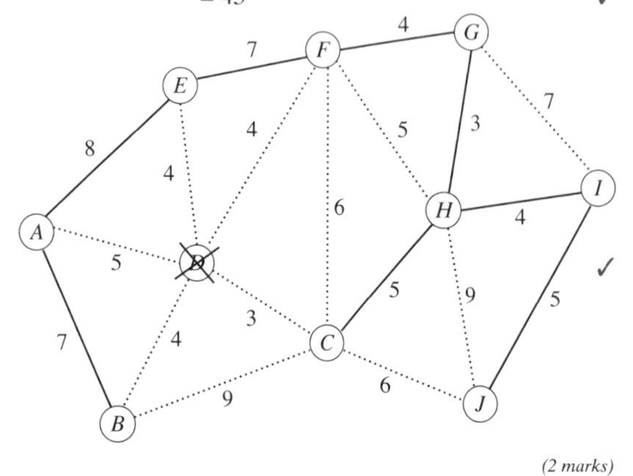

✓

(2 marks)

---

12 **a** $19 - 12 = 7$
$\therefore$ The float time is 7 days.

(1 mark)

b $BEHJKM$, a length of 28 days ✓✓

(2 marks)

c

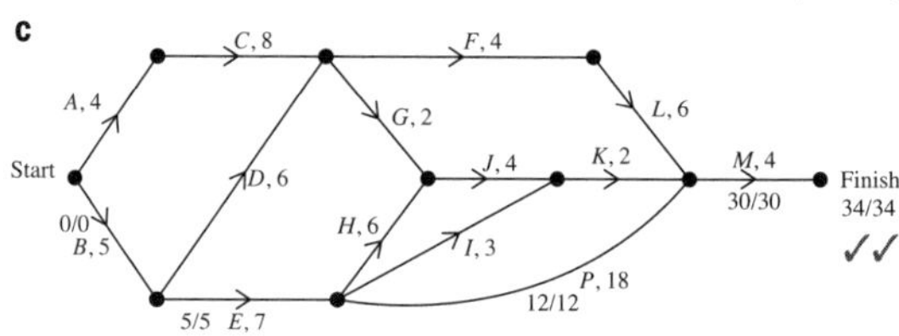

✓✓

(2 marks)

d The new critical path is $BEPM$ which takes 34 days. ✓ The project will now take 6 days longer to complete. ✓

(2 marks)

---

13 **a**

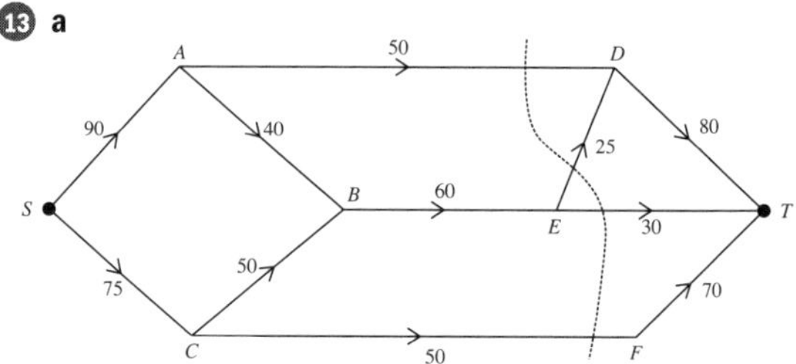

Maximum flow is
$50 + 25 + 30 + 50 = 155$
$\therefore$ 155 vehicles/minute

(1 mark)

b Increasing ED and ET to 40 vehicles/minute means that the cut used in part **a** gives 180 vehicles/minute. But another cut through AD, BE and CF gives 160 vehicles per minute. This is only a 5 vehicles/minute improvement. ✓

As $\frac{5}{155} \times 100\% = 3.2\%$, this is much lower than the engineer's expectations. ✓

(2 marks)

---

14 **a** 15 minutes

(1 mark)

b $225 + 15 = 240$
$\therefore$ 240 minutes

(1 mark)

c

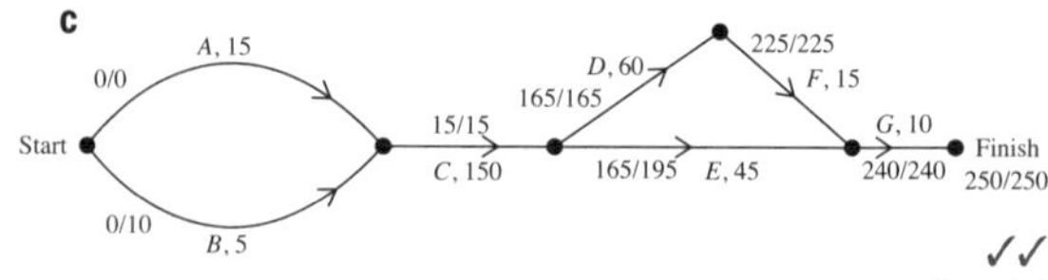

✓✓

(2 marks)

d $225 + 15 + 10 = 250$
$\therefore$ 250 minutes

(1 mark)

e $250 - 10 - 45 - 165 = 30$
$\therefore$ 30 minutes

(1 mark)

f An extra 15 minutes as the critical path is now $ACEG$ rather than $ACDFG$.
The shortest time is now 265 minutes.

(1 mark)

15

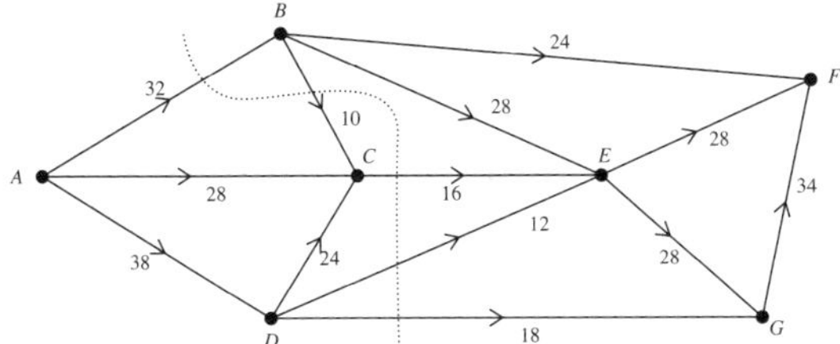

a F

(1 mark)

b $32 + 16 + 12 + 18 = 78$ (10 is not used as wrong direction)

$\therefore$ 78 kL/min ✓✓

(2 marks)

c Replacing CE with a new pipe means the previously calculated maximum flow is now 102 kL/min. However, the new minimum cut through BF, EF and GF gives a maximum flow of $24 + 28 + 34 = 86$. ✓

As $86 - 78 = 8$, the maximum flow only increases by 8 kL/min. ✓

(2 marks)

16 a 19 days

(1 mark)

b $CFIKL$

(1 mark)

c Activity H

(1 mark)

d F and K are on the critical path, so one day's reduction would cost \$2800 (by reducing K by one day).

(1 mark)

e

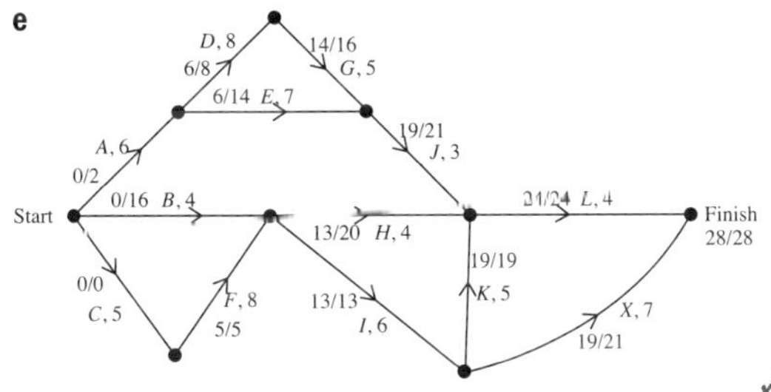

✓

$28 - 19 - 2 = 7$ ✓

$\therefore$ Activity X will take 7 days.

(2 marks)

17 a $11 + 9 + 12 + x = 43$

$\therefore x = 11$

(1 mark)

b

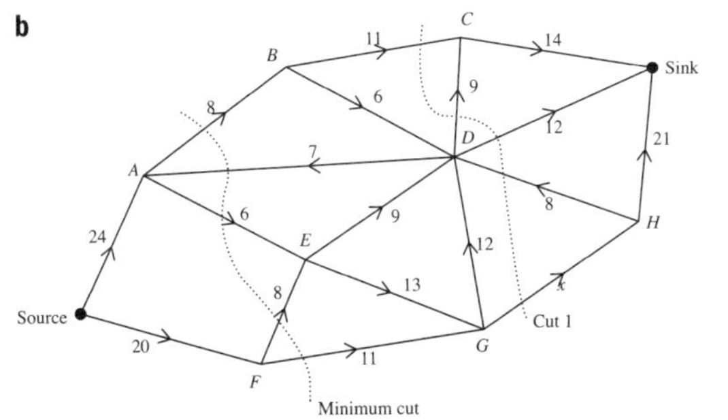

The maximum flow is $8 + 6 + 8 + 11 = 33$.

$\therefore$ 33 L/second

(1 mark)

18 a

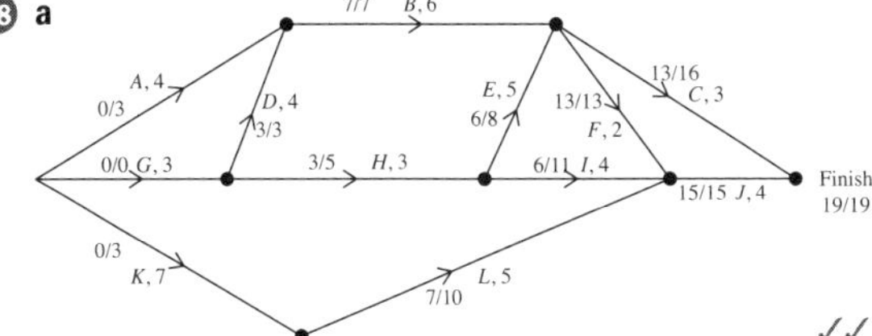

✓✓

(2 marks)

b $GDBFJ$, 19 hours ✓✓

(2 marks)

c Increasing C by 4 hours means the path involving C has a maximum of 20 hours ($GDBC$). Increasing H by 4 hours means the path involving H has a maximum of 21 hours ($GHEFJ$). Best to choose activity C to increase the project by just 1 hour. ✓✓

(2 marks)

19 a Minimum cut:

$40 + 30 + 50 = 120$

$\therefore$ maximum flow is 120 vehicles per minute.

(1 mark)

b As $80 + 60 = 140$ and $55 + 85 = 140$, the flow capacity from the source to the sink is 140. Building a new road linking A to E with a capacity of 20 vehicles per minute will ensure a maximum flow of 140 vehicles per minute. ✓✓

(2 marks)

20 a Six activities must be completed (A, B, C, D, E, F).

(1 mark)

b

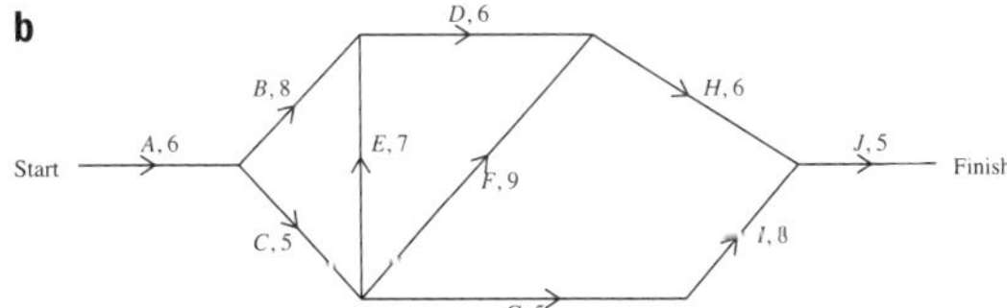

(1 mark)

c

Activity	Duration	EST	LST	Prerequisite
A	6	0	0	–
B	8	6	10	A
C	5	6	6	A
D	6	18	18	B, E
E	7	11	11	C
F	9	11	15	C
G	5	11	17	C
H	6	24	24	D, F
I	8	16	22	G
J	5	30	30	H, I

✓✓ *(2 marks)*

d The critical path is $ACEDHJ$ which is 35 hours.

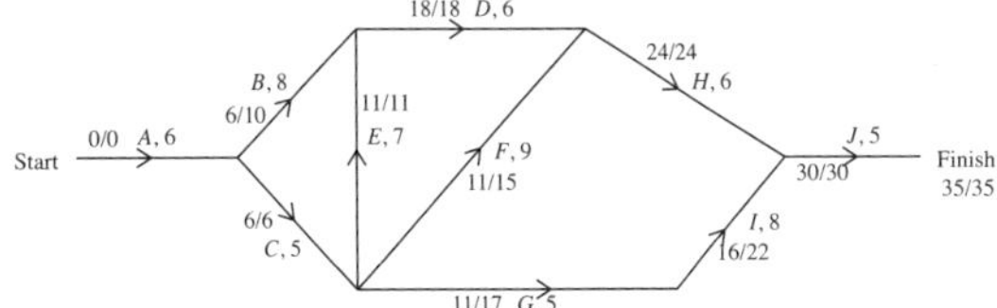

(1 mark)

e $6 + 5 + 12 + 5 + 8 + 5 = 41$ which means a new critical path. The 6-hour delay will cost $6 \times \$1500 = \$9000$.

(1 mark)

21 **a** EST = 13 days and ✓
LST = 14 days ✓

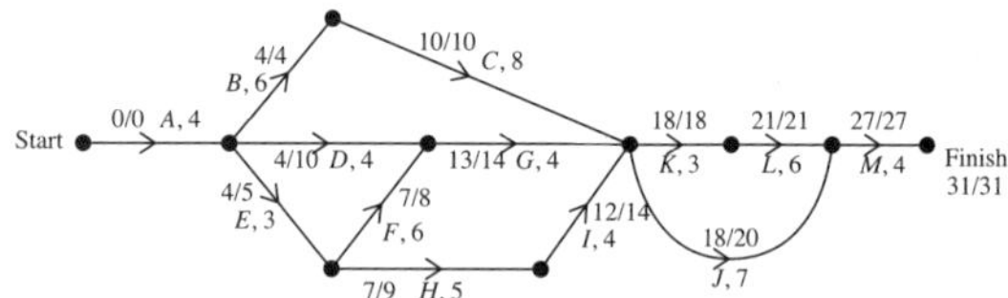

(2 marks)

b $ABCKLM$ = 31 days *(1 mark)*

c Reducing C and L by 2 days each means the $ABCKLM$ path is 27. However, this also means that the $AEFGKLM$ path is 28. The cost of the reduction by 3 days is 3 × \$14 500 = \$43 500. ✓✓✓ *(3 marks)*

22 **a** The arrow on CE means that water can only flow *from* Source 1. *(1 mark)*

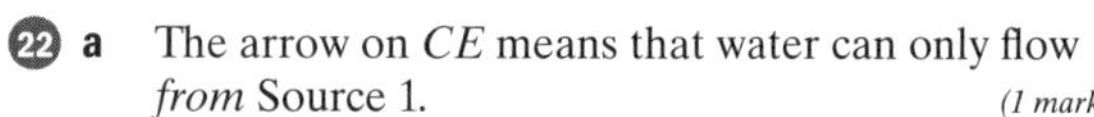

b 100 + 350 + 450 = 900
∴ 900 kL/min *(1 mark)*

c Increase capacity of CE to 150 means the minimum cut and hence maximum flow is 950 kL/min through Outlet Y.
∴ the new flow capacity should be 150 kL/min ✓✓ *(2 marks)*

23 **a**

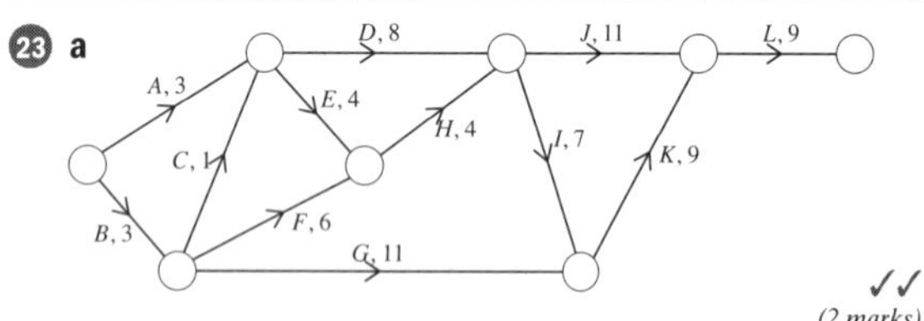

✓✓ *(2 marks)*

b 5 days *(1 mark)*

c $BFHIKL$

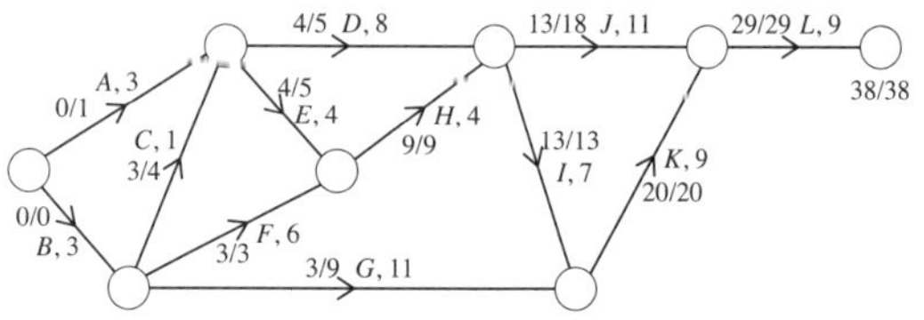

(1 mark)

d Activity I lies on the critical path. As 17 − 3 = 14, the project started on 14 July, and as 14 + 13 = 27, the latest starting date for activity I is 27 July. *(1 mark)*

24 **a**

Activity	Time (in days)	Prerequisite(s)
A	4	-
B	5	-
C	3	A
D	7	A
E	4	A
F	2	B, C
G	3	B, C
H	6	D, F
I	2	D, F
J	4	E, I
K	3	J
L	2	J
M	4	G, H
N	2	L, M
O	1	K, N

✓✓ *(2 marks)*

b

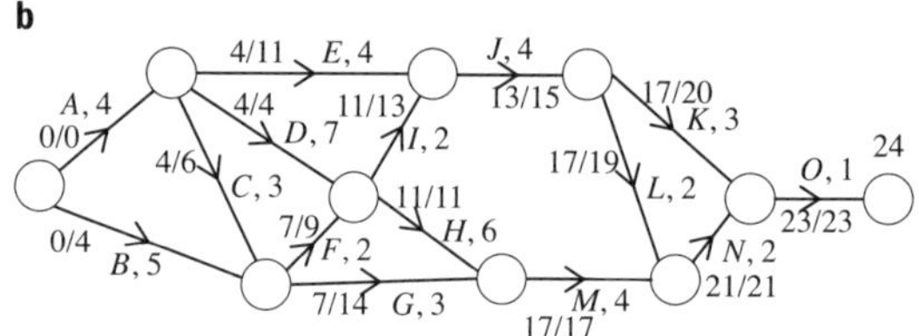

$ADHMNO$, 24 days ✓✓ *(2 marks)*

c 2 days *(1 mark)*

d Reducing activity H by 2 days means there are now two critical paths.
As 6000 × 2 = 12 000, the cost is \$12 000 to reduce the project's duration by 2 days. ✓ *(2 marks)*

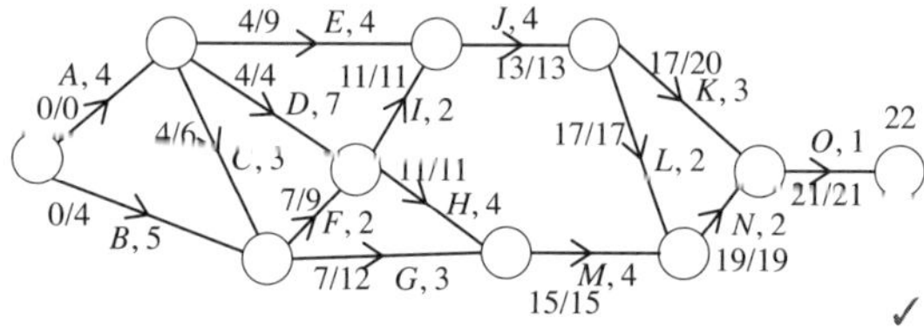

✓

25 **a**

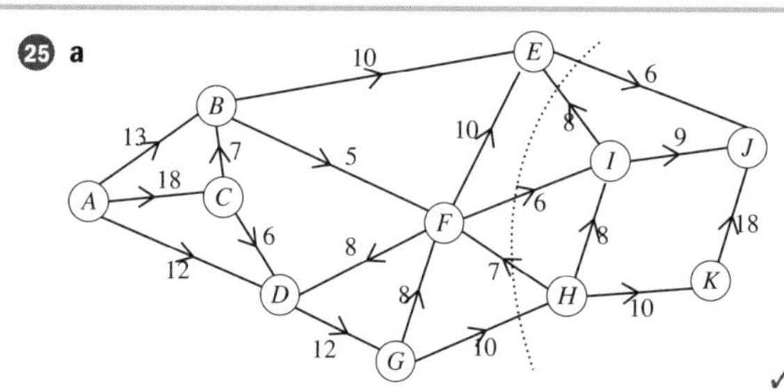

✓

Using the weights for the edges in the direction of Source to Sink:

Flow = 6 + 0 + 6 + 0 + 10
= 22

∴ the maximum flow is 22 kL/min. ✓ *(2 marks)*

b

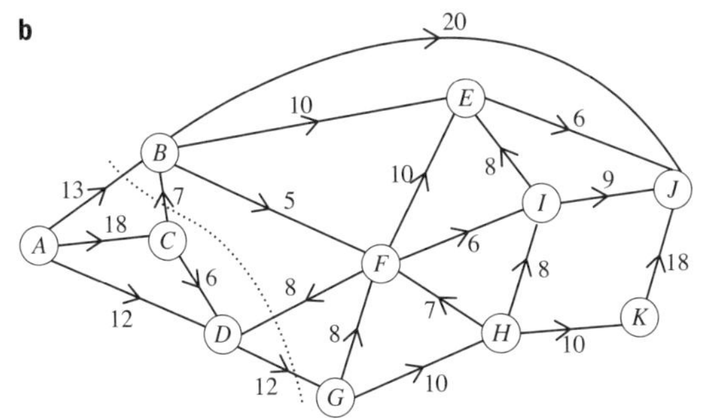

New maximum flow
$= 13 + 7 + 12$
$= 32$
Change $= 32 - 22$
$= 10$
$\therefore$ the maximum flow is increased by 10 kL/min.

(1 mark)

26

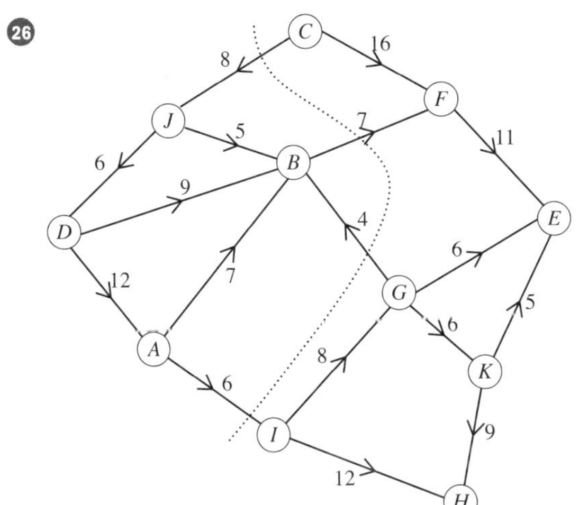

Minimum cut $= 7 + 6$
$= 13$
$\therefore$ the maximum flow is 13 L/second.

(1 mark)

27

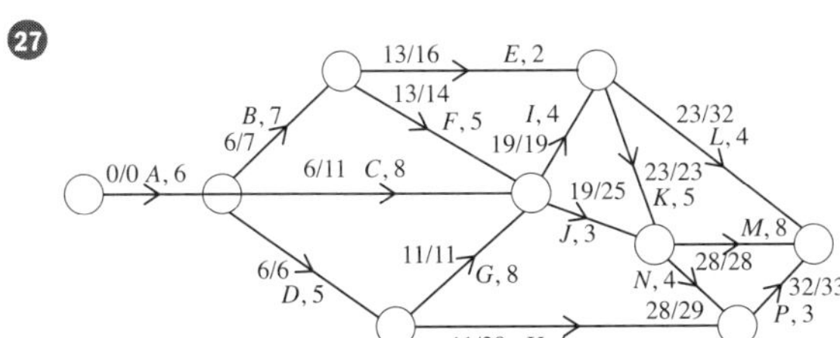

a $6 + 5 + 8 + 4 + 5 + x = 36$
$28 + x = 36$
$x = 8$

(1 mark)

b 19 days

(1 mark)

c $29 - 11 = 18$
$\therefore$ 18 days

(1 mark)

d As activity M takes 8 days and activities $N + P$ take 7 days, the manager should only reduce M by 1 day. This means a cost of \$2300.

(1 mark)

NSW Education Standards Authority

2021 HIGHER SCHOOL CERTIFICATE EXAMINATION

Mathematics Standard 1
Mathematics Standard 2

REFERENCE SHEET

Measurement

Limits of accuracy

Absolute error = $\frac{1}{2} \times$ precision

Upper bound = measurement + absolute error

Lower bound = measurement − absolute error

Length

$$l = \frac{\theta}{360} \times 2\pi r$$

Area

$$A = \frac{\theta}{360} \times \pi r^2$$

$$A = \frac{h}{2}(a + b)$$

$$A \approx \frac{h}{2}(d_f + d_l)$$

Surface area

$$A = 2\pi r^2 + 2\pi rh$$

$$A = 4\pi r^2$$

Volume

$$V = \frac{1}{3}Ah$$

$$V = \frac{4}{3}\pi r^3$$

Trigonometry

$$\sin A = \frac{\text{opp}}{\text{hyp}}, \quad \cos A = \frac{\text{adj}}{\text{hyp}}, \quad \tan A = \frac{\text{opp}}{\text{adj}}$$

$$A = \frac{1}{2}ab\sin C$$

$$\frac{a}{\sin A} = \frac{b}{\sin B} = \frac{c}{\sin C}$$

$$c^2 = a^2 + b^2 - 2ab\cos C$$

$$\cos C = \frac{a^2 + b^2 - c^2}{2ab}$$

Financial Mathematics

$$FV = PV(1 + r)^n$$

Straight-line method of depreciation

$$S = V_0 - Dn$$

Declining-balance method of depreciation

$$S = V_0(1 - r)^n$$

Statistical Analysis

An outlier is a score

less than $Q_1 - 1.5 \times IQR$

or

more than $Q_3 + 1.5 \times IQR$

$$z = \frac{x - \mu}{\sigma}$$

Normal distribution

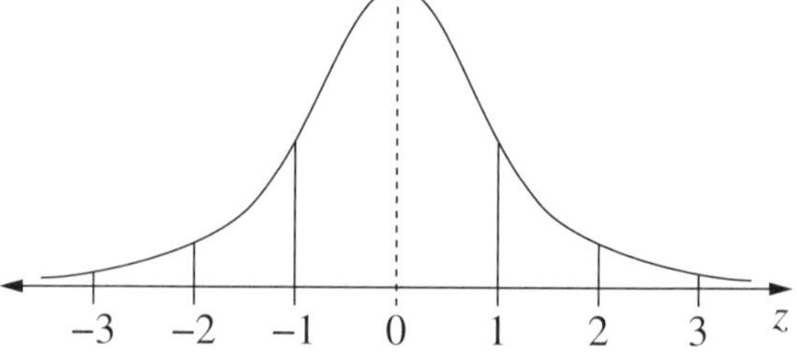

- approximately 68% of scores have z-scores between −1 and 1
- approximately 95% of scores have z-scores between −2 and 2
- approximately 99.7% of scores have z-scores between −3 and 3